Fluid Mechanics and Its Applications

Founding Editor

René Moreau

Volume 139

Series Editor

André Thess, German Aerospace Center, Institute of Engineering Thermodynamics, Stuttgart, Germany

The purpose of this series is to focus on subjects in which fluid mechanics plays a fundamental role. As well as the more traditional applications of aeronautics, hydraulics, heat and mass transfer etc., books will be published dealing with topics, which are currently in a state of rapid development, such as turbulence, suspensions and multiphase fluids, super and hypersonic flows and numerical modelling techniques. It is a widely held view that it is the interdisciplinary subjects that will receive intense scientific attention, bringing them to the forefront of technological advancement. Fluids have the ability to transport matter and its properties as well as transmit force, therefore fluid mechanics is a subject that is particulary open to cross fertilisation with other sciences and disciplines of engineering. The subject of fluid mechanics will be highly relevant in such domains as chemical, metallurgical, biological and ecological engineering. This series is particularly open to such new multidisciplinary domains. The median level of presentation is the first year graduate student. Some texts are monographs defining the current state of a field; others are accessible to final year undergraduates; but essentially the emphasis is on readability and clarity.

Springer and Professor Thess welcome book ideas from authors. Potential authors who wish to submit a book proposal should contact Dr. Mayra Castro, Senior Editor, Springer Heidelberg, e-mail: mayra.castro@springer.com

Indexed by SCOPUS, EBSCO Discovery Service, OCLC, ProQuest Summon, Google Scholar and SpringerLink

Hiroshi Kawamura · Koichi Nishino ·
Satoshi Matsumoto · Ichiro Ueno · Taishi Yano
Editors

Thermocapillary Convection in Microgravity

Thermohydrodynamic Experiment in Kibo
Aboard International Space Station

Editors
Hiroshi Kawamura
Tokyo University of Science and Suwa
University of Science
Chino, Nagano, Japan

Satoshi Matsumoto
Human Spaceflight Technology Directorate
Japan Aerospace Exploration Agency
Tsukuba, Ibaraki, Japan

Taishi Yano
Department of Mechanical Engineering
Kanagawa University
Yokohama, Kanagawa, Japan

Koichi Nishino
Department of Mechanical Engineering,
Materials Science, and Ocean Engineering
Yokohama National University
Yokohama, Japan

Ichiro Ueno
Department of Mechanical and Aerospace
Engineering
Tokyo University of Science
Noda, Chiba, Japan

ISSN 0926-5112　　　　　　　　ISSN 2215-0056　(electronic)
Fluid Mechanics and Its Applications
ISBN 978-981-96-2990-9　　　　ISBN 978-981-96-2991-6　(eBook)
https://doi.org/10.1007/978-981-96-2991-6

© The Editor(s) (if applicable) and The Author(s) 2025. This book is an open access publication.

Open Access This book is licensed under the terms of the Creative Commons Attribution 4.0 International License (http://creativecommons.org/licenses/by/4.0/), which permits use, sharing, adaptation, distribution and reproduction in any medium or format, as long as you give appropriate credit to the original author(s) and the source, provide a link to the Creative Commons license and indicate if changes were made.

The images or other third party material in this book are included in the book's Creative Commons license, unless indicated otherwise in a credit line to the material. If material is not included in the book's Creative Commons license and your intended use is not permitted by statutory regulation or exceeds the permitted use, you will need to obtain permission directly from the copyright holder.

The use of general descriptive names, registered names, trademarks, service marks, etc. in this publication does not imply, even in the absence of a specific statement, that such names are exempt from the relevant protective laws and regulations and therefore free for general use.

The publisher, the authors and the editors are safe to assume that the advice and information in this book are believed to be true and accurate at the date of publication. Neither the publisher nor the authors or the editors give a warranty, expressed or implied, with respect to the material contained herein or for any errors or omissions that may have been made. The publisher remains neutral with regard to jurisdictional claims in published maps and institutional affiliations.

This Springer imprint is published by the registered company Springer Nature Singapore Pte Ltd.
The registered company address is: 152 Beach Road, #21-01/04 Gateway East, Singapore 189721, Singapore

If disposing of this product, please recycle the paper.

Preface

The International Space Station (ISS), a massive crewed experimental facility in space, orbits at an altitude of 400 km above the Earth. One of its components is Japan's experimental module *'Kibo,'* which means *'hope'* in English. The 'Kibo' module began to be attached to the ISS in March 2008, enabling experiments inside the pressurized module starting in August 2008.

The first call for experiment themes for 'Kibo' was in 1992. Our group, in collaboration with some other scientists, applied with a proposal focused on thermocapillary convection under microgravity. In August 1993, 50 candidate themes were selected, including our proposed experiment. Our proposed experiment was realized in 2008, after 15 years, as the first scientific experiment carried out on 'Kibo'.

Looking back at that era, rapid advancement in semiconductor technology caused significant interest in the production of large, high-quality single crystals. The process includes melting of polycrystalline materials, creating a temperature gradient that leads to "thermocapillary convection" caused by surface tension gradient due to temperature distribution. This convection tends to be unstable and may impact the crystal structure. However, the gravity of Earth obscures and complicates these phenomena. Consequently, our proposal aimed to experimentally investigate the phenomena related to surface tension, particularly thermocapillary convection, under microgravity conditions.

Besides the crystal-growth technology, the thermocapillary convection plays significant roles in technically important phenomena such as mass and energy transport in microfluidics, the wetting/dewetting of liquid films and droplets, bubble behavior in boiling, and dropwise/filmwise condensation. More recently, as technology expands further into micro and nano realms—such as Lab-on-a-Chip, drug delivery, materials processing, coating, and cleaning—there is growing interest and demand in microhydrodynamics and microfluidics. The thermocapillary effect must also play a governing role in developing chemical reaction devices for environmental control and life support systems (ECLSS) under micro- and low-gravity conditions.

The convection caused by the difference in surface tension occurs not only due to the temperature differences but also due to the concentration differences of liquid components. A familiar example in our daily life can be seen in the 'tears of wine'

Fig. 1 Tears of wine. Photo by the author

as shown in Fig. 1. This phenomenon is called "solutocapillary" convection caused by the evaporation of alcohol in the wine. However, recent studies have shown that the thermocapillary convection, caused by the temperature decrease due to the evaporation of alcohol, also contributes to the emergence of tears (see the first chapter "Surface Tension, Thermocapillary Convection, Microgravity and 'Kibo' Aboard International Space Station").

Returning to the topic of preparations on the ISS, we were gradually informed around 2001–2002 that the orbital experiment would be feasible in approximately three years. However, in early 2003, the Space Shuttle Columbia disaster tragically took place and the lives of seven astronauts were lost. We were deeply saddened by this incident, but through this tragedy, we were strongly reminded that experiments conducted in space can only be carried out through the generous and dedicated support of astronauts, satellite and facility operators, project supervisors, and the general public.

Experiments in orbit were remotely operated from the Tsukuba Space Center (TSC). The orbital experiment commenced on the morning of August 13, 2008, 15 years after we first proposed this research theme. The initial operation, open to the media, proceeded smoothly with the formation of a liquid column. Upon applying the specified temperature difference, thermocapillary convection was successfully induced, allowing us to announce to the public a smooth start of this series of experiment in orbit.

However, such smooth progress did not last long. Since that afternoon, we were seriously troubled with appearance of many small bubbles inside the liquid bridge, resulting from absorbed gas in the test liquid. After serious struggle with bubbles for a week, we developed a method to eliminate bubbles remotely in microgravity with use of the thermocapillary effect. More detailed description in this respect is given in Sect. 3 of Chapter "Microgravity Experiments in Kibo Onboard the International Space Station." Since then, our experiment has proceeded almost smoothly.

Figure 2 depicts examples of straight-shaped liquid bridges we were able to form in space, thanks to weightlessness. For comparison, the rightmost image is one we usually employ on-ground with the maximum height for this diameter. The inset in the top left is the official mission patch of our experiment MEIS (Marangoni Experiment In Space), which was named after consultation with JAXA and NASA.

The invaluable results obtained from microgravity research focusing on thermocapillary convection have already been presented at various international conferences and in scientific journals. The main purpose of publishing this book in addition is to provide a comprehensive record of the scientific results obtained, as well as the experimental apparatus and measurement methods developed, such as three-dimensional Particle Tracking Velocimetry (PTV). Additionally, significant emphasis will be placed on introductory sections to assist young researchers in university and industry labs who are developing an interest in this field and also those looking forward to conducting experiments or engaging in activities in space.

In the first chapter "Surface Tension, Thermocapillary Convection, Microgravity and 'Kibo' Aboard International Space Station", we introduce surface tension as it appears in nature and daily life. In addition, another main theme of this book; that is, microgravity, the International Space Station (ISS), and Kibo, will be introduced. Following in the second chapter "Thermodynamic and Molecular Aspects of Surface Tension", the genesis of surface tension and its temperature dependence are explained from both a macroscopic thermodynamic perspective and a microscopic molecular dynamics viewpoint.

In the third chapter "Thermocapillary Convection in an Infinite Liquid Layer and in an Infinite Liquid Column", we focus on the main subject of this book: the thermocapillary convection. Initially, we present the fundamental equations and dimensionless numbers that describe this convection. The chapter then discusses the instability threshold of thermocapillary convection in two canonical configurations: an infinitely wide liquid film layer and an infinitely long liquid cylinder. A particularly effective analytical method for this problem is Linear Stability Analysis (LSA).

In the fourth chapter "Thermocapillary Convection in Liquid Bridges of Finite Length", we address thermocapillary convection in finite-length liquid bridges, a configuration employed in the orbital experiments on Kibo. The chapter begins with early time research in this field, followed by descriptions of key phenomena,

Fig. 2 Liquid bridges build in microgravity with a diameter (D) of 30 mm with various heights (H). Right most: a liquid bridge on the ground with D = 5 mm deformed due to gravity. *Photo courtesy of JAXA, All rights reserved.* Inset/Top Left: Official mission patch of the experiment MEIS (Marangoni Experiment in Space). Own work, designed by the author

including the conditions for transition to oscillatory flows, mode numbers, and Particle Accumulation Structures (PAS). These findings are primarily derived from experimental and numerical studies conducted in preparation for orbital experiments. Furthermore, the chapter discusses general research advancements in this area.

The fifth chapter "Effect of Heat Exchange, Control and Suppression of Thermocapillary Convection" then shifts focus to the substantial effects of heat exchange with the surrounding atmosphere, a crucial factor influencing the transition conditions. Additionally, methods for controlling or suppressing the transition to oscillatory flows will also be discussed.

The sixth chapter "Microgravity Experiments in Kibo Onboard the International Space Station" details the experiments conducted within Kibo, the primary focus of this book. It includes descriptions of the experimental apparatuses, conditions, and methods, along with measurement techniques such as three-dimensional particle tracking velocimetry, the photochromic dye activation method, and the micro-imaging displacement meter. The chapter also covers the transition conditions to oscillatory flow, the spatiotemporal structures of velocity and temperature fields, the effects of heat exchange through the free surface, the dynamic behavior of the liquid bridge due to thermocapillary convection and the residual acceleration of gravity on the ISS (i.e., g-jitter), and the transition process to chaotic or turbulent states.

The seventh chapter "Surface-Tension Related Flows in Microgravity and Microscale: Liquid Bridges" illustrates some unique phenomena observed in liquid bridge configurations driven by the thermocapillary effect. Coherent structures formed by small particles are observed in hydrothermal wave instability in the liquid bridge geometry. It is followed by insights gained from microgravity experiments in the series named MEIS and Dynamic Surf. Results from microgravity experiments and numerical simulations demonstrate how thermocapillary convection affects the shrinkage and pinching-off of liquid bridges, leading to the formation of accompanying the so-called satellite droplets. Control of the residual droplet volume after pinching-off is introduced as an example of practical applications in microfluidics

In the eighth chapter "Surface-Tension Related Flows in Microgravity and Microscale: Hanging Droplet, Thin Films, and Positive Surface Tension Temperature Coefficient (Self-rewetting Fluids)", research findings related to various configurations, such as droplets and thin films, are presented, with a special focus on microhydrodynamics. The behaviors of tiny particles due to the thermocapillary effect are illustrated through ground-based and microgravity experiments. The non-uniform distribution of surface tension also plays a crucial role in technologically fundamental processes such as coating and heat exchange. The distribution of surface tension significantly affects the quality of coatings, impacting both the bulk and the periphery of the edge. The solutocapillary effect induced by the evaporation of solvent is explored. Additionally, a unique feature of surface tension is introduced; some liquids exhibit a positive temperature coefficient of surface tension, leading to enhanced heat transfer characteristics in heat exchange devices.

In the final ninth chapter "Development of Fluid Dynamics Experiments in Kibo Aboard International Space Station and Beyond", we provide an overview of the history, current status, and future plans of the ISS and Kibo. Additionally, the chapter discusses future objectives, including the return of humans to the Moon and missions aimed at Mars. In such longer and more distant activities, the utilization of surface tension, an intrinsic property of liquids, becomes increasingly important. Therefore, the chapter concludes with a discussion on the current state and future prospects of active development to utilize phenomena involving surface tension. Space exploration is entering a new phase, with the focus shifting towards the Moon and Mars, which will involve experiencing a variety of gravitational environments, from microgravity in space transport vehicles to 1/6G on the Moon and 1/3G on Mars. Liquid handling technology is indispensable in these explorations. When used effectively, it serves as a passive pump that operates with minimal external energy by fully leveraging the capillary, thermocapillary, and solutocapillary forces.

In closing this Preface, as the author of this Preface, I would like to express my heartfelt gratitude to the co-editors of this book, K. Nishino, S. Matsumoto, I. Ueno, and T. Yano, for their significant contributions not only to the publication of this book but also to the microgravity experiment on Kibo. Regarding the publication, we approached Springer, renowned for its scientific publishing, to undertake this project. Springer Nature's Tokyo office provided us with comprehensive support for publication. Furthermore, JAXA enabled us to share the current achievements more widely with society by publishing this work as Open Access.

Finally, from a broader perspective, the present experiments in space made possible thanks to the support of JAXA, international collaboration with NASA, the dedicated efforts of astronauts and ground stuff, and the wide support from the public. Such collaboration and support made it possible to carry out our experiments in space and thus enabled us to publish this book.

Suwa, Japan Hiroshi Kawamura
Early Spring 2024

Acknowledgements

We would first like to thank André Thess, editor of the book series "Fluid Mechanics and Its Applications," in which this volume appears. We also thank A. Tokuno of Springer Nature for his helpful advice and support. T. Sato provided detailed preliminary review of our manuscripts and gave us helpful advice. We also thank S. Selvam and D. Varadharajan for their efforts in making this publication a reality.

We are deeply grateful to the contributors, Y. Yamaguchi, K. Fujimura, D. Schwabe, Y. Kamotani, M. Kudo, J. Shiomi, M. Onishi, S. Uemura, S. Shiratori, A. Cecere, R. Savino and Y. Abe, for providing their valuable manuscripts. Their contributions have significantly enriched the content of this book, covering a wide range of topics in the related field of capillarity.

The present book is dedicated to a space experiment project carried out through international cooperation and excellent teamwork. Members of the operation, user integration, researcher teams, as well as the experimental facility developers fulfilled their respective roles with dedication.

We would like to thank the advisory group members H. Azuma, K. Ito, N. Imaishi, H. Ota, H. Kawamura, M. Suzuki, K. Nishino, and S. Maruyama for their invaluable role in evaluating the specifications, discussing issues, and providing advice on the equipment, which led to the development of an experimental device with excellent basic performance. The advisory group of the Fluid Physics Experiment Facility (FPEF) whose members included A. Hirata, H. Azuma, H. Kawamura, T. Hibiya held lively discussions on the specifications of the equipment and installation of 3D-PTV as a novel observation technique for liquid bridge, which yielded numerous fruitful experimental results. Among them, T. Hibiya, also a scientist of the thermocapillary flow of low Prandtl number fluid, gave us strong encouragement for publishing the outcomes of the space experiment.

M. Ohnishi, M. Kawaji, Y. Takeda, M. Sakurai, S. Yoda, Y. Kamotani, A. Komiya, and N. Imaishi participated in the research team as principal investigators or collaborators. They shared responsibilities for planning of the research, the development of the instruments, the preliminary experiments on the ground, and the experiments on "Kibo."

For the preparatory work on the space experiment involving "Boundary-Condition Susceptibility and Controllability of Instability Mechanisms in Thermocapillary Convection", we have engaged in suggestive discussions with an international team, including V. Shevtsova, A. Mialdun, D. Melnikov, I. Ryzhkov, H. Kuhlmann, D. Schwabe, J. M. Montanero, M. Lappa, Y. Kamotani, M. Kawaji, S. Matsumoto, K. Nishino, I. Ueno, T. Yano.

Numerous students from university laboratories at Tokyo University of Science, Yokohama National University, University of Tsukuba and Tohoku University joined the operation team and played a pivotal role in the execution of the experiment experiencing both the tension and pleasure of being integrated in a space mission. We are certain that without student assistance, this space experiment would not have been performed, and at the same time we believe that their valuable experience gained here will significantly contribute to their future careers. In particular, doctoral students, S. Tanaka, S. Uemura, M. Kudo, K. Motegi and T. Yano, made substantial contributions to the research outcomes.

The role of User Integration is to bridge the gap between the research team and the operation team, and we would like to thank K. Enokido, Y. Kinefuchi, and Y. Toyoshima of Japan Manned Space Systems Corporation (JAMSS), and K. Kogure, N. Sakurai, and T. Shimaoka of Japan Space Forum (JSF), Y. Watanabe and E. Yoda of Advanced Engineering Services (AES) played an important role in the development of the instruments and in supporting space experiments. Without the user integrators' efforts, the space experiment would not have been achieved.

We would like to thank the JEM Flight Director, who is responsible for coordinating "Kibo" operations and control, as well as the Planner, the Experimental Operations Team for their joint efforts in conducting the experiment. The JEM Flight Director coordinated Kibo's operations with the domestic and international operations personnel. K. Sakagami, the Thermocapillary Experiment Lead, led the operations team during the Thermocapillary experiment and helped to create a comfortable environment for the research team to conduct the experiment. The FISICS team provided the commanding during the experiment. It was a very stressful experiment with a lot of commands. In particular, K. Noguchi, Y. Konisho, T. Sato, and Y. Ohkawa of Thermocapillary Experiment Lead made a great contribution.

Astronaut G. Chamitoff assembled the payloads at the start of the experiment in 2008, and subsequently, numerous astronauts have served on the thermocapillary convection experiment. Astronaut D. Pettit, who is also a chemical scientist, was greatly interested in the thermocapillary convection experiment and gave us fruitful discussions. Japanese astronauts K. Wakata, S. Noguchi, and S. Furukawa worked steadily guided our experiment towards success and also served as spokespeople, reaching out to the general public and promoting the experiment. The astronauts, M. Mohri, C. Mukai, and T. Doi gave us a lot of advice from the preparation stage and during the space experiment, which encouraged greatly us and the participating students as well.

In the present experiment, the liquid bridge is held between the two end disks only by the surface tension. It was conducted during night-time on the ISS to avoid the rupture due to the vibration caused by the astronauts activities. To avoid such accidents, Japanese astronauts actively requested cooperation of other crew in nighttime to provide an excellent microgravity environment in Kibo.

Engineers at IHI AEROSPACE Co., Ltd. were in charge of the facility development and manufacture for space experiment. In overcoming the challenges encountered during development, they successfully produced an excellent experimental facility for the fluid physics experiment. After the space experiments began, some malfunctions occurred with the experimental equipment. However, with limited tools and under the constraints of orbit, they developed a skillful repair technique. In cooperation with the astronauts, they were able to resume the experiments after completing the repairs.

The ISS international partners -NASA, ESA and CSA- were very cooperative in the implementation of the space experiment and in the coordination of the experiment phases to ensure the success of the experiment. We heartily appreciate their outstanding collaboration with these partners.

JAXA was the organization responsible for the overall promotion of the present experiment. In particular, S. Yoda, as a former team leader, motivated us toward this publication. We thank T. Ishikawa for his great efforts to implement the experiment during the preparatory stage. We would also like to thank S. Ogawa, the director supervising the relevant activities, and other administrators for their promotion of the space experiment, publicity activities, and the present publication.

JAXA enabled us to publish the current achievements as Open Access, allowing broader public access.

University members were supported by Grants-in-Aid for Scientific Research in conducting their research activities.

2024

Hiroshi Kawamura
Koichi Nishino
Satoshi Matsumoto
Ichiro Ueno
Taishi Yano

Contents

Surface Tension, Thermocapillary Convection, Microgravity and 'Kibo' Aboard International Space Station 1
Hiroshi Kawamura and Satoshi Matsumoto
1   Surface Tension, Thermocapillary Convection and Their
     Appearances in Nature and Daily Life 2
2   Microgravity, International Space Station (ISS), and "Kibo" 10
References 14

Thermodynamic and Molecular Aspects of Surface Tension 17
Yasutaka Yamaguchi and Hiroshi Kawamura
1   Thermodynamic Aspect of Surface Tension 17
2   Introduction to Molecular-Scale Understanding of Surface Tension 22
3   Concluding Remarks 44
References 45

Thermocapillary Convection in an Infinite Liquid Layer and in an Infinite Liquid Column 47
Hiroshi Kawamura and Kaoru Fujimura
1   Thermo-Hydraulic Equations and Nondimensional Numbers
     Related to the Thermocapillary Convection 48
2   Thermal Convection in an Infinite Liquid Layer with a Finite Depth
     Subjected to a Temperature Gradient Perpendicular to the Surface 55
3   Thermal Convection in a Thin Infinite Liquid Layer
     and an Infinitely Long Liquid Cylinder with a Temperature
     Gradient Along the Surface 69
4   Linear Stability Analysis of Thermocapillary Convection in Liquid
     Layer and Bridge 84
References 100

Thermocapillary Convection in Liquid Bridges of Finite Length 103
Hiroshi Kawamura and Dietrich Schwabe
1   Introduction 103
2   First Experiments on Time-Dependent Thermocapillary Flow 106
3   On Ground Experiments and Numerical Analyses of Hydrothermal
    Convection in Liquid Bridges with Finite Length 125
References 149

**Effect of Heat Exchange, Control and Suppression
of Thermocapillary Convection** 155
Koichi Nishino, Yasuhiro Kamotani, Masaki Kudo, and Junichiro Shiomi
1   Introduction 156
2   Effect of Interfacial Heat Loss/Gain on Stability of Thermocapillary
    Flow of High Prandtl Number Fluids 157
3   Control and Suppression of Oscillatory Thermocapillary
    Convection in a Half-Zone Liquid Bridge 164
4   Thermocapillary Convection in a Full Zone Liquid Bridge 176
References 190

**Microgravity Experiments in Kibo Onboard the International
Space Station** 193
Taishi Yano, Koichi Nishino, Satoshi Matsumoto, Ichiro Ueno,
Hiroshi Kawamura, and Yasuhiro Kamotani
1   Outline of Microgravity Experiment 194
2   Experiment Cell and Measurement Apparatus 200
3   Experimental Procedure 205
4   Instability of Thermocapillary Convection—Effects of Geometrical
    Conditions, Prandtl Number, and Temperature Control Method 210
5   Flow Measurement with Contactless Techniques 232
6   Heat Transfer Through the Liquid-Gas Interface 248
7   Dynamic Free-Surface Deformation 270
8   Transition to Chaos and Turbulence 281
References 288

**Surface-Tension Related Flows in Microgravity and Microscale:
Liquid Bridges** 297
Ichiro Ueno, Mitsuru Ohnishi, Satoshi Matsumoto, and Suguru Uemura
1   Experimental Study on Coherent Structures by Particles Suspended
    in Half-Zone Thermocapillary Liquid Bridges 297
2   Dynamics of Liquid Bridge 314
3   Nano Liter Droplet Formation and Its Volume Control
    by Thermocapillary Effect 318
References 325

Surface-Tension Related Flows in Microgravity and Microscale: Hanging Droplet, Thin Films, and Positive Surface Tension Temperature Coefficient (Self-rewetting Fluids) 331
Ichiro Ueno, Suguru Shiratori, Anselmo Cecere, Raffaele Savino, and Yoshiyuki Abe
1 Thermocapillary-Driven Convection in Hanging Droplet 332
2 Thin Liquid Film 345
3 Positive Surface Tension Temperature Coefficient (Self-rewetting Fluids) 355
References 369

Development of Fluid Dynamics Experiments in Kibo Aboard International Space Station and Beyond 377
Satoshi Matsumoto
1 International Space Station and Kibo Module 377
2 Thermocapillary Experiment in Microgravity 378
3 Experimental Facility 383
4 Experimental Operation 387
5 Perspectives of Kibo Utilization 387
6 Use of Surface-Tension-Driven Flows in Future Human Activities in Space 389
References 390

Index 391

Surface Tension, Thermocapillary Convection, Microgravity and 'Kibo' Aboard International Space Station

Hiroshi Kawamura and Satoshi Matsumoto

Abstract Surface tension is a phenomenon that we often encounter in our everyday lives. In this chapter, we present several examples from daily life, such as water droplets and soap films to illustrate the concept. We will also explain the balance of forces related to surface tension using an example of an air bubble and capillary rise in a thin tube. Furthermore, we will introduce and explain familiar examples, such as the tears of wine to illustrate thermocapillary and solutocapillary flows, which are induced by temperature and concentration gradients in liquids. When comparing the forces acting on fluid, the surface tension is usually smaller than buoyant or gravitational forces. Therefore, to conduct experiments eliminating the effects of gravity, we applied to a public call for experiment proposals for a scientific experiment in the Japanese Experiment, Module 'Kibo,' "Hope" in Japanese, on the International Space Station (ISS), and our proposal was selected. In the latter half of this chapter, we will introduce the ISS and Kibo, including a brief history of their construction, their structure, and major experimental facilities in Kibo. Even in ISS, variations in gravitational acceleration are unavoidable due to the operation of equipment and the activities of astronauts (known as g-jitter). Our experience of the g-jitter experienced in our Kibo experiment is also described.

H. Kawamura (✉)
Professor Emeritus, Tokyo University of Science, Tokyo, Japan
e-mail: kawa@rs.sus.ac.jp

Professor Emeritus, Suwa University of Science, Chino, Nagano, Japan

Principal Investigator of Thermocapillary Convection in "Kibo", Tsukuba, Japan

S. Matsumoto
Japan Aerospace Exploration Agency (JAXA), Tsukuba, Japan
e-mail: matsumoto.satoshi@jaxa.jp

© The Author(s) 2025
H. Kawamura et al. (eds.), *Thermocapillary Convection in Microgravity*,
Fluid Mechanics and Its Applications 139,
https://doi.org/10.1007/978-981-96-2991-6_1

1 Surface Tension, Thermocapillary Convection and Their Appearances in Nature and Daily Life

Surface tension is a phenomenon that we encounter frequently in our everyday lives. In Fig. 1a, we see a photograph of water droplets on the leaves of lotus plants in a pond at an ancient temple. The smaller water droplets maintain a nearly spherical shape, whereas the larger ones become flatter. This occurs because water droplets tend to maintain their spherical shape to minimize their surface area. Whereas, as they grow in size, the force of gravity begins to dominate over surface tension, causing them to deform.

Figures 1b provides example of how we can manipulate surface tension (σ) for practical purposes in our daily routines; that is, in car wash (b), a special coating is applied to the vehicle's body to reduce its surface energy. This coating helps prevent water from spreading across the surface to minimize their total surface energy. Another example of manipulating surface tension in daily life can be seen in cosmetics. Here, it is crucial for the surface tension to be low, as this allows them to spread smoothly and evenly over the skin.

Figure 1c depicts astronaut Noguchi creating a water droplet and allowing it to float within the Japanese Experiment Module "Kibo". An almost perfectly spherical water droplet is successfully suspended in space

Figure 2a and b are the drawings with which introductory explanations on the surface tension are often made. First, Fig. 2a illustrates a soap film formed in a rectangular wire frame in which the right side is movable. The movable frame is pulled to the right by a distance Δx (m) to increase the area of the soap film. A force F (N) is required because a force is acting on the surface of the soap film perpendicular to the edge and toward inside of the film. The reason for this force is that molecules inside of the film must be brought out to the surface to increase the surface area. The physical reason for this force is discussed briefly later in this section and will be discussed in more detail in Chap. 2 from both thermodynamic and molecular aspects. Returning to Fig. 2a, since this force is proportional to width ℓ of the soap film, the required force can be expressed as $F = 2\sigma\ell$, where σ is

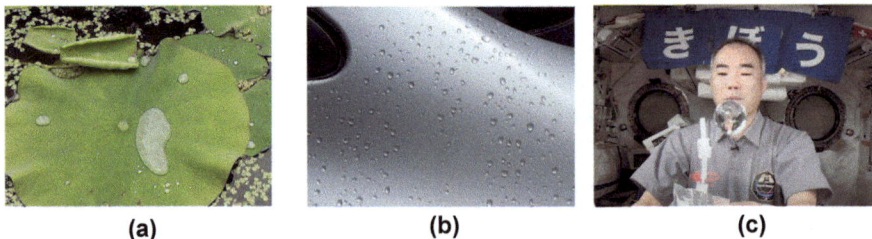

(a) (b) (c)

Fig. 1 Scenes of droplet in daily life and space; **a** Water droplets over lotus leaf, **b** Droplets over water-repellent coating, **c** Large spherical droplet floated in "Kibo" by astronaut Noguchi. **a** and **b**: photo. by Kawamura, **c** © JAXA. All rights reserved

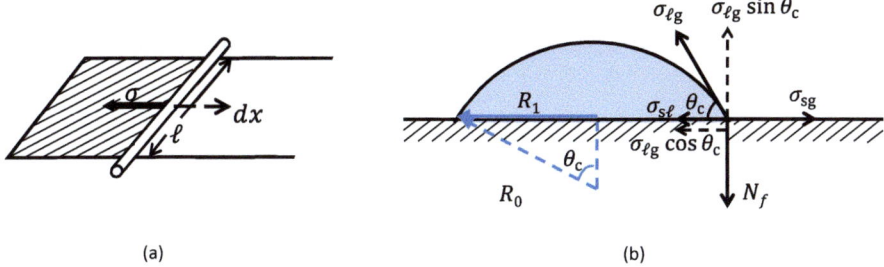

Fig. 2 Surface tension over **a** a thin film and **b** a partial droplet

a proportionality constant and the "2" arises because the film has both sides. The constant σ is called the surface tension representing the force per unit width (N/m).

There is another way of interpreting the surface tension. That is, for the frame of Fig. 2(a) to be moved by a distance Δx, a work of $F\Delta x = 2\sigma \ell \Delta x$(J) had to be done. The work required to create the unit surface area of the soap film becomes $F\Delta x/\Delta A = 2\sigma \ell \Delta x/2\ell\Delta x = \sigma$ (J/m^2), where ΔA is the increased surface area, $\Delta A = 2\ell\Delta x$(m^2). Accordingly, the surface tension σ (J/m^2) can be interpreted as an amount of energy to increase unit surface area of the soap film. Accordingly, a term "surface energy" is sometimes assigned to this quantity σ.

Reason why the surface formation requires energy can be outlined as follows. Molecules or atoms (called "molecules" for the brevity hereafter) in the bulk liquid are surrounded by many molecules and attractive forces act mutually between the adjacent molecules, which causes so-called "favorable" energy condition for them. The molecules over the liquid–vapor interface are, however, surrounded by smaller number of molecules, so that they are in "less favorable" state than the inner ones. This means that to bring the inner molecules to the interface requires an extra energy. This is an origin of the "surface energy". A point to be noticed here is that, according to the above explanation, the force acting to the surface molecules must be inward, whereas, in reality, the surface tension acts "tangentially" along the surface. This is seemingly mysterious and will be explained in Sect. 2 of Chapter "Thermodynamic and Molecular Aspects of Surface Tension" by Yamaguchi Y with the assistance of the molecular dynamics.

In the above example, we discussed the surface formation between liquid and gas. The surface tension of this case is denoted as σ_{lg}. It should be noted that the energy is also required to create interfaces between solid and air, as well as solid and liquid. Their interface tensions are usually represented by σ_{sg} and σ_{sl}, respectively.

Figure 2(b) depicts a small amount of liquid (e.g. water) placed on a flat solid surface. Ignoring the effect of the gravity, the water droplet will form a partial sphere, which may be called simply "hemisphere" sometimes. In this example, three kinds of surface (interfacial) tensions, σ_{lg}, σ_{sg} and σ_{sl} take place in the respective directions illustrated in Fig. 2(b). The combination of these three σ's determines the shape of the hemisphere, which can be characterized by the contact angle θc. That is, if θc is close to 180°, the droplet will be almost spherical, and if close to 0 then the droplet will spread over the surface. The force balance in the horizontal direction over the surface gives $\sigma_{lg}\cos\theta_c + \sigma_{sl} = \sigma_{sg}$ and thus we obtain the contact angle as

$$\cos\theta_c = \frac{\sigma_{sg} - \sigma_{sl}}{\sigma_{lg}}, \qquad (1)$$

which is called "Young's equation".

If $\sigma_{sg} - \sigma_{sl} \geq \sigma_{lg}$ i.e., $\cos\theta_c \geq 1$, then the contact angle θ_c can be regarded as zero, which means the liquid will spread over the solid surface. In contrast, if $\cos\theta_c \leq -1$, the surface is called "non-wetting" or "hydrophobic". We have seen the nearly non-wetting examples of water droplet in Fig. 1a. In this case, however, the surface is not perfectly non-wetting, but micro roughness over the surfaces helps the non-wettability.

The contact angles are rather sensitive to the surface contamination and other conditions. Accordingly, Kirby [3] gives the contact angles not as a definite value but a range of angles, such as 0–30° for water/glass, 40–70° for water/poly (methyl methacrylate) and 105–120° for water/Teflon.

A question raised often is about the vertical component of $\sigma_{\ell g}$. Not many textbooks describe balancing of this component. Shikhmurzaev [9] mentions that this vertical component balances with the reaction force N_f exerted on the contact line by the solid.

In this context, another less-discussed force balance, a vertical force equilibrium with respect to the partial sphere may be interesting to be discussed. Since a force of $\sigma_{\ell g}\sin\theta_c$ per unit length acts vertically upwards along this circular contact line, a total upward force along the entire length of the contact line becomes $2\pi R_1$ times $\sigma_{\ell g}\sin\theta_c$; that is, $2\pi R_0 \sigma_{\ell g}\sin^2\theta_c$. On the other hand, as we will find immediately below in Eq. (3), the pressure inside a sphere of radius R_0 is higher than the surroundings with an amount of $2\sigma_{\ell g}/R_0$. By multiplying it to the area of the contacting circle $\pi(R_0\sin\theta_c)^2$, we get the downward force of $2\pi R_0 \sigma_{\ell g}\sin^2\theta_c$, which is equal to the upward force along the circular contact line obtained above.

Next, let us discuss a water sphere (radius R) floating in the air and imagine a virtual rectangular parallelepiped which includes the center of the sphere in one of its surfaces (Fig. 3a and b). The pressure of the ambient gas is P_{out} and inside the water sphere P_{in}. These two pressures are not equal and the difference $P_{\text{in}} - P_{\text{out}}$ can be derived as follows.

That is, in Fig. 3b, the force acting the rectangular plane (area A_0) from right to left is $P_{\text{out}} A_0$, while the force pushing from left to right is $P_{\text{out}}(A_0 - \pi R_0^2) + P_{\text{in}}\pi R_0^2$. In addition to these two, the circular ring of the radius R_0 is pulled towards left by the surface tension σ(N/m). Accordingly, the force balance over the rectangular surface becomes

$$P_{\text{out}}(A_0 - \pi R_0^2) + P_{\text{in}}\pi R_0^2 - 2\pi R_0 \sigma = P_{\text{out}} A_0. \qquad (2)$$

Thus, we get

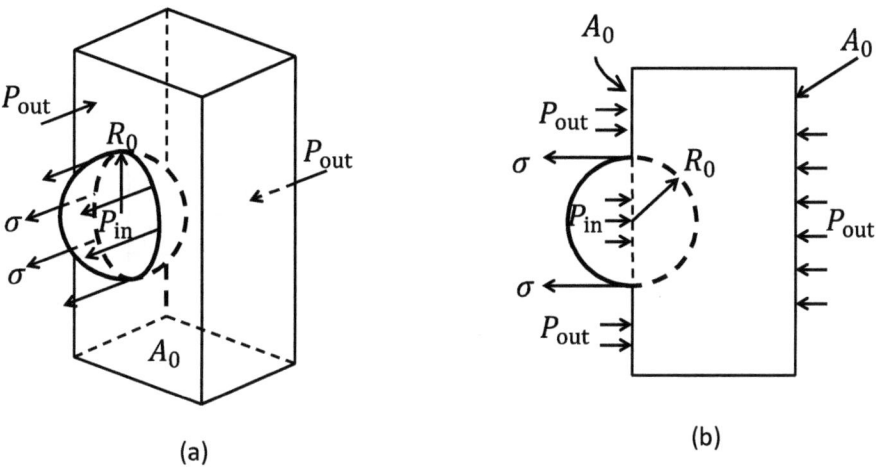

Fig. 3 Surface tension in a spherical liquid droplet; force balance across a virtual parallelepiped

$$P_{\text{in}} - P_{\text{out}} = \frac{2\sigma}{R_0}. \tag{3}$$

This means that the internal pressure P_{in} in the sphere is larger than the external pressure P_{out} with an excess of $2\sigma/R_0$. Note that this relationship holds true if liquid is inside and air is outside of the sphere. Accordingly, the side of the higher or lower pressure is determined solely by the geometry, that is, the pressure inside is higher than outside if its outer boundary surface is "convex", and lower if it is "concave." Accordingly, the pressure inside of a bubble or droplet is higher than its surroundings.

Another well-known visible appearance of the surface tension is so-called "capillary rise", that is, the rise of liquid column in a thin tube. The cause of this capillary rise is often called "capillarity". Two snapshots of such the capillary rise are shown in Fig. 4. Liquids used are water (natural mineral water) and liqueur (about 40 volume % alcohol) available in daily life. The thin tubes are made of borosilicate glass with an inner diameter of 2 mm. Remember that this experiment is not for an accurate measurement but for demonstration.

Here we will obtain the height of the capillary rise using Fig. 4c. We will assume the pressure in the liquid depends on the depth but that of the air is uniform irrespective of the position. Since the pressure is lower in the side whose boundary surface is concaved, the liquid pressure just beneath the meniscus P'_{liq} must be lower than the atmospheric one P_{air}. This is the reason why the liquid column rises up. Letting the radius of the capillary R_1 and remembering $R_1 = R_0 \cos\theta_c$, Eq. (3) gives

$$P'_{liq} = P_{air} - \frac{2\sigma}{R_0} = P_{air} - \frac{2\sigma \cos\theta_c}{R_1}. \tag{4}$$

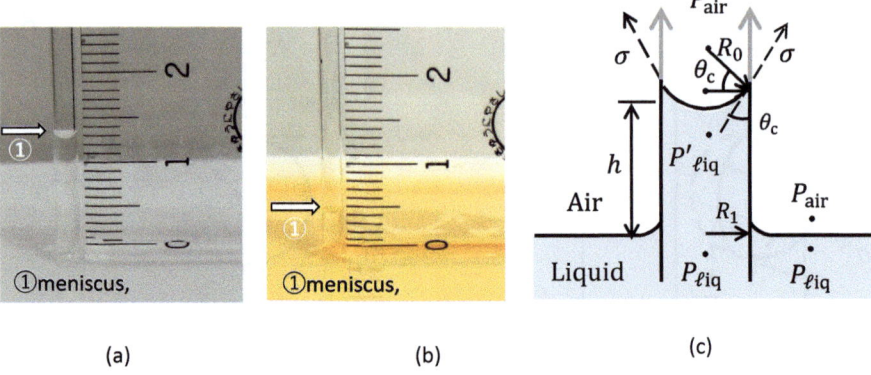

Fig. 4 Capillary rise in a thin tube (2 mm in diameter): **a** water, **b** liqueur, **c** force balance. **a** and **b**: photo by Kawamura

Since we can assume the pressure of the air over the liquid surface is uniform and thus $P'_{liq} + \rho_{liq}gh = P_{air}$, so we get

$$h = \frac{2\sigma \cos\theta_c}{\rho_{liq}gR_1}. \tag{5}$$

Then with use of $\sigma = 72$ mN/m (Table 1) for the water and $\theta_c = 25°$ for water versus glass and $R_1 = 1$ mm, we get $h \approx 13$ mm, which is fairly in good agreement with the observation in Fig. 4a.

Another photo of the capillary rise shown in Fig. 4b is that of liqueur. In case of the liqueur, its capillary rise is much smaller than that of water, with $h \approx 4 - 5$ mm. If the wetting condition and the density are assumed unchanged from those of water, the surface tension of this liqueur can be obtained as about 25 mN/m, which is nearly equal to that of alcohol given in Table 1, although one should note that the assumptions above are rather crude.

One of the widely-used technical applications of the capillarity is the heat pipe, which is now a common cooling device in laptops and other various types of computers. Figure 5(c) is an inside view of a laptop computer. The two copper

Table 1 Surface tension and its temperature dependence of some fluids. Produced referring to Shikhmurzaev [9]: Table D.2, p.413

	Surface tension (σ: mN/m) at several temperatures		
Liquid	25 °C	50 °C	75 °C
Water	71.99	67.94	63.57
Ethanol	21.97	19.89	–
Mercury	488.48	480.36	475.23

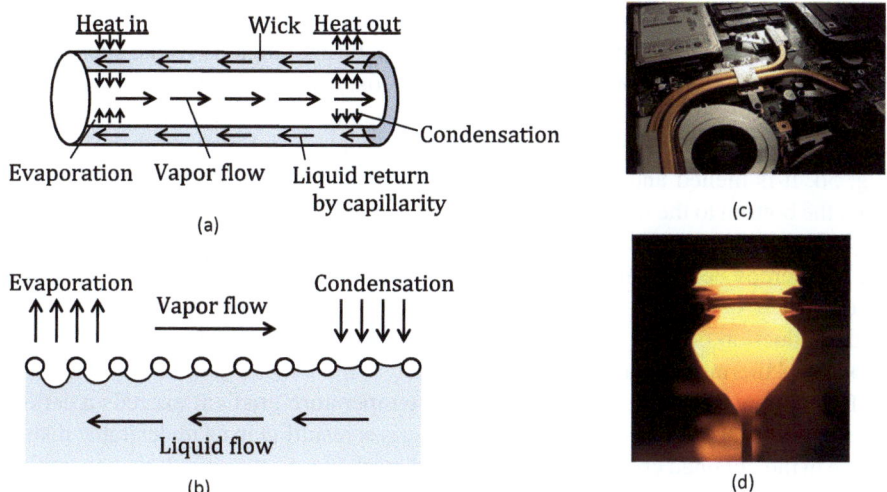

Fig. 5 Examples of capillary effect in technology. **a** and **b** Operating principle of heat pipe, **c** Heat pipe in laptop computer, **d** Crystal growth with use of surface tension effect. **a, b** Own drawing referring to Peterson [8] and Ochterbeck [7]. **c** from "Heat pipe" at Wikipedia CC-BY-SA attribution Kristoferb. **d** from "Floating-zone silicon" at Wikipedia CC-BY-SA attribution Marathoni62

tubes are the heat pipes designed to transport heat from a CPU to a heat sink like a cooling fan.

A typical heat pipe structure is schematically depicted in Fig. 5a, referring to Peterson [8] and Ochterbeck [7]. The heat pipe consists of a sealed metal tube, typically made of copper, a working fluid such as water for heat transport, and capillary material for wicking. Heat is introduced to the left side of the pipe, causing the working fluid to evaporate. The vapor moves through the central passage to right (the cooling section), where it condenses, releasing heat.

The condensed liquid must return to the left (hotter end). Figure 5b simplifies the fundamental process. That is, the evaporation in the heated section (left side) makes the meniscus of the liquid–vapor interface more concaved into the liquid layer. This concaved meniscus keeps the liquid side pressure lower than that of the vapor side, as discussed in relation to the capillary rise. In the condenser region (right), supply of liquid through the condensation makes the meniscus less concaved than in the evaporation region. Accordingly, the liquid pressure is kept higher in the condenser region than in the evaporating part and thus the working liquid can flow back to the evaporation side, closing the heat transport cycle.

The heat pipes are now widely used in various fields of applications such as cooling of semiconductor devices, solar thermal utilization, cooking, airplane, and spacecraft. Those interested more in the heat pipe technology may refer to Peterson [8].

Another example of the presently major technology involving the effect of the surface tension is crystal growth for semiconductor, which is now indispensable products in modern society, with a growing demand for high-purity, defect-free single crystals. There are various methods for manufacturing single crystals, such as the Czochralski (CZ) and floating zone (FZ) methods. Concise summary of the crystal growth methods will be described in Sect. 1 of Chapter "Thermocapillary Convection in Liquid Bridges of Finite Length". As an example, the FZ method is shown in Fig. 5d. It is melted and grown into a single crystal by moving the heated portion from the bottom to the top using a ring-shaped heater. In this method, surface tension plays a crucial role in two aspects. Firstly, it suspends the molten liquid column in the air, thereby determining the shape of the column. Secondly, due to a temperature gradient across the melt surface, a distribution of surface tension occurs, leading to the induction of melt flow over the liquid column surface. This phenomenon is known as "thermocapillary convection".

In this type of convection, if the surface temperature gradient exceeds a certain threshold, a time-dependent fluctuating flow arises resulting in an undesirable disturbance in the obtained crystal structure. Such disturbances are particularly concerning since semiconductor silicon for integrated circuits requires homogeneous doping. A similar process may emerge in other types of crystal growth techniques, such as the Czochralski (CZ) method. Therefore, research on flow stability in molten material was strongly motivated and initiated. This phenomenon, thermocapillary convection, is the primary focus of this book and will be described further in the following chapters.

Thermocapillary and solutocapillary convections

As mentioned above, the surface tension of liquid depends upon the fluid temperature. Some examples are given in Table 1. In most liquids, their surface tension decreases with increasing temperature. Accordingly, if a temperature distribution exists over a surface, flow will be induced from the hotter to the colder side. In case of some multicomponent liquid, however, its surface tension depends on its composition and in some exceptional cases of liquid mixtures their surface tension behaves inversely, that is, increases with increasing temperature. It causes interesting and sometimes useful effects, which will be treated in more detail later in Sect. 3 of Chapter "Surface-Tension Related Flows in Microgravity and Microscale: Hanging Droplet, Thin Films, and Positive Surface Tension Temperature Coefficient (Self-rewetting Fluids)."

The flow induced by the temperature difference is called "thermocapillary flow" and the one by the concentration difference, "solutocapillary flow" or "Marangoni flow", which is named after the Italian physicist Carlo Marangoni [5], is often used as a general term for the flows caused by the gradient of the surface tension.

According to Loglio [4], Carlo Marangoni was born at Pavia in 1840 and studied physics there. He obtained the university degree in physics through a thesis on the phenomena occurring in liquids due to the surface tension. Since 1869 until 1910, Carlo Marangoni continuously held the Chair of Physics at the Liceo Classico Dante (classical high school) at Florence, Loglio [4].

The solutocapillary flow is easier to be observed in daily scene than the thermocapillary flow. Soap, for example, has much less surface tension than water to spread

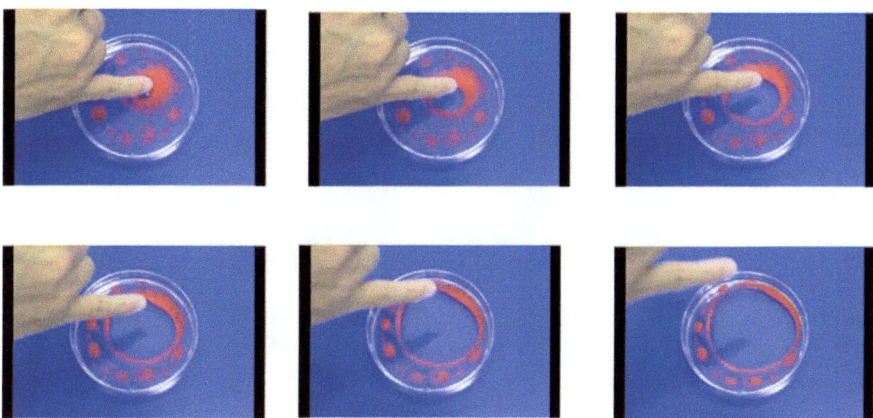

Fig. 6 Solutocapillary flow. Dropping a small amount of soapy water at the center results in a rapid outward flow caused by the solutocapillarity. Interval of every other frame:15 fps. Photo by Matsumoto

well over a variety of surfaces. In Fig. 6, a small amount of soap is applied to a fingertip and immersed at the center of the water surface. We can see that the water spreads instantly from the center to the circumferential area. Every other frame of the movie is displayed with the frame rate of 15 frames/s; so the induced flow is very quick.

"Tears of wine" is one of the widely known examples of the Marangoni flow. When a clean wine glass is partially filled with alcohol-rich wine or liqueur, a chain of droplets are formed over the inner wall of the glass, causing the droplets to slide down intermittently along the glass surface. Figure 7 is a picture of the tears of wine and enlarged photos of the movement of a droplet. This photograph was taken by the authors group with use of the same liqueur which was used in the capillary rise demonstration of Fig. 4.

This phenomenon, the tears of wine, was first scientifically discussed in 1855 by James [10], an elder brother of William Thomson (Lord Kelvin). He stated in his article that "the thin film adhering to the inside of the glass must very quickly become more water rich than the rest, on account of the evaporation of the alcohol contained in it being more rapid than the evaporation of the water". As stated here, the surface tension of the water-rich part is greater than that of the original liqueur, the bulk liquid is continuously sucked up along the side wall. Then the water portion becomes accumulated and drops down along the glass surface. So, the tears of wine have been regarded as a typical example of the solutocapillary flow.

Recently, however, Venerus and Simavilla [11] conducted experiments using the infrared thermography and found that the temperature of the rising water film and the falling droplets was lower than that of the bulk liquid due to the evaporation of the alcohol component. They also made estimates of temperature and concentration gradients and concluded that both temperature and concentration gradients contribute to this phenomenon.

Fig. 7 Tears of wine. From Author's LabDataArchive. Photo by Kawamura and Tanaka

2 Microgravity, International Space Station (ISS), and "Kibo"

Satoshi Matsumoto

We are usually conducting our research in a nominal gravity environment (1G or 1 g) on the ground and observing phenomena on that basis. In such cases, buoyancy, natural convection, deformation due to self-weight, and other factors brought about by gravity may be undesirable for elucidating the phenomena. By creating a microgravity environment, it is expected to simplify the phenomena and contribute to a better understanding of the essence of these phenomena.

A microgravity environment can be obtained in space. Strictly speaking, however, gravity still acts on the International Space Station (ISS), but due to the balance between gravity and centrifugal force, weightlessness or microgravity is effectively achieved in the ISS. We have conducted space experiments in microgravity environments using space platforms such as Drop Shaft, parabolic flights by aircraft, sounding rockets, Space Shuttles, and the International Space Station to elucidate a variety of phenomena.

The International Space Station (ISS) continues to orbit about 400 km above the Earth. Floating in space, the ISS offers several unique environments that are difficult to achieve on Earth. Microgravity, cosmic radiation, high vacuum with vast space, and direct exposure to solar energy are typical examples. By making good use of this environment, space experiments in various fields, including physics and life science experiments, are conducted on the ISS.

While the ISS orbits at breakneck speed, the interior of the ISS, on the other hand, is in a very calm and fascinating state. It is a microgravity environment. We call it

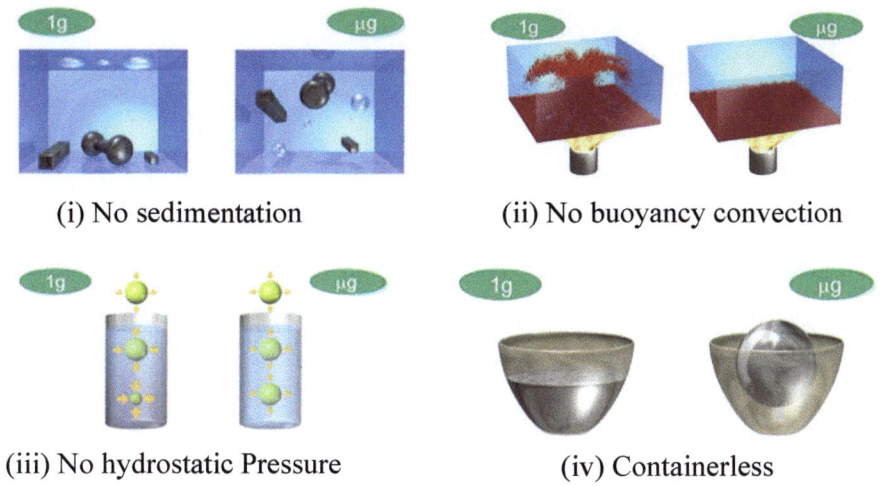

Fig. 8 Characteristics of microgravity environment. © JAXA. All rights reserved

microgravity rather than weightlessness because the acceleration inside is about one millionth (10^{-6} g) of the acceleration of gravity on the ground (1 g). In a microgravity environment, phenomena can be very different from those on the ground. These phenomena in the microgravity can be characterized as "no sedimentation," "no buoyancy convection," "no hydrostatic pressure," and "containerless levitation" (Fig. 8). These phenomena often contradict our common sense based on earthly experiences. In addition, phenomena which are hidden behind gravity on the ground often appear prominently and surprise us. These phenomena are caused by the fact that the vertical symmetry, which is restricted by the downward gravity on the ground, is freed from the constraint in space and becomes three-dimensional symmetry, resulting in perfect spatial symmetry.

In physical experiments, space experiments are conducted mainly from two perspectives: (1) from an applied perspective by utilizing the "microgravity environment" as a means of solving problems related to the manufacture of new materials and processes, and (2) from the perspective of contributing to the advancement of basic science through the understanding of physical and chemical phenomena. Through space experiments, theories have been proven, and unknown phenomena and mechanisms have been elucidated, and the obtained results have been applied to applied technology on the ground.

In life science experiments, research is conducted in environments such as "microgravity," "space radiation," and "outer space with a closed environment" contributing to the expansion of the sphere of human activity and the exploration of the universal laws of life.

The ISS is the largest floating structure in space in human history with an overall length of about 110 m, a solar array length of about 75 m, and an overall mass of about 450 tons, orbiting the Earth at an altitude of about 400 km every 90 min (Fig. 9).

Fig. 9 International Space Station (left) and Japanese Experiment Module "Kibo" (right). © JAXA. All rights reserved

It is intended to provide the capability for a variety of space activities without being limited to any specific field. For this reason, the ISS project has greatly expanded the scope of its use to include communications, broadcasting, meteorology, and earth observation, as well as the use of the microgravity environment and vastness of space in space and the convenience provided by astronauts' activities.

Construction of the ISS in space began with the basic module, the Zarya Control Module, launched in 1998, and has now been completed. The ISS program is the first international cooperative space project that mankind has ever experienced with 15 countries participating. Importantly, six astronauts are permanently on board and working on the ISS.

In March 2008, the Space Shuttle Endeavour (STS-123) carrying Astronaut Doi on the 1 J/A mission delivered the Experiment Logistics Module-Pressurized Section to the ISS, and in June of the same year, the Discovery (STS-124) with Astronaut Hoshide on the 1 J mission carried the Pressurized Module and docked with the ISS. This allowed the inclusion of *Japan's own space facility*, the Japanese Experiment Module "Kibo," on the International Space Station (ISS). Later, in July 2009, during the STS-127 (2 J/A) mission with Astronaut Wakata aboard, the Exposed Facility was attached to the Pressurized Module, completing Kibo as an on-orbit experiment facility (Fig. 9).

The Pressurized Module, serving as an experimental laboratory, has an outer diameter of approximately 4.45 m, an inner diameter of approximately 4.22 m, and a total length of approximately 11.68 m. Its mass in orbit is approximately 15.7 tons, excluding the experimental payload and other Kibo elements.

Inside the pressurized section, four equipment mounting structures, known as "Racks," are arranged along the same circumference. A total of 23 racks can be mounted in six rows in the pressurized section along the machine axis, with five rows on the ceiling side. Of these, 12 racks are designated for experimental payloads, two of which are used as storage space for experiments.

The experimental payload is mounted in the pressurized section in the form of an integrated rack. The mechanical attachment interface between the rack and the pressurized module is common for all ISS pressurized modules except for the module in charge of Russia. The coupling interface of the standard resource lines supplied to the experimental payload rack is also standardized, and the experimental payload rack that conforms to this standard is called the International Standard Payload Rack (ISPR). The thermocapillary convection experiment was conducted in the "Ryutai Rack" located at the aft3 position, which indicates a specific position toward the rear of the designated area in the pressurized module.

When conducting experiments in space, researchers may plan their experiments under the assumption of "zero" gravity. However, in reality, experimental systems are exposed to complex acceleration environments, including residual acceleration and fluctuations in acceleration referred to as 'g-jitter'. Researchers have come to recognize that the space experiment environment is not "ideally zero gravity" but rather "microgravity (μg)".

There are many sources of disturbance in an orbiting platform, and forces of various directions and magnitudes are exerted. As a result, a complex combination of acceleration disturbances is generated.

The causes of gravity modulation in a microgravity environment include:

(a) Disturbances (Earth's gravity gradient, atmospheric drag, solar wind and cosmic dust impacts, docking)
(b) Actions from the platform (propulsion rockets, mass damping, band antenna vibration, structural vibration)
(c) Inertial movement within the platform (crew activity, back-and-forth and rotational movement of onboard equipment).

The gravitational environment caused by these factors can be broadly classified as follows.

(1) quasi-steady acceleration (<0.01 Hz), g_{qs}
(2) transient acceleration, g_t
(3) periodic acceleration, g_p.

Acceleration measurements have been made on the International Space Station (ISS). The data and analytical results in the typical events are published on the website by NASA Principal Investigator Microgravity Services (PIMS). Acceleration is measured using MMA (Microgravity Measurement Apparatus) and SAMS (Space Acceleration Measurement System) aboard the Japanese Experiment Module "Kibo".

Researchers have acknowledged the significant impact of residual acceleration on certain space experiments, Zhao and Alexander [13]. The residual acceleration acting in fluids with density gradients introduces buoyant forces and results in the occurrence of convection in continuous fluids. Effect of residual acceleration on diffusion coefficient measurement was analyzed by numerical simulation Matsumoto and Yoda [6]. Lower frequencies below 0.1 Hz may affect the accuracy of measurements because buoyancy convection is induced more strongly than at higher frequencies.

The other is the problem of deformation of the gas–liquid interface due to forced oscillations. In the case of a liquid bridge configuration, the residual acceleration sometimes causes serious issues. If the amplitude of the periodic acceleration, which corresponds to the natural frequency of the liquid bridge, is large, the liquid bridge will swing significantly, and in the worst case it will breakup. Our experiments were conducted at nighttime of the activity in the ISS. Accordingly, the gravity acceleration was relatively quiet, but on rare occasions the liquid bridge vibrated considerably so that the experiment had to be interrupted. Yano et al. [12] discussed the g-jitter characteristics observed during their thermocapillary experiment in a liquid bridge configuration.

Since Japanese astronauts began staying on the ISS, they helped to spread awareness of high sensitivity to vibrations of our experiments. This led to greater caution with activities in the Kibo module by the astronauts during our experimental operations, resulting in an improved environment for our experiment. We were grateful for the cooperation of the astronauts, which contributed to the successful completion of our experiment.

References

1. Bruus H (2008) Capillary effects. Theoretical microfluidics. Oxford Univ. Press, Oxford, pp 123–140
2. Fisher GR et al (2012) Silicon crystal growth and wafer technologies. Proc IEEE 100:1454–1474
3. Kirby BJ (2010) Miro- and nanoscale fluid mechanics. Cambridge University Press, Cambridge, pp 22–23
4. Loglio G (2006) Carlo Marangoni and the laboratory of physics at the high school "Liceo Classico Dante" in Firenze. In: Gade M et al (eds) Marine surface films. Springer, Berlin, Heidelberg. https://doi.org/10.1007/3-540-33271-5_2
5. Marangoni CGM (1871) Über die Ausbreitung der Tropfen einer Flüssigkeit auf der Oberfläche einer anderen. Ann Phys Chem (Poggendorf) 143(7):337–354
6. Matsumoto S, Yoda S (1999) Numerical study of diffusion coefficient measurements with sinusoidal varying accelerations. J Appl Phys 85:8131–8136. https://doi.org/10.1063/1.370651
7. Ochterbeck JM (2003) Heat pipes. In: Kraus AD, Bejan A (eds) Heat transfer handbook, Wiley, pp 1181–1230
8. Peterson GP (1998) Heat pipes. In: Rohsenow WM et al (ed) Handbook of heat transfer, 3rd ed. McGraw-Hill, pp 12.1–12.20
9. Shikhmurzaev YD (2008) Capillary flows with forming interfaces. Chapman & Hall/CRC, pp 79–83
10. Thomson J (1855) On certain curious motions observable at the surface of wine and other alcoholic liquors. Phil Mag Sci 10(67):330–333
11. Venerus DC, Nieto Simavilla D (2015) Tears of wine: new insights on an old phenomenon. Sci Rep 16162. https://doi.org/10.1038/srep16162
12. Yano T, Nishino K, Matsumoto S, Ueno I, Komiya A, Kamotani Y, Imaishi N (2018) Overview of "dynamic surf" Project in Kibo–dynamic behavior of large-scale thermocapillary liquid bridges in microgravity. Int J Micrograv Sci Appl 35:350102. https://doi.org/10.15011/jasma.35.1.350102
13. Zhao Y, Alexander ID (2003) Effects of G-jitter on experiments conducted in low-earth orbit: a review, In: 41st Aerospace sciences meeting and exhibit. https://doi.org/10.2514/6.2003-994

Open Access This chapter is licensed under the terms of the Creative Commons Attribution 4.0 International License (http://creativecommons.org/licenses/by/4.0/), which permits use, sharing, adaptation, distribution and reproduction in any medium or format, as long as you give appropriate credit to the original author(s) and the source, provide a link to the Creative Commons license and indicate if changes were made.

The images or other third party material in this chapter are included in the chapter's Creative Commons license, unless indicated otherwise in a credit line to the material. If material is not included in the chapter's Creative Commons license and your intended use is not permitted by statutory regulation or exceeds the permitted use, you will need to obtain permission directly from the copyright holder.

Thermodynamic and Molecular Aspects of Surface Tension

Yasutaka Yamaguchi and Hiroshi Kawamura

Abstract In this chapter, we describe the fundamental features of surface tension from both thermodynamic and molecular perspectives. In the first section, we briefly introduce the fundamentals of the thermodynamics, such as the Gibbs and Helmholtz free energy, and then discuss their increments with respect to changes in surface area, which causes the surface tension. In the second section, first a brief history of the research on the surface tension is introduced, and then we provide the microscopic understanding of the surface tension, which needs basic knowledge of thermodynamics, statistical mechanics as well as continuum mechanics. By introducing the intermolecular interaction potential and temperature definition, and by showing conceptual pictures including some results obtained by molecular dynamics simulations, the authors hope that the target readers of undergraduate level students would find fascinating aspects of surface tension as the boundary of macroscopic and microscopic physics.

1 Thermodynamic Aspect of Surface Tension

Hiroshi Kawamura

As mentioned in the previous Chapter, "Surface Tension, Thermocapillary Convection, Microgravity and 'Kibo' Aboard International Space Station", surface tension can be characterized from two perspectives: the tensile force and the excess energy over the liquid surface. In this section, we will delve into the concept of surface

---

Y. Yamaguchi (✉)
Department of Mechanical Engineering, The University of Osaka, Yamadaoka, Suita, Japan
e-mail: yamaguchi@mech.eng.osaka-u.ac.jp

H. Kawamura
Professor Emeritus, Tokyo University of Science, Tokyo, Japan

Professor Emeritus, Suwa University of Science, Chino, Nagano, Japan

H. Kawamura
e-mail: kawa@rs.sus.ac.jp

© The Author(s) 2025
H. Kawamura et al. (eds.), *Thermocapillary Convection in Microgravity*,
Fluid Mechanics and Its Applications 139,
https://doi.org/10.1007/978-981-96-2991-6_2

energy in some detail with the aid of thermodynamics. The molecular aspect, which is more closely related to forces, will be discussed in the next Sect. 2.

In the example of a thin liquid film given in Chapter, "Surface Tension, Thermocapillary Convection, Microgravity and 'Kibo' Aboard International Space Station", increase/decrease of the surface energy $\gamma \Delta A$ due to the change of ΔA was considered, while a change of its volume was neglected. Here, we will take both changes of the volume V and the surface area A with an amount of ΔV and ΔA into account. In this case, we must consider also the mechanical work done by the expansion $P \Delta V$ under the external pressure P. Please note that the surface tension is expressed as γ in this chapter, rather than σ as in the preceding and forthcoming chapters, for editorial convenience.

The first law of thermodynamics can be expressed as

$$\delta Q = \Delta U + \delta W. \qquad (1)$$

Here Q is the heat transferred to the system concerned, U stands for the internal energy of the system and W denotes the work done by the system to surroundings. The symbol of increment Δ is referred to as an 'exact differential,' which applies to changes in 'state functions.' These changes depend solely on the initial and final states of the system. On the other hand, δ is treated as an 'inexact differential' for 'path functions,' where its value depends not only on the initial and final states but also on the transition path of the system. In order for these terms to be mathematically integrable, the inexact differentials must be expressed using the exact ones.

Accordingly, when both the volume and the surface area change in an amount of ΔV and ΔA, the work performed by the concerned system to the surroundings is

$$\delta W = P \Delta V - \gamma \Delta A. \qquad (2)$$

The negative sign arises because the work $\gamma \Delta A$ must be stored within the system's surface to increase its area ($\Delta A > 0$); thus it cannot be transferred to the surroundings. As for another inexact differential δQ, the second law of thermodynamics indicates the well-known relation $\delta Q = T \Delta S$, where S represents the entropy. Accordingly, Eq. (1) can now be expressed solely in terms of the state functions as:

$$T \Delta S = \Delta U + P \Delta V - \gamma \Delta A. \qquad (3)$$

Since the surface tension is thermodynamically well defined under constant temperature and pressure, the pressure P can be often assumed constant and Eq. (3) becomes

$$\Delta(U + PV - TS) = \gamma \Delta A. \qquad (4)$$

Here it is convenient to introduce an energy defined as

$$G = U + PV - TS, \qquad (5)$$

which is called the Gibbs free energy. The product TS indicates a part of the internal energy, which cannot be taken out of the system as a work and is called "bound energy." Accordingly, we can now express the surface tension in terms of Gibbs free energy as

$$\Delta G = \gamma \Delta A \tag{6}$$

or

$$\gamma = \left(\frac{\partial G}{\partial A}\right)_{p,T}. \tag{7}$$

Analogously, in case of constant temperature and volume, we can derive a relation between γ and the Helmholtz energy defined as:

$$F = U - TS. \tag{8}$$

$T\Delta S = \Delta U - \gamma \Delta A$

In this case, Eq. (3) becomes, and thus $\Delta(U - TS) = \gamma \Delta A$. Accordingly, we get the surface tension γ in terms of the Helmholtz energy F as $\Delta F = \gamma \Delta A$ and

$$\gamma = \left(\frac{\partial F}{\partial A}\right)_{V,T}. \tag{9}$$

The well-known Young's correlation, Eq. (1.1), which gives the contact angle θ_c between liquid and solid has been obtained in Chap. 1 based on the force balance among the related surface tensions, γ_{sg}, γ_{sl} and γ_{lg}, which denote the values of γ between solid–gas, solid–liquid and liquid–gas, respectively. Hereafter we will rederive Young's correlation from the aspect of the free energy referring to [2, 9].

Let a system concerned be composed of several free surfaces of area A_i with combinations of liquid, gas and solid. Further, we assume for the simplicity that the system is in equilibrium under constant pressure and temperature and only slight changes of the surface areas are considered. Since the state is in equilibrium, the Gibbs free energy must be at the minimum and thus its change ΔG must also be zero for any small perturbation of system parameters. Accordingly,

$$\Delta G = \sum_i \gamma_i \Delta A_i = 0, \tag{10}$$

where γ_i is the surface tension of the interface A_i.

Figure 1 is a modification of an illustration given in both [2, 9]. Liquid contacts with a solid surface with a contact angle θ and their initial interface is in equilibrium. The system is assumed two dimensional with a depth of unit length. Then a small perturbation is added to move the contact line with a small distance of ΔL_{sl} and a new liquid–gas interface of ΔL_{lg} is produced. A crucial point here is that if the liquid moves only *parallel* to the solid surface, no new liquid–gas interface ΔL_{lg} is produced. As for this point, Bruus [2] mentions that "we must consider that the liquid/gas interface is tilted for an infinitesimal angle $\Delta \theta$ around an axis which is parallel to the surface and placed far away from the interface." [9] further states that "the location of the interface far away from the solid surface is tacitly fixed, so this is actually a rotation of the interface." To explain this point explicitly, the tilting of the interface with an angle of $\Delta \theta$ is depicted schematically in Fig. 1. Then the length of the newly produced liquid–gas interface ΔL_{lg} can be expressed as

$$\Delta L_{lg} = \Delta L_{sl} \cos(\theta - \Delta \theta) = \Delta L_{sl} \cos \theta, \text{ as } \Delta \theta \to 0, \tag{11}$$

where the last equality is resulted from the Taylor series expansion

$$f(x + \Delta x) = f(x) + f'(x)\Delta x + \frac{1}{2}f''(x)\Delta x^2 + \cdots.$$

That is, $\cos(\theta - \Delta \theta)$ can be approximated as $\cos\theta + \sin\theta \Delta\theta + O(\Delta\theta^2)$, and as $\Delta\theta \to 0$ it tends to $\cos\theta$ since its residual error is an order of $\Delta\theta^2$.

Because the newly wetted length ΔL_{sl} was previously in contact with the gas, the relation Eq. (10) becomes

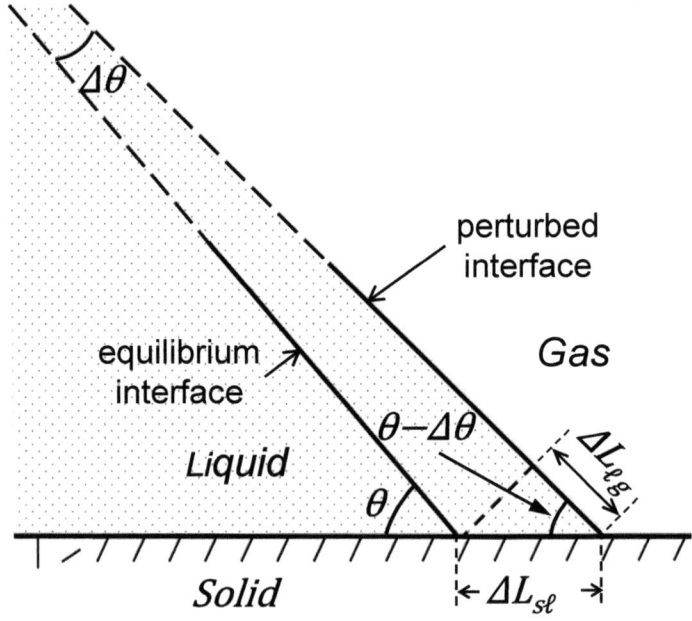

Fig. 1 Derivation of Young's equation from the thermodynamic point of view

$$\Delta G = \gamma_{sl}\Delta L_{sl} - \gamma_{sg}\Delta L_{sl} + \gamma_{lg}\Delta L_{lg} = (\gamma_{sl} - \gamma_{sg})\Delta L_{sl} + \gamma_{lg}\Delta L_{sl}\cos\theta = 0. \tag{12}$$

Accordingly, we obtain

$$\cos\theta = \frac{\gamma_{sg} - \gamma_{sl}}{\gamma_{lg}}, \tag{13}$$

which is the Young's equation Eq. (1.1) itself.

Since this book is much concerned with the "temperature coefficient" of the surface tension, its thermodynamical aspect will be discussed briefly below. The temperature coefficient of the surface tension γ_T is expressed as

$$\gamma_T = -\frac{d\gamma}{dT}, \tag{14}$$

where the sign of minus is introduced to make γ_T positive for most liquids. It is interesting to note here that the dimension of γ_T is [N/m·K] = [N·m/m²·K] = [J/K/m²], which is dimensionally equal to the entropy per unit area. This point will be discussed below, referring to [3].

We are now concerned with a very thin surface layer with a several-molecule thickness, in which molecules behave rather differently from those in the bulk fluid. Thus, we will introduce surface values of the thermodynamic quantities such as G_s, and S_s. In addition, the volume of the surface layer is neglected, because it is so thin. Instead, a term γdA of the surface excess energy will be added.

Now, small increments of G and U are shortly recalled as

$$dG = dU + PdV + VdP - TdS - SdT$$
$$dU = TdS - PdV. \tag{15}$$

Thus, we obtain $dG = VdP - SdT$. Accordingly, if we neglect the volume of the surface layer and designate an increment of the Gibbs free energy of the surface layer by the subscript s, it becomes

$$dG_s = -S_s dT + \gamma dA, \tag{16}$$

where the surface energy of γdA is added.

If the surface area is increased from an infinitesimally small value to A with an assumption of the constant temperature ($dT = 0$), Eq. (16) can be integrated into

$$G_s = \gamma A. \tag{17}$$

With use of Eqs. (15) and (17) and the relation $d(\gamma A) = Ad\gamma + \gamma dA$, we get

$$Ad\gamma = -S_s dT. \tag{18}$$

Introducing the entropy per unit surface area as $s_s = S_s/A$, one obtains

$$s_s = -\left(\frac{\partial \gamma}{\partial T}\right)_A. \tag{19}$$

This means that the temperature coefficient of the surface tension corresponds to (minus of) the entropy per unit surface area [4].

From the analogy with the definition of the Gibbs free energy Eq. (5) and neglecting the volume of the thin surface layer, we may express the Gibbs free energy per unit surface area g_s as

$$g_s = u_s - Ts_s. \tag{20}$$

This expression indicates that the Ts_s is the "bound energy," which is "not" available as the free energy.

In this layer, molecules exhibit behavior different from those in the bulk fluid. This topic will be described in more detail from molecular perspectives in the next section by Prof. Yamaguchi on the request of the present editorial committee; so the readers are strongly recommended to visit the upcoming section.

2 Introduction to Molecular-Scale Understanding of Surface Tension

Yasutaka Yamaguchi

2.1 Introduction

Liquids are supposed to be incompressible due to their small compressibility, e.g., its value of water is about 5×10^{-10} Pa^{-1}. Thus, when liquids are in an open space or in a closed space with a volume larger than theirs, they dispose solid–liquid and/or solid–gas interfaces, hence, liquids we see in our daily life almost always have interface. Interfacial tensions are termed as the force exerted on such interfaces.

At present, the existence of molecules is commonly accepted, and the mechanism of the liquid–vapor (LV) or liquid–gas (LG) surface tension, known as a tensile force at the interface, is usually explained by using a schematic as exemplified in Fig. 2. As in this figure, the molecules around the liquid–vapor interface have less partners to interact with than those in liquid bulk, and the liquid–vapor interface is generally disadvantageous to liquid bulk. This seems to be simple and intuitively understandable; however, as [11] pointed out, Fig. 2 with the arrow directing outward

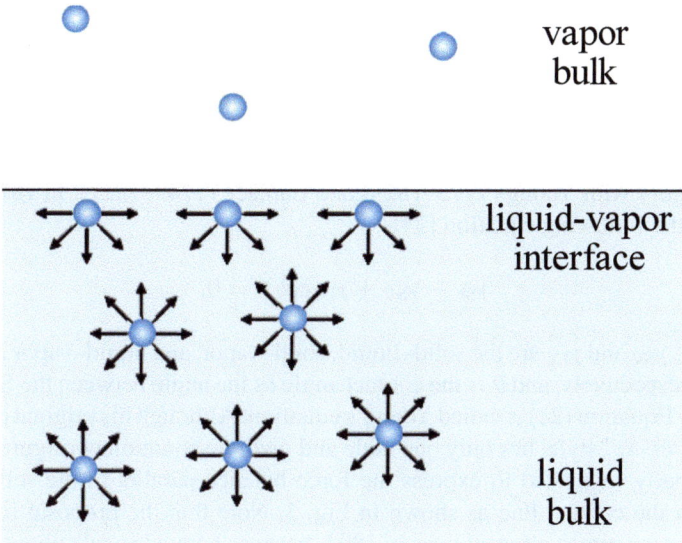

Fig. 2 Schematic for the molecular scale explanation of the surface tension

the molecules gives an impression that molecules in the liquid bulk and at the liquid–vapor interface 'pull' each other, and the molecule at the interface is subject to a net force along the direction perpendicular to the interface, not a force parallel to the interface. This kind of confusing questions are often posed especially for interfacial phenomena, and indeed, I personally think this is one of the fascinating aspects of interface physics which includes macroscopic and microscopic features.

In this section, basic pictures for understanding of the surface tension are provided with some simulation results. This concept is similar to [7, 11] basically targeting undergraduate level students, contrary to many other textbooks on this subject; however, I believe that basic but wide and throughout outlook of thermodynamics, statistical mechanics and continuum mechanics is needed to understand the physics of interface, and more importantly, to avoid misleading which also have trapped professional scientists [6]. Partly related to this point, molecular dynamics (MD) simulations have become common especially with the development of high-performance computers and simulation packages including LAMMPS, [18] and GROMACS from the beginning of the twenty-first century, and we can easily have access to the movies of various (colorful) molecules. The great scientists in the history could not have such movies, and they had to imagine basically from macroscopic experimental results; however, this does not mean that we will not be trapped by the misunderstanding. To advance science, we have to make use of both the robust theories constructed by the great scientists in the history and computer simulations to extend our understanding.

2.2 Brief History

Although there are several opinions about the origin of the study of capillarity or surface tension, [7, 12, 15], scientific and mathematical approach toward its understanding through the modeling of wetting started from the beginning of the nineteenth century with Young (1773–1829) and Laplace (1749–1827). In 1805, Young proposed the following equation [24].

$$\gamma_{SL} - \gamma_{SV} + \gamma_{LV}\cos\theta = 0, \tag{21}$$

where γ_{SL}, γ_{SV} and γ_{LV} are the solid–liquid, solid–vapor, and liquid–vapor interfacial tensions, respectively, and θ is the contact angle as the angle between the SL and LV interfaces. Equation (21) is called Young's equation. Although his original article [8], indeed in "essay" style, has only one table and gives no equation nor figure, Eq. (21) was originally suggested to express the force balance parallel to the solid surface exerted on the contact line as shown in Fig. 3. Note that the proposal of Young's Eq. (21) is sometimes referred to as in 1804, because he made a talk about this topic in this year. Young's Eq. (21) was extended to evaluate wettability of a liquid on a solid surface, known as the Young-Dupré equation:

$$S = \gamma_{LV}(\cos\theta - 1), \tag{22}$$

where the spreading coefficient S given by

$$S = \gamma_{SV} - (\gamma_{SL} + \gamma_{LV}) \tag{23}$$

expresses the wettability: $S = 0$ gives $\cos\theta = 1$ in Eq. (22) and positive S means that the liquid completely covers the solid surface.

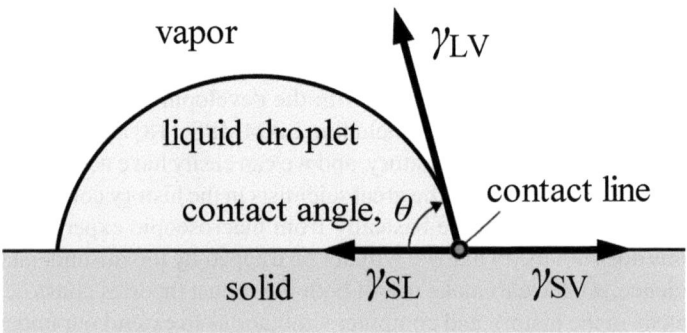

Fig. 3 Schematic of Young's equation proposed as the surface lateral force balance of interfacial tensions exerted on the contact line

The year of 1804 or 1805 was before the establishment of thermodynamics: Carnot (1796–1832) was under ten, and Joule (1818–1881), Thomson (Lord Kelvin, 1824–1907) Clausius (1822–1888), and Gibbs (1839–1903) were not born, and this clearly means that Young did not and could not bring the concept of thermodynamics nor molecular interaction into his modelling of wetting and surface tension as shown in Fig. 2, [6]. Note that the primary hydrodynamic description summarized in "Hydrodynamica" by Bernoulli (1700–1782), which was based on Newtonian mechanics by Newton (1642–1727), was already published at that time, and the Euler Equations about ideal fluids by Euler (1707–1783) as the mathematical framework of fluid mechanics as well as the Lagrangian mechanics by Lagrange (1736–1813) were already available, [5]. Probably, continuum mechanics including the concept of stress tensor was not available considering the ages of Cauchy (1789–1857), Navier (1785–1836) and Poisson (1781–1840).

Anyway, by the frontiers including the above-mentioned scientists, thermodynamics was established in the nineteenth century, and based on it, the van der Waals equation of state expressing the phase coexistence was proposed by van der Waals (1837–1923). He also wrote an article about surface tension in 1893, van der Waals JD (1893) translation by [14], and introduced the concept of dividing surface and formulated a thermodynamic framework with a liquid–vapor interface. Just after the tragedy of Boltzmann (1844–1906), Perrin (1870–1942) experimentally proved the existence of molecules inspired by the idea of Einstein (1879–1955), in [5], and this enabled the above-mentioned explanation of the surface tension through the interaction potential between molecules interacting with a potential well depth, i.e., with a moderate attraction force for long intermolecular distance range and a strong repulsion for short distance range described in detail below. Based on this interaction potential model, G. Bakker (1856–1938), Tolman (1881–1948), Kirkwood (1907–1959), Buff (1924–2009), etc. constructed the theoretical framework of the surface tension in thermodynamic equilibrium. Note that Japanese scientists including Ono (1918–1995), Kondo (1922–2014), Harashima (1908–1986), etc. largely contributed to the development. Especially, Ono and Kondo wrote a great review, [13], and they categorized the approaches of surface tension into thermodynamic and quasi-thermodynamic ones, statistical mechanical ones and hydrodynamic (also mentioned as "mechanical" in [12]) ones. Ono also left an instructive textbook in Japanese [12], and he wrote in the introduction of the book: "surface is very thin but still has thickness, and it is a pity that its physics is difficult to understand because of this fact." This sentence clearly points out the difficulty of surface, and also shows the fascinating feature of surface physics. Another great textbook [16] is also available; indeed, the history in the present article mentioned above is based on these books [12, 16].

As described above, it is basically necessary to include the microscopic concept to fully explain or understand the mechanism of surface tension and wetting. On the other hand, as we see water droplets on solid surfaces almost every day, such a liquid motion affected by the surface tension is very common and macroscopic, i.e., visible by bare eyes. As Young modelled, it is possible to understand and predict the macroscopic liquid behavior by simply considering the surface tension as the force to reduce the surface area without the above-mentioned microscopic knowledge.

Indeed in the middle of the nineteenth century, an experimental approach to measure the surface tension was proposed by Wilhelmy (1812–1864). In this method called the Wilhelmy plate method, [21], which is commonly used also at present, a solid plate or a cylinder is vertically immersed into a liquid pool, and the surface tension is evaluated by measuring the contact angle and the vertical force exerted on the solid. This standpoint is equivalent to the above-mentioned hydrodynamics approach that Ono and Kondo categorized, in which the microscopic physics of interface with non-zero thickness is integrated into the surface tension on a surface with zero thickness. As the name indicates, this hydrodynamic implementation matches well with the governing equations of fluid dynamics, and is especially familiar with the numerical simulation technique called the computational fluid dynamics (CFD). In practice, thermodynamic and statistical mechanical approaches cannot be applied for dynamic systems in principle because these approaches assume static thermodynamic equilibrium. Anyway, CFD simulations of liquid flow with moving interface are common at present owing to the development of high-speed computers from 1980s.

Parallel to the CFD, the (classical) molecular dynamics (MD) method, which solves the motion of constituent molecules of fluids governed by inter-molecular potential functions based on the Newtonian mechanics, has also been developed, and this method enabled the simulations of liquid behavior with interfaces as a microscopic approach. Indeed, one can obtain an equilibrium liquid–vapor coexistence system with an interface by locating a certain number of molecules in a simulation cell, whose intermolecular force is given by a potential function exemplified in Fig. 4, and by controlling the temperature of the system. As pointed out in [16], this solves the main difficulty in the statistical mechanical approach of how to analytically obtain the equilibrium density distribution in such liquid–vapor coexistence system from the microscopic intermolecular potential function.

2.3 Why Phase Separation Happens and Why Interface is Formed—from Thermodynamics

2.3.1 Ideal Gas Model and Temperature

In order to understand the properties of the liquid–vapor interface, a primitive question of why the constituent molecules try to gather to form liquid and vapor phases having different densities as illustrated in Fig. 2 instead of forming a single homogeneous phase with a uniform density. Indeed, a fluid which always stays single homogeneous phase exists as a conceptual one: an ideal gas. The principal (macroscopic) definition of ideal gases is that they always obey the equation of state:

$$pV = nRT, \tag{24}$$

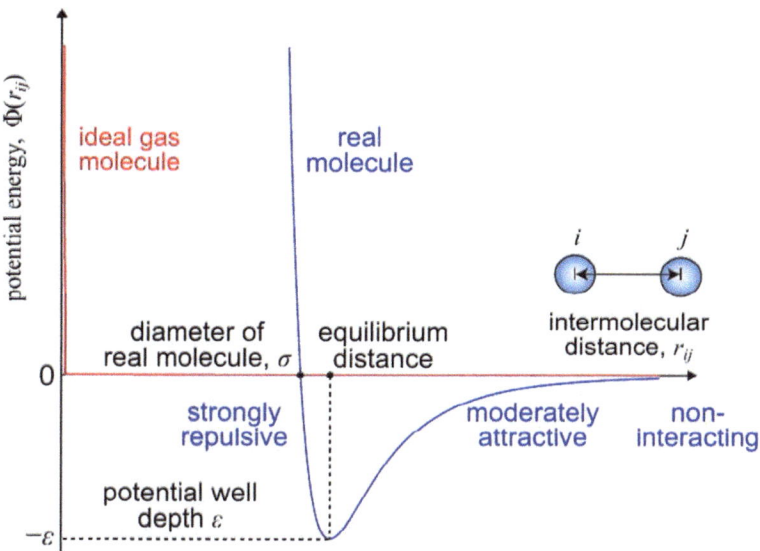

Fig. 4 Intermolecular interaction between ideal gas molecules and between real molecules. No intermolecular forces act between ideal gas molecules and they have no volume. On the other hand, real molecules are generally attracted by each other for an intermediate interatomic distance and repelled for a smaller distance around the molecular diameter, and they have almost no interaction for a longer distance. The minimum point of the potential energy is the equilibrium distance

where p, V, n, T and R denote the pressure, the volume, the amount of substance of the gas in moles, the absolute temperature, and the universal gas constant, respectively.

Indeed, Eq. (24) holds under high temperature and low pressure conditions for real gases, and an ideal gas model is an extended imaginary substance which obeys Eq. (24) irrespective of the temperature and pressure. D. Bernoulli was the first to propose the basis for the kinetic theory of gases in his book of Hydrodynamica, [1], in the eighteenth century. In this book, he argued that a gas consists of a huge number of molecules moving in all directions, and their impact on a surface causes the pressure of the gas, and thus their average kinetic energy determines the temperature of the gas under the following assumptions:

- The constituent molecules of the gas are infinitesimally small hard spheres, besides they are subject to elastic collisions among each other and with the surroundings (container wall).
- There are no attractive or repulsive forces between the molecules apart from those that determine their point-like collisions, i.e., the only forces between the gas molecules and the surroundings are impulsive force upon the point-like collisions.
- The molecules are constantly moving, and as a result of multiple random collisions, their velocity distribution reaches a certain isotropic random direction as an equilibrium state.

Almost after a century, the idea was revisited in the nineteenth century by Clausius and Maxwell (1831–1879), etc. From the assumption above, it is possible to relate the mean square velocity of the constituent molecules $\langle |v|^2 \rangle$ with the pressure p with the density ρ of the gas in three-dimension as

$$p = \frac{\rho \langle |v|^2 \rangle}{3} \tag{25}$$

by considering the sum of impulse exerted from the molecules on the container wall of unit area per unit time, where $\langle |v|^2 \rangle$ is defined as the space, temporal and molecular average by

$$\langle |v|^2 \rangle \equiv \lim_{t_\infty \to \infty} \frac{1}{t_\infty} \int_0^{t_\infty} \left[\frac{1}{N} \sum_{i=1}^{N} |v_i|^2 \right] dt \tag{26}$$

for N molecules with the velocity of i-th molecule being v_i. Note that v_i satisfies

$$\frac{1}{N} \sum_{i=1}^{N} v_i = 0. \tag{27}$$

For present equilibrium gases, e.g., confined in a closed container. In case the container is moving at a constant velocity, $\langle |v|^2 \rangle$ must be defined by using the molecular velocities relative to the group motion. By the way, the relation in Eq. (25) derived by Clausius is amazing because the speed of invisible molecules can be estimated only by the two measurable macroscopic values of the pressure p and the density ρ, [5]. By inserting Eq. (25) into Eq. (4), it follows

$$M \langle |v|^2 \rangle = 3RT, \tag{28}$$

where M denotes the mass of gas per mole. Then, let m and k_B be the mass of single molecule and the Boltzmann constant, respectively defined by

$$m \equiv \frac{M}{N_A}, \quad k_B \equiv \frac{R}{N_A} = 1.38 \times 10^{-23} \text{ J/K} \tag{29}$$

using the Avogadro number N_A, we obtain a fundamental relation between the kinetic energy of the constituent molecule and the absolute temperature:

$$\frac{1}{2} m \langle |v|^2 \rangle = \frac{3}{2} k_B T. \tag{30}$$

Equation (30) can be extended to real molecules as the microscopic definition of temperature T.

2.3.2 Intermolecular Potential

The intermolecular interaction between ideal gas molecules can be schematically illustrated in Fig. 4 with the red line. The horizontal line for overall distance except at zero distance shown with the vertical line indicates that the molecules have no interaction force except the hard point-like collision. In contrast to this ideal gas feature, the interaction potential between real molecules can be modeled as the blue line in Fig. 4. Two real molecules within an intermediate intermolecular distance attract each other whereas a strong repulsive force acts between the two within a smaller distance around the molecular diameter, and they have almost no interaction for a longer distance. The distance giving the minimum potential energy of $-\varepsilon$ is the equilibrium distance.

Now we try to intuitively understand phase change from this interaction potential of real molecules with a potential well depth and the definition of absolute temperature T in Eq. (30). Figure 5 shows the schematic for the understanding of the temperature-dependent phase change from the intermolecular interaction between real molecules. Suppose two molecules vibrating around the potential well at a low temperature without total translational motion as in the left panel, i.e., satisfying Eq. (27). Then, as indicated by Eq. (30), the two molecules have certain relative average kinetic energy with a scale of $k_B T$.

At a low temperature with the corresponding kinetic energy sufficiently smaller than the potential well depth ε, the molecules are trapped in the potential well. In that case, molecules rarely change their interaction pairs and if there more than two molecules, they try to maximize the number of neighboring pairs to minimize the

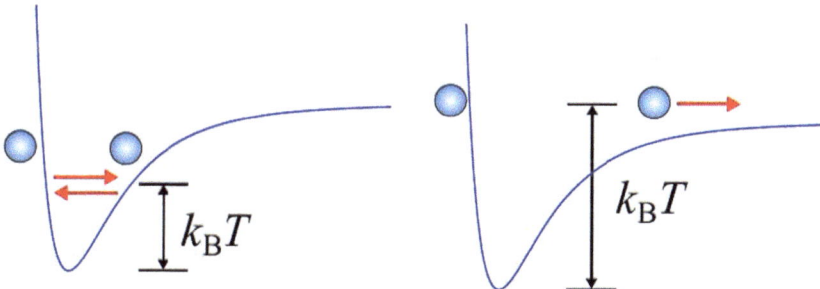

Solid or Liquid: molecules are trapped in the potential well and they stay near at low T.

Gas (critical state for high density): molecules are apart from each other and have no interaction at high T & low ρ (low p).

Fig. 5 Schematic of the intuitive understanding of the temperature-dependent phase change from intermolecular interaction between real molecules. (Left): molecules are trapped in the potential well because the average kinetic energy corresponding to $k_B T$ is not sufficiently large. (Right): molecules can escape from the potential well with the kinetic energy and freely moving in a ballistic manner. The left corresponds to the solid or liquid phase whereas the right is the gas phase

potential energy, and they eventually form a solid crystal typically with a closed packing structure. Note that thermal expansion of solid can also be understood from the asymmetric vibration feature around the equilibrium point; the intermolecular distance at $T = 0$ is apparently at the minimum of the potential well while the time averaged mean distance at $T > 0$ becomes longer as the temperature increases. This expansion is not expected if the intermolecular interaction is modeled by a simple harmonic spring connecting the two molecules, which gives a symmetric vibration around the point of minimum potential energy.

At a moderate temperature, molecules can frequently escape from the potential well and change the pairs without having certain fixed structures. This can be understood as liquid. At high temperature as shown in the right panel of Fig. 5, molecules are not trapped in the potential well due to their sufficient kinetic energy and freely moving in space. This corresponds to the gas phase.

Hence, the key features of real molecules can be summarized as follows:

- There are short-range repulsive and long range attractive forces between the molecules to form a potential well.
- The molecules are constantly moving, and as a result of multiple random collisions, their velocity distribution reaches a certain isotropic random direction as an equilibrium state.

The second feature is the same as ideal gases and the kinetic energy of the random motion is equally distributed to each degree of freedom: the equipartition under thermal equilibrium as a primitive basis of statistical mechanics. It is easy to imagine why Eq. (4) holds under high temperature and low pressure conditions for real gases from the intermolecular interaction potential of real molecules; however, the history of thermodynamics and statistical mechanics tells that constructing it from ideal gas potential is not that easy. Indeed, the outstanding idea of D. Bernoulli proposed in the first half eighteenth century that gases, as fluids consist of huge number of infinitesimally small molecules, was not further investigated for about a century until Clausius and Maxwell, [5].

Anyway, now we go back to the interpretation of the surface tension in Fig. 2. This figure does not mean that the molecular pairs in the liquid bulk pull the partner molecule each other but they are at a certain mean inter-molecular distance as indicated in the right panel of Fig. 4. Molecules around the interface have less partners to interact with than those in the liquid bulk, and those in the vapor bulk have almost no partners. As mentioned above, with the present computers, MD simulations can be easily run even with laptop computers. Figure 6 shows the snapshot of equilibrium systems consisting of molecules with their intermolecular interaction described by the Lennard–Jones potential popularly used as a simple model expressing the above-mentioned feature:

$$\Phi_{LJ}(r_{ij}) = 4\varepsilon \left[\left(\frac{\sigma}{r_{ij}} \right)^{12} - \left(\frac{\sigma}{r_{ij}} \right)^{6} \right], \tag{31}$$

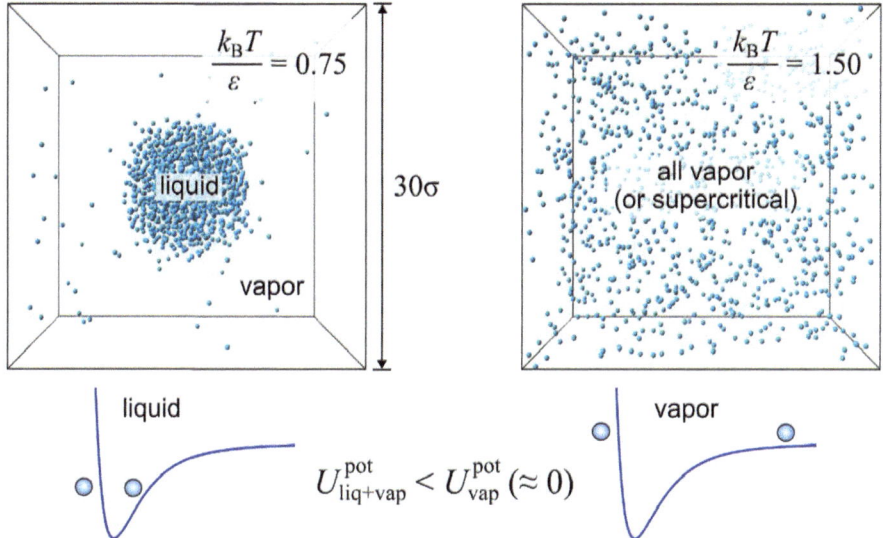

Fig. 6 MD simulation results of simple Lennard–Jones molecules confined in a calculation cell

where ε and σ denote the potential well depth and diameter shown in Fig. 4. One thousand molecules are confined in a cubic box sized $30\sigma \times 30\sigma \times 30\sigma$ with the time-averaged system temperatures, i.e., kinetic energy given by Eq. (30), set at $k_{\mathrm{B}}T/\varepsilon$ equal to 0.75 and 1.5 in the left and right systems, respectively.

Note that the molecules in the systems are constantly moving as described above, and even if one starts a simulation from arbitrary initial configuration, for instance, with locating the molecules at grid points, the molecules spontaneously form corresponding equilibrium states after a certain time as in the snapshots. This term 'equilibrium' in the microscopic scale roughly means that the molecules are moving with the positions and momenta of the constituent molecules changing with time, but their apparent feature as a group is unchanged. For instance, most of the constituent molecules are gathered to form a spherical liquid droplet with remaining molecules flying as vapor around the droplet in the left system as an equilibrium state; the constituent molecules of liquid and vapor change but the volume as well as the spherical structure of the droplet are unchanged. Similarly, the positions of momenta of the molecules change, but they always keep vapor phase in the right system after reaching the equilibrium state spontaneously achieved after a certain time irrespective of the initial condition. As described later, this spontaneous change of the system is due to the minimization of the Helmholtz free energy of the system.

As clearly observed from the figure, liquid–vapor coexistence kept at the lower temperature of $k_{\mathrm{B}}T/\varepsilon = 0.75$ in the left panel is lost with the temperature rise, and the system is filled with vapor (indeed supercritical phase) at the higher temperature of $k_{\mathrm{B}}T/\varepsilon = 1.5$. This phase change can be intuitively understood from the schematic in Fig. 5.

We further think about why the droplet in the left panel takes the spherical structure to minimize the surface area from an energetic point of view. Let the potential energy per molecule in the vapor bulk, at the LV interface, and in the liquid bulk be e_V, e_{LV} and e_L, respectively. Then, they satisfy

$$e_L < e_{LV} \ll e_V \approx 0 \tag{32}$$

considering the schematic in Fig. 2 and intermolecular potential in Fig. 5, which indicates that e_V and e_{LV} are basically negative whilst e_L is almost zero for molecules in the vapor phase without partners to interact with. The first inequality implies that the number of neighbouring molecules is smaller at the interface than in the liquid bulk. Consequently

$$e_V - e_{LV} > 0 \tag{33}$$

holds. In addition, let the number of corresponding molecules and volumes being N_L, N_{LV} and N_V, and V_L, V_{LV} and V_V, respectively. Then, the number of molecules N and volume V of the system are

$$N = N_L + N_{LV} + N_V \tag{34}$$

and

$$V = V_L + V_{LV} + V_V, \tag{35}$$

respectively. In the macroscopic scale, we assume $N_{LV} = 0$ and $V_{LV} = 0$ while the LV interface is a region in the molecular scale and molecules indeed exist there. We define the corresponding volumes per molecule as

$$v_L = \frac{V_L}{N_L}, \ v_{LV} = \frac{V_{LV}}{N_{LV}}, \ v_V = \frac{V_V}{N_V}. \tag{36}$$

Assuming that v_{LV} is approximately the same as v_L, it follows for the volume per molecule that

$$0 < v_L \approx v_{LV} \ll v_V. \tag{37}$$

Then, Eq. (35) is rewritten as

$$V = N_V v_V + + N_{LV} v_{LV} + N_V v_L \approx (N_L + N_{LV}) v_L + N_V v_V. \tag{38}$$

On the other hand, the internal energy U of a system is separated into the kinetic and potential contributions U^{kin} and U^{pot}, respectively as

$$U = U^{\text{kin}} + U^{\text{pot}}. \tag{39}$$

Using Eq. (30), the former is given by

$$U^{\text{kin}} \equiv \sum_{i=1}^{N} \frac{1}{2} m |v_i|^2 = \frac{3}{2} N k_B T, \qquad (40)$$

which is constant under the constant temperature condition. The latter of potential term is formally given by

$$U^{\text{pot}} \equiv \sum_{i=1} \sum_{j(>i)} \Phi(r_{ij}). \qquad (41)$$

This can be approximated using the mean potential energy per molecule as

$$\begin{aligned} U^{\text{pot}} &\approx N_V e_V + N_{\text{LV}} e_{\text{LV}} + N_L e_L \\ &= (N_L + N_{\text{LV}}) e_L + N_{\text{LV}} (e_{\text{LV}} - e_L). \end{aligned} \qquad (42)$$

From Eqs. (38)-(42) and Inequality (33), it is shown that with keeping $N_L + N_{\text{LV}}$ constant, reducing the internal energy U in Eq. (42) is possible by decreasing N_{LV} without changing the volume V in Eq. (38). This means that the system internal energy can be reduced by decreasing the surface area A_{LV} because N_{LV} is apparently proportional to A_{LV}. This leads to the formation of the spherical droplet in the left panel of Fig. 6 as an equilibrium shape with the smallest surface area in the three-dimensional space.

2.3.3 Free Energy and Entropy

The explanation above looks reasonable; however, another question arises why the system prefers to take the 'all vapor' state with less internal energy U at higher temperatures. The trap model in Fig. 5 merely indicates that the molecular pairs easily dissociate from each other, but they can have a lower internal energy even at a higher temperature if they form more intermolecular pairs. To answer this question, the free energy instead of the internal energy must be considered. The second law of thermodynamics tells us that the Helmholtz free energy F of a system given by

$$F \equiv U - TS \qquad (43)$$

decreases until the system reaches the equilibrium state, i.e.,

$$dF \leq 0 \text{ (nonequilibrium)}, \quad F = F_{\min}(\text{equilibrium}) \qquad (44)$$

holds for F for a system under constant volume V and constant temperature T condition. This corresponds to a closed container in contact with a constanttemperature heat bath. The key point is that the Helmholtz free energy in Eq. (43) includes the entropy of the system S.

In classical statistical mechanics, the entropy S is related to the "number of possible equivalent microstates corresponding to a certain macroscopic state" Ω as follows[1]:

$$S = k_B \text{ in } \Omega, \qquad (45)$$

where a microstate in 3-dimensional system of N-molecules is determined by giving $3N$-positions and $3N$-momenta of constituent molecules. From Eq. (45), it is possible to show that the position contribution to entropy S is the largest for a system with homogeneous density. We will see that in the following example:

(Example) Suppose a closed equilibrium system with ideal gas molecules in a box at a constant temperature, and let N_{left} and N_{right} be the numbers of gas molecules in the left-half and right-half of a box, respectively, and denote the case by $[N_{\text{left}}, N_{\text{right}}]$ ($N = N_{\text{left}} + N_{\text{right}}$). Evaluate the number of possible cases and compare the two for the following three cases, respectively:

1. Case of small total number $N_{\text{total}} = 200$: [99,101] versus [100,100]
2. Case of a larger total number $N_{\text{total}} = 2000$: [990, 1010] versus [1000, 1000]
3. Case of a huge (Avogadro scale) total number of $N_{\text{total}} = 2 \times 10^{23}$: $[0.99N_A, 1.01N_A]$ versus $[N_{\text{left}} = N_A, N_{\text{right}} = N_A]$

(Answer) The numbers of cases Ω to separate $N_{\text{left}} + N_{\text{right}}$ into left and right are given by

1. $N_{\text{total}} = 200$:

$$\Omega_{[99,101]} = \frac{200!}{99!101!}, \ \Omega_{[100,100]} = \frac{200!}{100!100!}$$

$$\frac{\Omega_{[99,101]}}{\Omega_{[100,100]}} = \frac{100!100!}{99!101!} = \frac{100}{101}.$$

2. $N_{\text{total}} = 2000$:

$$\Omega_{[990,1010]} = \frac{2000!}{990!1010!}, \ \Omega_{[1000,1000]} = \frac{2000!}{1000!1000!}$$

$$\frac{\Omega_{[990,1010]}}{\Omega_{[1000,1000]}} = \frac{1000!1000!}{990!1010!} = \frac{991}{1001} \cdot \frac{992}{1002} \cdots \frac{1000}{1010}.$$

The ratio above satisfies

$$\left(\frac{99}{100}\right)^{10} < \frac{\Omega_{[990,1010]}}{\Omega_{[1000,1000]}} < \left(\frac{100}{101}\right)^{10} \ \therefore \ \frac{\Omega_{[990,1010]}}{\Omega_{[1000,1000]}} \approx \left(\frac{99}{100}\right)^{10}.$$

---

[1] Boltzmann's tombstone bears the inscription of '$S = k \log W$'.

3. $N_{\text{total}} = 2 \times 10^{23}$:

$$\Omega_{[0.99N_A, 1.01N_A]} = \frac{2N_A!}{(0.99N_A)!(1.01N_A)!}, \quad \Omega_{[N_A, N_A]} = \frac{2N_A!}{N_A!N_A!}$$

$$\frac{\Omega_{[0.99N_A, 1.01N_A]}}{\Omega_{[N_A, N_A]}} = \frac{N_A!N_A!}{(0.99N_A)!(1.01N_A)!} \approx \left(\frac{99}{100}\right)^{0.01N_A} > 0. \tag{46}$$

Equation (46) indicates that the probability to see 2% density difference (0.99 ρ_0 and 1.01 ρ_0) between the left and right for ideal gas systems approaches to zero for $N_A \gg 1$. In addition, by inserting Eq. (45) into Eq. (46), it follows for the entropy difference ΔS between $S_{[N_A, N_A]}$ and $S_{[0.99N_A, 1.01N_A]}$ that

$$\Delta S \equiv S_{[N_A, N_A]} - S_{[0.99N_A, 1.01N_A]} \approx 0.01 N_A k_B \ln\left(\frac{100}{99}\right) > 0. \tag{47}$$

This indicates that the entropy S is the largest for uniform density, and also that S is proportional to the number of molecules N_A.

Now we go back to the comparison of the Helmholtz free energy in Fig. 6. The entropy S is larger for a system with homogeneous density than for an inhomogeneous system, meaning that the right system is advantageous from the entropy aspect. At low temperature, U contribution in Eq. (43) is dominant and the system tries to decrease U to minimize F, whereas at high temperature, the entropy contribution TS overcomes the potential minimization effect. This balance between the potential energy and the entropy governs the basic mechanism of phase separation as well as the surface tension. Note that for ideal gas systems, the potential contribution is constant (zero for the potential in Fig. 4, and the gas molecules tries to fill the system with a homogeneous density to maximize entropy S. Note as well that if a solid surface exists near the droplet at a temperature for the liquid–vapor phase separation, and the droplet is thermodynamically more stable on the surface, a hemispherical droplet is formed on the solid surface, and the equilibrium shape, i.e., the contact angle θ in Fig. 3, is determined so that the total free energy of the system may become minimum depending on the solid–fluid interaction strength.

2.4 Surface Tension

2.4.1 Bakker's Equation

As seen in the above example, we have to inevitably introduce the relation between the free energy of a fluid with interface and the local force to quantitatively evaluate the surface tension. For that purpose, we set a thought experiment with a flat liquid–vapor interface shown in Fig. 7 for the basic connection between the thermodynamic work as energy and the surface tension. Indeed, such flat interface can easily be achieved by MD simulations by using the periodic boundary condition in the surface lateral directions. In this thought experiment, one piston is set normal to a flat liquid-vapor interface, and it covers from $z = z_L$ to $z = z_V$ at the liquid and vapor bulk regions, respectively, across the plane of the liquid–vapor interface. Another piston parallel to the interface is set in the vapor bulk far from the interface. Through simultaneous virtual infinitesimal displacements of the pistons, only the interface area can be changed without changing the liquid and vapor volumes, V_L and V_V, respectively. Note that the change of bottom area in the figure is not considered. Let l be the depth normal to the xz-plane, and δx be the corresponding displacement of the side piston, the change of interface area A_{LV} is expressed by

$$\delta A_{LV} = l\delta x. \qquad (48)$$

In order to let the internal energies before and after the displacement unchanged both for the liquid and vapor parts, this displacement must be done under constant temperature. Then, the net minimum mechanical work δW exerted from the top and side pistons required for this change with quasi-static process can be associated with the change in the Helmholtz energy F. Let γ_{LV} be the LV interfacial energy per area, it follows for the quasi-static change that

$$\delta F = \gamma_{LV}\delta A_{LV}. \qquad (49)$$

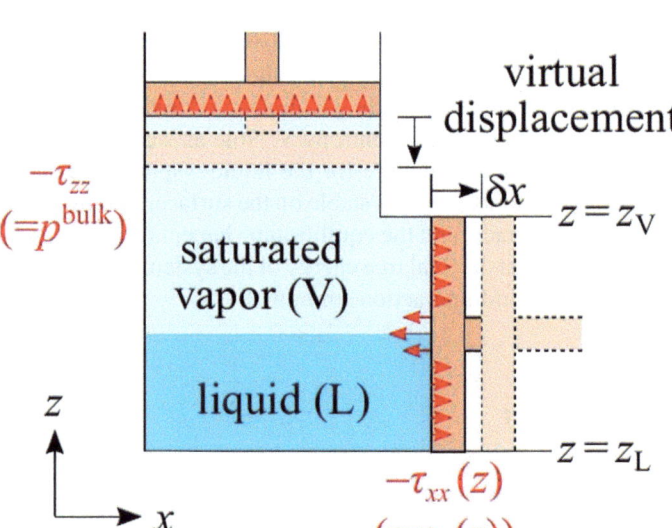

Fig. 7 Thought experiment of Bakker's equation for a flat liquid–vapor interface. The red arrows denote the pressure, i.e., normal fluid stress with its sign inverted, acting on the piston. Reprinted with permission from [23]. Copyright 2019 Author(s), CC BY 4.0

Thus, γ_{LV} is written by

$$\gamma_{LV} = \left(\frac{\partial F}{\partial A_{LV}}\right)_{N,V_L,V_V,T}. \tag{50}$$

This means that surface tension is intrinsically a thermodynamic force.

We further relate the mechanical stress changing as a continuous function with the LV interfacial tension γ_{LV}. The fluid stress tensor is not isotropic at phase interfaces even at equilibrium as shown with the red arrows. Due to the one-dimensional feature of the system along the z-direction, the symmetric stress tensor τ is expressed by

$$\tau = \begin{pmatrix} \tau_{xx} & 0 & 0 \\ 0 & \tau_{yy} & 0 \\ 0 & 0 & \tau_{zz} \end{pmatrix}, \tag{51}$$

where the surface-normal components τ_{zz} is constant in the entire region because of the force balance in the z-direction to be satisfied in the static equilibrium system for the present z-normal flat LV interface system. This constant value is equal to the saturated vapor pressure p^{bulk} with its sign inverted, i.e.,

$$\tau_{zz} = -p^{bulk}. \tag{52}$$

On the other hand, the surface-lateral diagonal components τ_{xx} and τ_{yy} satisfy

$$\tau_{xx}(z) = \tau_{yy}(z) \equiv -p_T(z), \tag{53}$$

where the surface tangential pressure is denoted by $p_T(z)$, which is a unique function of position z and is equal to the isotropic vapor pressure p^{bulk} in the liquid and vapor bulks. We set the liquid and vapor bulk positions at $z = z_L$ and $z = z_V$, respectively. Using these values, the changes of the volume and the Helmholtz free energy as the work upon the quasi-static displacement δx of the side piston are expressed by

$$\delta V = l\delta x \int_{z_L}^{z_V} dz \tag{54}$$

and

$$dF \equiv \delta W = p^{bulk}\delta V + l\delta x \int_{z_L}^{z_V} p_T(z) dz, \tag{55}$$

respectively. By inserting Eqs. (48) and (52)-(55) into Eq. (50),

$$\gamma_{LV} = \lim_{\delta x \to 0} \frac{dF}{dA_{LV}}\bigg|_{N,V_L,V_V,T}$$

$$= p^{\text{bulk}} \int_{z_\text{L}}^{z_\text{v}} dz - \int_{z_\text{L}}^{z_\text{v}} p_T(z) dz = \int_{z_\text{L}}^{z_\text{v}} [p^{\text{bulk}} - p_T(z)] dz,$$

thus

$$\gamma_{\text{LV}} = \int_{z_\text{L}}^{z_\text{v}} \left[\frac{\tau_{xx}(z) + \tau_{yy}(z)}{2} - \tau_{zz} \right] dz \qquad (56)$$

is derived. Equation (56) is called Bakker's equation, which serves as the basic connection between the microscopic anisotropic stress distribution and the surface tension. Note that the integration range in Eq. (56) is sometimes denoted by $\int_{-\infty}^{\infty}$; however, the necessary condition for this range is that z_L and z_V cover the entire range of the interface with anisotropic fluid stress, i.e., between the liquid and vapor bulks.

2.4.2 Molecular Dynamics Simulations

As mentioned in Sect. 2, molecular dynamics analysis is a powerful choice to overcome the theoretical difficulty of obtaining the equilibrium density distribution in liquid–vapor coexistence systems and resulting stress distribution to calculate the surface tension by Bakker's equation. The left panel of Fig. 8 shows a MD simulation system of Lennard–Jones (LJ) molecules (argon: $\sigma = 0.34$ nm, $\varepsilon = 1.65 \times 10^{-21}$ J in Eq. (31) and $m = 39.948$ g/mol) with flat solid–liquid (SL) and liquid–vapor (LV) interfaces, where 2000 argon molecules were confined in a rectangular simulation cell of $4 \times 4 \times 20$ nm^3. The periodic boundary conditions were imposed in the surface lateral x-and y-directions and the mirror boundary condition was set on the top boundary. In addition, a solid wall modelled through an integrated LJ potential was set on the bottom of the system to let the LJ-liquid be adhered on the solid. The system was equilibrated at a temperature of $T = 100$ K ($k_\text{B} T/\varepsilon = 0.835$) until the apparent feature of the system did not change. With such setting, one can easily realize a quasi-one-dimensional equilibrium MD system with flat LV and SL interfaces normal to the z-direction.

It is easy to calculate the density of this equilibrium system by taking the time average of the mass in each flat bin volume set normal to the z-direction with a small thickness of Δz. On the other hand, to calculate the stress, a method called the Method of Plane (MoP) is used, where one sets control surfaces in the system instead of the bin volumes, [19, 22]. The local fluid stress tensor $\tau(x, z)$ is calculated by setting z-normal and x- or y-normal flat bin faces in the system in the Cartesian coordinate.

Figure 9 shows the schematics of the stress tensor calculation by the MoP. The fluid stress tensor component $\tau_{\alpha\beta}$, which expresses the stress in β-direction exerted on a surface element with an outward normal in α-direction, is given by kinetic term $\tau_{\alpha\beta}^{\text{kin}}$ and inter-molecular interaction term $\tau_{\alpha\beta}^{\text{int}}$ as

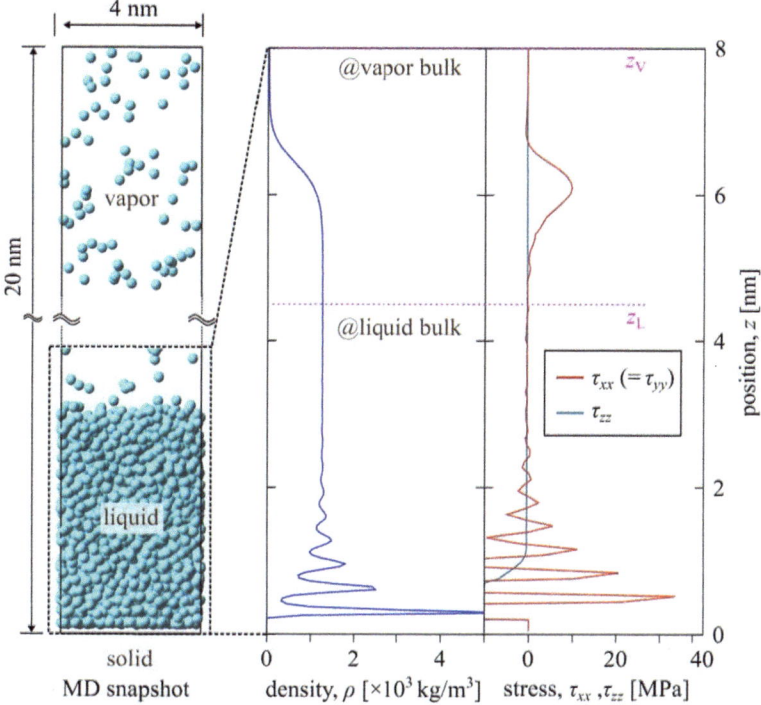

Fig. 8 (Left) snapshot of an equilibrium molecular dynamics simulation system of Lennard–Jones molecules (argon) with flat solid–liquid and liquid–vapor interfaces. (Middle) corresponding density distribution. (Right) distributions of stress diagonal components τ_{xx} and τ_{zz}

$$\tau_{\alpha\beta} = \tau_{\alpha\beta}^{\text{kin}} + \tau_{\alpha\beta}^{\text{int}}. \tag{57}$$

In the MoP, the kinetic term on an α-normal bin face with an area A_α is calculated by

$$\tau_{\alpha\beta}^{\text{kin}} = -\frac{1}{A_\alpha} \left\langle \sum_{i \in \text{fluid}, \delta t}^{\text{across } A_\alpha} \frac{(2\Theta(\mathbf{v}_i \cdot \mathbf{e}_\alpha) - 1) m_i \mathbf{v}_i \cdot \mathbf{e}_\beta}{\delta t} \right\rangle, \tag{58}$$

where m_i and $\mathbf{v}_i$ denotes the mass and velocity vector of i-th fluid molecule, and $\mathbf{e}_\alpha$ and $\mathbf{e}_\beta$ are the unit vectors in α- and β-directions, respectively. The angle brackets means the time average, and the summation $\sum_{i,\delta t}^{\text{across} A_\alpha}$ is taken for every fluid molecule i passing through the bin face within a time interval of δt, which is equal to the time increment for the numerical integration. A switching function $2\Theta(\mathbf{v}_i \cdot \mathbf{e}_\alpha) - 1$, which gives ± 1 depending on the sign of $\mathbf{v}_i \cdot \mathbf{e}_\alpha$ implemented through the Heaviside step function Θ, is included in the RHS of Eq. (58). The meaning of the kinetic stress tensor in Eq. (58) is shown with an example case with $\alpha = x$ and $\beta = y$. Suppose that molecule i with its x-directional velocity v_i^x passes through a x-normal plane

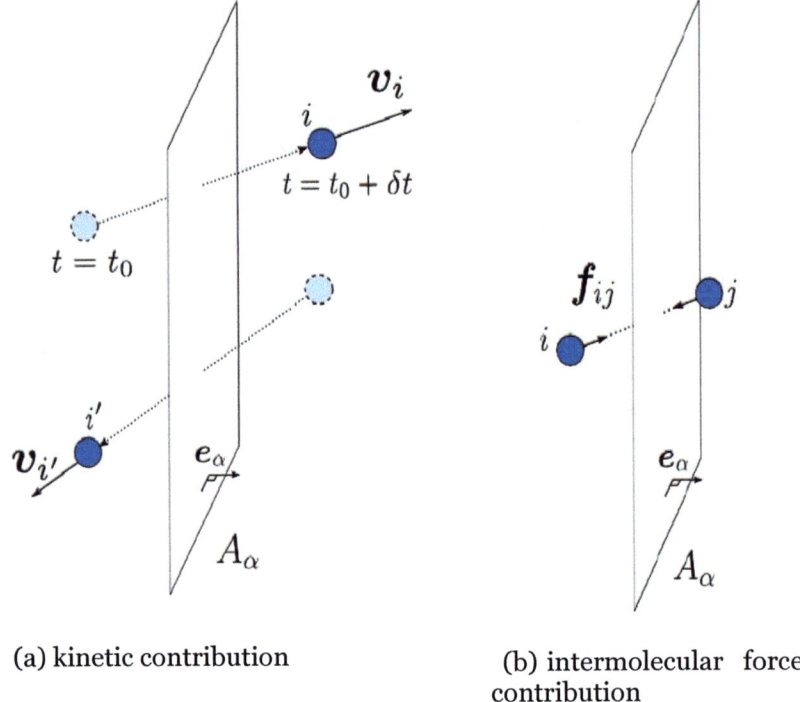

(a) kinetic contribution　　　(b) intermolecular force contribution

Fig. 9 Schematics of the stress tensor calculation based on the method of plane (MoP)

from left to right within a time interval δt as in the upper molecule in Fig. 9a. This is only achieved for positive $v_i^{x(=\alpha)}$ satisfying $\mathbf{v}_i \cdot \mathbf{e}_\alpha > 0$ with $\mathbf{e}_\alpha = (1, 0, 0)$, which corresponds to $2\Theta(\mathbf{v}_i \cdot \mathbf{e}_\alpha) - 1 = 1$. Then, the y-directional momentum of the fluid in the left is reduced by $m_i v_i^y$, i.e., is increased by $-m_i \mathbf{v}_i \cdot \mathbf{e}_\beta$ with $\mathbf{e}_\beta = (0, 1, 0)$. This change of momentum is equivalent to the impulse $\tau_{xy} A \delta t$ with the stress tensor. If the passage is in the opposite direction from right to left as in the lower molecule in Fig. 9a, then, $2\Theta(\mathbf{v}_i \cdot \mathbf{e}_\alpha) - 1 = -1$ and a momentum of $m_i \mathbf{v}_i \cdot \mathbf{e}_\beta$ is given to the fluid in the left. From this formulation, it is known that the kinetic contribution to surface normal stress $\tau_{\alpha\alpha}$ is always negative, i.e., results in positive pressure, and this indeed corresponds to the ideal gas pressure p given in Eq. (25), which is proportional to the density ρ on the face, and is also proportional to the temperature T from Eq. (30).

On the other hand, the intermolecular interaction term in Eq. (57) is given by

$$\tau_{\alpha\beta}^{\text{int}} = -\frac{1}{A_\alpha} \left\langle \sum_{(i,j) \in \text{fluid}}^{\text{across } A_\alpha} (2\Theta(\mathbf{r}_{ij} \cdot \mathbf{e}_\alpha) - 1) \mathbf{f}_{ij} \cdot \mathbf{e}_\beta \right\rangle, \qquad (59)$$

where $\mathbf{r}_{ij}$ and $\mathbf{f}_{ij}$ denote the relative position vector $\mathbf{r}_j - \mathbf{r}_i$ and force vector exerted on molecule j at position $\mathbf{r}_j$ from molecule i at $\mathbf{r}_i$, respectively, and the summation

$\sum_{(i,j)\in \text{fluid}}^{\text{across}} A_\alpha$ was taken for all line segments between $\mathbf{r}_i$ and $\mathbf{r}_j$ which crossed the bin face. In contrast to the kinetic contribution, this interaction contribution may give positive surface normal stress $\tau_{\alpha\alpha}$ on the bin face with an attractive interaction f_{ij}.

The middle and right panels of Fig. 8 show the density and surface-normal stress distributions, respectively. As expected, the LV interface with a large change in the density has a certain thickness, and in this region with non-constant density, the normal stress τ_{xx} tangential to the interface is different from the bulk stress τ_{zz}, which is constant in the whole system except near the solid at which the fluid is subject to an external force from the solid. Note that a remarkable wiggling structures in the density and stress at bottom is specific to the solid–liquid interface, and this indeed is an interesting topic especially related to the wetting; however, the effect vanishes as leaving away from the solid surface, and we mainly focus on the liquid–vapor interface. It is really amazing that the frontiers could expect the stress distribution at least around the LV interface without MD simulations. Also note that even by such great scientists, quantitatively evaluating the density and stress distributions only from the intermolecular potential was still difficult, and that we have a powerful tool of MD simulations to complement the missing piece. Related to this point, calculating the SL (and SV) interfacial tension from the stress distribution obtained through the MD simulations toward the understanding of wetting is a hot topic at present [10, 17, 23] however, I will not describe about this in detail considering the scope of this article.

A very simple and intuitive model to explain the stress anisotropy at the liquid vapor interface, i.e., $\tau_{zz} < \tau_{xx}$, is given here. Suppose z-normal and x-normal planes on a z-normal flat LV interface for the calculation of the normal stress components τ_{zz} and τ_{xx}, respectively based on the MoP as shown in Fig. 10. In this model, we assume average densities ρ_L and $\rho_V (< \rho_L)$ and $\frac{\rho_L + \rho_V}{2}$ in the liquid and vapor bulks and on the interface, respectively for simplicity. In addition, we introduce a mean-field like approach to the MoP here; instead of counting the molecules passing through the MoP plane for the kinetic stress term τ^{kin} in Eq. (58) or summing up the intermolecular interaction force for the line segment between two molecules crossing the MoP plane for the interaction stress term τ_{int} in Eq. (59), we evaluate τ^{kin} with the density on the point of interest (black cross) on the MoP plane, and τ^{int} with the two densities on the points which sandwich the point of interest (double-headed red arrow) in Fig. 10. As easily imagined, to correctly calculate the latter interaction term of τ^{int}, complicated space integration for the two points using a position- and direction-dependent radial distribution function is needed, and this is indeed the difficulties that the great scientists could not solve analytically, but here we try to represent the interaction term by the most probable single distance in the present simplest model. The kinetic terms $\boldsymbol{\tau}^{\text{kin}}$ is isotropic in equilibrium systems and is given by

$$\boldsymbol{\tau}^{\text{kin}} = -\frac{Nk_B T}{V}\mathbf{I} = -\rho\frac{k_B T}{m}\mathbf{I}, \tag{60}$$

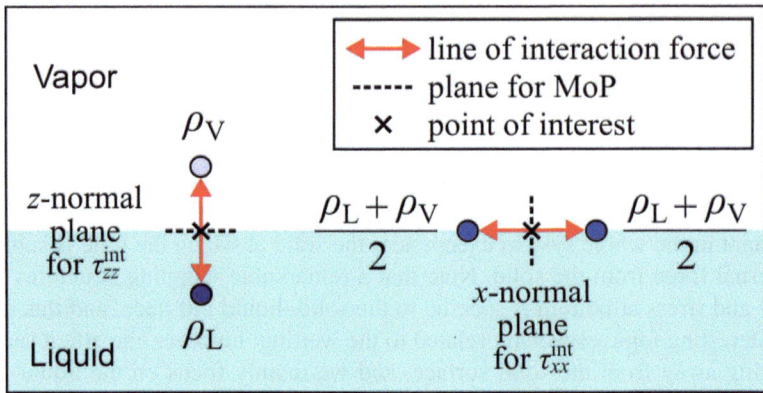

Fig. 10 The simplest model to explain the stress anisotropy at the interface. The interaction terms τ_{zz}^{int} and τ_{xx}^{int} on the point of interest (black crosses) are represented by the interaction between two points having densities (ρ_L, ρ_V), and $\left(\frac{\rho_L+\rho_V}{2}, \frac{\rho_L+\rho_V}{2}\right)$, respectively connected by the double-headed red arrows

where I denotes the identity tensor. Equation (60) corresponds to the ideal-gas pressure, which can be interpreted with Eqs. (30) and (58) as follows: fast molecules giving a momentum of $m v_i$ in Eq. (58) pass through the plane more with a chance proportional to $|v_i|$. From Eq. (60), τ_{zz}^{kin} and τ_{xx}^{kin} are identical and those at the LV interface at $z = z_{LV}$ are expressed by

$$\tau_{zz}^{kin}(z_{LV}) = \tau_{xx}^{kin}(z_{LV}) = -\frac{\rho_L + \rho_V}{2}\frac{k_B T}{m}. \tag{61}$$

On the other hand, we assume that the interaction term is represented by a single line of interaction force of the most probable distance (double-headed red arrows) which gives an intermolecular force $f_{rep}(z)$, then the interaction term $\tau_{\alpha\alpha}^{int}$ is written as

$$\tau_{\alpha\alpha}^{int}(z) \equiv -\rho_1 \rho_2 c(z) f_{rep}(z), \tag{62}$$

where ρ_1 and ρ_2 are the density of the two points sandwiching the point of interest, and $c(> 0)$ is a certain positive coefficient for the model due to this representation. Note that negative $f_{rep}(z)$ corresponds to attractive force across the MoP plane to give positive normal stress. Since τ_{zz} must satisfy Eq. (52), $\tau_{zz}(z_{LV})$ at the interface $z = z_{LV}$ given by

$$\begin{aligned}\tau_{zz}(z_{LV}) &= \tau_{zz}^{kin}(z_{LV}) + \tau_{zz}^{int}(z_{LV}) \\ &= -\frac{\rho_L + \rho_V}{2}\frac{k_B T}{m} - \rho_V \rho_L c(z_{LV}) f_{rep}(z_{LV})\end{aligned} \tag{63}$$

with $(\rho_1, \rho_2) = (\rho_L, \rho_V)$ in Eq. (62), is equal to $\tau_{zz}(z_V)$ in the vapor bulk at $z = z_V$, corresponding to the saturated vapor pressure. This is given by

$$\tau_{zz}(z_V) = \tau_{zz}^{\text{kin}}(z_V) + \tau_{zz}^{\text{int}}(z_V)$$
$$= -\rho_V \frac{k_B T}{m} - \rho_V^2 c(z_V) f_{\text{rep}}(z_V)$$
$$\approx -\rho_V \frac{k_B T}{m}, \qquad (64)$$

where the interaction term is assumed to be zero for the last approximation considering that the intermolecular interaction in the vapor is negligible. Thus, it follows

$$c(z_{LV}) f_{\text{rep}}(z_{LV}) = -\frac{\rho_L - \rho_V}{2} \frac{k_B T}{m} < 0, \qquad (65)$$

indicating that $f_{\text{rep}}(z_{LV})$ is negative, i.e., the representative interaction across the MoP plane at the LV interface is attractive. Assuming as well that $c(z_{LV}) f_{\text{rep}}(z_{LV})$ is independent of the direction at z_{LV}, then, the normal stress lateral to the interface is written as

$$\tau_{xx}|_{z=z_{LV}} = -\frac{\rho_L + \rho_V}{2} \frac{k_B T}{m} - \left(\frac{\rho_V + \rho_L}{2}\right)^2 c(z_{LV}) f_{\text{rep}}(z_{LV}). \qquad (66)$$

Using the relation between the arithmetic and geometric means expressed by

$$\rho_L \rho_V < \left(\frac{\rho_L + \rho_V}{2}\right)^2, \qquad (67)$$

it follows for $\tau_{zz}(z_{LV})$ and $\tau_{xx}(z_{LV})$ in Eqs. (63) and (66), respectively that

$$\tau_{zz}(z_{LV}) < \tau_{xx}(z_{LV}). \qquad (68)$$

This inequality is a simple model to explain the stress anisotropy shown in Fig. 8, due to the density anisotropy, and this is the fundamental basis of Bakker's equation (56).

The left panel of Fig. 11 shows the distributions of the normal stress τ_{xx} tangential to the interface and density ρ at three different temperatures $T = 90, 95$ and 100 K enlarged around the LV interface. As the temperature increases, the stress difference between the values of $\tau_{xx}(z)$ at the interface and away from the interface becomes smaller. The latter away from the interface is equal to and τ_{zz}, which is equal to the constant stress value away from the LV interface as shown in Fig. 8. From Bakker's Eq. (56), this means that the LV interfacial tension γ_{LV} decreases with the temperature rise. Note also that the bulk liquid density decreases with the temperature rise, and this

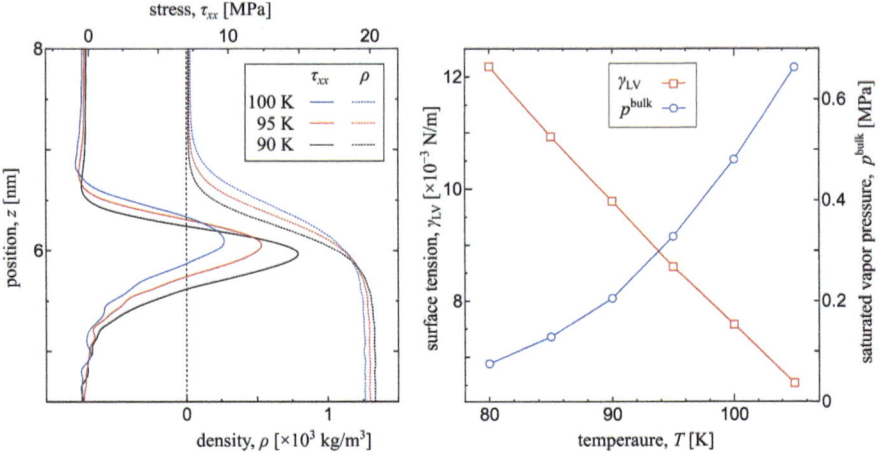

Fig. 11 (Left) distributions of the normal stress τ_{xx} tangential to the interface and density ρ at three different temperatures $T = 90, 95$ and 100 K enlarged around the LV interface. (Right) liquid–vapor interfacial tension γ_{LV} and p^{bulk} at different temperature T

is related to the increase of the average distance upon temperature rise indicated in Fig. 5. In addition, the bulk stress decreases with the temperature rise, and this means that the saturated vapor pressure p_V^{bulk} given by Eq. (52) increases with temperature. Such basic features are indeed realized in the MD systems.

The right panel of Fig. 11 shows the LV interfacial tension γ_{LV} and p^{bulk} at different temperature T. As expected from the right panel, γ_{LV} decreases and p^{bulk} increases with the temperature increase. The vapor pressure is above the atmospheric pressure, and the extraordinary high anisotropic local stress τ_{xx} gives the interfacial tension around 10×10^{-3} N/m, and the resulting temperature dependence of about -0.2×10^{-3} N/m $\cdot$ K. Of course, the absolute value and the coefficient of the temperature dependence depend on the molecular type and temperature range, but in general, if a temperature is anisotropic at an interface, this nonnegligible temperature dependence gives a local stress gradient around the interface known as the thermocapillary effect, which drives the liquid flow from a higher-temperature region to a lower-temperature region.

3 Concluding Remarks

Surface is physically as well as theoretically at the boundary of macroscopic and microscopic physics, understanding of surface tension intrinsically needs multiple knowledge of thermodynamics, statistical mechanics as well as continuum mechanics. In this chapter, the thermodynamic aspect of the surface tension was first described briefly. Then after introducing the interpretation of the phase change, liquid–vapor coexistence and interfacial tension from these points of view, molecular dynamics results were shown which indicate the existence of microscopic stress anisotropy at the interface as the origin of the macroscopic stress surface tension tangential to the interface. In addition, a simple model to explain the stress anisotropy due to the density anisotropy was provided. At present, molecular dynamics as a powerful tool is available, and we hope that unsolved issues about this fascinating physics, especially those related to the Marangoni effect and wetting, will be elucidated.

Acknowledgements The simulation results in Section 2 were provided by Minori Shintaku and Yule Ding, former members of the present author Yamaguchi's group. He would sincerely appreciate their cooperation.

References

1. Bernoulli D (1738) Hydrodynamica, sive de viribus et motibus fluidorum commentarii: Opus Academicum
2. Bruus H (2008) Theoretical Microfluidics. Oxford Univ Press: 123–134
3. Castellan GW, (1983) Physical chemistry, Addison-Wesley, Massachusetts: 420–422
4. Erbil Y (2006) Surface chemistry of solid and liquid interfaces, Blackwell: 90–97
5. Fujiwara K, Hyodo T (1995) Introduction to Thermal Physics. University of Tokyo Press, Tokyo (in Japanese), From a Macroscopic to Microscopic Approach
6. Gao L, McCarthy TJ (2009) Wetting 101. Langmuir 25:14105–14115
7. de Gennes P-G., Brochard-Wyart F, Quere D (2003) Capillarity and wetting phenomena: Drops, Bubbles, Pearls, Waves Springer, Berlin
8. Gromacs.org. http://www.gromacs.org. Accessed 22 Sept 2023
9. Kirby BJ, (2010) Micro- and Nanoscale Fluid Mechanics. Cambridge University Press, Cambridge, 21–24.
10. Kusudo H, Omori T, Yamaguchi Y (2021) Local stress tensor calculation by the method-of plane in microscopic systems with macroscopic flow: A formulation based on the velocity distribution function. J Chem Phys 155: 184103
11. Marchand A, Weijs JH, Snoeijer JH, Andreotti B (2011) Why is surface tension a force parallel to the interface? Am J Phys 79:999–1008
12. Ono S (1980) Surface tension. Kyoritsu, Tokyo (in Japanese)
13. Ono S, Kondo S (1960) Molecular theory of surface tension in liquids. Springer, Berlin, pp 134–280
14. Rowlinson JS (1979) Translation of van der Waals JD. The thermodynamic theory of capillarity under the hypothesis of a continuous variation of density. J Stat Phys 20:197–200
15. Rowlinson JS (2002) Cohesion: a scientific history of intermolecular forces. Cambridge University Press
16. Rowlinson JS, Widom B (1982) Molecular Theory of Capillarity. Dover, New York
17. Surblys D, Yamaguchi Y, Kuroda K, Kagawa M, Nakajima T, Fujimura H (2014) Molecular dynamics analysis on wetting and interfacial properties of water-alcohol mixture droplets on a solid surface. J Chem Phys 140:034505
18. Thompson AP et al (2022) LAMMPS—a flexible simulation tool for particle based materials modeling at the atomic, meso, and continuum scales. Comp Phys Comm 271:108171
19. Thompson SM, Gubbins KE, Walton JPRB, Chantry RAR, Rowlinson JS (1984) A molecular dynamics study of liquid drops. J Chem Phys 81:530–542

20. van der Waals JD (1893) Thermodynamische theorie der capillariteit in de onderstelling van continue dichtheidsverandering. Verhandel, Konink, Akad, Weten, Amsterdam (Sect. 1), 1(8)
21. Wilhelmy L (1863) Über die Abhängigkeit der Capillaritäts-Constanten des Alkohols von Substanz und Gestalt des benetzten festen Körpers. Ann Phys 195:177–217
22. Yaguchi H, Yano T, Fujikawa S (2010) Molecular Dynamics Study of Vapor-Liquid Equilibrium State of an Argon Nanodroplet and Its Vapor. J Fluid Sci Tech 5:180–191
23. Yamaguchi Y, Kusudo H, Surblys D, Omori T, Kikugawa G (2019) Interpretation of Young's equation for a liquid droplet on a flat and smooth solid surface: Mechanical and thermodynamic routes with a simple Lennard-Jones liquid, J Chem Phys 150: 044701
24. Young T (1805) An essay on the cohesion of fluids. Philos Trans R Soc Lond 95:65–87

Open Access This chapter is licensed under the terms of the Creative Commons Attribution 4.0 International License (http://creativecommons.org/licenses/by/4.0/), which permits use, sharing, adaptation, distribution and reproduction in any medium or format, as long as you give appropriate credit to the original author(s) and the source, provide a link to the Creative Commons license and indicate if changes were made.

The images or other third party material in this chapter are included in the chapter's Creative Commons license, unless indicated otherwise in a credit line to the material. If material is not included in the chapter's Creative Commons license and your intended use is not permitted by statutory regulation or exceeds the permitted use, you will need to obtain permission directly from the copyright holder.

Thermocapillary Convection in an Infinite Liquid Layer and in an Infinite Liquid Column

Hiroshi Kawamura and Kaoru Fujimura

Abstract In this chapter, we will describe the thermocapillary convection, which is induced by a gradient of the surface tension. The first section will be devoted to mathematical background, and in the subsequent sections, development of our understanding of the thermocapillary convection will be described with aid of historical viewpoints. Already since the early twentieth century the buoyancy-driven convection in liquids caused by the thermal expansion was carefully observed and analyzed by the scientists. Later in the mid-twentieth century, the convection due to gradient of the surface tension was noticed in industrial field and scientific analyses were intensively performed. Among them, the emergence of the traveling waves of velocity and temperature fluctuations was observed experimentally and then confirmed theoretically, leading to the designation "hydrothermal wave," which has attracted both scientific and practical interest. The major tool of the analysis in this field has been, and remains, the Linear Stability Analysis (LSA), whose fundamental aspects will be explained in this chapter with respect to basic configurations, such as infinitely large planar liquid layers and infinitely long cylindrical columns, as canonical examples. Furthermore, topics like the transition conditions to the oscillatory flows, traveling direction of the waves, and the influence of the heat exchange with environment are also discussed.

H. Kawamura (✉)
Professor Emeritus, Tokyo University of Science, Tokyo, Japan
e-mail: kawa@rs.sus.ac.jp

Professor Emeritus, Suwa Science University, Chino, Nagano, Japan

K. Fujimura
Professor Emeritus, Tottori University, Tottori, Japan
e-mail: kaoru@tottori-u.ac.jp

© The Author(s) 2025
H. Kawamura et al. (eds.), *Thermocapillary Convection in Microgravity*,
Fluid Mechanics and Its Applications 139,
https://doi.org/10.1007/978-981-96-2991-6_3

1 Thermo-Hydraulic Equations and Nondimensional Numbers Related to the Thermocapillary Convection

Hiroshi Kawamura

Figure 1(a) and (c) illustrate two major coordinate systems treated in this book, that is, the rectangular and cylindrical coordinates. First, the rectangular system of Fig. 1(a) is treated. A liquid layer with a depth of d is assumed and the deformation of the surface is neglected. The fundamental equations consist of the continuity, the momentum, and the energy equations along with necessary boundary conditions. One unique feature of the thermocapillary convection is that the driving force, i.e., the thermocapillarity, appears only in the boundary condition and not in the momentum equation itself. In contrast, the buoyant force influences directly the momentum equation for the buoyant flow.

Let u, v and w be the x, y and z components of the velocity, and p the pressure. In the vector form,

$$\boldsymbol{u} = (u, v, w), \tag{1}$$

$$\nabla = \left(\frac{\partial}{\partial x}, \frac{\partial}{\partial y}, \frac{\partial}{\partial z}\right), \tag{2}$$

$$\nabla^2 = \frac{\partial^2}{\partial x^2} + \frac{\partial^2}{\partial y^2} + \frac{\partial^2}{\partial z^2}, \tag{3}$$

$$\frac{D}{Dt} = \frac{\partial}{\partial t} + \boldsymbol{u} \cdot \nabla = \frac{\partial}{\partial t} + u\frac{\partial}{\partial x} + v\frac{\partial}{\partial y} + w\frac{\partial}{\partial z}. \tag{4}$$

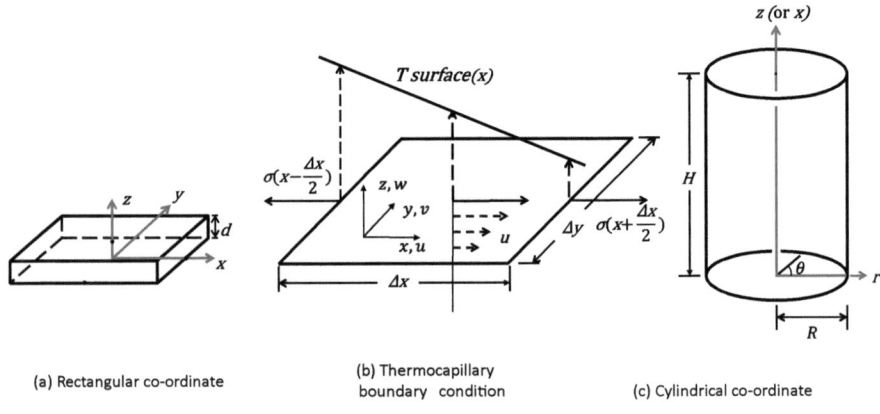

Fig. 1 Coordinate system and thermocapillary boundary condition

The physical properties are assumed constant in general, but dependencies of the surface tension and the density on the temperature are considered when their effects are significant.

With these simplifications, the continuity equation becomes

$$\frac{\partial u}{\partial x} + \frac{\partial v}{\partial y} + \frac{\partial w}{\partial z} = 0, \tag{5}$$

or in the vector form

$$\nabla \cdot \boldsymbol{u} = 0. \tag{6}$$

The momentum equations are

$$\rho \left(\frac{\partial u}{\partial t} + u\frac{\partial u}{\partial x} + v\frac{\partial u}{\partial y} + w\frac{\partial u}{\partial z} \right) = -\frac{\partial p}{\partial x} + \mu \left(\frac{\partial^2 u}{\partial x^2} + \frac{\partial^2 u}{\partial y^2} + \frac{\partial^2 u}{\partial z^2} \right), \tag{7}$$

$$\rho \left(\frac{\partial v}{\partial t} + u\frac{\partial v}{\partial x} + v\frac{\partial v}{\partial y} + w\frac{\partial v}{\partial z} \right) = -\frac{\partial p}{\partial y} + \mu \left(\frac{\partial^2 v}{\partial x^2} + \frac{\partial^2 v}{\partial y^2} + \frac{\partial^2 v}{\partial z^2} \right), \tag{8}$$

$$\rho \left(\frac{\partial w}{\partial t} + u\frac{\partial w}{\partial x} + v\frac{\partial w}{\partial y} + w\frac{\partial w}{\partial z} \right) = -\frac{\partial p}{\partial z} + \mu \left(\frac{\partial^2 w}{\partial x^2} + \frac{\partial^2 w}{\partial y^2} + \frac{\partial^2 w}{\partial z^2} \right), \tag{9}$$

or

$$\rho \frac{D}{Dt}\boldsymbol{u} = -\nabla p + \mu \nabla^2 \boldsymbol{u}, \tag{10}$$

where ρ is the density (kg/m^3), μ is the dynamic viscosity (Ns/m^2), and ∇^2 is the Laplacian operator, $\nabla^2 = \frac{\partial^2}{\partial x^2} + \frac{\partial^2}{\partial y^2} + \frac{\partial^2}{\partial z^2}$. The energy equation is

$$\rho c_p \left(\frac{\partial T}{\partial t} + u\frac{\partial T}{\partial x} + v\frac{\partial T}{\partial y} + w\frac{\partial T}{\partial z} \right) = \lambda \left(\frac{\partial^2 T}{\partial x^2} + \frac{\partial^2 T}{\partial y^2} + \frac{\partial^2 T}{\partial z^2} \right), \tag{11}$$

or

$$\rho c_p \frac{D}{Dt} T = \lambda \nabla^2 T, \tag{12}$$

where c_p is the heat capacity at the constant pressure (J/kgK) and λ the thermal conductivity (W/mK).

As for the boundary conditions, at a rigid wall ($z = 0$ in this example (a)), all the velocity components are zero, $u = v = w = 0$. Over a free surface ($z = d$), the usual condition is zero shear stress, meaning that the normal derivatives of velocity components are zero, that is, $\frac{\partial u}{\partial z} = \frac{\partial v}{\partial z} = \frac{\partial w}{\partial z} = 0$.

Next, boundary conditions related to the thermocapillarity will be discussed. First, we will consider a simplified example where a small rectangular element $\Delta x \cdot \Delta y$ is applied on the free surface at $z = d$ and a temperature gradient is applied in x direction as depicted in Fig. 1(b). Since the surface tension depends on the temperature, the surface tensions at both sides of the element are not balanced. Their difference is $\Delta \sigma = \sigma\left(x + \frac{\Delta x}{2}\right) - \sigma\left(x - \frac{\Delta x}{2}\right) = \frac{\partial \sigma}{\partial x} \Delta x$ (N/m). This unbalanced net force must balance with the viscous shear stress $\tau = \mu \frac{\partial u}{\partial z}$ (N/m^2) at $z = d$. Considering the force balance over a small surface area of $\Delta x \cdot \Delta y$ and reminding the surface tension is the force per unit length, we obtain

$$\mu \frac{\partial u}{\partial z} \cdot \Delta x \cdot \Delta y = \Delta \sigma \cdot \Delta y, \tag{13}$$

which results in $\mu \frac{\partial u}{\partial z} \Delta x = \Delta \sigma$ and $\mu \frac{\partial u}{\partial z} = \frac{\partial \sigma}{\partial x}$. Accordingly, the velocity and the temperature gradients $\frac{\partial u}{\partial z}$ and $\frac{\partial T}{\partial x}$ can be correlated as

$$\mu \frac{\partial u}{\partial z} = \frac{\partial \sigma}{\partial x} = \frac{\partial \sigma}{\partial T} \cdot \frac{\partial T}{\partial x} = -\sigma_T \frac{\partial T}{\partial x} \text{ at } z = d, \tag{14}$$

where σ_T is the temperature coefficient of the surface tension. If the temperature gradient exists also in the y-direction, the same procedure gives

$$\mu \frac{\partial v}{\partial z} = \frac{\partial \sigma}{\partial y} = \frac{\partial \sigma}{\partial T} \cdot \frac{\partial T}{\partial y} = -\sigma_T \frac{\partial T}{\partial y} \text{ at } z = d. \tag{15}$$

The boundary condition for the temperature over the surface ($z = 0$ or d) is usually expressed as

$$q(z = 0, d) = -\lambda_f \frac{\partial T}{\partial z}\bigg|_{z=0, d} = h\{T(z = 0, d) - T_\infty\} \text{ at } z = 0, d \tag{16}$$

where q is the heat flux (W/m^2), λ_f the thermal conductivity (W/mK) of the fluid, h the heat transfer coefficient (W/m^{2}K) and T_∞ the environmental temperature. If $h = 0$, the surface is called "insulating or adiabatic", while if $h \to \infty$, then $q/h \to 0$ and thus the surface temperature is held equal to the environmental temperature T_∞.

Now, we are going to nondimensionalize these equations. First, we will introduce specific quantities for the physical quantities such as the length, temperature difference, velocity, pressure, and time. Choosing the characteristic quantity is often self-evident, and there is no general rule; thus it must be specified individually. Some examples are shown below. The specific quantities will be designated by a subscript 0, but "0" will be omitted sometimes for simplicity.

(1) Length: L_0 (Thickness of liquid layer, radius, diameter, or height of cylinder)
(2) Temperature difference: ΔT_0 (Temperature difference to characterize the temperature fields such as the temperature difference across the thickness of a liquid layer or the specific length L_0 times the temperature gradient, $(L_0 \cdot dT/dx)$
(3) Marangoni velocity:

$$U_0 = u_M = \frac{\sigma_T \Delta T_0}{\mu} \qquad (17)$$

(4) Time: $t_0 = L_0/U_0$
(5) Pressure: $P_0 = \rho U_0^2$

With use of these quantities, variables are nondimensionalized as

$$u^* = \frac{u}{U_0},\ v^* = \frac{v}{U_0},\ w^* = \frac{w}{U_0},\ x^* = \frac{x}{L_0},\ y^* = \frac{y}{L_0},\ z^* = \frac{z}{L_0},\ p^* = \frac{p}{P_0},\ t^* = \frac{t}{t_0}. \qquad (18)$$

The asterisks indicate the nondimensionalized variables. Then, Eq. (7) becomes

$$\frac{\partial u^*}{\partial t^*} + u^*\frac{\partial u^*}{\partial x^*} + v^*\frac{\partial u^*}{\partial y^*} + w^*\frac{\partial u^*}{\partial z^*} = -\frac{dp^*}{dx^*} + \frac{1}{\mathrm{Re}_M}\left(\frac{\partial^2 u^*}{\partial x^{*2}} + \frac{\partial^2 u^*}{\partial y^{*2}} + \frac{\partial^2 u^*}{\partial z^{*2}}\right), \qquad (19)$$

where $\mathrm{Re}_M = \frac{u_M L_0}{\nu} = \frac{\sigma_T \Delta T_0 L_0}{\mu \nu}$ is called the Marangoni-Reynolds number and $\nu = \mu/\rho$ is the kinematic viscosity. The procedures for the other components v^* and w^* are the same.

The boundary conditions Eqs. (14) and (15) become

$$\frac{\partial u^*}{\partial z^*} = -\frac{\partial T^*}{\partial x^*} \text{ and } \frac{\partial v^*}{\partial z^*} = -\frac{\partial T^*}{\partial y^*} \text{ at } z^* = 1.$$

The nondimensionalization of the energy equation is analogous. With use of a reference temperature T_{ref}, which represents such as an ambient temperature T_∞ or a boundary temperature of the liquid layer or cylinder, the nondimensional temperature $T^*(x,y,z) = (T(x,y,z) - T_{\mathrm{ref}})/\Delta T_0$ is defined and we get

$$\frac{\partial T^*}{\partial t^*} + u^*\frac{\partial T^*}{\partial x^*} + v^*\frac{\partial T^*}{\partial y^*} + w^*\frac{\partial T^*}{\partial z^*} = \frac{1}{\mathrm{Ma}}\left(\frac{\partial^2 T^*}{\partial x^{*2}} + \frac{\partial^2 T^*}{\partial y^{*2}} + \frac{\partial^2 T^*}{\partial z^{*2}}\right), \qquad (20)$$

where $\mathrm{Ma} = \mathrm{Re}_M \mathrm{Pr} = \frac{\sigma_T \Delta T_0 L_0}{\mu \kappa}$ and is called the Marangoni number, where $\kappa = \lambda_f/\rho c_p$ is the thermal diffusivity and $\mathrm{Pr} = \nu/\kappa$ is the Prandtl number.

The boundary condition Eq. (16) becomes

$$-\frac{\partial T^*}{\partial z^*} = \text{Bi} \cdot T^* \text{ at } z^* = 1 (z = d), \tag{21}$$

where $\text{Bi} = \frac{hd}{\lambda_f}$ is the Biot number.

Figure 1(c) is the cylindrical coordinate. A liquid column formed between two circular end plates is often referred to as 'liquid bridge' or 'liquid column.' Although an extremely long liquid bridge cannot be formed because of the Plateau–Rayleigh instability (e.g., [5], pp. 22–27), an infinitely long liquid column is often assumed to investigate the fundamental nature of the thermocapillary convection in a liquid bridge with a finite length. The temperature gradient is typically applied in the z-direction of the cylindrical coordinate, and it causes a basic (mean or unperturbed) flow in the axial direction, although radial and azimuthal components may also be induced due to the flow instability.

The fundamental equations for the cylindrical coordinate are given below in the nondimensionalized forms, although the asterisks are eliminated for brevity. Let the radial, azimuthal and axial velocity components u_r, u_θ and u_z. Note that the axial direction is represented by z in this section in accordance with convention, but in subsequent sections, it may be denoted by x to align with the direction of the basic flow in the flat liquid layer.

Then the continuity equation becomes

$$\frac{1}{r}\frac{\partial (r u_r)}{\partial r} + \frac{1}{r}\frac{\partial u_\theta}{\partial \theta} + \frac{\partial u_z}{\partial z} = 0. \tag{22}$$

The momentum equations are

$$\frac{\partial u_r}{\partial t} + u_r \frac{\partial u_r}{\partial r} + \frac{u_\theta}{r}\frac{\partial u_r}{\partial \theta} - \frac{u_\theta^2}{r} + u_z \frac{\partial u_r}{\partial z}$$
$$= -\frac{\partial p}{\partial r} + \frac{1}{\text{Re}_M}\left[\frac{1}{r}\frac{\partial}{\partial r}\left(r\frac{\partial u_r}{\partial r}\right) + \frac{1}{r^2}\frac{\partial^2 u_r}{\partial \theta^2} - \frac{2}{r^2}\frac{\partial u_\theta}{\partial \theta} - \frac{u_r}{r^2} + \frac{\partial^2 u_r}{\partial z^2}\right], \tag{23}$$

$$\frac{\partial u_\theta}{\partial t} + u_r \frac{\partial u_\theta}{\partial r} + \frac{u_\theta}{r}\frac{\partial u_\theta}{\partial \theta} + \frac{u_r u_\theta}{r} + u_z \frac{\partial u_\theta}{\partial z}$$
$$= -\frac{1}{r}\frac{\partial p}{\partial \theta} + \frac{1}{\text{Re}_M}\left[\frac{1}{r}\frac{\partial}{\partial r}\left(r\frac{\partial u_\theta}{\partial r}\right) + \frac{1}{r^2}\frac{\partial^2 u_\theta}{\partial \theta^2} + \frac{2}{r^2}\frac{\partial u_r}{\partial \theta} - \frac{u_\theta}{r^2} + \frac{\partial^2 u_\theta}{\partial z^2}\right], \tag{24}$$

$$\frac{\partial u_z}{\partial t} + u_r \frac{\partial u_z}{\partial r} + \frac{u_\theta}{r}\frac{\partial u_z}{\partial \theta} + u_z \frac{\partial u_z}{\partial z} = -\frac{\partial p}{\partial z} + \frac{1}{\text{Re}_M}\left[\frac{1}{r}\frac{\partial}{\partial r}\left(r\frac{\partial u_z}{\partial r}\right) + \frac{1}{r^2}\frac{\partial^2 u_z}{\partial \theta^2} + \frac{\partial^2 u_z}{\partial z^2}\right]. \tag{25}$$

The energy equation is

$$\left(\frac{\partial T}{\partial t} + u_r \frac{\partial T}{\partial r} + \frac{u_\theta}{r}\frac{\partial T}{\partial \theta} + u_z \frac{\partial T}{\partial z}\right) = \frac{1}{\text{Ma}}\left[\frac{1}{r}\frac{\partial}{\partial r}\left(r\frac{\partial T}{\partial r}\right) + \frac{1}{r^2}\frac{\partial^2 T}{\partial \theta^2} + \frac{\partial^2 T}{\partial z^2}\right]. \tag{26}$$

The thermocapillary boundary conditions are

$$\frac{\partial u_z}{\partial r} = -\frac{\partial T}{\partial z} \text{ and } r\frac{\partial}{\partial r}\left(\frac{u_\theta}{r}\right) = -\frac{1}{r}\frac{\partial T}{\partial \theta} \text{ at } r = 1 \text{ or dimensionally } r = R. \quad (27)$$

The thermal boundary condition is

$$-\frac{\partial T}{\partial r} = \text{Bi} \cdot T \text{ at } r = 1 \text{ or dimensionally } r = R. \quad (28)$$

Nondimensional numbers

Here, we will derive several nondimensional numbers related with the contents of this book. Figure 2 shows conceptual drawings to explain these derivations, in which l represents a characteristic length of an object currently under consideration and is denoted as L_0 below.

(1) Bond number (Bo): Fig. 2a

This number represents the relative importance between gravity and capillarity (surface tension); that is

$$\text{Bo} = \frac{\text{gravity}}{\text{capillarity}} = \frac{\rho g \ell}{\sigma/\ell} = \frac{\rho g L_0^2}{\sigma}. \quad (29)$$

With increase of Bo, the gravity prevails the surface tension.

(2) Capillary number (Ca): Fig. 2b

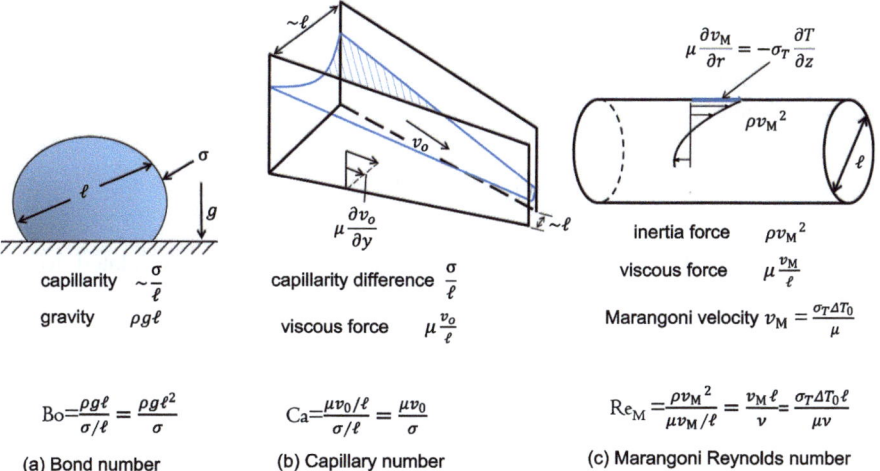

Fig. 2 Non-dimensional numbers and their features

This number is the ratio of the viscous drag force to the force caused by the capillarity difference.

$$\text{Ca} = \frac{\text{viscous drag}}{\text{capillarity difference}} = \frac{\mu v_0/L_0}{\sigma/L_0} = \frac{\mu v_0}{\sigma}. \tag{30}$$

As a related quantity, one may pay attention here to the Capillary length $Lc = \sqrt{\frac{\sigma}{\rho g}}$, which has a dimension of the length. If a diameter of a liquid droplet is smaller than its Lc, it tends to keep a spherical shape. As examples, Lc of the water is 2.7 mm, while that of the mercury is even smaller and 1.9 mm.

(3) Reynolds number (Re) and Marangoni Reynolds number (Re$_M$): Fig. 2c

These numbers are ratio of the inertia force to the viscous one, where the former tends to destabilize the flow while the latter to stabilize. Accordingly, this is a measure of stability of the flow and can be expressed as

$$\text{Re} = \frac{\text{inertia force}}{\text{viscous force}} = \frac{\rho v_0^2}{\mu v_0/L_0} = \frac{v_0 L_0}{\nu}. \tag{31}$$

The thermocapillary flow may also be unstable (often called "oscillatory") if a driving force (i.e., temperature difference) exceeds a certain threshold, which can be expressed by the Marangoni Reynolds number composed of the Marangoni velocity u_M Eq. (17) as the specific velocity v_0; That is,

$$\text{Re}_M = \frac{u_M l}{\nu} = \frac{\sigma_T \Delta T_0 L_0}{\mu \nu}. \tag{32}$$

(4) Marangoni number (Ma):

This is intrinsically the same as the Marangoni Reynolds number. Both are used to describe the transition condition from steady to oscillatory flows of the thermocapillary convection.

$$\text{Marangoni number: Ma} = \text{Re}_M \text{Pr} = \frac{\sigma_T \Delta T_0 L_0}{\mu \kappa}, \tag{33}$$

where Pr is the Prandtl number. The nondimensional numbers related to the thermocapillary convections are summarized in Table 1.

Table 1 Non-dimensional numbers related to the thermocapillary convection

Name	Definition	Note
Biot number	$Bi = \frac{hL_0}{\lambda_{s/f}}$	$L_0, \lambda_{s/f}$: characteristic length and thermal conductivity of internal solid (s) or fluid (f)
Bond number	$Bo = \frac{\rho g L_0^2}{\sigma}$	Refer to Eq. (29) g: gravity acceleration
Capillary number	$Ca = \frac{\mu U_0}{\sigma}$	Refer to Eq. (30)
Grashof number	$Gr = \frac{g\beta \Delta T L_0^3}{\nu^2}$	β: thermal expansion coefficient of fluid
Marangoni number	$Ma = \frac{\sigma_T \Delta T_0 L_0}{\mu \kappa}$	Refer to Eq. (33)
Marangoni Reynolds number	$Re_M = \frac{\sigma_T \Delta T_0 L_0}{\mu \nu}$	Refer to Eq. (32)
Nusselt number	$Nu = \frac{hL_0}{\lambda_f}$	L_0, λ_f: characteristic length and thermal conductivity of fluid
Prandtl number	$Pr = \frac{\nu}{\kappa}$	ν: kinematic viscosity κ: thermal diffusivity
Rayleigh number	$Ra = \frac{g\beta \Delta T L_0^3}{\kappa \nu} = GrPr$	L_0: thickness of fluid layer
Reynolds number	$Re = \frac{U_0 L_0}{\nu}$	L_0: characteristic length of fluid domain

2 Thermal Convection in an Infinite Liquid Layer with a Finite Depth Subjected to a Temperature Gradient Perpendicular to the Surface

2.1 Flow Instability Due to the Buoyancy

A thin liquid layer with an infinite horizontal surface area is a configuration, on which experimental and analytical studies of the thermocapillary convections have been performed for long period to explore new paradigm of the science on the thermally induced flow instabilities. We may assume fundamentally two types of temperature gradients in this configuration, that is, the gradient along or perpendicular to the surface.

Before discussing the thermocapillary driven convection, which is the main subject of this volume, a description will be provided on buoyancy driven one, the so-called Rayleigh-Bénard convection [24], which will be abbreviated as R-B convection hereafter. The research works on this configuration were started much earlier and a number of pioneering works were performed on the stability of the convections.

The first careful scientific experiment was made and reported by [1]. According to description by [24], Bénard kept a very thin layer of liquid about 1mm deep over a metallic plate maintained at a uniform temperature. The upper surface was in contact with the air at a lower temperature. Various kinds of liquid were employed including

Fig. 3 Bénard convection heated from below and cooled by the air of a low temperature. Reprinted from [9] with Permission from Elsevier

extremely high viscosity ones. With increase of the base plate temperature, the layer rapidly resolved into a number of cells. The fluid motion inside the cells was an ascension in the middle and a descension at the common boundary between a cell and its neighbors. Their forms were nearly regular convex polygons of, in general, 4 to 7 sides.

A photograph taken later by [9] is shown in Fig. 3. According to them, its test fluid was silicone oil, which was confined laterally by a Lucite ring of 133 mm in diameter over a copper heating plate. The fluid depth was 5.08 mm in most of their experiments. The photograph indicates that the fluid layer is filled with a number of regular hexagons, although some irregular ones are seen sparsely. In the following, we will describe an analysis of instability in the buoyancy-driven convections. This serves as an excellent example for understanding the fundamental characteristics of thermally activated flow instability. Furthermore, the reader will be introduced to the Linear Stability Analysis (LSA), one of the essential analytical tools in this field.

The configuration concerned is a horizontal liquid layer of a depth d as depicted in Fig. 4a. The coordinate z is directed from the bottom to the top surface. Temperatures of the bottom and top surfaces are kept at T_0 and T_1, respectively, and $T_0 > T_1$. The temperature difference is introduced as $\Theta = T - T_\infty$, where T_∞ is the environmental temperature. The bottom temperature is $\Theta_0 = T_0 - T_\infty > 0$. The assumed temperature profile in the liquid is linear with a gradient of $d\Theta/dz = dT/dz = (T_1 - T_0)/d = -\gamma$, where γ represents the temperature gradient and here $\gamma > 0$. Note that even if the bottom temperature is higher than the top, convection does not occur if its temperature difference is minimal, because the heat conduction through the fluid is capable of transferring heat to the environment. The purpose of the present analysis is to determine the critical conditions at which the stable stagnant state is lost, leading to the amplification of convective disturbances.

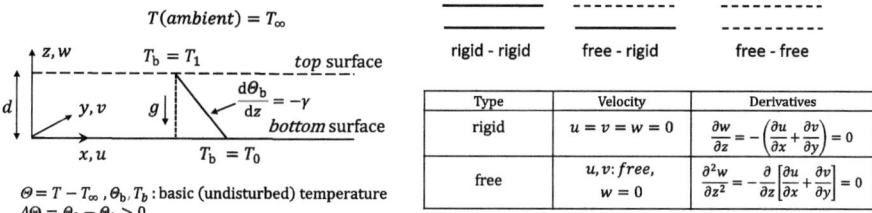

Fig. 4 Analysis of Rayleigh-Bénard convection (**a**) Coordinate and (**b**) Types of velocity boundary conditions

We will recall the continuity, momentum and energy equations in the vector representation as

$$\nabla \cdot \boldsymbol{u} = 0, \tag{34}$$

$$\rho \left[\frac{\partial}{\partial t} \boldsymbol{u} + (\boldsymbol{u} \cdot \nabla) \boldsymbol{u} \right] = -\nabla p + \mu \nabla^2 \boldsymbol{u} - \rho g \boldsymbol{i}_z, \tag{35}$$

$$\rho c_p \left[\frac{\partial}{\partial t} \Theta + (\boldsymbol{u} \cdot \nabla) \Theta \right] = \lambda \nabla^2 \Theta, \tag{36}$$

where the gravity force ρg is newly included in the momentum equation Eq. (35) using $\boldsymbol{i}_z = (0, 0, 1)$, a unit vector indicating the z axis and a negative sign is placed since gravity acts in the negative direction of z. The physical properties, ρ, μ, λ, c_p are dependent on the temperature, but they are assumed constant except the density ρ in the gravity force term of Eq. (35). This is called the Boussinesq approximation. The temperature dependence of the density ρ is simplified using the volumetric expansion coefficient β (>0) as

$$\rho(\Theta) = \rho_0[1 - \beta(\Theta - \Theta_0)], \text{ where } \rho_0 = \rho(\Theta_0). \tag{37}$$

Next, we will decompose the quantities $\boldsymbol{u}$, Θ and p into the basic (i.e., unperturbed) and small fluctuating parts with the designations of the subscript b and the superscript $'$, respectively; that is,

$$\boldsymbol{u} = \boldsymbol{u}_b + \boldsymbol{u}', \Theta = \Theta_b + \Theta' \text{ and } p = P_b + p'. \tag{38}$$

As for the basic condition, we assume $\boldsymbol{u}_b = 0$, $\Theta_b = \Theta_0 - \gamma z$, and from Eq. (37) $\rho_b(z)$ can be approximated as $\rho_b(z) = \rho_0(1 + \gamma \beta z)$.

Thus, the pressure gradient of the basic state balances with the gravity force, in which the density variation is caused by the basic temperature distribution $\Theta_b(z)$ as:

$$\frac{\partial P_b}{\partial z} = -\rho_b(z)g = -\rho_0(1+\gamma\beta z)g. \tag{39}$$

In the Linear Stability Analysis (LSA), we assume perturbations of the basic state are so small that the square products between the perturbations are negligible compared to the first order, i.e., linear terms. Then by substituting Eq. (38) to Eqs. (34)–(36) and using Eq. (39), the following equations are obtained for the perturbation quantities:

$$\nabla \cdot \boldsymbol{u}' = 0, \tag{40}$$

$$\rho_0 \left[\frac{\partial}{\partial t}\boldsymbol{u}'\right] = -\nabla p' + \mu \nabla^2 \boldsymbol{u}' + \rho_0 g \beta \Theta' \boldsymbol{i}_z, \tag{41}$$

$$\rho_0 c_p \left[\frac{\partial}{\partial t}\Theta' + w'\frac{\partial \Theta_b}{\partial z}\right] = \lambda \nabla^2 \Theta'. \tag{42}$$

Now we will nondimensionalize the above equations. The asterisks mean the nondimensional quantities;

coordinate: $\boldsymbol{x}^* = \boldsymbol{x}/d$, time: $t^* = t/(d^2/\kappa)$, velocity: $\boldsymbol{u}^* = \boldsymbol{u}/(\kappa/d)$,

pressure: $p^* = p/(\rho_0 \kappa^2/d^2)$, temperature: $\Theta^* = \Theta/(\gamma d) = \Theta/\Delta\Theta_0$.

The obtained nondimensional equations for the perturbations are presented below. Since in the rest of this subsection only the nondimensional quantities are treated, the asterisks will be omitted without confusion. In addition, the superscript prime ' indicating the perturbation will also be omitted and the nondimensionalized perturbations will be represented by the lowercase letters with "hat" such as $\hat{\boldsymbol{u}}(\hat{u}, \hat{v}, \hat{w})$, $\hat{\vartheta}$ and $\hat{p}$. With these notational simplifications, we get

$$\nabla \cdot \hat{\boldsymbol{u}} = 0, \tag{43}$$

$$\frac{\partial}{\partial t}\hat{\boldsymbol{u}} = -\nabla \hat{p} + \Pr \nabla^2 \hat{\boldsymbol{u}} + \text{RaPr}\,\hat{\vartheta}\,\boldsymbol{i}_z, \tag{44}$$

$$\frac{\partial}{\partial t}\hat{\vartheta} - \hat{w} = \nabla^2 \hat{\vartheta}, \tag{45}$$

where Pr and Ra are the Prandtl and Rayleigh numbers given in Table 1, the latter of which can be expressed using the present notations as

$$\text{Ra} = \frac{g\beta\gamma d^4}{\kappa\nu} = \frac{g\beta\Delta\Theta d^3}{\kappa\nu}, \tag{46}$$

where $\gamma = -d\Theta_b/dz$ and $\Delta\Theta = \Theta_0 - \Theta_1 > 0$.

We now discuss boundary conditions for the bottom and top surfaces. Two categories of boundary conditions must be considered; "velocity" and "temperature". First, as for the velocity boundary, we can classify them into two types: those are 'rigid' and 'stress-free' boundaries. The former refers to solid surfaces, while the latter to surfaces facing air.

[Rigid boundary]: All the velocity components (both basic and perturbation) vanish.

$$\hat{u} = 0 \text{ or } \hat{u} = \hat{v} = \hat{w} = 0. \tag{47}$$

As for the wall normal derivative of $\hat{w}$, since $\hat{u} = \hat{v} = 0$ everywhere over the x–y plane,

$$\frac{\partial \hat{w}}{\partial z} = -\left(\frac{\partial \hat{u}}{\partial x} + \frac{\partial \hat{v}}{\partial y}\right) = 0. \tag{48}$$

[Free boundary]

$$\hat{w} = 0, (\hat{u}, \hat{v} \neq 0) \text{ and } \frac{\partial \hat{u}}{\partial z} = \frac{\partial \hat{v}}{\partial z} = 0. \tag{49}$$

As for the derivative of $\hat{w}$, with use of the continuity equation, we obtain

$$\frac{\partial^2 \hat{w}}{\partial z^2} = -\frac{\partial}{\partial z}\left[\frac{\partial \hat{u}}{\partial x} + \frac{\partial \hat{v}}{\partial y}\right] = -\left[\frac{\partial}{\partial x}\left(\frac{\partial \hat{u}}{\partial z}\right) + \frac{\partial}{\partial y}\left(\frac{\partial \hat{v}}{\partial z}\right)\right] = 0. \tag{50}$$

These "rigid" and "free" boundary conditions are summarized in Fig. 4b.

As for the thermal (temperature) boundary, one may assume three kinds of conditions; those are, constant temperature (isothermal), insulating, and heat-transfer boundaries.

[Constant temperature (isothermal) boundary], called sometimes "conducting boundary", too.

$$\hat{\Theta} = 0. \tag{51}$$

[Insulating boundary]

$$\frac{\partial \hat{\Theta}}{\partial z} = 0. \tag{52}$$

[Heat-transfer boundary] With use of Eqs. (16) and (21), we obtain.

$$\frac{\partial \hat{\Theta}}{\partial z} = -\text{Bi}\hat{\Theta}, \tag{53}$$

where Bi is the Biot number, given in Table 1. Note that Bi = 0 and ∞ correspond to the insulating and isothermal boundaries, respectively. It should be noted here that these thermal conditions for the fluctuating part can be specified independently of those of the mean temperature at the boundary; that is, even if the mean one is constant temperature or insulating condition, the fluctuation part can be assumed to be the heat transfer one.

We have now three equations, the continuity (43), momentum (44) and the energy (45) equations corresponding to the three unknown variables, velocity $\hat{u}$, pressure $\hat{p}$ and temperature $\hat{\vartheta}$. Therefore, they are solvable. A standard method to solve this set of equations has been established and described in textbooks such as [2, 5]. Their basic procedure will be described briefly below.

We first eliminate the pressure to obtain a set of equations for the velocity and the temperature. Thus, we operate the rotation operator $\nabla \times$ to the momentum Eq. (44). With use of the continuity $\nabla \cdot \hat{u} = 0$ and the formula $\nabla \times \nabla \phi = 0$ for an arbitrary scaler ϕ, an equation for the vorticity $\hat{\omega} = \nabla \times \hat{u}$ can be obtained as

$$\frac{\partial}{\partial t}\hat{\omega} = \text{RaPr}(\nabla\hat{\vartheta} \times i_z) + \text{Pr}\nabla^2\hat{\omega}. \tag{54}$$

Now, we are able to eliminate the pressure, but the equation obtained is not for $\hat{u}$ itself but for the $\hat{\omega}$. Accordingly, we apply again $\nabla \times$ to the Eq. (54), and with use of $\nabla \cdot \hat{u} = 0$ and several vector operations such as

$$\nabla \times \hat{\omega} = \nabla \times \nabla \times \hat{u} = -\nabla^2\hat{u} + \nabla(\nabla \cdot \hat{u}) = -\nabla^2\hat{u},$$

$$\nabla \times (\nabla\hat{\vartheta} \times i_z) = -i_z\nabla^2\hat{\vartheta} + (i_z \cdot \nabla)\nabla\hat{\vartheta},$$

$$\nabla \times \nabla^2\hat{\omega} = \nabla \times \nabla \times \nabla^2\hat{u} = -\nabla^2\nabla^2\hat{u} + \nabla(\nabla \cdot \nabla^2\hat{u}) = -\nabla^2\nabla^2\hat{u},$$

we get

$$\frac{\partial}{\partial t}\nabla^2\hat{u} = \text{RaPr}(i_z\nabla^2\hat{\vartheta} - (i_z \cdot \nabla)\nabla\hat{\vartheta}) + \text{Pr}\nabla^2\nabla^2\hat{u}. \tag{55}$$

Next, we would obtain the solution of $\hat{\vartheta}$, by solving Eq. (45). For this purpose, we need the velocity component $\hat{w}$, which is the z component of Eq. (55). It can be expressed as

$$\frac{\partial}{\partial t}\nabla^2\hat{w} = \text{RaPr}\,\nabla^2_{xy}\hat{\vartheta} + \text{Pr}\nabla^2\nabla^2\hat{w}, \tag{56}$$

where ∇^2_{xy} is an operator indicating $\nabla^2_{xy} = \partial^2/\partial x^2 + \partial^2/\partial y^2$. Thus, we have obtained a set of the linearized equations; Eq. (56) for the momentum and Eq. (45) for energy. The latter is rewritten below in a transposed form as

$$\frac{\partial}{\partial t}\hat{\vartheta} - \nabla^2\hat{\vartheta} = \hat{w}. \tag{57}$$

These constitute a set of linearized equations, whose solutions are to be examined under proper boundary conditions. Since the linear equations and their boundary conditions are symmetric in x and y axis and are bounded between $z = 0$ and 1, the analysis can be made in terms of two-dimensional periodic waves with assigned wavenumbers. So, we describe the perturbations with the dependence of the form

$$\hat{w} = \tilde{W}(z)e^{i(lx+my)}e^{st}, \tag{58}$$

$$\hat{\vartheta} = \tilde{\Theta}(z)e^{i(lx+my)}e^{st}, \tag{59}$$

where $l, m \,(\in \mathbb{R})$ and resultant $a = \sqrt{l^2 + m^2}$ are the wavenumbers. The term e^{st} describes the temporal nature of the disturbance and s is a complex growth rate. An important point to be noticed here is that in this configuration "the limiting conditions of stability can be obtained when all the time variations are made zero"; that is $s = 0$, which was proved by [21].

[a]: Free-Free boundaries

As for the boundary conditions, we first select "free velocity" and "constant temperature" for both sides of the liquid layer, where the "free velocity" means that free surface is assumed for the velocity boundary condition. Although the "free velocity" for a bottom surface is rather unrealistic, this assumption is made because for these combinations we are able to obtain *analytically* the critical Rayleigh number, at which the disturbance starts to be amplified.

That is, with selection of these boundary conditions, we can easily find that

$$\tilde{W}(z) = w_0 \sin(n\pi z), \quad \tilde{\Theta}(z) = \theta_0 \sin(n\pi z) \tag{60}$$

are able to satisfy the velocity and temperature boundary conditions simultaneously for the top and bottom surfaces, i.e., Eqs. (49, 50 and 51).

Now we will insert Eqs. (58 and 59) into Eqs. (56 and 57). Although we know already the functional forms of $\tilde{W}(z)$ and $\tilde{\Theta}(z)$ as Eq. (60), we will use differential operators, $\mathrm{D}\Phi = \frac{\mathrm{d}}{\mathrm{d}z}\Phi$ and $\mathrm{D}^2\Phi = \frac{\mathrm{d}^2}{\mathrm{d}z^2}\Phi$ (Φ could be $\tilde{W}$ or $\tilde{\Theta}$) for convenience of later generalization. Then Eqs. (56) and (57) can be written as

$$s(\mathrm{D}^2 - a^2)\tilde{W} = -\mathrm{Ra}\,\mathrm{Pr}\,a^2\,\tilde{\Theta} + \mathrm{Pr}(\mathrm{D}^2 - a^2)(\mathrm{D}^2 - a^2)\tilde{W}, \tag{61}$$

$$\{s - (\mathrm{D}^2 - a^2)\}\tilde{\Theta} = \tilde{W}, \tag{62}$$

where $a^2 = l^2 + m^2$. By eliminating $\tilde{W}$, we get

$$\text{RaPr}\, a^2 \tilde{\Theta} = -(D^2 - a^2)\{\Pr(D^2 - a^2) - s\}\{(D^2 - a^2) - s\}\tilde{\Theta}. \qquad (63)$$

Note that the same equation is obtained for $\tilde{W}$, if we eliminate $\tilde{\Theta}$.

In the present case, since a simple trigonometric function of Eq. (60) can satisfy both boundary conditions at top and bottom surfaces, thus, we are able to perform the differentiation as $D^2\tilde{\Theta} = \theta_0 \cdot \frac{d^2}{dz^2}\sin(n\pi z) = -n^2\pi^2\theta_0$. Accordingly, Eq. (63) can be rewritten as

$$\text{RaPr}\, a^2 = (n^2\pi^2 + a^2)\{\Pr(n^2\pi^2 + a^2) + s\}(n^2\pi^2 + a^2 + s). \qquad (64)$$

As mentioned in relation to Eqs. (58 and 59), we may set $s = 0$ to obtain a marginal (neutral) Rayleigh numbers, the Ra number at which disturbance starts to grow for a given wavenumber a. Accordingly, we get the marginal Ra number as

$$\text{Ra} = (n^2\pi^2 + a^2)^3/a^2. \qquad (65)$$

Since it is obvious that $n = 1$ gives the lowest level of the marginal Ra number for a given wavenumber a, the marginal Rayleigh number can be obtained as

$$\text{Ra} = (\pi^2 + a^2)^3/a^2. \qquad (66)$$

Among the marginal Ra numbers, the minimum one with respect to the wavenumber a gives the critical Rayleigh number Ra_{cr}, at which the disturbance starts to grow if the Rayleigh number Ra is increased from a minimal level. The minimum point against the wavenumber a can easily be calculated as,

$$\frac{d}{da^2}\text{Ra} = \frac{(2a^2 - \pi^2)}{a^4}(\pi^2 + a^2)^2 = 0. \qquad (67)$$

Accordingly, the critical Rayleigh number Ra_{cr} can be obtained with use of $a^2 = \pi^2/2$ as

$$\text{Ra}_{cr} = \frac{27}{4}\pi^4 = 657.5 \text{ at } a_{cr} = \frac{\pi}{\sqrt{2}}; \lambda_{cr} = \frac{2\pi}{a_{cr}} = 2\sqrt{2} = 2.828, \qquad (68)$$

where a_{cr} and λ_{cr} are called respectively the critical wavenumber and the critical wavelength. They are nondimensionalized with the thickness d. (Note that although the nomenclature λ is used as the thermal conductivity in other parts of this chapter, there should be no confusion because of their significant difference.) It is interesting to note here that the critical Rayleigh number Ra_{cr} does not depend upon the Prandtl number. This is because Pr drops out of both sides when s is set to be zero in Eq. (64).

(b): Rigid-Rigid boundaries

The case of a fluid layer bounded by two rigid surfaces with bottom heating can be solved by means of the same principle as above. In this case, however, the functions corresponding to $\tilde{W}(z)$ and $\tilde{\Theta}(z)$ of Eqs. (58 and 59) cannot be a simple trigonometric function because the boundary condition of w is different from the former free-free case. Those who are interested in this mathematical treatment may refer to Chapter "Thermodynamic and Molecular Aspects of Surface Tension", Sect. 1. An approximate solution for the marginal Ra number in this rigid-rigid case is given by [2] as

$$\text{Ra} = \frac{(\pi^2 + a^2)^3/a^2}{1 - 16a\pi^2 \cosh^2(a/2)/\left\{(\pi^2 + a^2)^2 (\sinh a + a)\right\}}. \tag{69}$$

The resultant critical conditions are

$$\text{Ra}_{cr} \simeq 1715.08 \text{ at } a_{cr} = 3.117 \text{ and } \lambda_{cr} = 2.016, \tag{70}$$

which are in enough good agreement with more exact values given by [2].

[c]: Free-Rigid boundaries

This case is the closest to realistic one. According to [2], this case can be deduced from the rigid-rigid case. That is, $\tilde{W}$ of the lowest mode for the rigid-rigid one is symmetric with respect to the mid plane and has a peak at the center (called "even" solution). In case of the second lowest mode of this case, on the other hand, $\tilde{W}$ is antisymmetric and must be zero at the mid plane (called "odd" solution). Thus, if we pay attention to the second lowest mode (n = 2) of the Rigid-Rigid case, we find the boundary conditions for the free surface of the Free-Rigid one, Eqs. (49 and 50), can be satisfied at the mid plane $z = 1/2$ of the Rigid-Rigid one. Accordingly, we can utilize the critical Ra number for the second lowest mode of the Rigid-Rigid boundaries Ra_{cr2} with replacing d with $(1/2)d$.

As for the critical conditions for the relevant mode (n = 2) of the rigid-rigid case, an approximate expression for the so-called "odd (one free and one rigid)" case is given by [2] as

$$\text{Ra} = \frac{(4\pi^2 + a^2)^3/a^2}{1 - 64a\pi^2 \sinh^2(a/2)/\left\{(4\pi^2 + a^2)^2 (\sinh a - a)\right\}}. \tag{71}$$

Recalling Ra is proportional to d^4, the Ra_{cr} for the free-rigid surface is obtained as

$$\text{Ra}_{cr} \simeq 17803.2/2^4 = 1112.7 \text{ at } a_{cr} = 5.365/2 = 2.682, \lambda_{cr} = 2.342. \tag{72}$$

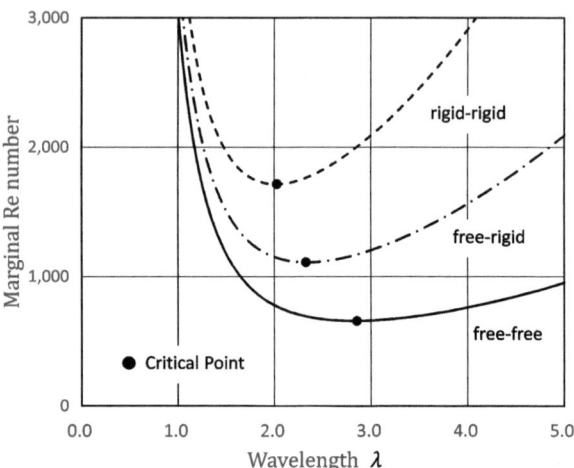

Fig. 5 Marginal Rayleigh number versus wavelength of Rayleigh-Bénard convection (Solid circle: Critical point)

The marginal Ra numbers for the three boundary conditions, i.e., "free-free", "free-rigid" and "rigid-rigid", are compared in Fig. 5 with use of the Eqs. (66), (69), and (71). The small black circles indicate the critical conditions at which disturbance of the most "dangerous" wavelength λ_{cr} starts to grow. The figure shows that Ra_{cr} is the smallest and λ_{cr} the largest for the "free-free" boundaries, while for the "rigid-rigid" ones, Ra_{cr} is the largest and λ_{cr} the smallest. Those of the "free-rigid" stay in between. These trends are quite understandable because the liquid motion is most constrained in the "rigid-rigid" boundaries while least in the "free-free" ones.

2.2 Flow Instability Due to the Surface Tension Gradient

In this subsection, we will explore the flow instability within the same configuration as the previous one. However, the present case is not due to the density difference but rather to the surface tension distribution induced by the temperature fluctuation across the surface.

Pearson JRA, a research scientist of a large British chemical industry, I. C. I. Ltd., was the first to notice and to analyze scientifically the cellular fluid motion in a thin liquid layer, which is due to the surface tension variation caused by the mean temperature and/or concentration gradients in the direction of depth. In his pioneering paper, Pearson [20] mentioned that his research was inspired by finding of an engineer at the I.C.I. Ltd, who found that drying paint films often exhibited steady cellular circulatory flows, similar to those observed in the fluid layers heated from below. Remarkably, these cellular motions emerged whether the free surface faced either topside or underside of the paint layer. This finding strongly suggested that gravity was not the cause; rather, the motions were likely initiated by differences in surface tension between the more and less volatile components of the paint.

We have already fundamental tools for the analysis of this type of instability, too. The coordinate of the present analysis is the same as the former one, Fig. 4, except that the upper surface must be free and the heat transfer boundary as Eqs. (16 and 21). The basic temperature gradient, $\Theta_b = \Theta_0 - \gamma z$, exists in this case, too, but the thermal expansion of the fluid is neglected for the simplicity; thus, no buoyancy force arises in the present analysis.

We will, as before, decompose the quantities $\boldsymbol{u}$, Θ and p into the basic and small fluctuating parts Eq. (38) and the superscript prime ' indicating the perturbation will be omitted. The continuity, momentum and energy equations are same as Eqs. (43)–(45) except that buoyancy is not included in this case. Then we get

$$\nabla \cdot \hat{\boldsymbol{u}} = 0, \tag{73}$$

$$\frac{\partial}{\partial t}\hat{\boldsymbol{u}} = -\nabla \hat{p} + \Pr \nabla^2 \hat{\boldsymbol{u}}, \tag{74}$$

$$\frac{\partial}{\partial t}\hat{\vartheta} - \hat{w} = \nabla^2 \hat{\vartheta}. \tag{75}$$

The equation for the z component of the velocity $\hat{w}$ is also the same as the z component of Eq. (44) with exclusion of the buoyancy term. After operating ∇^2, it becomes

$$\frac{\partial}{\partial t}\nabla^2 \hat{w} = \Pr \nabla^2 \nabla^2 \hat{w}. \tag{76}$$

As for the boundary conditions at the bottom surface, the rigid and the constant temperature are assumed:

$$\hat{u} = \hat{v} = \hat{w} = 0 \text{ and } \frac{\partial \hat{w}}{\partial z} = 0 \text{ at } z = 0, \tag{77}$$

$$\hat{\vartheta} = 0 \text{ at } z = 0. \tag{78}$$

At the top surface ($z = 1$), the wall normal velocity component becomes zero

$$\hat{w} = 0, \ (\hat{u}, \hat{v} \neq 0) \text{ at } z = 1, \tag{79}$$

and the heat transfer boundary is applied for the temperature

$$\frac{\partial \hat{\vartheta}}{\partial z} = -\text{Bi}\,\hat{\vartheta} \text{ at } z = 1. \tag{80}$$

The thermocapillary boundary condition coupling $\hat{\boldsymbol{u}}$ and $\hat{\vartheta}$ needs some additional explanation. Referring to Eqs. (14) and (15), the thermocapillary boundary conditions over the free surface can be expressed in the "dimensional" form as $\mu \frac{\partial u}{\partial z} = -\sigma_T \frac{\partial \Theta}{\partial x}$ and $\mu \frac{\partial v}{\partial z} = -\sigma_T \frac{\partial \Theta}{\partial y}$. In the nondimensional form, they become

$$\frac{\partial \hat{u}}{\partial z} = -\text{Ma}\frac{\partial \hat{\vartheta}}{\partial x} \quad \text{and} \quad \frac{\partial \hat{v}}{\partial z} = -\text{Ma}\frac{\partial \hat{\vartheta}}{\partial y} \quad \text{at } z = 1, \tag{81}$$

where $\text{Ma} = \frac{\sigma_T \gamma d^2}{\mu \kappa} = \frac{\sigma_T \Delta \Theta_b d}{\mu \kappa}$ is the Marangoni number in this case. As for the component $\hat{w}$, we get with use of the continuity equation

$$\frac{\partial^2 \hat{w}}{\partial z^2} = -\frac{\partial}{\partial z}\left(\frac{\partial \hat{u}}{\partial x} + \frac{\partial \hat{v}}{\partial y}\right) = -\left[\frac{\partial}{\partial x}\left(\frac{\partial \hat{u}}{\partial z}\right) + \frac{\partial}{\partial y}\left(\frac{\partial \hat{v}}{\partial z}\right)\right]. \tag{82}$$

Then we obtain

$$\frac{\partial^2 \hat{w}}{\partial z^2} = \text{Ma}\left(\frac{\partial^2 \hat{\vartheta}}{\partial x^2} + \frac{\partial^2 \hat{\vartheta}}{\partial y^2}\right) \quad \text{at } z = 1. \tag{83}$$

Since we have now the set of equations and necessary boundary conditions, we are able to analyze the growth of disturbances employing the same perturbation forms as Eqs. (58) and (59).

$$\hat{w} = \tilde{W}(z)e^{i(lx+my)}e^{st}, \tag{84}$$

$$\hat{\vartheta} = \tilde{\Theta}(z)e^{i(lx+my)}e^{st}, \tag{85}$$

although $\tilde{W}(z)$ and $\tilde{\Theta}(z)$ cannot be simple trigonometric functions as was in the previous Sect. 2.1. Then, the equations corresponding to Eqs. (61) and (62) are

$$s(D^2 - a^2)\tilde{W} = \text{Pr}(D^2 - a^2)(D^2 - a^2)\tilde{W}, \tag{86}$$

$$\{s - (D^2 - a^2)\}\tilde{\Theta} = \tilde{W}. \tag{87}$$

The boundary conditions at the rigid and conducting (constant temperature) bottom surface are

$$\tilde{W} = 0, \ D\tilde{W} = 0 \text{ and } \tilde{\Theta} = 0 \text{ at } z = 0. \tag{88}$$

At the top surface,

$$\tilde{W} = 0, \ D^2\tilde{W} = -a^2\text{Ma}\,\tilde{\Theta} \text{ and } D\tilde{\Theta} = -\text{Bi}\,\tilde{\Theta} \text{ at } z = 1, \tag{89}$$

where the thermocapillary condition Eq. (83) and the heat transfer boundary Eq. (80) are applied.

By setting $s = 0$ in Eqs. (86) and (87) as before, these set of equations allow us to obtain the marginal (threshold) condition for the onset of the thermocapillary convection in an infinite liquid layer. The solution was first obtained by Pearson [20] and confirmed by Nield [18] with use of the Fourier series expansion.

The marginal Marangoni number can be expressed in terms of a wavenumber a for a given Biot number as

$$\mathrm{Ma} = \frac{8a(a\cosh a + \mathrm{Bi}\sinh a)(a - \sinh a \cosh a)}{a^3 \cosh a - \sinh^3 a}. \tag{90}$$

This relation corresponds to Eq. (71) of the previous case of the R-B convection with the "free-rigid" boundaries. The critical Marangoni number $\mathrm{Ma_{cr}}$, that is, the minimum value against all wavenumbers a, were obtained by [18] and given in a tabular form of $\mathrm{Ma_{cr}}$ versus Bi. In the table, the critical Ra number, $\mathrm{Ra_{cr}}$, for the R-B convection between the "free-rigid" boundaries, is also given against Bi number for comparison.

Now, we are able to examine the effect of Bi on the critical Ra and Ma numbers for the "free-rigid" boundaries. Both critical numbers are plotted against the Bi number in Fig. 6. As for their boundary conditions, the heat transfer condition is applied for top surfaces of both cases, and the thermocapillary condition for the thermocapillarity-driven convection. On the bottom surfaces, the nonslip and isothermal conditions are applied in both cases. An aim of this figure is *not to compare their absolute magnitudes of the critical numbers* but to compare their behaviors against the Bi number. That is, we can notice in Fig. 6 that the two curves behave similarly for small Bi numbers but quite differently for larger ones. First for Bi⟶0, the critical Ra and Ma numbers approach their respective values; that is $\mathrm{Ra_{cr}} = 669$ and $\mathrm{Ma_{cr}} = 79.61$. The former was taken from the table of [18], and the latter can be obtained by inserting $\mathrm{Bi} = 0$ and $a_{cr} \simeq 2.0$ into Eq. (90).

When Bi increases, on the other hand, both curves behave quite differently. With increasing Bi, the magnitude of the surface temperature variation decreases in both flows. In case of the buoyancy driven flow, the critical $\mathrm{Ra_{cr}}$ must approach to the critical value for the isothermal top surface; that is, $\mathrm{Ra_{cr}} \simeq 1100$ from Eq. (72). If this $\mathrm{Ra_{cr}}$ is compared with that of the adiabatic case (Bi⟶0), it is certainly increased but the difference is rather moderate. This is because the driving force of buoyancy driven convection exists inside of the fluid, so the surface condition has only a secondary effect.

In case of the thermocapillary flow, on the other hand, Fig. 6 indicates that the critical $\mathrm{Ma_{cr}}$ increases indefinitely in proportion to Bi in the large Bi region. To obtain this asymptotic relation between the $\mathrm{Ma_{cr}}$ and Bi, we adopt Eq. (90) and retain only the term with Bi with neglecting the other terms. Then with use of an approximate value of $a_{cr} \simeq 3.0$, we obtain $\mathrm{Ma_{cr}} \simeq 32.1\mathrm{Bi}$, which is plotted in Fig. 6. This increasing trend of $\mathrm{Ma_{cr}}$ against Bi can be explained as follows, that is, an increase

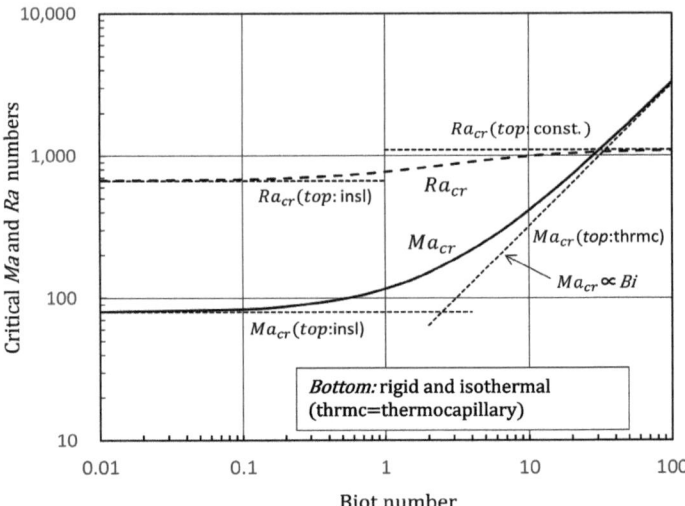

Fig. 6 Comparison of the effect of Bi number upon critical Ra and Ma numbers. Both critical numbers behave similarly for the decreasing Bi, while quite differently for the increasing Bi

in Bi tends to decrease the surface temperature fluctuations and thus reduces the motive force due to the thermocapillarity itself. Accordingly, an increasingly larger Ma number is required to excite the flow instability.

It should be noted here that the observed increase in Ma_{cr} with increasing Bi numbers is not a universal trend, but depends on geometry. Specifically, for a liquid bridge of finite length, this trend actually reverses within a practical range of Bi numbers. This difference arises because, in the current analysis, the basic temperature gradient is specified and unchanged, whereas for the finite-length liquid bridge, only the temperature difference between the hot and cold ends is specified. Consequently, the basic temperature profile is directly influenced by the Bi number. A more detailed explanation will be provided in Sect. 3 of "Thermocapillary Convection in Liquid Bridges of Finite Length".

At the end of this subsection, we would briefly discuss the history of interpretation of the Bénard's experiment. Gravity had long been believed to be the driving force behind formation of cells in his experiments. Pearson [20] threw doubt on this point with substantial analytical grounds. He pointed out that the liquid layer in Bénard's experiment was so thin, less than 1 mm, that its Ra value must not have exceeded the Ra_{cr} for "free-rigid" boundary with an insulating top and also that its Ma number must have been larger than the critical value of $Ma_{cr} \simeq 80$.

Finally, regarding the naming of this kind of convection, "Marangoni-Bénard convection" is rather often referred to. In this context, Koschmieder [10] mentioned that, since the Marangoni's work has already been justified as the "Marangoni number", it would be appropriate to call the surface-tension-driven convection the "Bénard-Pearson convection", as Davis [4] also proposed. Considering these discussions and also the current widespread acceptance of "Rayleigh-Bénard convection" for the buoyancy-driven convection, the present case will be referred to as "Pearson-Bénard convection" in this chapter, maintaining the sequence of "theoretical" and "experimental" works.

3 Thermal Convection in a Thin Infinite Liquid Layer and an Infinitely Long Liquid Cylinder with a Temperature Gradient Along the Surface

3.1 Configuration of Flow Field

The thermocapillary flow in the previous section was induced by a temperature gradient perpendicular to the surface, resulting in no net flow in any direction. From the following sections, we will focus on thermocapillary flows induced by temperature gradient along the fluid surface. Early research in this field, which included both analytical and experimental approaches, was stimulated by advancements in single-crystal growth techniques at the time. This is because the temperature gradient along fluid surface arises over the melt surfaces leading to the thermocapillary flow. In usual crystal growth methods, which are briefly introduced in Sect. 1 of "Thermocapillary Convection in Liquid Bridges of Finite Length". one side of polycrystal is heated to be melt, while the other side is cooled for single crystal to grow. Therefore, the thermocapillary convection occurs in a relatively simple geometry, which has attracted interest of not only the crystal growth but also the thermo-fluid scientists.

S. H. Davis's group conducted analytical studies on two canonical configurations with use of the Linear Stability Analysis (LSA). Those were an infinitely large liquid layer with a finite depth d and an infinitely long circular cylinder with radius R. Although the latter may seem unrealistic due to the length limitation of the liquid column resulting from the Plateau-Rayleigh instability, it still provides a valuable basis for fundamental understanding of the phenomena. The works related to these two canonical configurations will be described in some detail below.

3.2 Infinitely Large Flat Liquid Layer with a Finite Depth d

As for the infinitely large liquid layer, Smith and Davis [28] examined the stability conditions using the LSA. Later, [22, 23] reexamined the calculation and added new cases of the fixed bottom temperature and of higher Prandtl numbers.

In their analysis, an infinitely large flat liquid layer is assumed, as given in Fig. 7. A constant temperature gradient of the environmental temperature $T_\infty(x)$ is imposed along the x-axis; that is, $T_\infty(x) = T_{\text{ref}} - \gamma x$ and $dT_\infty/dx = -\gamma$ with $\gamma > 0$, where T_{ref} is a reference temperature and is arbitrary. The basic flow has only an x component $U_b(z)$ directed towards the positive x-axis and is a function of z.

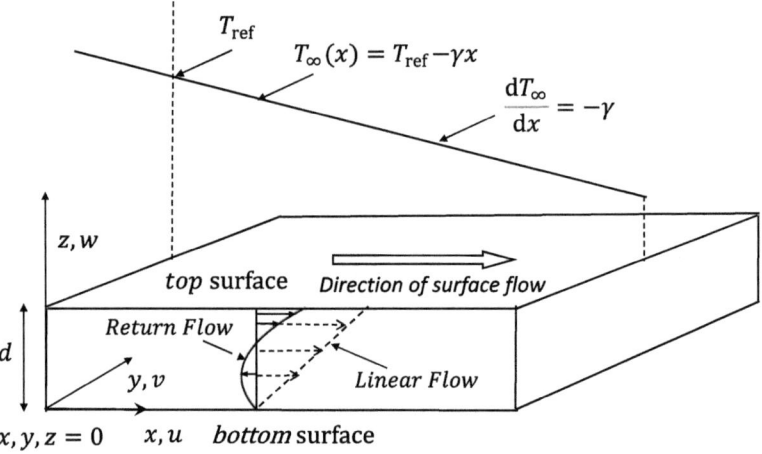

Fig. 7 Return and linear flows (Coordinate and basic velocity distribution)

We will now decompose the quantities u, T and p into the basic (i.e., unperturbed) and fluctuating parts. First we would introduce the temperature deviation from the ambient temperature $T_\infty(x)$, that is,

$$\Theta(x, y, z, t) = T(x, y, z, t) - T_\infty(x), \qquad (91)$$

which is more suitable for analytical treatment because its basic (unperturbed) part Θ_b depends on z but must not on x and y since we assume an infinitely wide area along the axis x and y.

In the followings, the basic part will be designated with the subscript b ($U_b(z)$, $\Theta_b(z)$ and $P_b(x)$), which are all time-independent. The fluctuating parts will be expressed with the lower case letters ($\hat{u}, \hat{v}, \hat{w}, \hat{\vartheta}$, and $\hat{p}$), which are all functions of x, y, z, and t.

The basic velocity is steady and has only an x component $U_b(z)$ as a function of z. Thus, the momentum equation for the basic velocity U_b becomes from Eq. (7)

$$-\frac{dP_b}{dx} + \mu\left(\frac{d^2 U_b}{dz^2}\right) = 0. \qquad (92)$$

The boundary condition over the free surface ($z = d$) is, referring to Eq. (14),

$$\mu \frac{dU_b}{dz}\bigg|_{z=d} = \frac{d\sigma}{dx}\bigg|_{z=d} = -\sigma_T \frac{dT_b}{dx}\bigg|_{z=d} = \sigma_T \gamma. \qquad (93)$$

At the bottom surface ($z = 0$), the non-slip condition is assumed; $U_b = 0$. The energy equation for the basic temperature T_b becomes using Eq. (11)

$$\rho c_p \left(U_b \frac{\partial T_b}{\partial x} \right) = \lambda \left(\frac{\partial^2 T_b}{\partial z^2} \right). \tag{94}$$

The boundary condition over the free surface is given using Eq. (16) as

$$-\lambda \frac{\partial T_b}{\partial z}\bigg|_{z=d} = -\lambda \frac{\partial \Theta_b}{\partial z}\bigg|_{z=d} = h\{T_b - T_\infty(x)\} = h\Theta_b(z=d) \text{ at } z = d. \tag{95}$$

At the bottom surface, the insulated condition is often assumed, that is,

$$-\lambda \frac{\partial \Theta_b}{\partial z}\bigg|_{z=0} = 0 \text{ at } z = 0. \tag{96}$$

Now, we will nondimensionalize these equations with use of the relation given by Eq. (18). The characteristic quantities for the length, temperature, velocity, pressure and heat flux are defined respectively,

$$L_0 = d, \ \Delta T_0 = \gamma d, \ U_0 = \frac{\sigma_T \Delta T_0}{\mu} = \frac{\sigma_T \gamma d}{\mu}, \ P_0 = \mu \frac{U_0}{d}, \tag{97}$$

$$t_0 = \frac{d}{U_0} \text{ and } q_0 = \lambda \frac{\Delta T_0}{d} = \lambda \gamma. \tag{98}$$

From now on in this section, *all quantities are nondimensionalized, but the same notations as above will be assigned*, because it would not lead to any confusion.

Then Eqs. (92) and (93) become

$$\frac{d^2 U_b}{dz^2} = \frac{dP_b}{dx}, \tag{99}$$

$$\frac{dU_b}{dz}\bigg|_{z=1} = 1 \text{ at } z = 1, \text{ and } U_b = 0 \text{ at } z = 0. \tag{100}$$

From the set of equations above, we obtain

$$U_b(z) = \frac{dP_b}{dx}\left(\frac{1}{2}z^2 - z\right) + z. \tag{101}$$

The pressure gradient dP_b/dx is determined depending on the assumed velocity field. Two typical cases are well known, the so-called "linear flow" and "return flow", which were first discussed and named by [28].

In the "linear flow", a null pressure gradient $dP_b/dx = 0$ is assumed, then we obtain a "linear" velocity profile of $U_b(z) = z$. This case, however, will not be treated here further because it is less related to the thermocapillary flow in a liquid bridge than that of the "return flow".

In the return flow, "zero-net flow rate" is assumed; that is, from Eq. (101)

$$\int_0^1 U_b(z)dz = \int_0^1 \left\{ \frac{dP_b}{dx}\left(\frac{1}{2}z^2 - z\right) + z \right\} dz = \frac{dP_b}{dx}\left(-\frac{1}{3}\right) + \frac{1}{2} = 0. \quad (102)$$

From the above equation, we find $dP_b/dx = 3/2$. Thus, the velocity profile of the "return flow" becomes

$$U_b(z) = \frac{3}{4}z^2 - \frac{1}{2}z, \quad (103)$$

which is depicted with a thick solid line in Fig. 8. The basic flow is towards the positive x in the upper part of the layer, while it "returns" towards the negative x in the lower part. If it is integrated from the bottom to the top, the net flow is zero. Note that the returning flow is derived by the positivity of dP_b/dx.

As for the temperature related equations, Eqs. (94), (95) and (96) become

$$\frac{\partial^2 \Theta_b}{\partial z^2} = -\text{Ma} \cdot U_b(z), \quad (104)$$

where $\text{Ma} = \frac{\sigma_T \Delta T_0 d}{\mu \nu}$ $\text{Pr} = \frac{\sigma_T \gamma d^2}{\mu \kappa}$.

With use of the velocity profile Eq. (103), we obtain

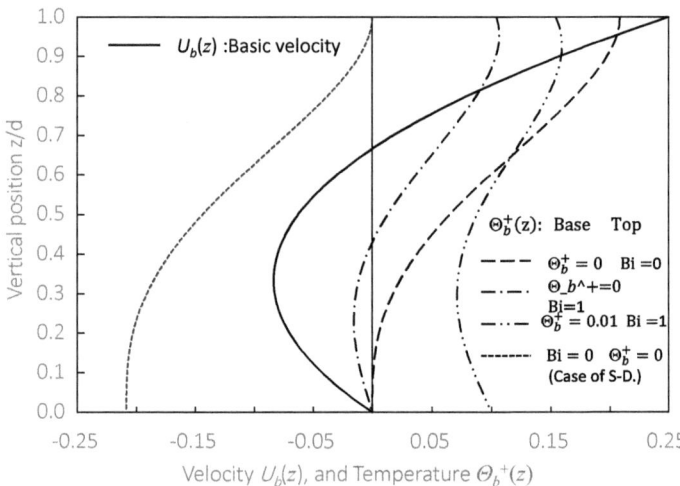

Fig. 8 Profiles of basic velocity $U_b(z)$ and temperature $\Theta_b^+(z)$ of the return flow. (*Notes* Value of the temperature profile $\Theta_b^+(z)$ is multiplied by a factor of 10. Temperature boundary conditions are $\Theta_b^+(0) = 0$ for dashed and dot-dashed lines, $\Theta_b^+(0) = 0.01$ for double-dot dashed line, and $\Theta_b^+(1) = 0$ for blue dotted line: the case of S-D)

$$\frac{\partial^2 \Theta_b}{\partial z^2} = -\text{Ma} \cdot \left(\frac{3}{4}z^2 - \frac{1}{2}z\right) \tag{105}$$

and

$$\Theta_b(z) = -\text{Ma} \cdot \left(\frac{1}{16}z^4 - \frac{1}{12}z^3\right) + C_1 z + C_2, \tag{106}$$

where C_1 and C_2 are the constants to be determined from the temperature boundary conditions. This process will be described below in some detail, because to the author's knowledge, it is not described in the beginner's level in former articles. First we will integrate Eq. (105) to examine relation of the heat fluxes between at the top and bottom surfaces.

$$\int_0^1 \frac{\partial^2 \Theta_b}{\partial z^2} dz = \left.\frac{\partial \Theta_b}{\partial z}\right|_{z=1} - \left.\frac{\partial \Theta_b}{\partial z}\right|_{z=0} = -\text{Ma} \int_0^1 \left(\frac{3}{4}z^2 - \frac{1}{2}z\right) dz = 0. \tag{107}$$

Thus, we find $\left.\frac{\partial \Theta_b}{\partial z}\right|_{z=1} = \left.\frac{\partial \Theta_b}{\partial z}\right|_{z=0}$, which means that the heat fluxes at the top and the bottom surfaces must be equal and there is no net heat flow in case of the "return flow". Accordingly, if one assumes an insulating surface on the bottom then the top surface becomes automatically insulating. On the other hand, if one side is a fixed temperature boundary then another side can be an insulating, a heat transfer or a fixed temperature boundary.

First, following Smith and Davis [28] (hereafter referred to as S-D), we will assume the insulating bottom boundary. Then

$$\left.\frac{\partial \Theta_b}{\partial z}\right|_{z=1} = \left.\frac{\partial \Theta_b}{\partial z}\right|_{z=0} = 0 \text{ at } z = 1 \text{ and } 0, \tag{108}$$

and thus we get $C_1 = 0$ from Eq. (106). To determine the remaining constant C_2, we must specify the temperature at either the top or bottom surfaces. S-D assumed the temperature of the top is equal to the ambient one $T_\infty(x)$, that is,

$$\Theta_{b,\text{S-D}}(z=1) = 0 \text{ at } z = 1, \tag{109}$$

which results in $C_2 = -(1/48)\text{Ma}$ and the temperature distribution becomes

$$\Theta_{b,\text{S-D}}(z) = -\frac{1}{48}\text{Ma} \cdot (3z^4 - 4z^3 + 1). \tag{110}$$

It is depicted in Fig. 8 with a thin blue dotted line. This seems to behave differently from others; this is only because S-D specified the top surface temperature to be zero.

We may specify the temperature at the bottom surface instead of the top. This approach was first employed by [22,23] with an assumption that the bottom temperature is equal to the ambient one;

$$T_b(x, y, z = 0) = T_\infty(x), \text{ i.e., } \Theta_b(z = 0) = 0 \text{ at } z = 0. \tag{111}$$

More generally we may assume that the bottom temperature is higher (or lower) than the ambient one, $T_\infty(x)$, with an amount of $\delta\Theta_0$, that is

$$T_b(x, y, z = 0) - T_\infty(x) = \Theta_b(z = 0) = \delta\Theta_0 \text{ at } z = 0. \tag{112}$$

In this case, we are able to apply the heat transfer condition between the top and the ambient temperature $T_\infty(x)$; that is,

$$\left.\frac{\partial \Theta_b}{\partial z}\right|_{z=1} = -\text{Bi}\{T(x)_{z=1} - T_\infty(x)\} = -\text{Bi} \cdot \Theta_b(z = 1) \text{ at } z = 1. \tag{113}$$

A set of equations, Eqs. (105), (112) and (113) can be solved as

$$\Theta_b(z) = -\frac{1}{48}\text{Ma} \cdot \left(3z^4 - 4z^3 + \frac{\text{Bi}}{1+\text{Bi}}z\right) + \left(1 - \frac{\text{Bi}}{1+\text{Bi}}z\right)\delta\Theta_0, \tag{114}$$

or with use of $\Theta_b^+ = \Theta_b/\text{Ma}$ and $\Delta\Theta_0^+ = \delta\Theta_0/\text{Ma}$,

$$\Theta_b^+(z) = -\frac{1}{48}\left(3z^4 - 4z^3 + \frac{\text{Bi}}{1+\text{Bi}}z\right) + \left(1 - \frac{\text{Bi}}{1+\text{Bi}}z\right)\Delta\Theta_0^+. \tag{115}$$

This solution for $\Delta\Theta_0^+ \neq 0$ was obtained in [26], although their nondimensionalization is different from the present ones.

Figure 8 shows profiles of the obtained solutions. The thick solid line (1) is the basic velocity $U_b(z)$. The blue thin dotted line is $\Theta_{b,\text{S-D}}(z)$ of Eq. (110) by S-D. The dashed and dot-dashed lines are for the fixed bottom temperature $\Theta_b(0) = 0$ with Bi = 0 and Bi = 1 at the top, respectively. The double-dot dashed line is for a non-zero bottom temperature $\Delta\Theta_0^+ = 0.01$ with Bi = 1. In case of Bi = 0, the temperature line intersects the boundaries perpendicularly, while in case of Bi $\neq$ 0, with a certain inclination. Note that the temperature inclination becomes equal at the top and bottom surfaces because the heat from the bottom must flow out of the top.

To examine the stability of the "return flow", it is useful to look back the instability analysis of the buoyancy-induced R-B convection described in Sect. 1 of Chapter "Thermodynamic and Molecular Aspects of Surface Tension", because the basics of the analysis is the same in both cases although the present one is more complexed.

First, we decompose the velocity, the temperature and the pressure into their basic and fluctuating parts as was in Eq. (38). The basic velocity has only the x component $U_b(z)$, in the present case, while the fluctuating parts are function of all the directions (x, y, z). Thus, the decomposition becomes

$$\boldsymbol{u} = \boldsymbol{U_b}(\boldsymbol{x}) + \hat{\boldsymbol{u}}(\boldsymbol{x}, t) = [U_b(z) + \hat{u}(\boldsymbol{x}, t), \hat{v}(\boldsymbol{x}, t), \hat{w}(\boldsymbol{x}, t)], \tag{116}$$

$$\Theta = \Theta_b(z) + \hat{\vartheta}(\boldsymbol{x}, t), \tag{117}$$

$$p = P_b(x) + \hat{p}(\boldsymbol{x}, t). \tag{118}$$

Now we will derive the equations for the fluctuating parts $\hat{\boldsymbol{u}}(\boldsymbol{x}, t)$, $\hat{\vartheta}(\boldsymbol{x}, t)$ and $\hat{p}(\boldsymbol{x}, t)$. As we did for the former R-B convection, we first substitute these equations into the momentum (velocity) and energy (temperature) equations, with proper nondimensionalizations. Next, we subtract the equations for the basic fields from the obtained equations. Then we get equations for the fluctuating parts of the velocity and temperature. These equations for the fluctuating parts correspond to Eqs. (55) and (57) in case of the R-B convection. There exists, however, a distinct difference from the former case. That is, in the R-B convection, the basic part exists for the temperature field but not for the velocity. While in the present case of the return flow, the basic parts arise in both the velocity and temperature fields.

Resulting equations for the fluctuating velocities $\hat{\boldsymbol{u}}(\boldsymbol{x}, t) = \{\hat{u}(\boldsymbol{x}, t), \hat{v}(\boldsymbol{x}, t), \hat{w}(\boldsymbol{x}, t)\}$, temperature $\hat{\vartheta}(\boldsymbol{x}, t)$ and pressure $\hat{p}(\boldsymbol{x}, t)$ become

$$\text{Re}\left(\frac{\partial \hat{u}}{\partial t} + U_b(z)\frac{\partial \hat{u}}{\partial x} + \hat{w}\frac{dU_b}{dz}\right) = -\frac{\partial \hat{p}}{\partial x} + \nabla^2 \hat{u}, \tag{119}$$

$$\text{Re}\left(\frac{\partial \hat{v}}{\partial t} + U_b(z)\frac{\partial \hat{v}}{\partial x}\right) = -\frac{\partial \hat{p}}{\partial y} + \nabla^2 \hat{v}, \tag{120}$$

$$\text{Re}\left(\frac{\partial \hat{w}}{\partial t} + U_b(z)\frac{\partial \hat{w}}{\partial x}\right) = -\frac{\partial \hat{p}}{\partial z} + \nabla^2 \hat{w}, \tag{121}$$

$$\text{Ma}\left(\frac{\partial \hat{\vartheta}}{\partial t} + U_b(z)\frac{\partial \hat{\vartheta}}{\partial x} + \hat{w}\frac{d\Theta_b}{dz}\right) = \nabla^2 \hat{\vartheta}, \tag{122}$$

where $\text{Ma} = \text{Re} \cdot \text{Pr}$ and the pair of Ma and Pr is often selected to specify the conditions.

Referring to Eqs. (81) and (80), the boundary conditions at top surface with the present normalization become

$$\text{Velocities:} \quad \frac{\partial \hat{u}}{\partial z} = -\text{Ma}\frac{\partial \hat{\vartheta}}{\partial x}, \quad \frac{\partial \hat{v}}{\partial z} = -\text{Ma}\frac{\partial \hat{\vartheta}}{\partial y} \text{ and } \hat{w} = 0 \text{ at } z = 1, \tag{123}$$

Temperature: $\dfrac{\partial \hat{\vartheta}}{\partial z} = -\text{Bi}\,\hat{\vartheta}$ at $z = 1$, (124)

where $\text{Bi} = 0$ is equivalent to the insulating boundary and $\text{Bi} = \infty$ to the fixed temperature.

At the bottom $z = 0$, the fluctuating velocities are zero $\hat{u}(x, t) = 0$, and the fluctuating temperature $\hat{\vartheta}$ is assumed either fixed $\hat{\vartheta}(z = 0, t) = 0$ (so-called conducting) or $\partial \hat{\vartheta}/\partial z = 0$ (insulating). Note that these boundary conditions on the fluctuating temperature $\hat{\vartheta}$ are applied irrespective of those for the basic temperature as noted already for the R-B convection.

These set of equations for the fluctuating components are solved as was done for the R-B convection described in Sect. 2.1. That is, we will describe all the fluctuating quantities with the following dependence in the form as

$$\hat{f}(x, t) = \tilde{f}(z)\exp[i(k_1 x + k_2 y)] \cdot \exp(st), \tag{125}$$

where $\hat{f}(x, t)$ represents either $\hat{u}(x, t)$, $\hat{v}(x, t)$, $\hat{w}(x, t)$, $\hat{p}(x, t)$ or $\hat{\vartheta}(x, t)$ and $\tilde{f}(z)$ is their amplitude function. The constant s can be complex and is expressed here as $s = s_r - i\omega$, where s_r and ω are both real constants and indicate the growth rate and the frequency of the fluctuation, respectively. Then, Eq. (125) can be rewritten as

$$\hat{f}(x, t) = \tilde{f}(z)\exp[i(k_1 x + k_2 y - \omega t)] \cdot \exp(s_r t), \tag{126}$$

where the growth rate s_r is the most critical factor affecting the flow stability. Specifically, the disturbance is either attenuated ($s_r < 0$) or amplified ($s_r > 0$) depending on the sign of s_r. Therefore, $s_r = 0$ corresponds to the marginal (or neutral) stability condition.

The equations given representatively by Eq. (126) correspond to those of Eqs. (58 and 59) of the R-B convection. Some differences, however, exist between these two cases. Firstly, in case of the R-B convection, the gradient of the basic pressure appears only in the vertical (z) direction, thus the flow field is homogeneous in horizontal x and y directions. Accordingly, the wavenumbers l and m was able to be reduced to one characteristic wavenumber $a = \sqrt{l^2 + m^2}$. In the present case, on the other hand, the basic flow exists along the x-axis. Thus, the wavenumbers k_1 and k_2 are not equivalent. Secondly, although the frequency ω in the R-B convection was theoretically found zero, it must be retained in the present case.

Detailed description on the solution method of this set of equations is beyond the scope of this section. Its description as well as some new computational results are given in the subsequent Sect. 4 of this Chapter, which is provided by Fujimura [8] of Tottori University, Japan, in response to a request from the editorial board of this book. Accordingly, only an outline of the solution method will be given below.

The governing equations and imposed boundary conditions with the nondimensional numbers, Ma, Pr, Bi and $\Delta\Theta_0^+$ together with the introduced constants, k_1, k_2, ω and s_r, constitute an eigenvalue problem. Among the nondimensional numbers, we first specify the property of the test fluid Pr and the temperature boundary conditions at top surface, Bi and at bottom, $\Delta\Theta_0^+$. Then, with use of a tentatively assumed set of constants k_1, k_2, the eigenvalue problem is solved to find marginal Marangoni numbers Ma_{mr} at which the neutral condition $s_r = 0$ is satisfied. Then this process is repeated using another set of the constants until we can attain the minimum among the Ma_{mr} numbers. The minimum one attained is the critical Marangoni number Ma_{cr}. Then, the relevant constants $k_{1\text{cr}}, k_{2\text{cr}}$ and ω_{cr} can also be determined at the same time.

As mentioned already, this stability problem was first solved by S-D for an insulated bottom and later by[22, 23], for various *Pr* numbers and several Bi's with an insulated bottom surface or a fixed temperature of $\Delta\Theta_0^+ = 0$. The obtained values of Ma_{cr} from the analysis of [22, 23] are shown in Fig. 9 against the Pr number for Bi $= 0$ and 0.1. Comparing the temperature conditions at the bottom surface, if Ma_{cr}'s with Bi $= 0$ are compared, it is larger for the fixed bottom temperature than for the insulated one. This is because if the bottom temperature is fixed the temperature fluctuation is more suppressed than with an insulated bottom. Additionally, as for the thermal condition at the top surface, a tendency is seen for Ma_{cr} to increase with increasing Bi. These effects of the boundary conditions are significant for the small Pr numbers and becomes less with increasing Pr.

A significant difference found by S-D from the aforementioned R-B and Pearson-Bénard convections is that in the present type of flow, a travelling wave of velocity and temperature appears in the flow and temperature fields. S-D called this wave the "hydrothermal wave" (often abbreviated as HTW) in their pioneering article of S-D (1983). A snapshot of such a traveling wave is shown in Fig. 10 from the present author's laboratory. The basic flow over the surface moves from the left to the right, while the HTW indicated by the high and low temperature stripes travels upstream against the basic flow with an inclination angle ψ', which is defined against the negative direction of *x*-axis traditionally since the work of S-D.

Fig. 9 The critical Marangoni number of the hydrothermal wave in the return flow with several thermal boundary conditions: Bottom: Insulating or fixed temperature, Top: Bi $= 0$ or 0.1: from [22, 23]

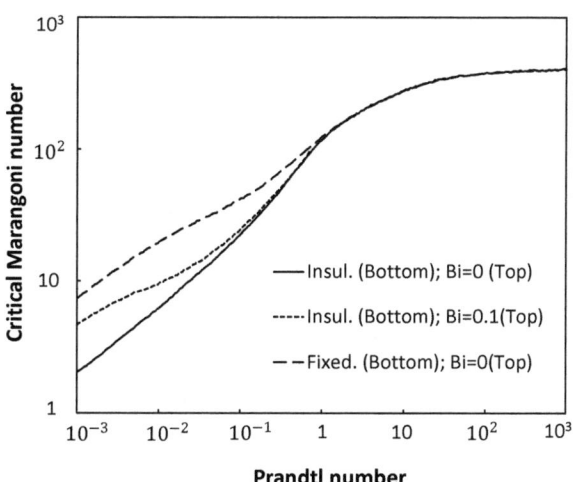

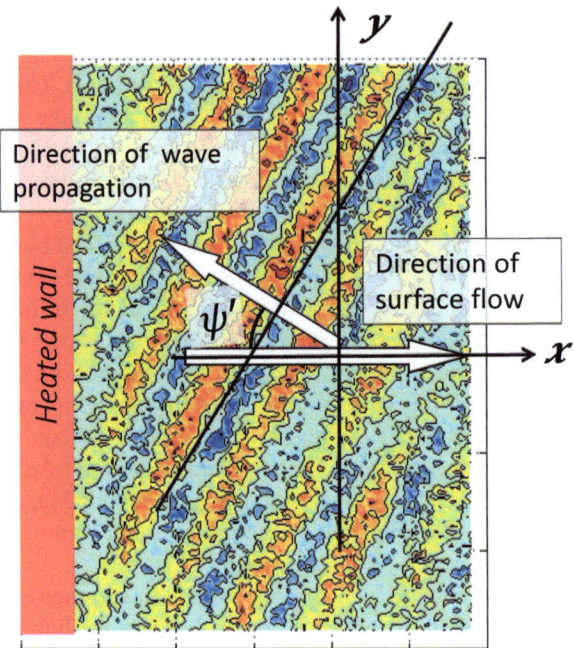

Fig. 10 A snapshot of a hydrothermal wave over a liquid layer with a temperature gradient imposed along the x-axis. The hydrothermal wave propagates obliquely upstream against the basic flow. Reproduced from Author's LabDataArchive, Produced by Ide T. (2003)

The relation among the angle ψ' and the constants, k_1, k_2 and ω is depicted in Fig. 11. Referring to Eq. (126), a wave front of the "crest or trough" can be designated by $k_1 x + k_2 y = \omega t_0$, which is depicted by the dot-dashed lines in Fig. 11. This line travels with increasing time $t_0 + \Delta t$ ($\Delta t > 0$) towards the direction depending on combinations among k_1, k_2 and ω. (Although the time increment Δt can be any positive (nonzero) value, it is more easy to understand if it is put equal to the one period of the oscillation, that is $\Delta t = 2\pi/\omega$.)

The signs of these constants are only significant in relation to each other. Here, we will take $k_2 > 0$ for consistency with the case of a cylinder, where k_2 corresponds to m, the azimuthal mode number, which is assigned usually to be positive.

In (a) and (b) of Fig. 11, the cases of $k_2 > 0$ and $\omega > 0$ are depicted. When $\omega/k_1 > 0$, the dot-dashed lines propagate towards the upper right of the figure (b). In this case of (b), the x-component of the propagation is in the same direction as the basic flow; thus, this propagation is often called the "*co-flow*". In the case of $\omega/k_1 < 0$ as in (a), on the other hand, their x-components propagate in the opposite direction to the basic flow, thus it is called the "*counter-flow*".

Although not shown for the simplicity, when ω is negative with the sign of ω/k_1 unchanged, the dot-dashed lines fold over to the negative y side of the x-axis, while their intersections with the x-axis remain unchanged, thus the conditions for the

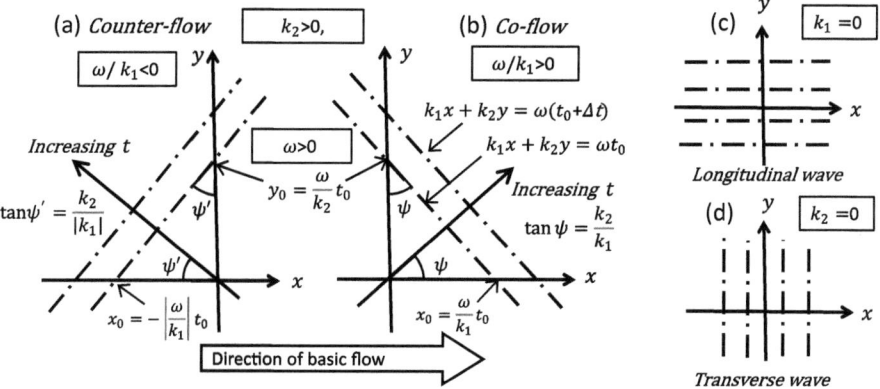

Fig. 11 Relation among the wavenumbers (k_1, k_2), the frequency (ω) and the propagation direction and angle ψ. As for k_2, $k_2 > 0$ in **a, b** and **c**, and $k_2 = 0$ in **d**; $\omega \neq 0$ in all cases

"co-flow" or the "counter-flow" also remain unchanged. Accordingly, whether the oblique wave propagation is the "co-flow" or the "counter-flow" depends on the sign of ω/k_1, regardless of the sign of ω.

In the above mentioned cases, neither k_1 nor k_2 is zero. On the other hand, the cases where either k_1 or k_2 is zero are depicted in figures (c) and (d), respectively. Specifically, in the case of $k_1 = 0$, the dot-dashed lines are aligned parallel to the basic flow and is called the "longitudinal wave". When $k_2 = 0$ they are perpendicular to the basic flow and is called the "transverse wave".

The propagation angle ψ' at the critical Ma number in the return flow is shown in Fig. 12 for various Pr numbers. The propagation angle ψ' is defined against the *negative* direction of x-axis, following the original definition by S-D. Figure 12 indicates that when the Pr number is small, the angle is close to 90°, so the hydrothermal wave (HTW) becomes similar to the longitudinal wave. As the Pr number increases, the angle becomes smaller, and the HTW becomes more like the transverse wave. At intermediate Pr values, the angle also takes intermediate values and does not exceed 90°. This means that the HTW in the return flow always propagates in the opposite direction to the basic flow.

In the experimental snapshot of Fig. 10, the angle ψ' is approximately 30°. Since Pr $\sim$30 with fixed bottom temperature in this experiment, the angle of about 20° is obtained from Fig. 12, which agrees fairly well with the experimental ψ'.

As for the non-zero bottom temperature, analysis was first made by [26] for positive $\delta\Theta_0'$ with a selected Pr of Pr $= 7$. In this book, a new work on this subject is presented by [8] as requested by the current editorial board. His work includes cases with both positive and negative $\delta\Theta_0'$s, for a higher Prandtl number of Pr $= 100$ and reveals the emergence of transverse waves and steady longitudinal rolls also in the return flow.

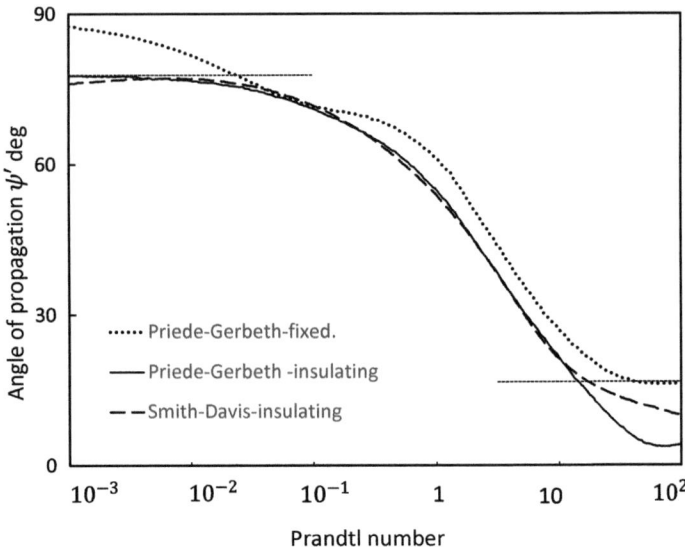

Fig. 12 Propagation angle ψ' of HTW at the critical point of the return flow. Top surface: Bi = 0. Bottom: insulating or fixed-temperature. Plotted from [22, 28]

3.3 Infinitely Long Cylinder with a Temperature Gradient Along Its Axis

The main focus of this book is a series of experiments on the thermocapillary convection in a finite-length liquid cylinder conducted under microgravity conditions. Since this system is three-dimensional and complex, we introduced the infinite plane liquid layer as one of the canonical configurations in the previous subsection. In this subsection, we will discuss another canonical system, an infinitely long cylinder with a temperature gradient along its axis, which is more directly related to the present main subject. Although this is unrealistic due to the length limit of the circular cylinder (the Plateau-Rayleigh instability), it is still worthy for the theoretical foundation. This configuration was initially examined by Xu and Davis [31] and later revised by [25].

An infinitely long cylinder with a radius R is assumed as shown in Fig. 13. As was in the previous section, a constant temperature gradient of the environmental temperature $T_\infty(x)$ is imposed along the x-axis; that is, $T_\infty(x) = T_{\text{ref}} - \gamma x$ and $dT_\infty/dx = -\gamma$ with $\gamma > 0$, where T_{ref} is an arbitrary reference temperature. The basic flow has only an x component with a function of r; that is, $U_b(r)$. We will derive equations of the basic velocity and temperature distributions analogously to the return flow of the plane layer.

The quantities and equations are nondimensionalized as the return flow using the same characteristic quantities as Eqs. (97 and 98) except L_0 is R instead of d.

The nondimensionalized momentum equation becomes

$$\frac{1}{r}\frac{d}{dr}\left(r\frac{dU_b(r)}{dr}\right) = \frac{dP_b}{dx}. \tag{127}$$

Referring to the thermocapillarity condition of Eq. (14), the boundary conditions at the surface and the center are

$$\left.\frac{dU_b}{dr}\right|_{r=1} = 1 \text{ (at } r = 1\text{) and } \left.\frac{dU_b}{dr}\right|_{r=1} = 0 \text{ (at } r = 0\text{)}. \tag{128}$$

From Eqs. (127) and (128), we get $\frac{dP_b}{dx} = 2$, that is, the pressure gradient is positive as was in the case of the return flow in the flat layer. Since these boundary conditions provide only the gradients of the velocity, we must specify an absolute value of the flow velocity in some way. Here we will assume a net flow, U_{net}, and impose it uniformly across the cylinder section. Then the basic velocity profile can be obtained as

$$U_b(r) = \frac{1}{2}\left(r^2 - \frac{1}{2}\right) + U_{net}. \tag{129}$$

If there is no net flow, this solution coincides with that of [25, 31].
We will introduce the temperature difference again as:

$$\Theta(r, \theta, x, t) = T(r, \theta, x, t) - T_\infty(x). \tag{130}$$

The basic (unperturbed) part Θ_b does not depend on x because we assume an infinitely long cylinder.

The energy equation can be nondimensionalized, again using the same characteristic quantities as Eqs. (97 and 98) with use of R instead of L_0, as

$$\frac{1}{r}\frac{d}{dr}\left(r\frac{d\Theta_b(r)}{dr}\right) = -\text{Ma}\, U_b(r), \tag{131}$$

where $\text{Ma} = \frac{\sigma_T \gamma R^2}{\mu \kappa}$. The minus sign on the right hand side arises from $\frac{dT_\infty(x)}{dx} = -1$ in the nondimensional form. With use of Eq. (129), we get

$$\frac{1}{r}\frac{d}{dr}\left(r\frac{d\Theta_b(r)}{dr}\right) = -\text{Ma}\left\{\frac{1}{2}\left(r^2 - \frac{1}{2}\right) + U_{net}\right\}, \tag{132}$$

which indicates that the added net flow term is equivalent to a uniform heat source or sink in the liquid bridge depending on the sign of U_{net}. Integration of Eq. (132) with reminding $\frac{d\Theta_b(r)}{dr} = 0$ at $r = 0$ results in

$$\Theta_b(r) = -\frac{\text{Ma}}{32}\left(r^4 - 2r^2 + 8U_{net}r^2\right) + C_\theta, \tag{133}$$

where C_θ is a constant to be determined from the boundary condition over the cylinder surface. That is,

$$-\left.\frac{\partial \Theta_b}{\partial r}\right|_{r=1} = \text{Bi}\{T_b - T_\infty(x)\} = \text{Bi}\Theta_b(r=1) \text{ at } r=1. \tag{134}$$

Then we obtain

$$\Theta_b(r) = -\frac{\text{Ma}}{32}(1-r^2)^2 + \frac{\text{Ma}}{4}(1-r^2)U_\text{net} + \frac{\text{Ma}}{2\text{Bi}}U_\text{net}. \tag{135}$$

If U_net is zero, the above equation coincides with the one obtained by [25, 31]. Here, one may notice that if U_net is non-zero and Bi tends to zero then $\Theta_b(r)$ becomes infinity. This is because we need heat release to environment since, as mentioned above, U_net is equivalent to a heat source. Thus, it is more reasonable to introduce a new variable $\delta\Theta_\text{surf} = \text{Ma}U_\text{net}/2\text{Bi}$, which is the (nondimensionalized) surface temperature at $r=1$. Accordingly, Eq. (135) can be rewritten as

$$\Theta_b(r) = -\frac{\text{Ma}}{32}(1-r^2)^2 + \left\{\frac{\text{Bi}}{2}(1-r^2) + 1\right\}\delta\Theta_\text{surf}. \tag{136}$$

The process of the linear stability analysis (LSA) is fundamentally same as the case of the return flow in the previous section. The main difference between the two cases is that, in the previous (return flow) case, two directions were infinitely large, while in the present case, the direction of the basic flow (x-axis) is infinite but the azimuthal direction (φ) is periodic.

Then, with use of the coordinate $\boldsymbol{x} = (r, \varphi, x)$, the equation corresponding to Eq. (126) becomes,

$$\hat{f}(\boldsymbol{x}, t) = \tilde{f}(r)\exp[i(k_1 x + m\varphi - \omega t)] \cdot \exp(s_r t), \tag{137}$$

where the major difference from Eq. (126) is the use of an azimuthal wavenumber m instead of k_2. The constant m represents the azimuthal wavenumber and is often called "mode number". That is, "m" refers to the number of periodicities exhibited by a rotating pattern during a rotation of 2π. It can be either positive or negative in the mathematical treatment; however, to correspond with the intuition associated with the term "mode number", it is assumed almost always to be positive, including in this article as well. In this relation, the wavenumber k_2 in the previous section was assigned positive.

The relation among the propagation angle ψ, the wavenumbers k_1 and m, and the frequency ω can be understood quite similarly as those depicted in Fig. 11. The direction of the basic flow is x in both cases, while the wavenumber k_2 should be replaced by m in the present case. The distinction of "co-flow" or "counter-flow" can be judged by the sign of ω/k_1 exactly as in the previous case.

The linear stability analysis (LSA) of this case was performed by [31] and [25]. Their results on the critical Marangoni number and the propagation angle are illustrated against the Prandtl number of the fluid in Fig. 14. The critical mode number was found $m = 1$ within the plotted conditions. Note that, the propagation angle ψ is defined here against the positive direction of the x axis; that is, $\tan \psi = m/k_1$. As for the propagation direction of the hydrothermal wave (HTW) with respect to the base flow, the "co-flow" and "counter-flow" are distinguished with use of the solid and open symbols, respectively.

The point to be noted here is that in the region where $Pr < 1.0$, the direction of the HTW changes from co-flow to counter-flow accompanied by the increase in the propagation angle crossing over 90° with decreasing Pr number. Instead, for high Pr of approximately over 20, direction of the HTW and its propagation angle both change discontinuously, which was not observed in case of the large liquid layer at least within the conditions examined in Figs. 12 and 13.

During the microgravity experiment of the HTW in a liquid bridge, we observed a change in the propagation direction of the HTW depending on the heat loss/gain from the liquid bridge. Accordingly, we asked [8] to perform an LSA on the HTW in the long liquid cylinder with the heat loss/gain; that is, the non-zero surface temperature difference, $\delta\Theta_{surf}$ of Eq. (136). His LSA revealed a significant effect of the $\delta\Theta_{surf}$ upon the propagation direction of HTW. The results are presented in the subsequent subsection; so, readers are recommended to refer to the subsequent Sect. 4.

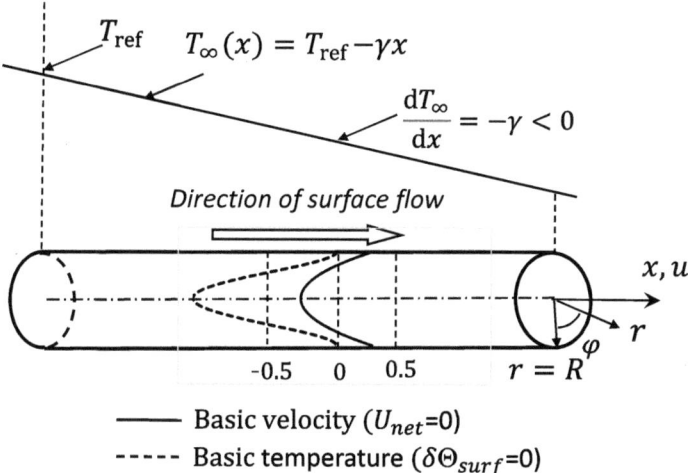

Fig. 13 Infinitely long cylinder in an environment with a constant temperature gradient. (Coordinate and the ambient temperature gradient)

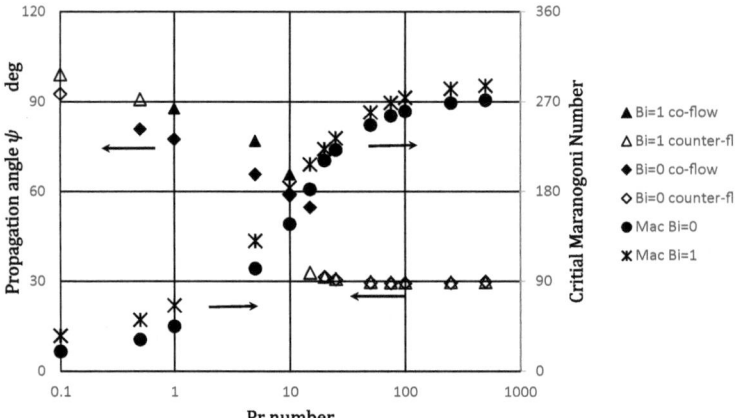

Fig. 14 Propagation angle ψ of HTW at the critical point and the critical Marangoni number of a long cylinder. Free surface: Bi = 0 and 1. Plotted from [25]

4 Linear Stability Analysis of Thermocapillary Convection in Liquid Layer and Bridge

Kaoru Fujimura

4.1 Physical Setup and Formulation of the Problem

In this subsection, we describe the analysis of the linear stability of thermocapillary convections and the stability characteristics of the following three specific problems:

P-I. a liquid layer formed on a horizontal flat plate of uniform depth d with an infinite extent where the ambient temperature $T_\infty(x)$, say, has a negative gradient in the x-direction $dT_\infty/dx = -\beta < 0$ as illustrated in Fig. 15(a).

P-II. a liquid column of radius d with an infinite extent in the x-direction where the ambient temperature T_∞ has a negative gradient $dT_\infty/dx = -\beta < 0$ as illustrated in Fig. 15(b).

P-III. a liquid bridge with radius R and height d, known as the half-zone model whose endplates (i.e., upper and lower rods) have uniform but different temperatures as illustrated in Fig. 15(c). The ambient temperature has a positive gradient $dT_\infty/dx = \beta > 0$. The aspect ratio of the liquid bridge is defined by $\Gamma = d/R$.

In all these problems, we assume that the free surface is non-deformable. The temperature gradient on the free surface causes an inhomogeneity of the surface tension and yields thermocapillary convection. In problems P-I and II, we further consider a situation where a heat transfer is imposed across the liquid layer, i.e., the

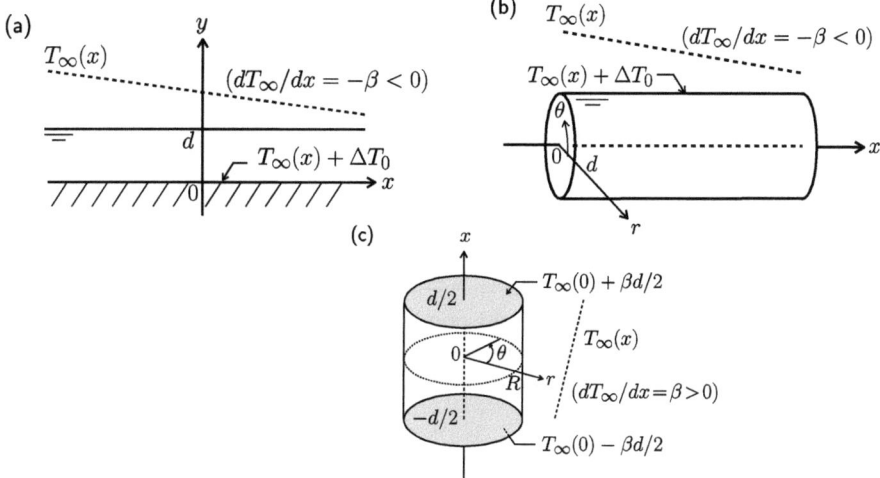

Fig. 15 Physical setup and coordinate systems. (a) problem P-I, (b) P-II, and (c) P-III. In (a) and (b), $dT_\infty/dx = -\beta < 0$ is assumed whereas $dT_\infty/dx = \beta > 0$ in (c)

temperatures on the bottom $y = 0$ in P-I and on the free surface $r = d$ in P-II are both given by $T_0 = T_\infty(x) + \Delta T_0$. In P-I, we take the Cartesian coordinate and define the position vector x by $x = (x, y, z)$ and velocity vector u by $u = (u, v, w)$. In P-II and P-III, we take the cylindrical coordinate and define x and u by $x = (x, r, \theta)$ and $u = (u_x, u_r, u_\theta)$, respectively.

We nondimensionalize physical variables by

$$x = dx^*, \ t = \frac{\mu}{\sigma_T \beta} t^*, \ u = \frac{\sigma_T \beta d}{\mu} u^*, \ p = \sigma_T \beta p^*, \ T - T_\infty(x=0) = \beta d T^*. \tag{138}$$

Here, the variables without asterisks denote dimensional and asterisked variables nondimensional. The characteristic length scale d is the depth of the liquid layer in P-I but is the radius of liquid column in P-II and the height of the liquid bridge in P-III. Here, σ_T is the temperature coefficient of the surface tension, and μ is the dynamic viscosity.

In what follows, we use only nondimensional variables by omitting the asterisks without confusion.

The governing equations for the velocity u, the pressure p, and the temperature T are the equation of continuity, the Navier–Stokes equations, and equation of energy. In nondimensional form, they are

$$\nabla \cdot u = 0, \tag{139}$$

$$\frac{\partial \boldsymbol{u}}{\partial t} + (\boldsymbol{u} \cdot \nabla)\boldsymbol{u} = \mathrm{Re}^{-1}(-\nabla p + \Delta \boldsymbol{u}), \tag{140}$$

$$\frac{\partial T}{\partial t} + (\boldsymbol{u} \cdot \nabla)T = \mathrm{Ma}^{-1}\Delta T \tag{141}$$

defined in the domain $\Omega = \{(x, y, z) | x \in (-\infty, \infty), y \in (0, 1), z \in (-\infty, \infty)\}$ for P-I, $\{(x, r, \theta) | x \in (-\infty, \infty), r \in [0, 1), \theta \in [0, 2\pi)\}$ for P-II, and $\{(x, r, \theta) | x \in (-1/2, 1/2), r \in [0, \Gamma^{-1}), \theta \in [0, 2\pi)\}$ for P-III. Here, Re is the Reynolds number defined by $\mathrm{Re} = \frac{\sigma_T \beta d^2}{\mu \nu}$, Ma is the Marangoni number defined by $\mathrm{Ma} = \mathrm{Pr}\mathrm{Re}$, and Pr is the Prandtl number defined by $\mathrm{Pr} = \frac{\nu}{\kappa}$ where ν is the kinematic viscosity and κ is the thermal diffusivity.

Boundary conditions on the free surface located at $y = 1$ in P-I, $r = 1$ in P-II, or $r = \Gamma^{-1}$ in P-III are

$$\boldsymbol{u} \cdot \boldsymbol{n} = 0, \; \boldsymbol{S} \cdot \boldsymbol{n} + (\boldsymbol{I} - \boldsymbol{nn}) \cdot \nabla T = 0, \; \boldsymbol{n} \cdot \nabla T + \mathrm{Bi}[T - T_\infty(x)] = 0, \tag{142}$$

where $\boldsymbol{S} = \nabla \boldsymbol{u} + (\nabla \boldsymbol{u})^T$ is the stress tensor, $\boldsymbol{I}$ is the 3×3 identity matrix, $\boldsymbol{n}$ is the outward unit vector normal to the free surface, T_∞ is the nondimensional ambient temperature ($= -x$ in P-I and II and x in P-III), and Bi is the Biot number defined by hd/λ with the heat transfer coefficient h and the thermal conductivity λ.

Boundary conditions on the solid surface in P-I are $\boldsymbol{u} = 0$ and $T = \delta\Theta_0 - x$ at $y = 0$ where $\delta\Theta_0$ is the nondimensional ΔT_0. In P-III, they are $\boldsymbol{u} = 0$ and $T = \pm\frac{1}{2}$ at $x = \pm\frac{1}{2}$.

4.2 Basic Field

To examine the stability of fluid motions, we first get the basic field which is a steady solution of Eqs. (139)–(141) and is independent of z or θ under the boundary conditions.

In P-I, an assumption of zero mass-flux across an arbitrary x-cross section yields a flow having the quadratic velocity profile. The closed form solutions for the basic flow $\overline{U}(y)$ and the temperature $\overline{T}(x, y)$ are

$$\overline{U}(y) = \frac{3}{4}y^2 - \frac{1}{2}y, \tag{143}$$

$$\overline{T}(x, y) = -\mathrm{Ma}\left(\frac{y^4}{16} - \frac{y^3}{12}\right) - \frac{\mathrm{Bi}\mathrm{Ma}\,y}{48(1+\mathrm{Bi})} + \left(1 - \frac{\mathrm{Bi}\,y}{1+\mathrm{Bi}}\right)\delta\Theta_0 - x. \tag{144}$$

For $\delta\Theta_0 = 0$, Eqs. (143) and (144) are the same as the basic field obtained by Priede and Gerbeth [22].

In P-II, the closed form solutions for the basic flow $\overline{U}(r)$ and the temperature $\overline{T}(x, r)$ are

$$\overline{U}(r) = \frac{1}{2}\left(r^2 - \frac{1}{2}\right) + \frac{2\mathrm{Bi}}{\mathrm{Ma}}\delta\Theta_0, \tag{145}$$

$$\overline{T}(x, r) = -\frac{\mathrm{Ma}}{32}\left[(r^2 - 1)^2 + \frac{16\mathrm{Bi}}{\mathrm{Ma}}\delta\Theta_0(r^2 - 1)\right] + \delta\Theta_0 - x. \tag{146}$$

We have Eqs. (145) and (146) in a situation where a uniform net flow $\frac{2\mathrm{Bi}}{\mathrm{Ma}}\delta\Theta_0$ is introduced to maintain a temperature difference on the free surface $r = 1$ at $\delta\Theta_0 - x$. For $\delta\Theta_0 = 0$, Eqs. (145) and (146) are the same as the basic field obtained by Xu and Davis [31] and Ryzhkov [25]. If $\delta\Theta_0 > 0$, the net flow is towards positive x; so the hotter fluid is conveyed from the upstream and the free surface ($r = 1$) temperature becomes higher than $T_\infty(x)$; while for $\delta\Theta_0 < 0$, the colder fluid is conveyed and the surface temperature decreases less than $T_\infty(x)$.

In contrast, in P-III, a closed-form solution of the basic field is hardly obtained. Because the basic field is axisymmetric, we may introduce Stokes' stream function and integrate the resultant equations numerically. As the Prandtl number increases, iso-thermal lines concentrate at the point $(x, r) = \left(-1/2, \Gamma^{-1}\right)$ so that a non-uniformity of the temperature gradient of the surface tension is localized near $x = -1/2$. This situation makes it difficult to get accurate basic fields. To avoid difficulties, Wanschura et al. [30] introduced a 'regularization'. Following them, one may replace the second boundary condition of Eq. (142) with $\mathbf{S}\cdot\mathbf{n} + f_\delta(x)(\mathbf{I} - \mathbf{nn})\cdot\nabla T = 0$. Here, the factor $f_\delta(x)$ has the form

$$f_\delta(x) = [1 - \cos(\delta\pi(x + 1/2))]^2/4 \text{ for } x \leq -1/2 + 1/\delta, \tag{147}$$

$$f_\delta(x) = 1 \text{ for } |x| < 1/2 - 1/\delta, \tag{148}$$

$$f_\delta(x) = [1 - \cos(\delta\pi(x - 1/2))]^2/4 \text{ for } x \geq 1/2 - 1/\delta. \tag{149}$$

Wanschura et al. [30] set $\delta = 10$. In Sect. 4.6, we use both $\delta = 10$ and 50.

4.3 Stability Analysis

To examine the stability of the basic field, we decompose $\mathbf{u}$, p, and T into the steady basic field with 'over-bar' and the disturbance with 'hat' as $\mathbf{u} = \overline{\mathbf{U}} + \hat{\mathbf{u}}$, $p = \overline{P} + \hat{p}$, and $T = \overline{T} + \hat{\vartheta}$. Substitution of these expressions into the governing Eqs. (139)–(141) and subtraction of the equations for $\overline{U}$, $\overline{P}$, and $\overline{T}$ yield the equations governing the disturbance. In the linear stability analysis, we ignore the nonlinear terms of the disturbance components therein. We further introduce the normal mode[1] such that

$$\text{P-I} \begin{pmatrix} \hat{u}(x,t) \\ \hat{p}(x,t) \\ \hat{\vartheta}(x,t) \end{pmatrix} = \begin{pmatrix} \tilde{u}(y) \\ \tilde{p}(y) \\ \tilde{\Theta}(y) \end{pmatrix} e^{i(k_1 x + k_3 z) + \tilde{\sigma} t}, \qquad (150)$$

$$\text{P-II} \begin{pmatrix} \hat{u}(x,t) \\ \hat{p}(x,t) \\ \hat{\vartheta}(x,t) \end{pmatrix} = \begin{pmatrix} \tilde{u}(r) \\ \tilde{p}(r) \\ \tilde{\Theta}(r) \end{pmatrix} e^{i(k_1 x + m\theta) + \tilde{\sigma} t}, \qquad (151)$$

$$\text{P-III} \begin{pmatrix} \hat{u}(x,t) \\ \hat{p}(x,t) \\ \hat{\vartheta}(x,t) \end{pmatrix} = \begin{pmatrix} \tilde{u}(x,r) \\ \tilde{p}(x,r) \\ \tilde{\Theta}(x,r) \end{pmatrix} e^{im\theta + \tilde{\sigma} t}. \qquad (152)$$

where $(\tilde{u}, \tilde{p}, \tilde{\Theta})$ are called the amplitude functions. The resultant equations for $(\tilde{u}, \tilde{p}, \tilde{\Theta})$ together with the boundary conditions compose a linear eigenvalue problem. After discretizing them, we get the linear dispersion relation[2]

$$D(\text{Re}, \text{Pr}, \text{Bi}, \boldsymbol{k}; \tilde{\sigma}^{(j)}) = 0, \; j = 1, 2, \ldots,$$

where the j-th linear eigenvalue $\tilde{\sigma}^{(j)} \in \mathbb{C}$ plays the role of a complex growth rate of the j-th eigenmode. Because there exist infinite discrete eigenvalues, we order the eigenvalues such that $\text{Re}[\tilde{\sigma}^{(1)}] \geq \text{Re}[\tilde{\sigma}^{(2)}] \geq \ldots$ where $\text{Re}[\tilde{\sigma}^{(j)}]$ stands for the real part of $\tilde{\sigma}^{(j)}$ indicating the linear growth rate of the j-th eigenmode. Hereafter, we denote $\tilde{\sigma}^{(1)}$ as $\tilde{\sigma}$ throughout in Sect. 4. In Sects. 4.4–4.6 we also let

$$\tilde{\sigma} = \text{Re}[\tilde{\sigma}] - i\omega \equiv \tilde{\sigma}_r - i\omega$$

with the frequency $\omega \in \mathbb{R}$. As for the discretization, the finite difference method, the finite element method, the finite volume method, expansion in Chebyshev polynomials associated with the tau-collocation method, and so on were adopted in past. In Sect. 4, all the presented results are due to the Chebyshev tau-collocation. In problems P-II and P-III, the disturbance should be finite at $r = 0$. Xu and Davis [31] clarified the asymptotic form of disturbance near $r = 0$, i.e.,

---

[1] In an initial value problem, the disturbance is expressed in terms of the inverse Fourier–Laplace transform $\hat{u}(t) = \frac{1}{(2\pi)^2} \int_{-\infty}^{\infty} d\boldsymbol{k} \left[\frac{1}{2\pi i} \int_{\varepsilon - i\infty}^{\varepsilon + i\infty} d\tilde{\sigma} \boldsymbol{u}^{\mathcal{FL}}(\boldsymbol{k}, \tilde{\sigma}) e^{i\boldsymbol{k}\cdot\boldsymbol{x} + \tilde{\sigma} t} \right]$. The linearity of the disturbance enables us to analyze each mode involved in the integrand $\boldsymbol{u}^{\mathcal{FL}}(\boldsymbol{k}, \tilde{\sigma}) e^{i\boldsymbol{k}\cdot\boldsymbol{x} + \tilde{\sigma} t}$, separately. This is called the normal mode.

[2] The linear dispersion relation is a functional relationship between the wavenumber and the frequency such that $\omega = \omega(\boldsymbol{k})$. In Sect. 4, we generalize its concept so as to include a dependence of $\tilde{\sigma}^{(j)} = \text{Re}[\tilde{\sigma}^{(j)}] - i\omega^{(j)}$ on Re, Pr, Bi, and $\boldsymbol{k}$ by denoting $D(\text{Re}, \text{Pr}, \text{Bi}, \boldsymbol{k}; \tilde{\sigma}^{(j)}) = 0$. In the parenthesis, we indicate the eigenvalue following the semicolon.

$$(\tilde{u}_x, \tilde{u}_r, \tilde{u}_\theta, \tilde{p}, \tilde{\Theta}) \sim (r^m, r^{m-1}, r^{m-1}, r^m, r^m) \text{ for } m > 0, \tag{153}$$

$$(\tilde{u}_x, \tilde{u}_r, \tilde{p}, \tilde{\Theta}) \sim (1, r, 1, 1) \text{ for } m = 0. \tag{154}$$

In the linear dispersion relation $D(\text{Re}, \text{Pr}, \text{Bi}, \boldsymbol{k}; \tilde{\sigma}) = 0$, the real part of $\tilde{\sigma}$ denotes the linear growth rate so that the neutral condition is given by $\text{Re}[\tilde{\sigma}] = 0$. For steady onset, i.e., bifurcation to a steady-state solution, we may set $\tilde{\sigma} = 0$ so that the neutral Reynolds number Re_n is obtained as an eigenvalue for prescribed Pr, Bi, and $\boldsymbol{k}$ such that $D(\text{Pr}, \text{Bi}, \boldsymbol{k}; \text{Re}_n) = 0$. In contrast, for Hopf onset, i.e., bifurcation to a time-periodic solution, an iterative procedure is unavoidable to get the neutral Reynolds number. To get the neutral condition (and the critical condition) in high accuracy, we adopted the Newton iteration in the problems P-I and P-II by assuming $\text{Re}[\tilde{\sigma}] = 0$ (with $\partial \text{Re}/\partial k = 0$). In P-III, we used numerical data of $\text{Re}[\tilde{\sigma}]$ to get the critical condition iteratively based on the secant method.

4.4 Problem P-I—Thermocapillary Convection in a Horizontal Liquid Layer

In P-I, the critical Marangoni number Ma_c is defined by $\text{Ma}_c = \text{Pr} \text{Re}_c$ with $\text{Re}_c = \min_{k_1, k_3} \text{Re}_n(\text{Pr}, \text{Bi}, \delta\Theta_0, \boldsymbol{k})$ where Re_n being the neutral Reynolds number given by the linear dispersion relation $D(\text{Re}_n, \text{Pr}, \text{Bi}, \delta\Theta_0, \boldsymbol{k}; \omega) = 0$ since $\text{Re}[\tilde{\sigma}] = 0$.[3] Here, we set $\tilde{\sigma} = \text{Re}[\tilde{\sigma}] - i\omega$ with frequency $\omega \in \mathbb{R}$. We define the direction of the wavevector $\boldsymbol{k} = (k_1, 0, k_3)$ measured from the positive k_1-axis on the $k_1 k_3$-plane by $\psi = \tan^{-1}(k_3/k_1)$. We write $e^{i\boldsymbol{k} \cdot \boldsymbol{x} + \tilde{\sigma} t}$ given in Eq. (150) as $e^{i\boldsymbol{k} \cdot \boldsymbol{x} + (\tilde{\sigma}_r - i\omega)t} \equiv e^{i\boldsymbol{k} \cdot \boldsymbol{x} + \tilde{\sigma}_r t - ikct} = e^{ik[x\cos\psi + z\sin\psi - ct] + \tilde{\sigma}_r t}$ where $\omega = kc$, k is the size of the wavevector defined by $k = |\boldsymbol{k}| = \sqrt{k_1^2 + k_3^2}$, $\tilde{\sigma}_r = \text{Re}[\tilde{\sigma}]$, and $c \in \mathbb{R}$. An oscillatory mode thus propagates in the direction of $(\cos\psi, \sin\psi)$ on the xz-plane with the phase velocity c.

i. $\delta\Theta_0 = 0$ case

Smith and Davis [28] investigated thermocapillary convection in a horizontal liquid layer with an infinite extent. They applied a uniform negative temperature gradient in the x-direction, imposed the insulating boundary condition on the bottom plate, and assumed the free surface was nondeformable. Under an assumption of zero mass-flux across an arbitrary x cross section, the basic flow has the quadratic profile, the so-called "return flow." If the zero mass-flux is violated, the basic flow becomes a linear function of y. They found that in the return flow, the oscillatory waves called the hydrothermal waves were the critical mode over the entire Prandtl number range. The increase in the Biot number exerts a stabilizing effect.

---

[3] On a closed disconnected neutral curve, the critical Reynolds number Re_c is also defined by $\text{Re}_c = \max_{k_1, k_3} \text{Re}_n(\text{Pr}, \text{Bi}, \delta\Theta_0, \boldsymbol{k})$. See Fig. 17(d).

Priede and Gerbeth [22] reinvestigated the return flow problem by carrying out an asymptotic analysis in a low Prandtl number limit as well as numerical analysis over the range $10^{-3} \leq \mathrm{Pr} \leq 10^3$. As the boundary condition on $y = 0$, they examined both the insulating and perfectly conducting boundaries, the latter of which means constant temperature over the relevant surface. They pointed out that in the case of the insulating lower boundary, critical wavelength $2\pi/k_c$ was much longer than the liquid layer thickness. Here, k_c is the critical wavenumber. In contrast, in the perfectly conducting case, $2\pi/k_c$ is comparable to the depth of the layer. The perfectly conducting lower boundary does not allow the longitudinal mode to be critical.

In this subsection, we impose the perfectly conducting boundary condition on $y = 0$. Figure 16 shows the critical curves for Ma_c, k_c, ω_c, and the direction of the wavevector $|\psi|$; that is, the basic flow becomes unstable if Ma exceeds the Ma_c for the given Pr and Bi. Both for Bi = 0 and 1, $\omega_c < 0$ holds so that the hydrothermal waves propagate at the angles $180° \pm \psi$ to the positive x-axis; that is, inversely to the mean flow. We call it the "counterflow".

ii. Effect of non-vanishing $\delta\Theta_0$

We show the effect of $\delta\Theta_0$ on the critical conditions for Pr = 100 and Bi = 1 in Fig. 17. Lower left part bounded by the critical curves in Fig. 17(a) is the region where the basic field is stable. If the parameter set $(\delta\Theta_0, \mathrm{Ma})$ crosses thick lines, the basic field loses its stability. Transverse waves T with $|\psi| = 0$ shown by the long-dashed line between the endpoint A and the crossover point C_1 give the critical

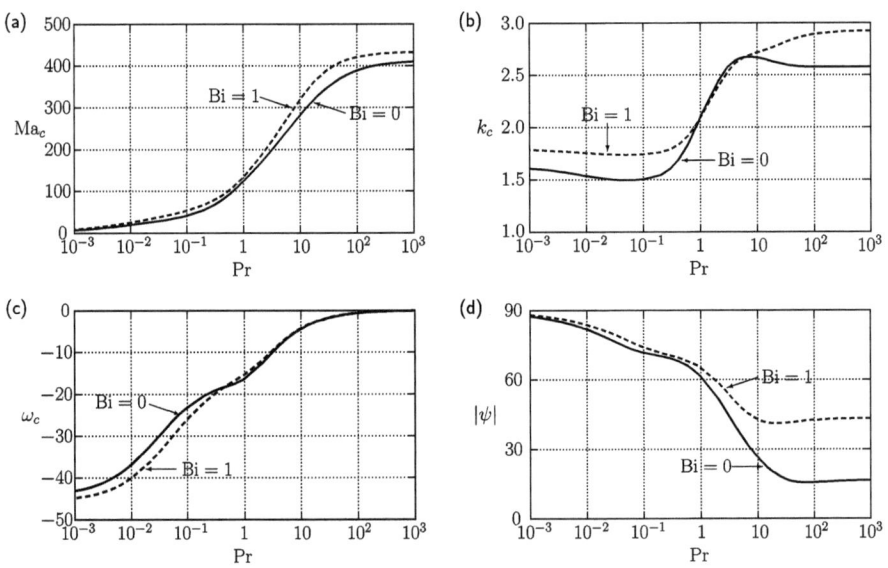

Fig. 16 Critical conditions in P-I for $\delta\Theta_0 = 0$ under the perfectly conducting boundary condition on $y = 0$. (a) critical Marangoni number, (b) critical wavenumber, (c) critical frequency, and (d) direction of the wavevector

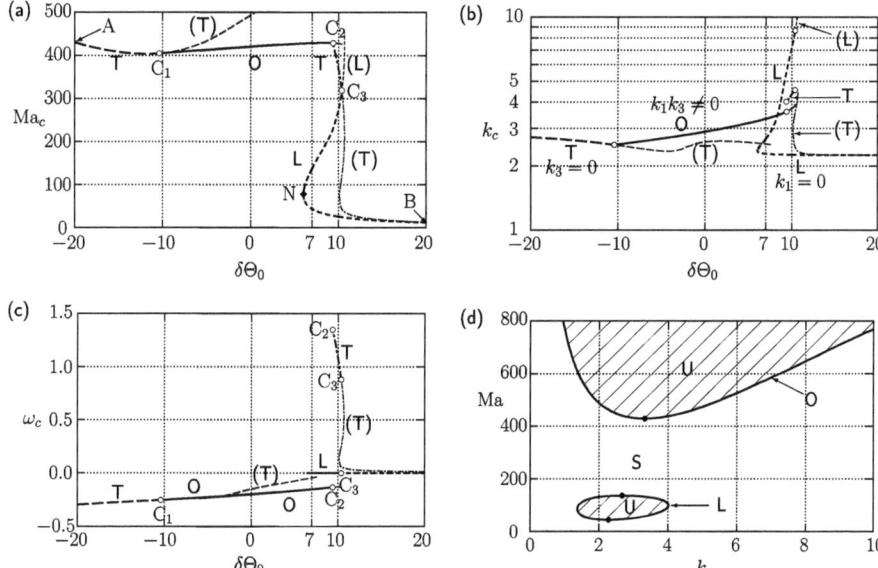

Fig. 17 Critical curves in P-I for various $\delta\Theta_0$ with $Pr = 100$ and $Bi = 1$. (a) critical Marangoni number, (b) critical wavenumber, (c) critical frequency, and (d) disconnected neutral curves at $\delta\Theta_0 = 7$. T: transverse waves, O: oblique waves, and L: longitudinal rolls. The letter in parentheses: non-critical mode. Thick lines: critical curves, thin lines: curves connecting non-critical local minima/maxima on the neutral curves. Open circles in (a-c): crossover points, filled diamond: the nose labeled N on the critical curve connecting the points C_3 and B. Filled circles in (d): critical points. Labels C_1-C_3 in (c) correspond to the crossover points in (a)

condition, while oblique waves shown by the solid line between C_1 and C_2 are the critical modes.[4] Across the crossover point C_1 located at $\delta\Theta_0 = -10.4$, there is no jump on Ma_c, k_c, ω_c, and $|\psi|$. The transverse waves again give the criticality in the range between C_2 and C_3 as the dash-dotted line shows. Along the short dashed line connecting C_3 and B via N, the longitudinal rolls give the criticality and are steady-state since $\omega_c = 0$ as seen in Fig. 17(c). Between the branches L and (T) in (a), branches of oblique waves are densely distributed. Although not shown in Fig. 17, the angle ψ of the wavevector to the k_1-axis is 0 along the long-dashed line in (a). As $\delta\Theta_0$ increases from the crossover point C_1 to C_2, $|\psi|$ is an increasing function from $0°$ to around $67°$. At C_2, $|\psi|$ jumps from $67°$ to $0°$, and at C_3, it jumps to $90°$. Across the crossover point C_2 at $\delta\Theta_0 = 9.4$, as Fig. 17(c) shows, the critical frequency jumps from negative to positive values, i.e., $\omega_c < 0$ holds below $\delta\Theta_0 = 9.4$ whereas $\omega_c > 0$ holds above 9.4.

Let us summarize the propagation direction of the critical modes. Between endpoint A and the crossover point C_1, the hydrothermal waves propagate in the

---

[4] In Fig. 17(a), we label the crossover points C_1, C_2, and C_3, the endpoints A and B, and the nose N located at C_1: $(\delta\Theta_0, Ma) = (-10.4, 404)$, C_2: $(9.37, 428)$, C_3: $(10.4, 318)$, A : $(-20, 430)$, B: $(20, 11.9)$, and N : $(6.02, 78.0)$.

pure counterflow ($-x$) direction, and as $\delta\Theta_0$ increases above $-10.4(C_1)$, the direction inclines and finally reaches $\pm 113°$ to the positive x-axis. At the crossover point C_2, the oblique waves change to the transverse waves again, and correspondingly, ω_c jumps from negative to positive meaning that the propagation direction changes from $\pm 113°$ to $0°$. This time, the transverse waves propagate in the pure coflow ($+x$) direction. Further increase of $\delta\Theta_0$ or decrease of Ma decelerates the waves that suddenly change to steady-state longitudinal rolls with $k_1 = 0$ at C_3 since $\omega_c = 0$ along the short dashed line connecting C_3, N, and B.

The longitudinal rolls are observed experimentally by [29], cited also in [11]. They carried out an experiment with rather high Pr numbers of Pr = 28 to 207 by imposing temperature gradients in the vertical as well as horizontal directions. Roughly speaking, for $\delta\Theta_0 \gtrsim 30$ and $\delta\Theta_0 \text{Ma} \gtrsim 80$, they observed Bénard-Marangoni convection in the form of hexagons. For $\delta\Theta_0 \simeq 30$, the hexagons became distorted by the basic flow, and the longitudinal rolls eventually emerged for $6 \lesssim \delta\Theta_0 \lesssim 30$ and $160 \lesssim \delta\Theta_0 \text{Ma} \lesssim 2500$. The existence region of the longitudinal rolls seems to be consistent with Fig. 17(a), at least qualitatively.

We show disconnected two neutral curves in Fig. 17(d) for $\delta\Theta_0 = 7$. Such curves exist for $6.02 < \delta\Theta_0 < 10.4$. The oval closed neutral-curve vanishes at the nose labeled N, $(\delta\Theta_0, \text{Ma}_c) = (6.02, 78.0)$, on the critical curve in Fig. 17(a). The disconnected two neutral curves merge into the primary neutral curve at the crossover point C_3, $(10.4, 318)$.

At the end of this subsection, we note that [17] examined the effect of an inclined temperature gradient on the stability characteristics of a two-layer problem of water and air, and later, [26] for a single layer of water. By applying the vertical as well as horizontal temperature gradients, these authors obtained similar critical curves to that connecting the point $(\delta\Theta_0, \text{Ma}) = (0, 420)$, C_2, C_3, N, and B in Fig. 17(a). The transverse waves between C_2 and C_3 are pointed out to be transverse rolls drifted by the basic flow. Since they did not examine the case corresponding to the negative $\delta\Theta_0$ in our problem, the transverse waves between A and C_1 were lacking. They showed the disconnected two neutral curves similar to Fig. 17(d).

4.5 Problem P-II—Thermocapillary Convection in a Liquid Column with an Infinite Extent

In P-II, the critical Marangoni number Ma_c is defined by $\text{Ma}_c = \text{Pr}\text{Re}_c$ with $\text{Re}_c = \min_{k_1,m} \text{Re}_n(\text{Pr}, \text{Bi}, \delta\Theta_0, \boldsymbol{k})$ where the neutral Reynolds number Re_n follows the linear dispersion relation $D(\text{Re}_n, \text{Pr}, \text{Bi}, \delta\Theta_0, \boldsymbol{k}; \omega) = 0$ since $\text{Re}[\tilde{\sigma}] = 0$.[5] Here, we set $\boldsymbol{k} = (k_1, 0, m)$ and $\tilde{\sigma} = \text{Re}[\tilde{\sigma}] - i\omega$ with frequency $\omega \in \mathbb{R}$. Like P-I in Sect. 4.4, $e^{i\boldsymbol{k}\cdot\boldsymbol{x}+\tilde{\sigma}t}$ in Eq. (151) is written as $e^{i\boldsymbol{k}\cdot\boldsymbol{x}+(\tilde{\sigma}_r-i\omega)t} \equiv e^{i\sqrt{k_1^2+m^2}[x\cos\psi+\theta\sin\psi-ct]+\tilde{\sigma}_r t}$ where

---

[5] On a closed disconnected neutral curve, the critical Reynolds number Re_c is also defined by $\text{Re}_c = \max_{k_1,m} \text{Re}_n(\text{Pr}, \text{Bi}, \delta\Theta_0, \boldsymbol{k})$. See Fig. 19d.

$\psi = \tan^{-1}(m/k_1)$ and $\omega = \sqrt{k_1^2 + m^2} \times c$. The hydrothermal waves thus propagate in the direction of $(\cos\psi, 0, \sin\psi)$ on the cylindrical surface $(x, 1, \theta)$ with the phase velocity c.

i. $\delta\Theta_0 = 0$ **case**

Xu and Davis [31] analyzed the linear stability of P-II by assuming a uniform negative temperature gradient in the x-direction and zero mass-flux across an arbitrary x-cross section for various Pr's and for Bi $= 0$ and 1. According to their result, the critical azimuthal wavenumber was basically $m_c = 1$. The $m_c = 0$ appeared only for high Pr's. In both $m_c = 1$ and 0 cases, $\omega_c \neq 0$ holds implying that the basic field is unstable to oscillatory hydrothermal waves. Ryzhkov [25] revisited this problem and found that Xu and Davis [31] overlooked another $m = 1$ mode that gave the criticality for a high Prandtl number range. As a result, the $m = 0$ mode is no longer critical, and the criticality is always due to the $m = 1$ modes for $10^{-2} \leq \text{Pr} \leq 10^3$.

Figure 18 shows the critical conditions for $\delta\Theta_0 = 0$ and $m = 1$. The open circles in (a–c) are the crossover points between different $m = 1$ modes and the filled circles in (d) are the critical points yielding the critical curves of (a-c). If $\omega_c > 0$ holds, the hydrothermal waves propagate in the $(\cos(\pm\psi), 0, \sin(\pm\psi))$-direction on the $(x, 1, \theta)$-cylindrical surface, where $\psi = \tan^{-1}(1/k_1)$ because $m = 1$ holds. In Fig. 18(b), we find that $k_{1c} < 0$ holds for $10^{-3} < \text{Pr} < 0.14$ with Bi $= 0$ and for $10^{-3} < \text{Pr} < 0.61$ with Bi $= 1$. Figure 18(c) shows that the critical frequency jumps and changes its sign at the crossover point C_1 located at $(\text{Pr}, \text{Ma}_c) = (20.0, 211)$ and corresponding (k_{1c}, ω_c) are $(0.776, 0.642)$ and $(1.64, -0.308)$ for Bi $= 0$ and C_2 located at $(\text{Pr}, \text{Ma}_c) = (10.2, 185)$ and (k_{1c}, ω_c) are $(0.457, 0.479)$ and $(1.38, -0.287)$ for Bi $= 1$. If we set Bi $= 0$, the propagation direction of the hydrothermal waves to the positive x-axis on the cylindrical surface is slightly counterflow and changes like $\pm 93.9° \rightarrow \pm 100° \rightarrow \pm 90°$ as Pr increases from 10^{-3} to 0.14 at which k_{1c} crosses zero in Fig. 18(b). Above 0.14, the direction is coflow and changes from $\pm 90°$ to $\pm 52.2°(=\pm\tan^{-1}(1/0.776) \times 180°/\pi)$, the latter of which is at C_1. Across the crossover point C_1, the direction jumps to the counterflow again with $\pm 149°(=\pm[\pi - \tan^{-1}(1/1.64)] \times 180°/\pi)$. As Pr $\rightarrow 10^3$, the direction tends to $\pm 151°$. For Bi $= 1$, a similar process takes place.

ii. **Effect of non-vanishing $\delta\Theta_0$**

In Fig. 19(a–c), we show the effect of $\delta\Theta_0$ on the critical conditions for Pr $= 100$ and Bi $= 1$. Figure 19(d) shows neutral curves at $\delta\Theta_0 = 6.5$. In (a), the basic field is stable in the lower left region bounded by the critical curves. If $(\delta\Theta_0, \text{Ma})$ crosses "thick" lines, the basic field loses stability, i.e., $m = 0$ mode for $\delta\Theta_0 < -16.9$ (shown by the dotted line between the endpoint A and the crossover point C_1), $m = 1$ mode with $k_{1c} \neq 0$ between the crossover points C_1 and C_2 shown by the solid line, $m = 1$ mode with $k_{1c} \neq 0$ between C_2 and C_3 shown by the long-dashed line, $m = 1$ mode

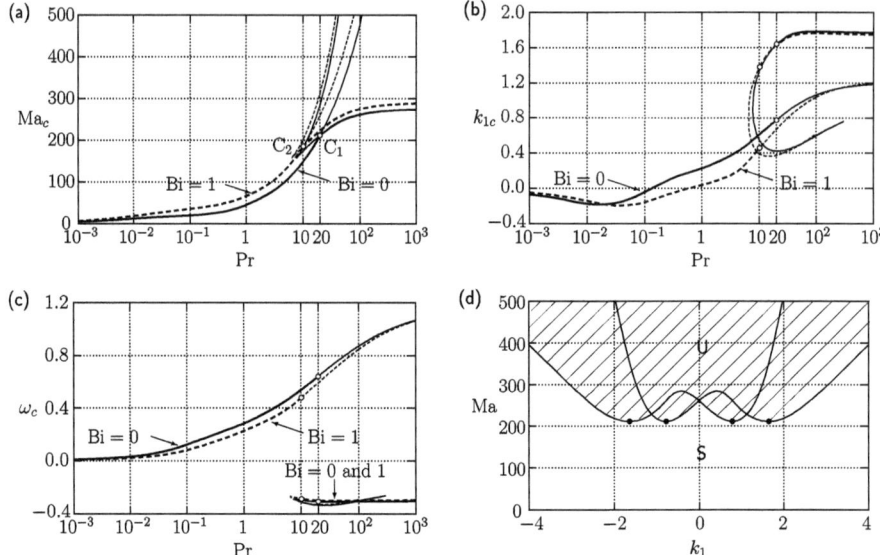

Fig. 18 Critical curves in the P-II for $\delta\Theta_0 = 0$ and $m = 1$. (a) critical Marangoni number, (b) critical axial wavenumber, (c) critical frequency, and (d) neutral curves at Pr = 20 and Bi = 0. In (a–c), thick lines: critical curves, and thin lines: curves connecting non-critical local minima/maxima on neutral curves. Open circles in (a–c): crossover points. Filled circles in (d): critical points on the neutral curves. The hatched part labeled U is a region where the basic field is unstable. The label S stands for the stable region

with $k_{1c} = 0$ between C_3 and C_4 shown by the short-dashed line, and $m = 1$ mode with $k_{1c} \neq 0$ between C_4 and the endpoint B shown by the dash-dotted line.[6]

The $m = 0$ mode shown by the thick dotted line is axisymmetric and oscillatory. It propagates in the counterflow $(-x)$ direction because of the negative ω_c. The thick solid line is due to the $m = 1$ oblique mode with $\omega_c < 0$ propagating in the "counterflow" $(\cos(180° \pm \psi), 0, \sin(180° \pm \psi))$-direction on the cylindrical surface $(x, 1, \theta)$ where $\psi = \tan^{-1}(m/k_1)$. Both the thick long-dashed line and the thick dash-dotted line represent modes with $m = 1, k_{1c} \neq 0$, and $\omega_c > 0$ propagating in the "coflow" $(\cos(\pm\psi), 0, \sin(\pm\psi))$ direction as shown in (c). The thick short-dashed line corresponds to the $m = 1$ and $k_{1c} = 0$ steady mode since $\omega_c = 0$. In summary, between point A and C_1, the axisymmetric waves propagate in pure counterflow direction, i.e., 180° to the positive x-axis on the cylindrical surface. At C_1, the propagation direction of hydrothermal waves jumps discontinuously to $\pm 155°$ and changes to $\pm 144°$ with the increase of $\delta\Theta_0$ before arriving at the crossover point C_2. At C_2, the direction jumps to $\pm 39.0°$, i.e., coflow direction, and across C_3, it jumps to $\pm 90°$. Along the critical curve, as the parameter set passes C_4, the direction

---

[6] In Fig. 19(a), we label the crossover points C_1, C_2, C_3, and C_4, the endpoints A and B, and the nose N whose coordinates on the $\delta\Theta_0$ Ma-plane are $C_1 : (-16.9, 626)$, $C_2 : (5.88, 272)$, $C_3 : (6.23, 75.5)$, $C_4 : (6.03, 20.5)$, A : $(-20, 706)$, B : $(20, 4.44)$, and N : $(5.02, 38.1)$.

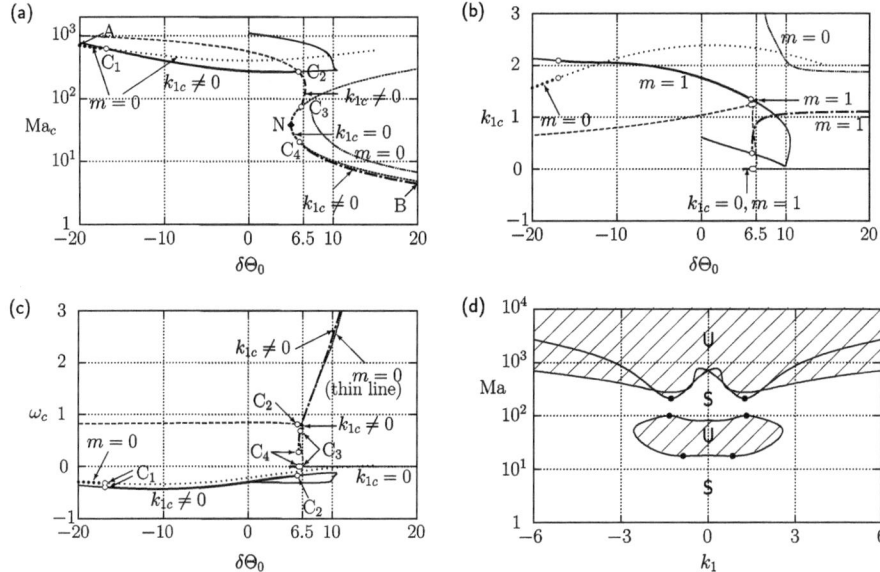

Fig. 19 Critical curves in the P-II for various $\delta\Theta_0$ with Pr = 100 and Bi = 1. (a) critical Marangoni number, (b) critical axial wavenumber, (c) critical frequency, and (d) disconnected neutral curves for $\delta\Theta_0 = 6.5$. The azimuthal wavenumber m is 1 if not designated as $m = 0$. In (a–c), thick lines: critical curves, and thin lines: curves connecting non-critical local minima/maxima on the neutral curves. Open circles in (a–c): crossover points. Labels C_1–C_4 in (c) correspond to the crossover points in (a). Filled diamond in (a): the nose labeled N on the critical curve between points C_3 and C_4. Filled circles in (d): critical points on the neutral curves. The hatched parts labeled by U are regions where the basic field is unstable. The basic field is stable in a region labeled by S outside the hatched regions

changes to $\pm 73.2°$ and reaches $\pm 42.2°$ at the endpoint B. Along the critical curve leaving C_2 and arriving at B via C_3 and C_4, the oscillatory modes, if exist, thus propagate in the coflow direction.

The $m = 0$ mode and $k_{1c} = 0$ mode correspond to the transverse waves and the longitudinal rolls, respectively, in Sect. 4.4. As above, on the left side of C_2 in Figs. 17(a) and 19(a), both the transverse waves and oblique waves propagate in the counterflow direction, and across the crossover point C_2, the propagation direction changes to the coflow. The steady longitudinal modes become critical at C_3. In this sense, there is a sort of 'similarity' between problems P-I and P-II. However, the thick dash-dotted line in Fig. 19(a) does not have its counterpart in Fig. 17(a).

Figure 20 summarizes the critical conditions in these figures. In the table, we simplify the evaluation process of Ma_c, i.e., we increase Ma until $Re[\tilde{\sigma}]$ vanishes for prescribed $\delta\Theta_0$ so as to regard Ma_c as a single valued function of $\delta\Theta_0$. The oblique hydrothermal waves on the line connecting C_1 and C_2 thus change abruptly to the longitudinal rolls. The snap through occurs from $(\delta\Theta_0, Ma) = (6.02, 421)$ to the nose N : (6.02, 78) in P-I and from (5.02, 270) to N:(5.02, 38) in P-II. The similarity mentioned above is clearly seen.

Problem	P-I		P-II		
Point	C_1	N	C_1	N	C_4
$\delta\Theta_0$	-10.4	6.02	-16.9	5.02	6.03
Ma_c	404	421|78	626	270|38	20.5
k_{1c}	$k_1 \neq 0$	$k_1 = 0$	$k_1 \neq 0$	$k_1 = 0$	$k_1 \neq 0$
k_{3c}, m_c	$k_3 = 0$	$k_3 \neq 0$	$m = 0$	$m = 1$	
ω_c	$\omega < 0$	$\omega = 0$	$\omega < 0$	$\omega = 0$	$\omega > 0$
Type	T	O	T	O	L	O
Direction	Counterflow	Steady	Counterflow	Steady	Coflow

Fig. 20 Summary of Figs. 17(a) and 19(a). Instead of tracing the critical curves on the $\delta\Theta_0$ Ma-plane, we here simplify a situation where Ma is increased with a given $\delta\Theta_0$ by simulating a usual experimental procedure. C_1 and C_4: crossover points, N: nose on the critical curve. T: transverse waves, O: oblique waves, L: longitudinal rolls. Direction: the propagation direction of the critical modes

Figure 19(d) shows disconnected two neutral curves spreading over $\pm k_1$ regions since the axial wavenumber k_1 may have both signs. Such neutral curves exist for $5.02 < \delta\Theta_0 < 6.71$. The closed disconnected neutral curve vanishes at $\delta\Theta_0 = 5.02$ while merging into the primary neutral curve at 6.71.

4.6 Problem P-III—Thermocapillary Convection in a Liquid Bridge with Finite Aspect Ratio

In the 1990s and 2000s, the linear stability of half-zone models was analyzed repeatedly. Concerning the analysis of non-deformable cylindrical liquid bridges with the aspect ratio $\Gamma = 1$, we cite [3, 13, 30] among others. For the influence of a liquid-bridge volume on the onset of instabilities, see [6, 19], for example.

i. **Linear stability results**

In P-III, the critical condition depends on the aspect ratio $\Gamma = d/R$ such that $Re_c(\Gamma) = \min_{m \in \mathbb{Z}} Re_n(Pr, Bi, m, \Gamma)$, where the neutral Reynolds number Re_n follows the linear dispersion relation $D(Re_n, Pr, Bi, m, \Gamma; \omega) = 0$.[7] Figure 21 for $Pr = 0.02$ and $Bi = 0$ demonstrates the Γ-dependence of Re_c over the range $0.25 \lesssim \Gamma^{-1} \lesssim 1.65$. In the figure, we also plotted the numerical data of Table V in [30] by open circles. As Γ^{-1} increases, the critical azimuthal wavenumber m increases from 1 through 3. All the plots are steady onset.

We summarize the stability characteristics of P-III in Fig. 22 for $\Gamma = 1$ and $Bi = 0$ with $\delta = 10$ and 50. For δ, see Eqs. (147)–(149). The solid line is the

---

[7] The upper branch of the critical curve connecting $(Pr, Re_c) = (0.0596, 7740)$ and $(0.0562, 12200)$ in Fig. 22 with $\delta = 50$ is due to the definition $Re_c(\Gamma) = \max_{m \in \mathbb{Z}} Re_n(Pr, Bi, m, \Gamma)$.

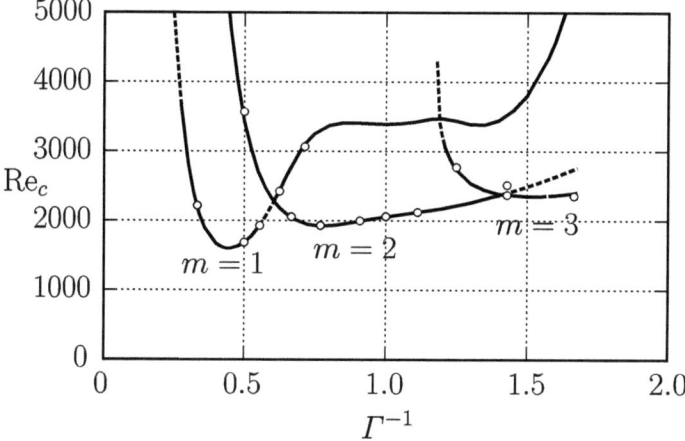

Fig. 21 Effect of the aspect ratio Γ on the critical Reynolds number Re_c for $\text{Pr} = 0.02$ and $\text{Bi} = 0$ in problem P-III. Open circles: Table V of [30]. Solid line: supercritical steady onset, dashed line: subcritical steady onset. $\delta = 50$

critical condition obtained with $\delta = 50$ while the dashed line is with $\delta = 10$. If we trace the critical curves from $\text{Pr} = 10^{-3}$ to 8, we find that the critical wavenumber changes from $m = 2$ to (1), 3, 2, 3, and 2. The $m = 1$ is only for $\delta = 10$ and $0.055 < \text{Pr} < 0.056$. The arc connecting $(\text{Pr}, \text{Re}_c) = (10^{-3}, 1800)$, $(0.0596, 7740)$, and $(0.0562, 12200)$ for $\delta = 50$ and the corresponding arc for $\delta = 10$ including the short arc with $m = 1$ represent the steady onset. All the rest are Hopf onset, i.e., the bifurcation to a time-periodic solution. Filled and hollow squares denote crossover points between different azimuthal wavenumbers m for $\delta = 50$ and 10, respectively. The results with $\delta = 50$ are qualitatively consistent with those with $\delta = 10$ except for the $m = 1$ steady onset for $\delta = 10$. Small open circles are the critical conditions listed in Table II of [13] based on the finite element method. They show good agreement with the solid line obtained with $\delta = 50$.

ii. **Selection between standing waves and rotating waves**

As mentioned above, the linear stability theory predicts the Hopf bifurcation with $m \neq 0$ for $\text{Pr} \gtrsim 0.056$. On a linear basis, however, we cannot conclude whether the realized wave pattern is rotating or standing even if the Hopf onset is accurately predicted. To conclude, we need to go a bit further, i.e., the weakly nonlinear stage.

In the P-I, the linear theory predicts that oblique hydrothermal waves with wavevector $\boldsymbol{k} = (k_1, 0, k_3)$ and $(k_1, 0, -k_3)$ become critical at $\text{Ma} = \text{Ma}_c$, simultaneously. On the weakly nonlinear basis, [27] examined the selection between the pure mode, i.e., pure oblique waves with either $(k_1, 0, k_3)$ or $(k_1, 0, -k_3)$, and the mixed mode of oblique pair. He clarified that the mixed mode was stable for $\text{Bi} = 1$ and $\text{Pr} \leq 0.01$, and the pure mode was stable, otherwise. He further examined the sideband instability of the realized wave pattern by taking account of spatial modulation. He then found that all these stable pure and mixed modes are unstable to the

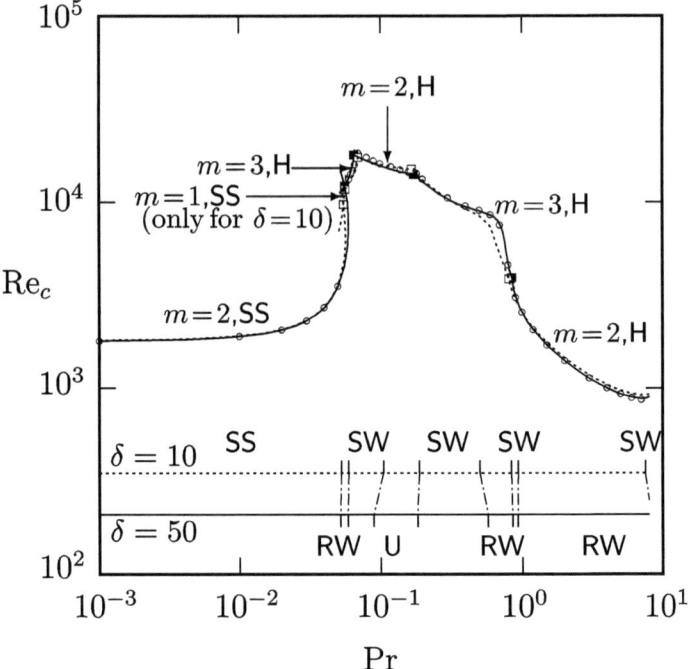

Fig. 22 Critical Reynolds number Re_c in P-III for $Bi = 0$ and $\Gamma = 1$. Solid line: $\delta = 50$, short-dashed line: $\delta = 10$, small open circles: Table 2 of [13], filled squares: crossover points with $\delta = 50$, and hollow squares: crossover points with $\delta = 10$. SS on the critical curve: onset of steady-state solution, and H: Hopf onset. The lower part of the figure summarizes weakly nonlinear characteristics. SS: stable steady-state solution, SW: stable standing waves, RW: stable rotating waves, and U: no stable finite solution

sidebands. To the author's knowledge, the selection between the pure mode and the mixed mode in P-II has not been examined on a weakly nonlinear basis, yet.

In P-III, the azimuthal wavenumber m can take a small integer value. There is no chance for the non-subharmonic sideband to grow. We thus ignore the effect of spatial modulation and reduce the governing nonlinear PDEs, i.e., infinite-dimensional dynamical systems, to one or two-dimensional complex dynamical systems. See [7] for the weakly nonlinear reduction in P-III.

Let us consider a situation where a simple eigenvalue satisfies $\text{Im}[\tilde{\sigma}] = 0$ as $\text{Re}[\tilde{\sigma}] \downarrow 0$. The disturbance is then expressed as $A(t)\tilde{\phi}(\check{x})e^{im\theta}+ $ c.c. $+$ h.o.t. where $A : \mathbb{R} \to \mathbb{C}$ is a complex function of t, $\tilde{\phi}(\check{x})$ denotes the linear eigenfunction of $\check{x} = (x, r)$, c.c. stands for complex conjugate of the preceding term, and h.o.t. higher order terms. The application of the weakly nonlinear reduction to the P-III yields an amplitude equation known as the Stuart-Landau equation

$$\dot{A} = A(\tilde{\sigma} + \lambda |A|^2) + \mathcal{O}(5), \quad \tilde{\sigma}, \lambda \in \mathbb{R} \tag{155}$$

for the steady-state bifurcation. Here, $\mathcal{O}(5)$ denotes nonlinear terms of the quintic order or higher. Truncated equation of (155) at the cubic order has a non-trivial solution $|A|^2 = -\tilde{\sigma}/\lambda$ if $\lambda < 0$ and $\tilde{\sigma} > 0$ or $\lambda > 0$ and $\tilde{\sigma} < 0$. We call the former situation supercritical bifurcation; the bifurcating nontrivial solution exists stably in the supercritical region $\tilde{\sigma} > 0$ where the trivial solution is unstable. The latter situation is called subcritical bifurcation where the bifurcating solution exists unstably for $\tilde{\sigma} < 0$.

If a complex conjugate pair of simple eigenvalues crosses the imaginary axis of the complex $\tilde{\sigma}$-plane and $\frac{\partial \tilde{\sigma}_r(\text{Re})}{\partial \text{Re}}\Big|_{\text{Re}=\text{Re}_c} \neq 0$ holds, Hopf bifurcation takes place. The group of symmetries of the problem P-III is O(2), i.e., rotations about the x-axis and reflections about a plane on which the x-axis lies. We may thus reduce the nonlinear PDEs to the amplitude equations having the form of

$$\dot{A} = A(\tilde{\sigma} + \lambda_1|A|^2 + \lambda_2|B|^2) + \mathcal{O}(5), \quad \dot{B} = B(\tilde{\sigma} + \lambda_2|A|^2 + \lambda_1|B|^2) + \mathcal{O}(5). \tag{156}$$

Here we write the disturbance as a sum of waves rotating in the $\pm\theta$-directions, i.e., $A(t)\tilde{\phi}_1(\vec{x})e^{i(-m\theta+\omega t)}$ + c.c. + $B(t)\tilde{\phi}_2(\vec{x})e^{i(m\theta+\omega t)}$ + c.c. + h.o.t., $A, B: \mathbb{R} \to \mathbb{C}$, and $\tilde{\phi}_{1,2}(\vec{x})$ denote the linear eigenfunctions. This time, $\tilde{\sigma} \in \mathbb{C}$ and $\lambda_{1,2} \in \mathbb{C}$. Truncated equations of Eq. (156) at the cubic order have two nontrivial solutions. They are $A \neq 0, B = 0$ (or $A = 0, B \neq 0$) exhibiting rotating waves and $|A| = |B| \neq 0$ standing waves. For rotating waves to exist stably, $\lambda_{1r}, \lambda_{2r}$, and Re$[\tilde{\sigma}]$ should satisfy $\lambda_{1r} < 0$, $\lambda_{1r} - \lambda_{2r} > 0$, and Re$[\tilde{\sigma}] > 0$ where $\lambda_{1r,2r} = \text{Re}[\lambda_{1,2}]$. In contrast, for standing waves to exist stably, $\lambda_{1r} + \lambda_{2r} < 0$, $\lambda_{1r} - \lambda_{2r} < 0$, and Re$[\tilde{\sigma}] > 0$. The inequality $\lambda_{1r} < 0$ states that the bifurcation of rotating waves is supercritical, and $\lambda_{1r} + \lambda_{2r} < 0$ states that the bifurcation of standing waves supercritical. This implies that a subcritical branch, if exists, is always unstable. Stability conditions for these solutions require opposite signs of $\lambda_{1r} - \lambda_{2r}$ so that if one of these solutions is stable, the other is unstable.

According to the weakly nonlinear analysis, in Fig. 21 for Pr = 0.02, a part of the critical curve shown by the dashed line is subcritical, but all the rest, solid lines, are supercritical. Typically, at $\Gamma = 1$, the steady-state solution with the azimuthal wavenumber $m = 2$ bifurcates supercritically.

In the lower part of Fig. 22, we summarize the weakly nonlinear characteristics for $\Gamma = 1$. The solid line is due to $\delta = 50$, and the short-dashed line is due to $\delta = 10$. The steady-state bifurcation along the arc connecting (Pr, Re$_c$) = $(10^{-3}, 1800)$, (0.0596, 7740), and (0.0562, 12200) is supercritical, consistent with the numerical results for Pr = 0.01 [12] and Pr = 0.02 [14]. The latter authors reported that under an initial condition composed of the standing waves, rotating waves were eventually realized in their full numerical simulation for Pr = 4. They also confirmed it based on the amplitude equations Eq. (156) whose coefficients were evaluated by fitting the data of their simulation. According to [15], rotating waves are stable up to Pr = 7.8 ± 0.1, and standing waves are stable above this Pr. Figure 22 shows that for $\delta = 10$, rotating waves are stable up to Pr = 7.5. For $\delta = 50$, we cannot specify the upper stability bound of the rotating waves decisively since no reliable data is available for Pr > 8. This is due to the lack of numerical accuracy for such a high Prandtl number region.

As mentioned above, the linear stability analysis predicts the critical condition above which the basic field loses its stability. The weakly nonlinear theory enables us to predict stable secondary solution branches bifurcating off the critical curve in slightly supercritical states. [16] applied the secondary instability analysis based on the Floquet theorem to examine the linear stability of the bifurcated solution branches with respect to the disturbance of the form $e^{\tilde{\sigma}^{(2)}t} \sum_{j=-\infty}^{\infty} \tilde{\phi}_j^{(2)}(\check{x}) e^{i(jm_c+m^{(2)})\theta}$ for $Pr \leq 0.02$ where $m^{(2)} \in \mathbb{Z}$ satisfying $-m_c/2 < m^{(2)} \leq m_c/2$. But we will not go further since the secondary instability analysis is beyond the scope of Sect. 4

Acknowledgements H. Kawamura, an author of this chapter, would extend his sincere thanks to co-author K. Fujimura for his significant contributions to Sect. 4 with the integration of his original research, as well as for his thorough proofreading of the entire chapter. Finally I (HK) would also like to express deep gratitude for our long-standing academic collaboration.

References

1. Bénard H (1901) Les tourbillons cellulaires dans une nappe liquide—Méthodes optiques d'observation et d'enregistrement, J Phys Theor Appl 10(1):254–266. HAL Id:jpa-00240502
2. Chandrasekhar S (1961) Hydrodynamic and hydromagnetic stability. Oxford Univ. Press, Oxford, Chapter. II:9–75
3. Chen G et al (1997) Bifurcation analysis of the thermocapillary convection in cylindrical liquid bridges. J Crystal Growth 180:638–647
4. Davis SH (1969) Buoyancy-surface tension instability by the method of energy. J Fluid Mech 39:347–359
5. Drazin PG, Reid WH (2004) Hydrodynamic stability, 2nd edn. Cambridge Univ. Press, Cambridge
6. Ermakov MK, Ermakova MS (2004) Linear-stability analysis of thermocapillary convection in liquid bridges with highly deformed free surface. J Crystal Growth 266:160–166
7. Fujimura K (2013) Linear and weakly nonlinear stability of Marangoni convection in a liquid bridge. J Phys Soc Jpn 82:074401-1-14
8. Fujimura K (2024) Linear stability analysis of thermocapillary convection in liquid layer and bridge In: Kawamura et al. (eds) Thermocapillary convection in microgravity, Springer, Heidelberg
9. Koschmieder EL, Pallas SG (1974) Heat transfer through a shallow horizontal convection fluid layer: Int J. Heat Mass Transf 17:991–1002
10. Koschmieder EL (1993) Bénard cells and Taylor vortices. Cambridge Univ. Press
11. Lappa M (2010) Thermal convection: patterns, evolution and stability. Wiley, pp 334–337
12. Levenstam M, Amberg G (1995) Hydrodynamical instabilities of thermocapillary flow in a half-zone. J Fluid Mech 297:357–372
13. Levenstam M et al (2001) Instabilities of thermocapillary convection in a half-zone at intermediate Prandtl numbers. Phys Fluids 13:807–816
14. Leypoldt J et al (2000) Three-dimensional numerical simulation of thermocapillary flows in cylindrical liquid bridges. J Fluid Mech 414:285–314
15. Leypoldt J et al (2001) Stability of hydrothermal-wave states. Z Angew Math Mech 81:785–786

16. Motegi K et al (2017) Floquet analysis of spatially periodic thermocapillary convection in a low-Prandtl-number liquid bridge. Phys Fluids 29:074104-1-14
17. Nepomnyashchy A et al (2001) Stability of thermocapillary flows with inclined temperature gradient. J Fluid Mech 442:141–155
18. Nield DA (1964) Surface tension and buoyancy effects in cellular convection. J Fluid Mech 19:341–352
19. Nienhüser Ch, Kuhlmann HC (2002) Stability of thermocapillary flows in non-cylindrical liquid bridges. J Fluid Mech 458:35–73
20. Pearson JRA (1958) On convection cells induced by surface tension. J Fluid Mech 4:489–500
21. Pellew A, Southwell RV (1940) On maintained convective motion in a fluid heated from below. Proc Roy Soc A 176:312–343
22. Priede J, Gerbeth G (1997a) Influence of thermal boundary conditions on the stability of thermocapillary-driven convection at low Prandtl numbers. Phys Fluids 9:1621–1634
23. Priede J, Gerbeth G (1997b) Hydrothermal wave instability of thermocapillary-driven convection in a coplanar magnetic field. J Fluid Mech 347:141–169
24. Rayleigh Lord (1916) On convection currents in a horizontal layer of fluid when the higher temperature is on the under side, Phil Mag 32:529–546
25. Ryzhkov II (2011) Thermocapillary instabilities in liquid bridges revisited. Phys Fluids 23:082103-1-6
26. Shklyaev OE, Nepomnyashchy AA (2004) Thermocapillary flows under an inclined temperature gradient. J Fluid Mech 504:99–132
27. Smith MK (1988) The nonlinear stability of dynamic thermocapillary liquid layers. J Fluid Mech 194:391–415
28. Smith MK, Davis SH (1983) Instabilities of dynamic thermocapillary liquid layers. Part 1. Convective instabilities. J Fluid Mech 132:119–144
29. Ueno I et al (2002) Thermocapillary convection in thin liquid layer with temperature gradient inclined to free surface. In Proc. IHTC12, 18–23 August 2002, Grenoble
30. Wanschura M et al (1995) Convective instability mechanisms in thermocapillary liquid bridges. Phys Fluids 7:912–925
31. Xu JJ, Davis SH (1984) Convective thermocapillary instabilities in liquid bridges. Phys Fluids 27:1102–1107

Open Access This chapter is licensed under the terms of the Creative Commons Attribution 4.0 International License (http://creativecommons.org/licenses/by/4.0/), which permits use, sharing, adaptation, distribution and reproduction in any medium or format, as long as you give appropriate credit to the original author(s) and the source, provide a link to the Creative Commons license and indicate if changes were made.

The images or other third party material in this chapter are included in the chapter's Creative Commons license, unless indicated otherwise in a credit line to the material. If material is not included in the chapter's Creative Commons license and your intended use is not permitted by statutory regulation or exceeds the permitted use, you will need to obtain permission directly from the copyright holder.

Thermocapillary Convection in Liquid Bridges of Finite Length

Hiroshi Kawamura and Dietrich Schwabe

Abstract This book is proposed by a group of scientists and a space-agency engineer who performed collaboratively a series of microgravity experiments on the thermocapillary convection in a liquid bridge in microgravity aboard the International Space Station. This chapter discusses advancements in research on thermocapillary convection in finite-length liquid bridges, which was initially motivated by the growth of single crystals. The chapter begins by introducing explanatory research in this field, followed by the descriptions of the formation of hydrothermal waves, including the condition of their transition, their course of development in the flow field, with a particular focus on the effects of dimensions, heat exchange with the surroundings, and gravity. Several pioneering microgravity experiments using sounding rockets are described. Additionally, we will focus on the particle accumulation structure (PAS), mentioning its discovery, the various patterns of its appearance, and subsequent related studies through both experimental and numerical analyses. Finally, recently developed applications related to microfluidics will be briefly introduced.

1 Introduction

Hiroshi Kawamura

In the previous chapter, we discussed the fundamentals of thermocapillary convection in canonical configurations, such as an infinite liquid layer and an infinitely long cylinder. Now, we shift our focus to the thermocapillary convection in a

---

H. Kawamura (✉)
Professor Emeritus, Tokyo University of Science, Tokyo, Japan
e-mail: kawa@rs.sus.ac.jp

Professor Emeritus, Suwa University of Science, Chino, Nagano, Japan

D. Schwabe
Professor Emeritus, Physics Institute, Justus-Liebig-University of Giessen, Giessen, Germany
e-mail: Dietrich.Schwabe@physik.uni-giessen.de

© The Author(s) 2025
H. Kawamura et al. (eds.), *Thermocapillary Convection in Microgravity*,
Fluid Mechanics and Its Applications 139,
https://doi.org/10.1007/978-981-96-2991-6_4

"liquid bridges of finite-length", which was the main subject of our in microgravity experiments in the Japanese Experiment Module "Kibo".

The present purpose is to publish a book describing its major scientific outcomes, experimental apparatus, procedures of the space experiments and so on. The transition threshold from the steady to unsteady thermocapillary flow, factors affecting the threshold and measured three-dimensional flow field, etc., will be reported. A peculiar phenomenon, named the Particle Accumulation Structure (PAS), where micro particles uniformly dispersed in a liquid bridge tend to accumulate by themselves along a three-dimensionally closed loop to travel inside the liquid bridge, was firstly found in the terrestrial studies and later emerged in the space experiments, too. The collaborations among the scientists, operation teams, Japanese and US space agencies, and terrestrial staffs and astronauts are also worth to be described. A call for scientific experiment proposals in the Japanese experimental module "Kibo" on the International Space Station took place in 1992, and its screened results released in 1993. If that era is looked back, there was a rapid development in semiconductor technology, which raised strong interests and demands in the production of large-sized high-quality single crystals.

In crystal growth of silicon for use as semiconductor substrates, the process involves melting polycrystalline raw material with n-doping (As) or with p-doping (B) and then to cool it directionally to grow a single crystal. As a result of the temperature gradients involved in the process, the hydrothermal convection occurs inevitably in the melt, and a time-varying fluctuating flow arises within the molten material. It will result in undesirable fluctuations of the growth velocity and this, in turn, will lead to disturbance of the crystal structure and especially in uneven distribution (striations) of the dopant. However, semiconductor silicon for integrated circuits needs to be homogeneously doped. Therefore, research on the flow stability in molten material was strongly motivated and initiated.

D. Schwabe, a physicist and a crystal growth scientist at the University of Giessen in Germany, initiated a series of early-stage experimental investigations on the flow induced by the thermocapillary effect at the gas–liquid (melt) interface. He conducted experiments using a transparent model liquid, $NaNO_3$ (melting point of 307 °C), to form a finite-length liquid column, demonstrating the emergence of thermocapillary flow, its transition to periodic oscillations, and the presence of a transition threshold. This discovery preceded the theoretical works on the hydrothermal waves, such as Smith and Davis [61], mentioned in the previous chapter. In the following Sect. 2, we will be introduced the explanatory early researches by D. Schwabe, who has provided us a manuscript in response to the request from the editorial committee.

Before discussing the thermocapillary convection in a liquid bridge in more detail, let us briefly outline the currently representative crystal growth methods from the melt. Among fairly large number of crystal growth methods, Fig. 1 depicts three typical examples of the methods, in which the hydrothermal wave may take place in their melts due to the temperature gradient over their melt surfaces. The figures are (a) the floating zone (FZ), (b) the Czochralski (CZ) and (c) the horizontal Bridgeman (HB) methods. In these illustrations, the molten regions are designated with 'Melt" where the temperature is higher than the neighbouring crystalline and polycrystalline

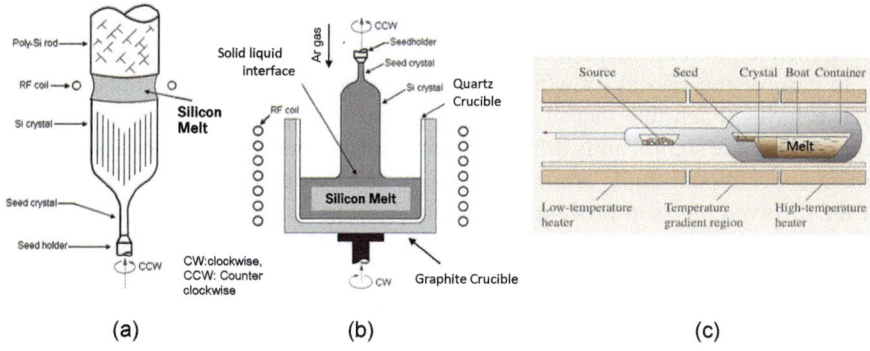

Fig. 1 Schematic drawings of some configurations of bulk crystal growth methods from melt. **a** Floating zone (FZ), **b** Czochralski (CZ) and **c** Horizontal Bridgeman. Reused (**a**, **b**) from Xiao and Xu [69], Permission of Tayl. Franc., **c** from Rudolph [37], Permission of Springer Nature

silicon regions. The temperature of the melt/crystal boundary is kept at the melting point and thus lower than the rest part of the melt. Accordingly, the thermocapillary convection takes place over the surface of the melt, because the melt surface faces to ambient gas (or vacuum) in these methods.

In the floating zone (FZ) method depicted in (a), a small part of the silicon rod is molten from above by RF (Radio Frequency) heating and solidified from below at the same time. This is achieved by moving the silicon rod downwards through the heated zone. The liquid silicon is thus processed without wall contact. The FZ-technique, though technically demanding, enables the production of silicon crystals with the highest purity.

The "liquid bridge with finite length", which is the primary focus of this chapter, is a simplified version of the FZ method and half of the FZ is modelled, that is, the test liquid is held by the surface tension between two end plates held at high and low temperatures. Thus, it is often called half zone liquid bridge (HZ).

The Czochralski method (CZ) depicted in Fig. (b) is currently the most widely used technique for the growth of large single crystal of silicon. The melt of silicon is held in a crucible and a seed crystal in immersed into the melt and slowly withdrawn under relative rotation. A very large single crystal can be formed as a result of gradual cooling.

Over the large rotating melt surface, there exists a temperature gradient and thus the thermohydraulic convection must emerge, which have been subjected to various experimental and numerical studies. Li et al. [25], for example, reported the appearance of the azimuthally travelling hydrothermal wave over the surface of a uniformly filled circular liquid layer, rotating in a disk pan heated from its outer periphery simulating the CZ method.

In the horizontal Bridgeman method (HB) in Fig. (c), the melt is held in a boat and pulled from high temperature to low temperature side. The melt surface is often open to the air for visual inspection, thus the thermocapillary flow may take place in this method, too. Those interested more in the crystal growth technology are recommended to refer to the textbooks in this field such as by Duffer [8].

In case of the crystal growth of silicon, the crystal is a semiconductor and its melt behaves much like that of metal. Its physical properties, such as the thermal conductivity, are significantly different from those of fluids in laboratory experiments, where conventional fluids are used normally. We can enhance the understandings of the involved physical processes and also develop analytical methods, especially numerical ones, and thus we are able to apply those outcomes to the process of semiconductors and metals.

In addition, the investigation of the thermocapillary driven flows will contribute also to the field of the micro hydrodynamics, as well as to applications such as the thin liquid film or droplet formations and handlings and other related phenomena. Some of these topics will be described in subsequent chapters of this book.

2 First Experiments on Time-Dependent Thermocapillary Flow

Dietrich Schwabe

2.1 The Technique of Floating Zone Crystal Growth and the Discovery of Time-Dependent Thermocapillary Flow

Around 1970 homogeneously doped silicon single crystals have been urgently asked for to build large integrated semiconductor circuits. But this material has not been available on the market because instable buoyant convection in the melt, from which silicon crystals were grown, resulted in dopant striations. One fantastic solution of the problem at this time was the growth of silicon crystals under microgravity without time-dependent buoyancy-driven convection, e.g. using the floating zone technique (FZ-technique) in a spacecraft. FZ-growth under microgravity could solve many problems because it is a containerless technique (oxygen-free) and it is feasible under microgravity in contrast to Czochralski growth from a quartz crucible. Crystal growers did neither count with thermocapillary convection nor with its possible time-dependence. In zone melting of silicon, both ends of a cylindrical rod of poly-crystalline Si are fixed and a short part between the ends is inductively heated by radio frequency to melt a "zone". The vertically oriented rods are moved with constant and small speed through the stationary radio frequency coil. This speed is the growth speed; moving downwards means that the rod is shifted slowly from above through the coil where it is molten. Material from the melt-zone crystallizes on the lower part of the rod after passing the RF-coil. Thus a molten zone of silicon is moved along the rod to become a single crystal. In floating zone growth of silicon, the highly reactive melt from Si has no contact with any other material but pure solid silicon and it therefore stays pure with a free melt surface. This is the advantage of the float-zone technique above crystal pulling from a crucible where the latter is contaminating the melt. An axial temperature difference is imposed in the FZ- technique from the hot centre of the liquid zone to both ends (Fig. 2a). This drives thermocapillary flow in the free surface from the (heated) hot part of the surface to both colder ends. This type of flow is called Marangoni convection or thermocapillary convection (TC), which was not known at that time.

When I entered the field in 1976, I was amongst all the crystal growers who knew nothing about Marangoni convection. However, Landau and Lifshitz [22] had already formulated the force-balance at a free liquid surface under a temperature gradient in their textbook about hydrodynamics. Ostrach [33] had published a non-dimensional analysis of thermocapillary flow in a 2D slot, pointing out its significance, as first analysed by Birikh [2]. Only one numerical paper existed [5] which was strongly debated and rejected because of an error in formulating the boundary conditions. I was happy to see their numerically calculated flow which looked much like that in our experiments with a liquid bridge from NaNO$_3$. No experimental work on TC was known at the time. I planned to investigate the rotationally driven (forced) convection in a floating zone and had chosen the transparent melt of sodium nitrate (NaNO$_3$, melting point 307 °C), as candidate material to see inside the zone.

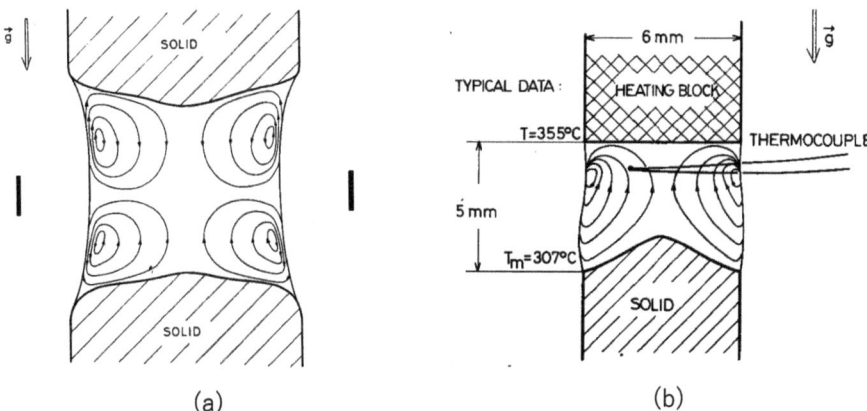

Fig. 2 a Sketch of the vertical cross-section of a floating zone from NaNO$_3$, displaying the streamlines of thermocapillary convection (TC) in a 6 mm-diameter floating zone. This picture is from the first experiments ever on thermocapillary flow. The ring heater from Pt (illustrated with two thick-dark segments) had a temperature of approximately 1000 °C. Reused from Schwabe [50] J. Crystal Growth with permission. **b** Vertical cross-section through a liquid bridge with 6 mm diameter, created by melting a solid NaNO$_3$ rod from above (This picture is from the first experiments ever on time-dependent (oscillatory) thermocapillary flow). The vortex centre of TC is situated very near the free surface and near to the heating block. The solid–liquid interface is shaped by convective heat transport. Reused from Schwabe et al. [50] J. Crystal Growth with permission

The flow was visualized by tracers. For a floating zone experiment under microgravity we needed dense, cylindrical solid rods from $NaNO_3$. We pressed such rods from granular material and melted a few millimetres of one end of the pressed rod from above by contacting it to an electrically heated platinum (Pt) sheet. A molten liquid zone was formed hanging on the Pt-sheet and supported by the solid part of the pressed cylindrical $NaNO_3$ rod. After crystallizing this liquid zone from below, the same procedure was executed from the opposite end. Early in 1977 we thus developed and worked with a precursor of a liquid bridge (floating half zone). We observed thermocapillary convection (TC, hereafter), and moreover, we discovered time-dependent (oscillatory) TC. I could see hot (lighter) melt with tracers flowing down from the place of the hot platinum sheet along the free surface of the zone, against the gravitational forces! Moreover, the flow inside the liquid zone and at the free surface of the melt was oscillating! I was startled by this observation of oscillatory flow in this configuration because I knew about the plans of NASA and others to grow doped, striation-free float-zone silicon crystals after the FZ-technique (Fig. 2a) under microgravity. This would not make sense after our discovery of oscillatory TC because oscillatory flow results in oscillating growth speed and this would result in inhomogeneities (dopant striations) in doped silicon! There is no advantage in growing silicon crystals after the FZ-technique in microgravity.

We improved our experiment based on this set-up; we replaced the platinum sheet by a temperature controlled electric heater, into which a cylindrical rod made of especially dense graphite (Fig. 2b) was screwed in. Initially, the lower end of the liquid zone was not fixed; the length L was determined by the position of the liquid–solid interface as in crystal growth, dependent on the temperature of the hot upper graphite heater and the heat transport by the flow (Fig. 2b).

We designed a tiny vacuum-tight experiment chamber with observation windows on 4 sides to be used later for experiments in sounding rockets. Disturbances of the TC due to convection in the surrounding air have been reduced by the design with a small volume of air around the zone (liquid bridge).

The TC was visualized by tracers and the motion of the tracer particles was observed by the aid of a stereo microscope. A very fine thermocouple (made of bare wires with a diameter of 0.05 mm) was inserted into the melt from the side. It served to measure the temperature oscillations which are coupled to the flow oscillations. The frequency and amplitude of the temperature oscillations were evaluated by hand from the paper of a chart recorder. First results have been presented already half a year later at the German Conference on Crystal Growth [53]. The results gained with this precursor configuration of a liquid bridge have been published by Schwabe et al. [50]. The results have been sensational because of the discovery of time-dependent thermocapillary convection in a small melt volume; thus time-dependent TC was expected to occur under microgravity. The temperature oscillations are sinusoidal with a well-defined frequency (e.g. with 2.16 Hz for a liquid bridge with 10 mm diameter). Figure 3 shows the oscillation periods in zones of different diameters d and of various lengths L. The oscillation period increases with both, the zone length L and the zone diameter d. This can be expected from theoretical hydrodynamics where the rule $f \sim V^{-3}$ is known, with the frequency f and the volume V of the liquid.

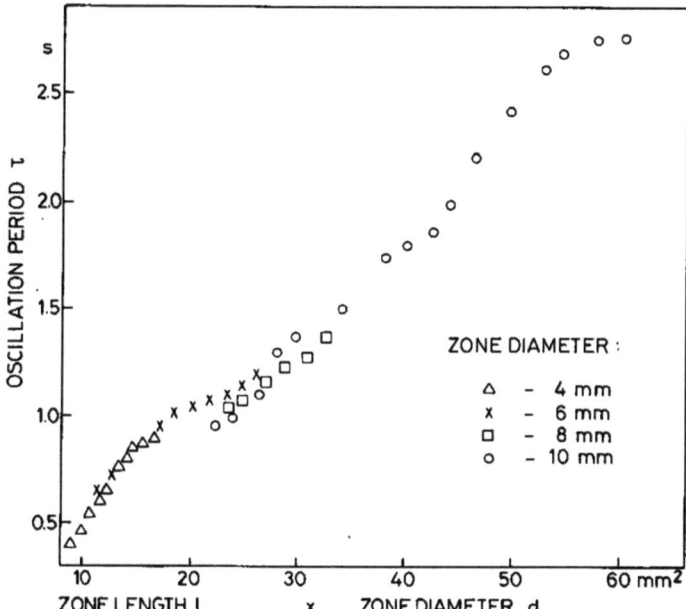

Fig. 3 Periods of the temperature oscillations of time-dependent thermocapillary convection measured in liquid bridges of various diameters d and various lengths L. The correspondence of the frequency to the dimension of the liquid bridge show that the observed oscillations come from TC. Reused from Schwabe et al. [50] J. Crystal Growth with permission of Elsevier

This experimental result confirmed to me that we had to do with a hydrodynamic instability originating from the flow in the liquid bridge and not from another source for the oscillations.

Conclusion: We performed the first investigation ever on TC and on its oscillatory state under normal gravity in a liquid bridge where L was not independently fixed.

2.2 The Advent of Liquid Bridges with Free Cylindrical Surface as a Paradigm for Thermocapillary Experiments

We finally received financial support from the German Ministry of Research and Technology to study oscillatory TC under microgravity during the 360 s of ballistic flight in the payload of a sounding rocket. To make the device and the handling as simple as possible, I had the idea to use a solidified sample of $NaNO_3$ that was held between two differentially heated graphite cylinders. We used a very dense graphite from Schunk Group (Heuchelheim, Germany) which was not wetted by $NaNO_3$-melt. This important feature of the graphite allowed us to form a liquid bridge of molten $NaNO_3$ between two differentially heated graphite rods. Thus the length of the liquid zone was fixed and its diameter was shaped as a cylinder. The sample could be melted with the help of TC in less than 1 min under microgravity to become a fluid sample. With this solidified $NaNO_3$-sample everything went without problems; tests on ground, transport to tests and to the launch site, no fluid handling, no cooling of the cold side needed, guaranteed no contamination of the sample surface by others because we delivered a sample contained in a closed cell. This concept worked fine and so did our experiment. It was developed in the beginning for the crystal grower community as a demonstration experiment and was now changed to become a real hydrodynamics experiment with fixed geometric boundaries (defined radius r and length L) and fixed thermal boundary conditions by setting well defined temperature differences ΔT between the upper and the lower graphite cylinder. With this we could assign a thermal Marangoni number Ma to each experimental situation, $Ma = \frac{|\sigma_T|\Delta T L}{\eta \chi}$ with temperature dependence σ_T ($\partial \sigma / \partial T$ of the surface tension σ), dynamic viscosity η, and thermal diffusivity χ, with zone length L and temperature difference ΔT between the upper and the lower graphite rod. The diameter of the liquid bridge (zone) was in most cases 6 mm (Fig. 4). One can combine the values of the physical properties of $NaNO_3$ in Ma into the factor $F^* = 1211$ cm^{-1} K^{-1}, writing $Ma = F^* \Delta T L$ for an easy conversion of the experimental parameters ΔT and L into the non-dimensional Marangoni number Ma.

Because of experimenting at elevated temperatures with a salt melt with melting point 307 °C in microgravity experiments, no active cooling of the cold side was needed in our set up because of sufficient cooling by heat flow from the hot zone to the colder structure of the payload.

This design, consisting of a cylindrical liquid sample between two differently heated cylinders (Fig. 4), became our working horse for many experiments under microgravity and in the laboratory [47]. It is now the paradigm for experiments on TC under microgravity. The only serious competitor in the field at that time was Chun [6]. I mention here that the second paradigmatic configuration for experiments on TC under microgravity is the annular gap [52].

To measure the temperature oscillations, we used fine naked thermocouples (Pallaplat from Heraeus with wire diameter 0.05 mm). The detection and the quantitative measurement of the oscillatory TC by fine thermocouples was a further important intentional simplification of our experiments compared to optical observation. To exclude any disturbance of the free surface, the thermocouples were introduced for the first experiments under microgravity from above through the upper heating block into the liquid zone.

The temperatures of both heating blocks were controlled and ramped according to a temperature–time programme. The data from the liquid bridge in the rocket were recorded by a chart recorder on board, or later, by transmission down to the ground station. At this time, it was normal to record temperature oscillations of low frequency by a chart-writer paper from the control centre. At the end of the experiments the experimenter received a piece of paper from control centre with the temperatures and temperature oscillations written over time. We could see the oscillatory state and could count the oscillation frequency from it.

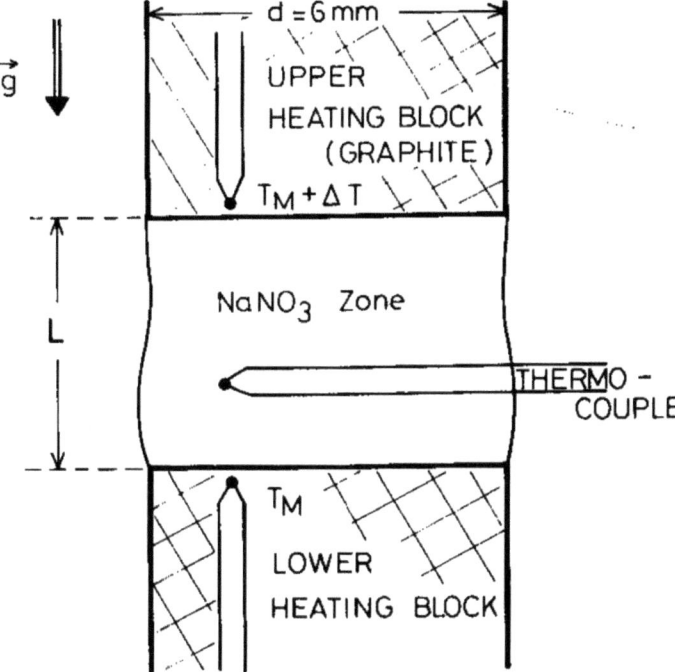

Fig. 4 Paradigm of experiments under microgravity on TC in liquid bridges from NaNO$_3$, which are hold by surface tension forces between two differentially heated cylindrical support rods (from special dense graphite with large thermal conductivity). Reused from Schwabe and Scharmann [47] J. Crystal Growth with permission of Elsevier

2.3 A Special Problem of Experiments with Thermocapillary Convection is the Need for a Clean Liquid Surface

TC is driven by a surface tension gradient which is due to a temperature gradient at the surface. The surface tension of a substance depends on its temperature but can also be influenced to a large extent by surface-active contaminations. Such liquids are not suited for experiments on TC. Examples are water with detergents or oxidized metal-melt surfaces. Water is totally unsuited for experiments on TC because many possible environmental contaminants reduce its surface tension in an unpredictable way. The lower surface tension of a locally contaminated liquid causes a flow on the surface to areas with higher surface tension. However, short-chain silicone oils have such a low surface tension that a contamination can hardly cause a significant reduction of the surface tension. Thus, the problem of contamination can largely be solved by using a silicone oil as test liquid. The disadvantage of silicone oils for thermocapillary experiments lies in handling-difficulties; they are wetting every surrounding wall and are not easily contained under microgravity.

We have found a different way with NaNO$_3$-melt; this material is highly oxidising above its melting point of 307 °C. Oil and other contaminations at the surface are readily oxidized. The oxygen for this reaction comes from the decomposition of NaNO$_3$ into NaNO$_2$ at temperatures above 320 °C. Fortunately NaNO$_2$ has very much the same surface tension as NaNO$_3$.

We have thus a continuous self-cleaning of the free surface of the NaNO$_3$-melt by oxidisation of organic dirt. The advantages of using NaNO$_3$ melt for experiments on thermocapillarity are its high surface tension and thus the high stability of the liquid bridge configuration. Interesting is the low vapour pressure and especially the smaller Prandtl number (Pr = 9) of NaNO$_3$ compared to that of silicone oils.

2.4 First Experiments During Sounding Rocket Flights to Clarify the Existence of Time-Dependent Thermocapillary Convection in Liquid Bridges Under Microgravity

We used the microgravity-time of approximately 6 min during the ballistic flight of the sounding rockets in the TEXUS programme to melt the solid NaNO$_3$-sample under microgravity (gravity level of approximately 10^{-4} G), to install a temperature difference ΔT between the two pre-heated cylindrical rods and for thermalization of the system. We measured and recorded continuously the temperatures in the liquid melt, in both heating blocks near the solid–liquid boundary, and on the surrounding walls of the experiment chamber.

Temperature oscillations are always coupled to convective oscillations in thermal convection of medium Prandtl-number liquids. This is known from literature and by own experience with NaNO$_3$-liquid bridges. Therefore, one can derive the existence and features of oscillatory flow from the existence of temperature oscillations and vice versa. There is no need in our case to visualize the flow and to analyse tracer motion for detecting oscillatory flow. Flow visualization would pose much larger difficulties than the measurement of temperature oscillations. Moreover, the complete optical observation of the flow by tracer motion in a 6 mm thick liquid bridge from the side is not possible because of optical problems, posed by the cylindrical free surface. Our first experiment to observe oscillatory thermocapillary flow under microgravity in a liquid bridge was on the sounding rocket TEXUS IIIa, launched on April 28, 1980. Our experiment worked fine but the rocked failed to de-spin [49]. The liquid bridge was thus exposed by chance to 0.5 g in radial direction. It remained stable and we registered temperature-oscillations. This, at least, confirmed the stability of the liquid bridge from NaNO$_3$ under microgravity and our experimental approach.

The results of the first completely successful experiment under microgravity to verify the oscillatory TC have been on TEXUS IIIb and are displayed in Fig. 5a, b [46]. After melting the liquid bridge under microgravity, we applied two temperature differences (40–80 K) between to the graphite rods for 120 s each. Rather regular temperature oscillations of 1.8 K peak-peak amplitude occurred for $\Delta T = 40$ K (Fig. 5a). The nature and the strength of the temperature oscillation signal indicate that it is from a flow state just above the threshold of the transition from steady to oscillatory flow. We interpreted this signal as being caused by an azimuthally travelling disturbance wave as formulated for the azimuthal component v_φ of the flow velocity by Schwabe et al. [46].

For $\Delta T = 80$ K more irregular temperature oscillations appear with amplitude up to 5 K peak to peak (Fig. 5b). This can be interpreted by the occurrence of more than one travelling wave.

The signals during both temperature differences show by their constancy in time that we approached steady state conditions.

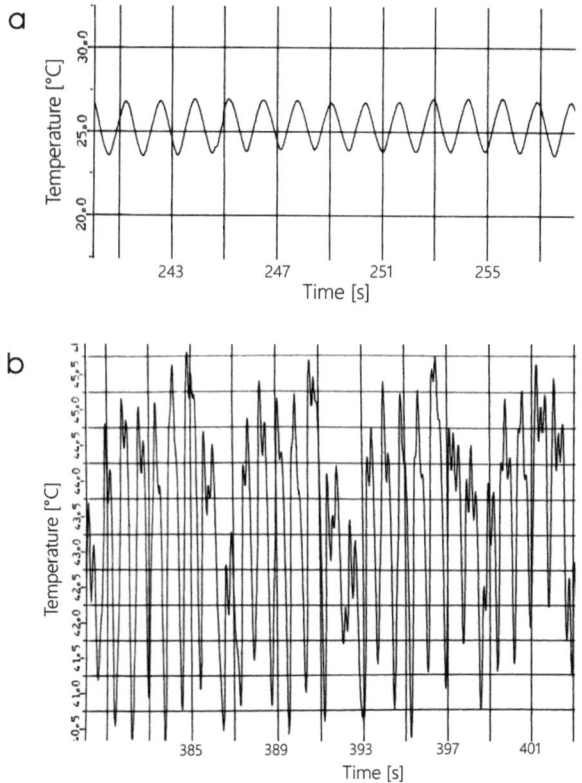

Fig. 5 a, b From the sounding rocket flight of the rocket TEXUS III b, April 30th, 1981. **a** Temperature oscillations under microgravity (10^{-4} G) at Ma $= 1.6 \times 10^4$ ($\Delta T = 40$ K). **b** Temperature oscillation from the same experiment at Ma $= 3.2 \times 10^4$, $\Delta T = 80$ K. The temperatures shown are reduced by 300 °C to compensate for the high working temperature (melting point of $NaNO_3$ is 309 °C). Reused from Schwabe et al. [46] Acta Astronautica, with permission of Elsevier

Conclusion: We could demonstrate the existence of time-dependent TC under microgravity. The critical ΔT for our experiment is smaller than 40 K. The time-dependence can be interpreted as being due to azimuthally travelling waves in the liquid bridge. TC does not need the coupling to gravity to become time-dependent.

2.5 The Measurement of the Critical Marangoni Number Ma_c of the Transition to Time-Dependent Thermocapillary Flow in Liquid Bridges Under Microgravity

An interesting point for crystal grower is the threshold in ΔT the transition from steady to oscillatory flow. Temperature differences above this transition point need to be avoided. We measured this threshold, under microgravity represented by the critical Marangoni number three times in total. The first measurement on TEXUS 5 was by ramping up ΔT linearly in time through the expected range of the transition temperature difference. The experiment was disturbed by vibrations from a cine camera of another experiment in the payload. The repetition on TEXUS 8 had better conditions. The onset of temperature oscillations under microgravity is at experiment time t = 270 s and at t = 340 s for the same type of run under normal gravity (Fig. 6a, b). The oscillations have a frequency near 1 Hz and the amplitude grows with ΔT. We calculated a Ma_c under microgravity of 8.5×10^3 (critical temperature difference ΔT_c = 22.7 K, Fig. 6a). The critical temperature difference ΔT_c under normal gravity was larger than under microgravity. Does gravity stabilize the flow because of heating from above? As well the different thermal environment could make a difference between an experiment under microgravity compared to one under normal gravity (Fig. 6b).

What about the influence of measuring during heating up (non-stationary conditions compared to quasi stationary conditions)? This must result in the measurement of a larger Ma_c. We found a lag between the real temperature and the programmed one [48]. The measured ΔT_c could be larger by 0.6 K and Ma_c by ramping ΔT up is between 2–3% larger than the value which could be gained in a perfect "quasi stationary" measurement. For a short-time experiment during sounding rocket flight this result is not too bad.

2.6 The Structure of Time-Dependent Thermocapillary Flow Under Microgravity in Liquid Bridges Depends on the Aspect Ratio

We had two similar experiment chambers on board of the free flying satellite SPAS-01 with $NaNO_3$ zones of 6 mm diameter and length $L_1 = 4.0$ mm and $L_2 = 4.8$ mm,

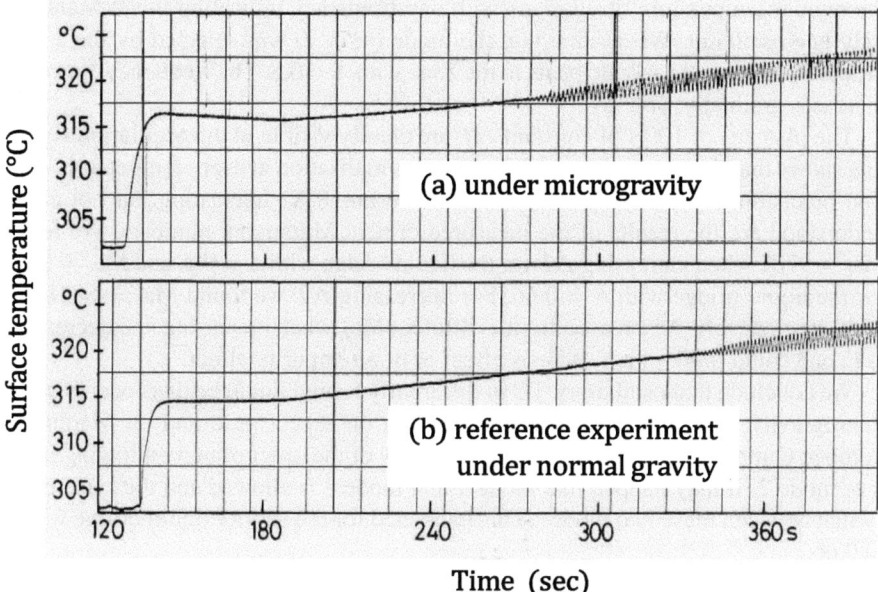

Fig. 6 From the ballistic flight of TEXUS 8, May 13th 1983. Temperature—time trace in a liquid bridge from NaNO$_3$ with 6 mm diameter, length $L = 3.2$ mm during increase of the temperature difference ΔT. With oscillatory thermocapillary flow at higher ΔT: **a** during increase of ΔT under microgravity; **b** reference experiment at normal gravity, during increase of ΔT. Reused from Schwabe and Scharmann [48] (Z. Flugwiss. Weltraumforsch with permission of Springer Nature)

processed at the same time. The aspect ratio A (= length (L)/diameter (d)) is 0.66 and 0.8, respectively. The temperature differences have been ramped up and later down in small steps of $\Delta T = 0.36$ K every 100 s, [54]. Originally, we aimed with this experiment mainly at a more accurate measurement of Ma_c. Besides this, our attention was attracted by beats in the temperature oscillation spectrum in the liquid bridge with smaller aspect ratio $A = 0.66$. The beats are indicating the existence of two waves with slightly different frequencies.

We received oscillation signals of very good quality for the duration of 5.28×10^3 s from both liquid bridges. We discuss some of the more interesting features of the oscillation signals from this experiment comparing the influence of the aspect ratios. In the liquid bridge with $L = 4.0$ mm ($A = 0.66$) in Fig. 7, we observe beats of the frequency starting at the very beginning of the oscillations until the end of the experiment after more than one hour. At higher Ma we observe stronger beats in this liquid bridge with $A = 0.66$, followed by periods of no beats. This behaviour can be explained by the existence of two oscillations with two different frequencies (mode $m = 1$ and mode $m = 2$) coexisting at different, but not at all ΔT. This is in contrast to the behaviour of the flow in the liquid bridge with $A = 0.8$. The zone with $A = 0.8$ shows continuous increase and decrease of the oscillation frequency during ramping up and ramping down of ΔT. This zone shows for more than 1 h and for various ΔT

the regular temperature oscillations with one frequency indicating the existence of only one oscillator. We assume that the mode ($m = 1$) was selected by the aspect ratio $A = 0.8$. We observe no beats in the zone with $A = 0.8$. The frequency, however, increases gradually with ΔT.

The plateaus of 100 s of constant ΔT are clearly visible in the oscillation signal; this shows that there was plenty of time for thermalisation at every temperature step. The precision of measuring ΔT_c is therefore ± 0.18 K. Interesting but not easily understood are the results of the measured critical Marangoni numbers. We found $Ma_c = 9702$ when increasing ΔT for the liquid bridge with $A = 0.8$, and $Ma_c = 9444$ for the liquid bridge with $A = 0.66$. For decreasing ΔT we found $Ma_c = 8848$ and 8340, respectively. As a reason for the different Ma_c when increasing and decreasing ΔT, one could think of a hysteresis effect or of an impurity effect.

We conclude that oscillatory TC in differently heated liquid bridges occurs under microgravity with different modes, depending on aspect ratio and on Marangoni number. Our aspect ratios $A = L/d$ are generally of the order of unity allowing mode 1 or mode 2. It may happen that mode 1 and mode 2 is allowed and the system can switch between these two modes. This happened for the shorter liquid bridge with $A = 0.66$.

A different scenario holds for the liquid bridges with very large aspect ratio, with e.g. $d = 20$ mm and $L = 2$ mm, not investigated by us under microgravity but under normal gravity. Here only modes much higher than $m = 2$ are allowed and a travelling wave with such a high mode can be more easily accommodated without the competition of two modes without disturbing the travelling waves; Particle accumulation structures (PAS) as described later Sect. 2.11 can develop more easily under such conditions [55].

2.7 Structure of Time-Dependent Thermocapillary Flow Under Microgravity; Transition to Chaotic Flow

The experiment Fig. 8 [56] was performed on a liquid bridge with $L = 4.00$ mm and $A = 0.66$ in a "getaway special" on space shuttle STS-89 during a docking phase to the Russian space station MIR in January 1998 [44]. We aimed for high ΔT to realize chaotic TC. The study of time-dependent convection in small liquid bridges is interesting because the flow is two-dimensional in the steady state and stays as such in the average in the time-dependent state. The system is "hard" against transition to chaos because the driving force for the flow is located mainly in the free surface. The transition to chaos in thermocapillary liquid bridges has been extensively studied on earth in the authors group by Schwabe et al. [51] and by Frank and Schwabe [10]. Chun [6] claimed to have reached a turbulent flow state of TC in a sounding rocket experiment, but he did not reach an equilibrium state.

In the experiment MAUS G 141 we equipped two of our experiment chambers with 6 mm diameter $NaNO_3$ samples, with length $L_1 = 2.5$ mm ($A = 0.41$) and L_2

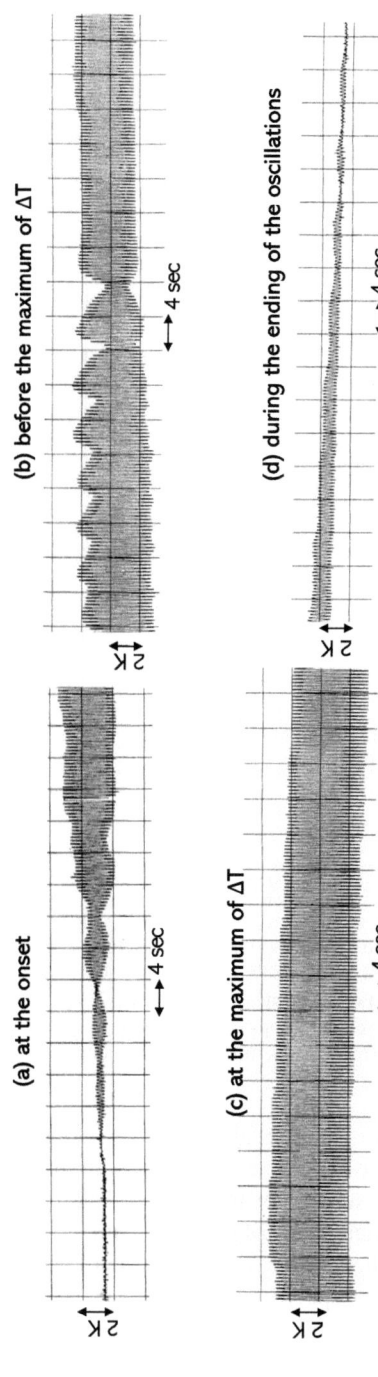

Fig. 7 Temperature oscillations of TC under microgravity from an experiment on the free flying satellite SPAS-01 (June 22.1983 together with STS 7, Challenger). TC in the liquid bridge from NaNO$_3$ with 6.0 mm diameter, length $L = 4.0$ mm, showing beat frequencies for this aspect ratio $A = 0.66$) during increase and during decrease of ΔT; **a** at the onset, **b** before the maximum of ΔT, **c** at the maximum of ΔT, and **d** during the ending of the oscillations. A liquid bridge with $A = 0.8$ processed in the same way showed no beat frequencies at all. Reused from Schwabe and Scharmann ESA SP-222 [54], with permission of ESA

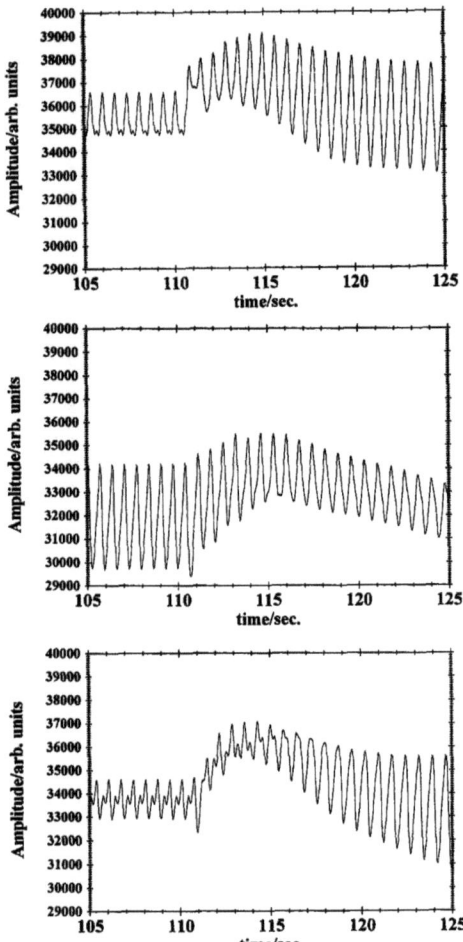

Fig. 8 The temperature oscillation spectra from oscillatory TC under microgravity in the 2.5 mm liquid bridge in the experiment MAUS G 141 on STS 89, January 1998. The temperature signals come from the three thermocouples positioned at different azimuth (72° apart). All time scales are shifted in time in the same way. The correlated jumps of the amplitudes indicate a standing wave. Reused from Schwabe [56] Adv. Space Res, with permission of Pergamon

= 4.5 mm ($A = 0.75$). The longer liquid bridge aimed to establish a mode $m = 1$ (hydrothermal wave with one wavelength around the circumference of the liquid bridge). The shorter liquid bridge aimed for mode $m = 2$ and the results from it are described here. We applied increasing Marangoni numbers by increasing the temperature differences between the graphite rods in steps of 2 K until $\Delta T = 120$ K [44]. The background in the Fourier spectrum was less than 10^{-3} of the main peak around 1.2 Hz for the lower ΔT until $\Delta T = 100$ K, where it increased to a level of

10^{-2} for 102 K, and further to 10^{-1} for $\Delta T = 106$ K. This sudden increase of noise in the Fourier spectrum is interpreted as the point of transition to chaos.

Results: We could verify that the instability develops as travelling wave (especially clearly in the long liquid bridge) and we could show the transition to chaotic flow at very high temperature differences $\Delta T = 106$ K, near Ma $= 5.6 \times 10^4$. Before the onset of chaotic flow, the peak of the main frequency in the Fourier spectrum was always higher by a factor of 10^3 compared to that of the background noise. A sudden increase of the background noise by 1–2 orders of magnitude occurred in the Fourier spectra when increasing ΔT in steps of 2 K from $\Delta T = 100$ K to 106 K. I interpret this sudden increase of the background noise as the transition to chaotic flow.

2.8 Standing Thermocapillary Waves in Liquid Bridges Under Microgravity

In the experiment MAUS G141 we tried to verify azimuthally travelling waves in the oscillatory state by positioning 3 thermocouples at the same axial position but at different azimuthal positions and by measuring the azimuthal phase differences between the signals from these three thermocouples [56]. However, we found standing thermocapillary waves in the liquid bridge with $L = 4.5$ mm, $(A = 0.75)$. The reason for this finding is not really known except that standing waves have been found in thermocapillary liquid bridges as well by numerical simulation by Leypoldt [24] and that standing waves of TC can be provoked by deliberate addition of impurities [45].

We describe and interpret in the following jumps of the oscillation amplitude which occurred at the same moment for all 3 thermocouples in the liquid bridge with $L = 4.5$ mm (Fig. 8 b, c). These jumps of the signal amplitude happened only 3 times during the experiment, namely at 19,500 s experiment time, 23,000 s, and 26,400 s during the 23,000 s long measuring phase [56]. The signal amplitude is constant before and after the jump. Such correlated jumps of the oscillation amplitude from thermocouples, which are azimuthally differently positioned, can originate only from standing waves. They are likely to occur in the small-aspect-ratio zones, where neither the aspect ratios A_1 nor A_2 are exactly within the existence range of mode 1 or of mode 2. The hydrothermal waves can develop as standing waves because of this mismatch [45]. Standing hydrothermal waves have an azimuthal point of origin (source) and an azimuthal point where the waves meet again (sink). Source and sink are fixed under constant conditions but can move from time to time during a short moment, e.g. when Ma is changed as in our experiment. The movement of the origin of a standing hydrothermal wave will be in the direction of better matching of the wave to its existence range in the given aspect ratio. This movement of the origin of the standing wave results in the synchronous jumps of the oscillation amplitudes at the measurement points. The scenario for standing waves described above could well explain the synchronous jumps in oscillation amplitude; source and sink of the standing wave move as indicated by the jump of the signal-amplitude from the 3 thermocouples. The amplitudes at all three thermocouples change at the same time. Figure 8 from Schwabe [56] show the synchronous jump of the oscillation amplitude of the three thermocouples around time $t = 111$ s with large time-resolution. It is important to note that none of the three thermocouples was in one of the nodes of the standing wave; the nodes develop at their position independently from a possible disturbance by the thermocouples.

Conclusion: We found indications that time-dependent TC under microgravity favours the occurrence of standing waves in zones with aspect ratio around 0.75. This is in accordance with the numerical work by Leypoldt et al. [24].

2.9 The Hydrothermal Waves in a Liquid Bridge with Length Near the Raleigh-Limit Display a Critical Marangoni Number Near the Theoretical Value and Indicate Inclined Travelling Waves

To establish a long liquid bridge under microgravity I adopted the technique from Kawamura lab, working with silicone oil as the test fluid and using a transparent sapphire rod as one end for the view from one end into the liquid bridge. I established a liquid bridge with $A = 2.5$, length $L = 15.0$ mm and radius $r = 3.0$ mm (Fig. 9a) from 2 cSt silicone oil (Pr $= 28$) by telecommand. This was done under microgravity during the flight of the sounding rocket MAXUS 4 (launched from ESRANGE in Kiruna, April 29, 2001). This bridge length was dangerously close to the Raleigh-limit, which is in our case $2\pi r = 18.84$ mm.

The main purpose of the experiment was to measure the axial component of the hydrothermal wave what is only possible for large L. Five fine thermocouple tips have been placed near the surface of the liquid bridge in different axial but in the same radial position. Four more thermocouples had the same axial but different azimuthal positions near the free surface. We could thus measure the orientation and travelling speed of the hydrothermal wave at various Marangoni numbers from the phase shifts between the thermocouple signals [57]. Some of the four thermocouples at the same axial position obviously touched and deformed the free surface of the liquid bridge to relatively large degree (Fig. 9a). This strong deformation comes from the united action of the 5 thermocouples with the same azimuthal position. The deformation of a bridge from silicone oil due to normal gravity would already be large for a bridge-length $L = 4$ mm, which hinders the exact measurement of the azimuthal component, though [29] claimed a measurement of the axial component under gravity with the help of an infrared camera. At a liquid bridge with $L = 15$ mm, which is only stable under microgravity, one can more easily measure the axial component of the travelling wave. We applied altogether 4 different temperature differences ($\Delta T = 7$ K, 9 K, 10 K, and 12 K) and registered the amplitude of the corresponding temperature oscillations. From their squared amplitudes over ΔT (Fig. 9b), assuming a Hopf

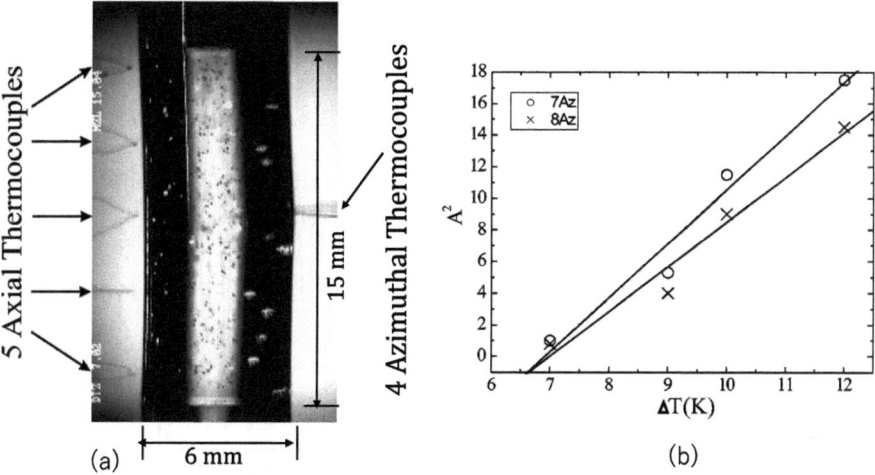

Fig. 9 MAXUS 4, (April 29 2001). **a** Photograph with background illumination of the liquid bridge from 2cSt silicone oil with 6 mm diameter and 15 mm length under microgravity. Five thermocouples (TCs) are coming from the left in this figure, with their tips very near the free surface of the liquid bridge, not touching it. Four TCs are coming from the right at $L/2$, disturbing the shape of the free surface, especially because they are touching the free surface all at the same axial position. These four disturbances are adding. (Advances in Space Research 36 (2005), **b** Squared amplitudes of the temperature oscillations measured by two thermocouples, plotted over the temperature difference ΔT applied to the 15 mm long liquid bridge from 2 cSt oil under microgravity. Reused from Schwabe [57], Phys. Fluids with permission of AIP

bifurcation, I could extract a critical temperature difference $\Delta T_c = 5.2$ K (Ma$_c$ = 952), which is much smaller than that measured at the short liquid bridges on ground. This can be explained by a significant drop of the applied ΔT in the thermal boundary layers at the end-rods with higher thermal conductivity than that of the liquid. This temperature-drop near the ends of the liquid bridge is approximately the same for all bridge lengths, but it counts more for the short liquid bridges. The definition of a critical Marangoni number is therefore meaningless without the information about the bridge-length L. Our value of the Ma$_c$ measured in this experiment with a bridge length $L = 15$ mm, $r = 3$ mm is already near the theoretical value of an infinitely long zone, Xu and Davis [70].

From the temperature–time traces of the axially positioned thermocouples (Fig. 10) and of the azimuthally positioned ones I could derive that the wave was with mode $m = 1$, travelling counter-clockwise from the cold side to the hot side with an angle of 47° against the applied temperature gradient. It had a wavelength of 24 mm and a period of 6.52 s (frequency $f_1 = 0.153$ Hz). The angle between the temperature gradient and the travelling direction is very close to the theoretical one in extended liquid layers [70]. The good agreement between our experimental and the theoretical values comes from our rather long liquid bridge, which is close to the theoretical model of an infinitely long liquid bridge.

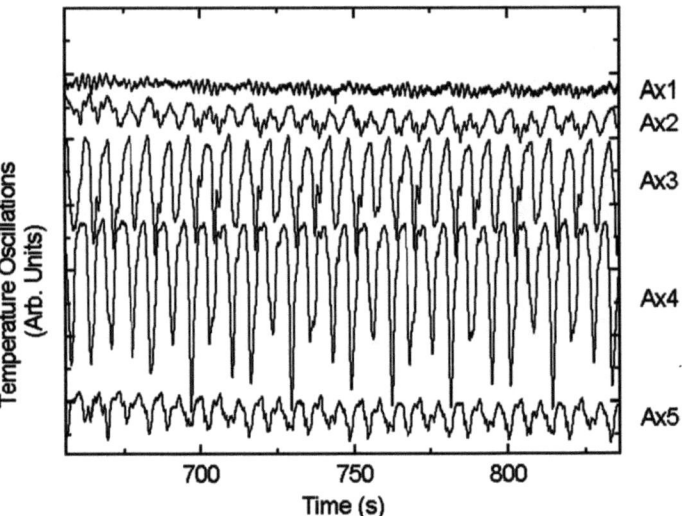

Fig. 10 A result from MAXUS 4, April-29 2001. Temperature oscillation signals from thermocouples with different axial position. The axial component of the hydrothermal wave can be derived from the phase shifts between the thermocouple signals. Reused from Schwabe [43] Adv. Space Res. with permission of Elsevier

2.10 Bénard Cells, Drifting with the Thermocapillary Surface Flow, have been Found Under Microgravity on the Surface of a Long Differentially Heated Liquid Bridge. They Originated from Cooling of the Free Surface by Cold Air Advected by Thermocapillarity

We found in the experiment on MAXUS 4 two further well-defined higher frequencies ($f_2 = 0.56$ Hz, and $f_3 = f_1 + f_2$) besides that of the hydrothermal wave with $f_1 = 0.154$ Hz (Fig. 11a, b), Schwabe [43]. We assign them to convection cells that drift in the thermocapillary surface flow, with their cold cell boundary touching the thermocouple tips when passing by. These cells are Bénard-like with hot up-flow in the middle and colder down-flow in the cell boundaries [3]. We could measure the flow speed in the free surface of the liquid bridge from particle motion in a video to be $v = 4.3$ mm/s for $\Delta T = 12$ K. We can assume the existence of two Bénard cells in axial direction over the length $L = 15$ mm of the liquid bridge. Note that the free surface of the liquid bridge has approximate rectangular extension ($L \approx 2\pi r$), giving unrestricted the space for 4 Bénard cells. The drift velocity of 4.3 m/s gives for the frequency f' of passing by cell boundaries, the value: $2*(15 \text{ mm})^{-1}*34 \text{ mm s}^{-1} = 0.57$ Hz. This value is very near the directly with the thermocouples measured frequency of $f' = 0.57$ Hz (Fig. 11a, b). For $\Delta T = 7$ K we found from the tracer drift velocity $f = 0.4$ Hz compared to the value from the Fourier analysis with $f' = 0.38$ Hz. The more detailed analysis of the cooling of the free surface can be found in Schwabe [43]. We note that not all of the azimuthally distributed thermocouples show the strong signal of Bénard cell-boundaries passing by the thermocouples; the two thermocouples Az6 and Az9 show this signal only as a weak decoration. We note from thermocouple 4Ax in Fig. 10 and from TZ9R in Fig. 11a the larger and more pronounced deflection of the temperature signals towards the colder temperature. This is the well-known signature of the cold cell boundary passing by. Drifting Bénard cells have been observed later by the author and his co-worker in layers of liquid under the action of an inclined temperature gradient [28]. This type of inclined T-gradient seems to be as well present in the long liquid bridge with its primary axial temperature gradient driving thermocapillary convection and the radial cooling of the free surface by air-convection.

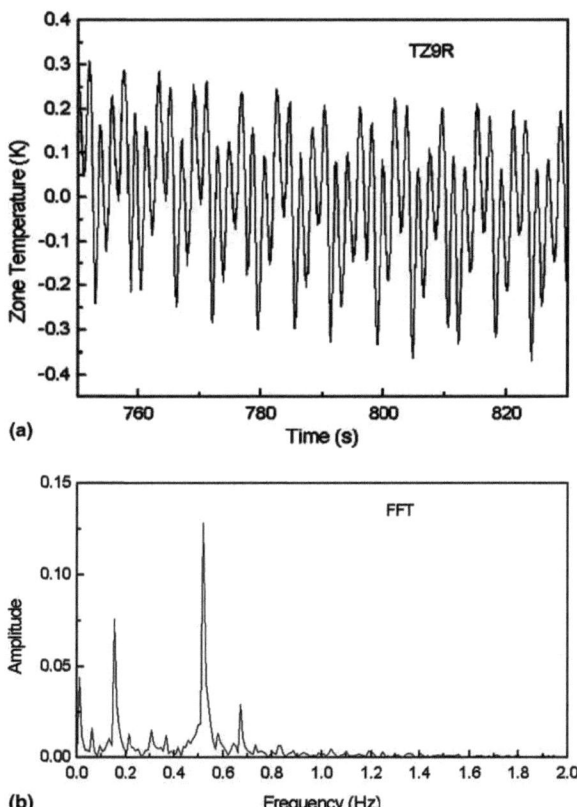

Fig. 11 From MAXUS 4, April-29 2001; **a** Complex temperature signal from TZ9R at $\Delta T = 12$ K and **b** Fourier analysis of this signal as example for oscillations of the hydrothermal wave with $f = 0.154$ Hz plus a higher frequency $f' = 0.56$ Hz, and $(f + f')$ Hz. The new frequency f' is interpreted as generated by Bénard-Marangoni cells drifting in the thermocapillary flow, passing by the thermocouple tips with their cold cell-boundaries. Reused from Schwabe [43] Adv. Space Res. with permission of Elsevier

The analysis of the thermal environment around the liquid bridge under microgravity showed that evaporative cooling alone would not be sufficient to develop the Bénard instability. The formation of Bénard cells can be explained only by air convection driven by TC (motion of the free surface of the liquid bridge). The circulating air contacts colder metal parts of the cold side, is cooled down there and the cool gas is transported back to the free surface of the liquid bridge, cooling it [43]. This is an inherent mechanism for a thermocapillary liquid bridge.

2.11 Discovery of Travelling Waves with Particle Accumulation Structures (PAS)

Particle accumulation structures (PAS) have been discovered in our laboratory under normal gravity by my diploma student Schwabe et al. [55] when experimenting with liquid bridges with small aspect ratio (typically 20 mm diameter, 2.5 mm length). These liquid bridges with large diameter can always accommodate a large mode m which is not disturbed by mode transitions. And they allow direct view from the side with a stereo microscope into its volume near the free surface. The larger diameter allowed us to observe the thermocapillary return flow with PAS. We could observe particle accumulation taking place there and particle movement as in travelling waves (called "travelling stripes" by P. Hintz). This is only mentioned here because PAS under microgravity is treated in Sect. 3 of this Chapter by Kawamura H. We could observe PAS in our ground based experiments only with selected graphite rods of 20 mm diameter, presumably because of difficulties to achieve perfect rotational symmetry of the temperature distribution in the graphite rods, which are screwed into the heaters. To my understanding, the perfect rotational symmetry of the liquid bridge is required to create a perfect rotationally travelling wave, which is needed for PAS to form.

Conclusions and Acknowledgements

We report on the results of short-time experiments on time-dependent thermocapillary flow in small liquid bridges under microgravity. For the conditions of microgravity, we mainly used the time provided by the sounding flight of rockets payloads (6 min for TEXUS and 12 min for MAXUS). This research took place partly before long-duration experiments in Spacelab became feasible. Long-term experiments were possible on the free flying satellite SPAS-01 and on the payload MAUS G 141 of Spacelab D-2, when coupled to the MIR space station. Our interest was.

(1) in the time-dependent thermocapillary convection in liquid bridges;
(2) in the existence of this state under microgravity,
(3) in the critical Marangoni number of the transition to oscillations,
(4) in the influence of the aspect ratio on the mode number,
(5) in the various manifestations of the oscillations (standing and travelling hydrothermal waves),

(6) in the transition to chaotic states at high Marangoni number,
(7) in the angle of the travelling wave to the thermal gradient,
(8) in the influence of the bridge length, on the critical Marangoni number, and.
(9) in Bénard cells due to cooling of the free surface of the liquid bridge by air-convection driven by thermocapillarity.

The short experiment-time under microgravity of 5–8 min, provided by the sounding rockets, forced us to investigate time-dependent thermocapillary flow in small fluid samples (typically 6 mm diameter, 3.5 mm long) to reach convective equilibrium of the flow state in less than one minute. Fortunately, the oscillation periods are of the order of one second in these samples, long enough for a meaningful analysis. We applied the uncommon technique in most cases to launch the samples in their solid state, melting them (melting point 309 °C) under microgravity and to have a liquid with Prandtl-number Pr = 9. This has the advantage, amongst others, of no need for cooling and no need for the complicated fluid handling under gravity.

We consider our experiments as precursors because repetition-experiments are missing in most cases because of the high costs. We could, however, indicate many features of hydrothermal thermocapillary waves in liquid bridges under microgravity.

D. Schwabe would thank the German Ministry for Research and Technology and the European Space Agency for the continuous financial support of my research under microgravity. The engineers from Astrium Space (Bremen) I thank for their professional support and the final conduct of the flight experiments. Great thanks are to Prof. A. Scharmann, Director of the 1. Physics Institute at Justus-Liebig-Universität Giessen, who gave me the possibility to pursue my research on this new field of thermocapillary flow and its instabilities. My friend and colleague Robert Oeder assisted me with thorough proofreading in finalizing the manuscript.

3 On Ground Experiments and Numerical Analyses of Hydrothermal Convection in Liquid Bridges with Finite Length

Hiroshi Kawamura

3.1 Introduction

In the previous section, we were introduced to the early research of thermocapillary convection experiments with liquid bridges on ground as well as under microgravity by D. Schwabe. In this subsection, we will describe the experiments and numerical analyses primarily conducted by the lead author (H. K.) and his fellow students in the laboratory during the preparation for the 'Kibo' experiment. The application and the ensuing 15-year preparation period for the Kibo experiment were described in the Preface. Additionally, we will provide a brief overview of related research developments during the relevant period and in more recent years.

In the "Kibo" experiment, a "liquid bridge" was created. The liquid bridge is a cylindrical liquid column that bridges the gap between the endfaces of a pair of disks, simulating the containerless crystallization method (see Sect. 1). For example, we conducted various preparatory on-ground experiments using various length of liquid bridges within the range possible on the ground and examined the transition of flow field from steady to unsteady flows up to the transition to the chaotic state. Effects of the surrounding gas and environmental temperature using actual experimental conditions with argon gas have been studied. Alongside, considerable focus was given to fundamental research. In this regard, the rapid advancement of high-performance computers, greatly aided the experimental endeavors.

This chapter will describe mainly the results obtained in these preparatory works for the microgravity experiment obtained by the present author H.K.'s group during the preparatory research for the microgravity experiment in "Kibo." Those interested in a broader perspective on research in this field are recommended to refer to a comprehensive monograph by Lappa [23].

3.2 Outline of On-Ground Experiments of Liquid Bridge with a Finite Length

Figure 12 illustrates the arrangement of standard equipment employed in ground-based experiments. In these experiments, we mostly utilized liquid bridges with a diameter (D) of approximately 2 to 5 mm and an aspect ratio ($\Gamma = H/R$, or $A = H/D$) ranging from $\Gamma \simeq 0.1$ to 2.0, where R is the radius and D the diameter. While the small aspect ratios can be realized by increasing the bridge diameter, there are inherent limitations to increasing the aspect ratio. When attempting to increase the length of the liquid bridge beyond a certain point, the liquid starts dripping out from the gap between the solid cylinders.

Shapes of liquid bridges created on the ground and in space have been compared in Fig. 2 of the Foreword of this book. A shape of a liquid bridge formed on the ground, with a diameter of 5 mm and a height of 4.5 mm, is shown in the rightmost photo. It is significantly deformed by gravity, representing the practical limit of the aspect ratio for this diameter in ground experiments. In contrast, the series of images on the left depict liquid bridges formed under microgravity in space, where the largest one, with a diameter of 50 mm and a height of 60 mm, demonstrates a nearly perfect cylindrical shape.

The top rod was made of sapphire (Al_2O_3) because it is transparent and has a high thermal conductivity, so that, we were able to observe the flow from the top as well as to heat the liquid bridge by coiling an electric heater around the top rod. This heating method using the transparent top rod has been proven highly effective

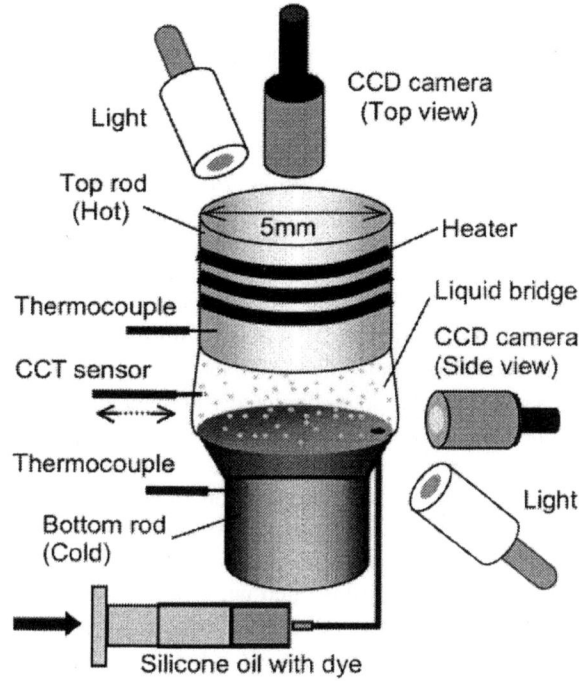

Fig. 12 Arrangement of on-ground experiment. The test fluid is silicone oil in which polystyrene particles are dispersed to visualize the flow field. Reproduced from Tanaka [65] with permission

in understanding the three-dimensional flow field within the liquid bridge. I would like to note and pay respect here that we adopted this method after the work of the late Professor Akira Hirata at Waseda University, Hirata et al. [12].

The bottom rod was made of aluminum and features a sharp edge cut at the periphery to prevent dripping of the liquid from the edge. It was cooled by the cold water fed from a regulated cold bath. For illumination, optical fibers were used; one from the top end face of the top rod and the other from the side of the liquid bridge. In addition, the temperatures of necessary parts were measured using thermocouples, and the flow state was recorded by the motion of the tracer particles from the top and side using CCD cameras.

Silicone oil was often used as the test liquid. While water has a high surface tension, it has been well known that it does not exhibit stable thermocapillary flow, because of its property to dissolve various substances and to adsorb them onto its surface. The advantages of the silicone oil include its stable properties, the availability of a wide range of viscosity grades and its relatively easy accessibility. In our on-ground experiments, we used silicone oils with kinematic viscosity of 1, 2, or 5 cSt.[1] The corresponding Prandtl numbers for the silicone oils are approximately 16, 28, and 68, respectively. It is worth noting that the viscosity decreases appreciably with

---

[1] The unit cSt represents the dynamic viscosity coefficient, where 1 cSt = 1 mm^2/s. Although not part of the SI unit system, it is commonly used in this field since the kinematic viscosity of water at 20 °C is approximately 1 cSt.

increasing temperature; so, its Prandtl number is also temperature-dependent. Its density ρ given in its catalogue is 0.87 g/cm^3 and almost irrespective of the viscosity.

In order to visualize the flow fields, tracer particles were suspended in the fluid. Polystyrene particles were mainly used as the particles because their density is relatively close to that of the test fluid, the silicone oil. According to the catalogue, the ratio of the densities $\rho_{particle}/\rho_L$ is about 1.2. The diameter was about 10–20 µ. Since various sizes of particles were mixed in the as-purchased particles, they were sifted through a sieve, but a certain amount of unexpected size of particles remained unavoidably. For some experimental purposes, smaller and/or heavier particles were also used.

During the experiment, a desired liquid bridge was formed by injecting the test liquid using a syringe, increasing the spacing between the upper and lower rods up to the target aspect ratio, as depicted schematically in Fig. 13a. Usually, the temperature of the low-temperature rod is held constant, and that of the high-temperature one is gradually increased, because the hotter temperature is easier to be controlled. With increase of the temperature difference, the thermocapillary convection occurs along the cylindrical liquid surface from the high- to the low-temperature side. Accordingly. returning flow from the low-to the high-temperature side takes place within the liquid bridge as illustrated with small arrows in Fig. 13b, by numerical analysis of Kousaka and Kawamura [20]. In this analysis, the heat transfer between the liquid surface and the surrounding air is expressed with use of the Bi number (Eq. 21 in chapter "Thermocapillary Convection in an Infinite Liquid Layer and in an Infinite Liquid Column"), and the environmental temperature T_∞ is given by linear interpolation between the hot and cold temperatures of the both end plates (Fig. 13a). In more detailed analyses, the flow induced in the surrounding air was also calculated simultaneously as depicted in Fig. 13d, by Kawame [18]. In this case, one need not specify the Bi number.

Figure 13c presents the temperature profile along the liquid surface as determined by numerical analysis. Two key observations are highlighted: First, the surface temperature profile is non-linear, with steeper gradients near both ends and a flatter profile in the central region. Second, an increase in the Biot number (Bi) leads to reduced temperature gradients near the ends, while the central region shows a more pronounced gradient. This effect is attributed to a higher Biot number causing the surface temperature to more closely align with the ambient temperature, which is presumed to be linear in this analysis. Whereas, in the simulation depicted in Fig. 13d, Bi is not used; instead, the air field is solved simultaneously, with top and bottom boundaries of the air region being bounded by hypothetical horizontal plates for simplicity. The obtained critical Marangoni numbers will be compared with experiments later in Fig. 14a. The effect of heat exchange with the surrounding gas was found significant during the preparatory research phase and a more detailed description will be presented in chapters "Effect of Heat Exchange, Control and Suppression of Thermocapillary Convection" and "Microgravity Experiments in Kibo Onboard the International Space Station".

As the temperature difference is increased, the flow instability emerges at a certain threshold, similar to the canonical case of an infinite long liquid bridge mentioned in

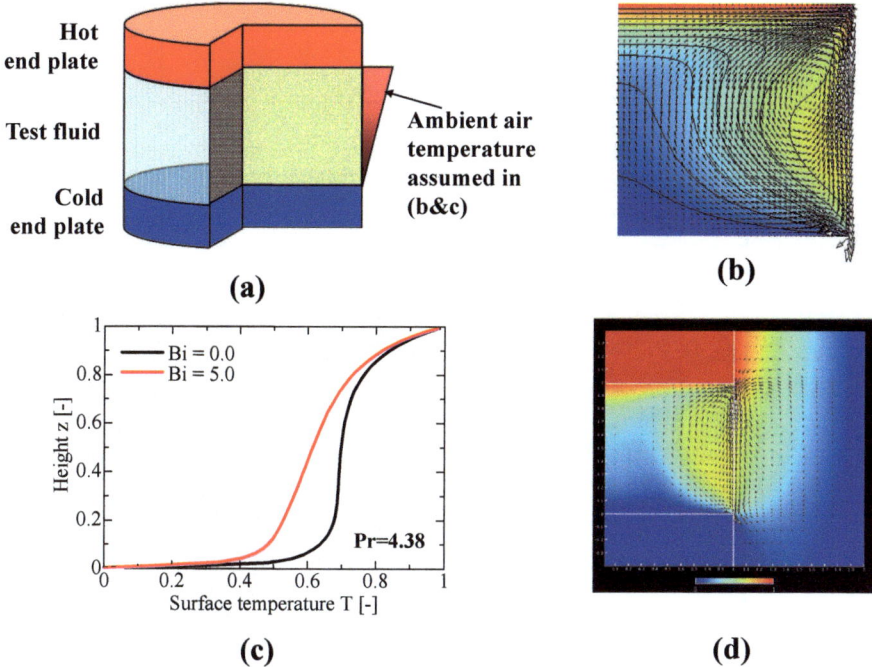

Fig. 13 Numerical analysis of thermocapillary convection in a liquid bridge. **a** Liquid filled between hot and cold end plates, and an ambient temperature profile assumed often in numerical analysis, **b** Example of temperature and velocity distributions **c** Example of surface temperature distributions obtained by numerical simulation with adiabatic surface (Bi = 0) and with a nonvanishing heat exchange (Bi ≠ 0), **d** internal liquid flow and induced surrounding air flow solved simultaneously: Reproduced from Author's LabDataArchive; **a, b, c** produced by Kousaka and Kawamura [20] and **d** by Kawame [18]

section "Infinitely Long Cylinder with a Temperature Gradient Along its Axis". The Marangoni number, Ma, is the major non-dimensional parameter that determines this threshold. The Marangoni number for a finite-length liquid bridge is defined as.

$$\mathrm{Ma} = \frac{\sigma_T \Delta T L}{\rho \nu \kappa}. \tag{1}$$

Here, ΔT is the temperature difference between top and bottom surfaces, σ_T, ρ, ν and κ are the liquid properties; those are, the temperature coefficient of surface tension, density, kinematic viscosity and the thermal diffusivity, respectively. The quantity L is a characteristic length and can be the height (H), the radius (R) or the diameter (D) and their Ma is expressed as Ma_H, Ma_R or Ma_D, respectively. The Ma number at the threshold is called the critical Marangoni number and often symbolled as "Mac".

Since the height of the liquid bridge is finite, the aspect ratio, $A = H/D$ or $\Gamma = H/R$ becomes an important additional condition to determine the Mac. Similarly to the canonical cases in the previous chapter "Thermocapillary Convection in an Infinite Liquid Layer and in an Infinite Liquid Column", the Mac depends on the heat transfer

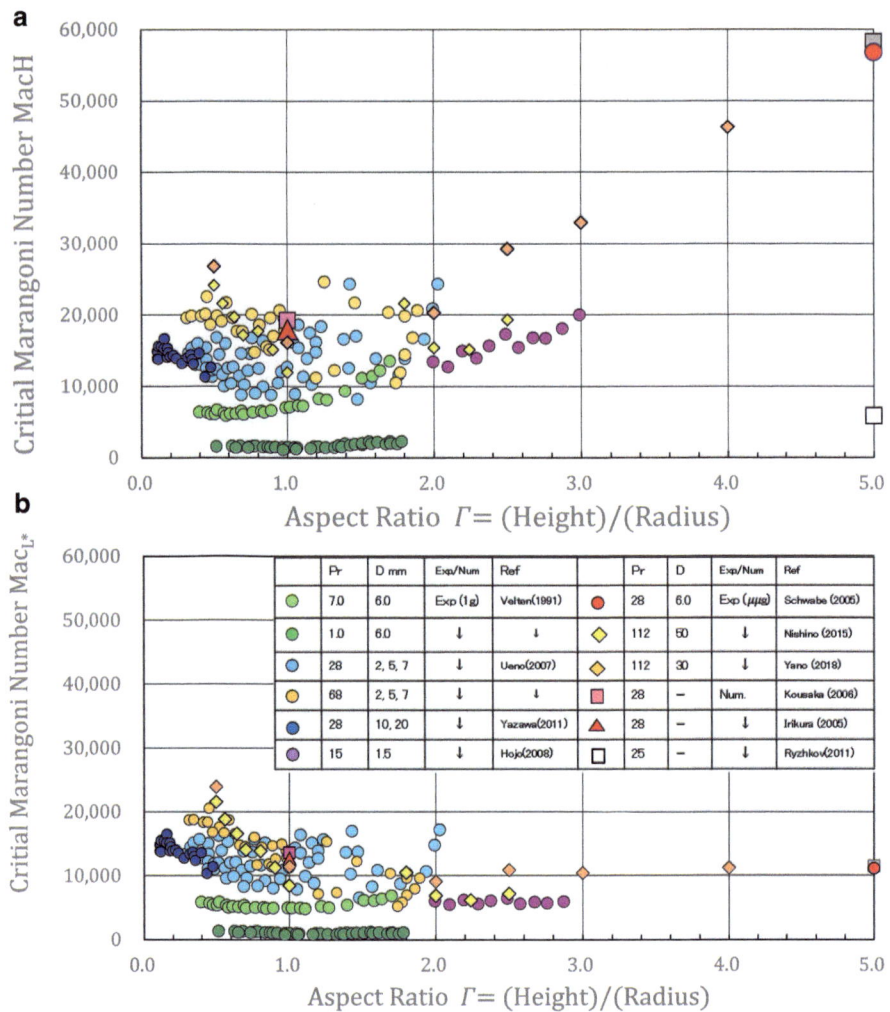

Fig. 14 a Summary of critical Ma numbers based on height H by experiment and numerical analyses (Legend: given in Fig. 14b). **b** Critical Ma numbers based on effective length L^*; inset: legend for Fig (**a**) and (**b**). References: Velten et al. [68], Ueno et al. [67], Yazawa [72], Hojo [15], Schwabe [57], Nishino et al. [32], Yano et al. [71], Kousaka and Kawamura [20], Irikura and Arakawa [17], Ryzhkov [38]

at liquid surface, which will be described in more detail in the next chapter "Effect of Heat Exchange, Control and Suppression of Thermocapillary Convection".

Since the critical Marangoni number, Mac, is one of the most interested issue in this field, we would like to overview the critical Marangoni number (Mac), obtained experimentally and numerically for liquid bridges with finite-lengths. To this end, Fig. 14a corrects the experimental results of Mac on the ground primarily from the present author's (H. K.) group. Some characteristic data are included from the Schwabe's group, too. The symbols are given in an inset of Fig. 14b. Vertical axis represents Ma_H, and the horizontal one is the aspect ratio $\Gamma = H/R$. Some numerical results on Mac by Irikura-Arakawa [17] are also included, where surrounding air motion was solved simultaneously.

Looking first at the results with larger aspect ratios, the experiment with the largest aspect ratio in this field is the small rocket experiment, the red circle, by Schwabe [57], as already mentioned in the previous Sect. 2 of this Chapter.

From our microgravity experiment in Kibo, the Mac obtained in a series with a medium Prandtl number $Pr = 112$ are represented here by the yellow and dark yellow diamonds, which will be detailed in chapter "Microgravity Experiments in Kibo Onboard the International Space Station". Despite the difference in Prandtl numbers, these points and the datum of Schwabe [57] align well, almost on a straight extension of one another.

An interest here lies in the comparison with the Mac for an infinitely long cylinder described in section "Thermal Convection in a Thin Infinite Liquid Layer and an Infinitely Long Liquid Cylinder with a Temperature Gradient Along the Surface". The Mac for an infinitely long cylinder was defined as $Ma_{Long} = \frac{\sigma_T \gamma R^2}{\mu \kappa}$, where $\gamma = \left|\frac{dT}{dz}\right|$. Although the surface temperature gradient of finite length liquid bridges is not linear (see Fig. 13c); nevertheless, if we approximate it linear as $\gamma \simeq \Delta T/H$, then we get the Mac in the present case as $Mac_H = Ma_{Long} \Gamma^2$. Accordingly, using the Ma_{Long} by Ryzhkov [38] for $Pr = 25$, $Bi = 1.0$, an obtained value at $\Gamma = 5.0$ becomes $Mac_H \simeq 5,800$, which is plotted with a white square at $\Gamma = 5.0$ in Fig. 14a for reference. This value is much smaller than the experimental one even for a finite length close to the Rayleigh limit (red circle), which indicates that the non-linearity of the surface temperature profile is always significant in the liquid bridge of a finite length.

This value is considerably lower than the experimental one by the small rocket experiment indicated by a red circle. The difference is due to that in a finite-length liquid bridge, the temperature gradient in a significant portion of surface is much smaller than the one linearized as $\Delta T/H$, because a large temperature gradient exists near both ends of the bridge, see Fig. 13c. To obtain a value close to the experiment (red circle), an empirical factor of 10.0 must be multiplied (grey square).

The aspect ratio Γ attainable in ground experiment is typically up to about 1.5 to 2. Beyond that, by reducing its diameter and with particular careful treatments. Hojo [14, 15] in our group achieved an aspect ratio up to approximately $\Gamma \simeq 3.0$ using n-decane as the test with thin rods of $D = 1.5$ mm. Those results are depicted between $\Gamma = 2.0 \sim 3.0$ by violet circles in Fig. 14a. The results align well with other results although their Macs are relatively low because of its smaller Prandtl number ($Pr = 15$).

We next turn our attention to the regime with medium to small aspect ratios. In this region, plenty of experimental data are available and they exhibit considerable scatter. This could be due to several reasons, including insufficient control of heat exchange with the ambient environment, surface contamination of the liquid, and sensitive shift in the mode number among these aspect ratios.

In terms of the Prandtl number's influence, a general trend can be observed where the Mac values increase with the increasing Pr number. This trend is shown clearly in Fig. 24 in chapter "Microgravity Experiments in Kibo Onboard the International Space Station" of this book, in which Mac values are collected from experiments, numerical simulations and linear stability analyses. The general increase of Mac with Pr is obvious; however, a closer inspection of Fig. 24 in chapter "Microgravity Experiments in Kibo onboard the International Space Station" reveals that the increase relative to Pr is not monotonous. This is likely due to the transition in the mode number and/or the type of waves, as also indicated in section "Linear Stability Analysis of Thermocapillary Convection in Liquid Layer and Bridge" of this book by Fujimura K.

Next, we examine below the dependency of the critical Marangoni number (Mac) on the decreasing aspect ratio. Referring to Fig. 14a, there appears to be rather large scattering among the Mac's compared. When considering the reduction of the aspect ratio $\Gamma = H/R$, it is common to decrease the height H while maintaining radius R constant. Alternatively, one may increase R while keeping H constant. In the latter scenario, the flow fields in the central region of the bridge must have a minimal impact on the stability of the surface flow. Therefore, it is conceivable that the critical state at a small aspect ratio will primarily be influenced by the channel width H and the temperature difference ΔT, regardless of changes in R. This concept was already presented in an early time by Preisser et al. [34]. This scenario suggests that Ma_H based on the height H will asymptotically converge to a constant value, as Γ decreases. Indeed, this trend can be seen for several data sets in Fig. 14a, but an increasing trend in the Mac is also observed with decreasing Γ in cases such as the current microgravity experiments; indicating a discrepancy that remains to be elucidated.

Finally in relation to the expression for the Mac, we would propose a practical empirical formula. As previously discussed in relation to the limit of a small aspect ratio, it can be understood that the smaller of these two lengths, H and R, must play a more dominant role in deciding the critical condition.

Therefore, we may devise an effective characteristic length L^* as.

$$\frac{1}{L^{*2}} = \frac{1}{H^2} + \frac{1}{R^2}. \tag{2}$$

With this definition, L^* is closer to the smaller of H and R; that is $L^* \simeq H$ for small Γ, while $L^* \simeq R$ for large one.

The Marangoni number Mac_{L*}, with use of this L^* as the characteristic length, is plotted against the aspect ratio $\Gamma = H/R$ in Fig. 14b. As expected, Mac_{L*} stays approximately constant with respect to Γ within this range of the aspect ratio. One should however reminded that the Mac must tend to Ma_{Long} as Γ further increases, although hypothetical. The dependence upon the Pr number remains still.

3.3 Evolution of the Flow Regime and Surface Temperature Fluctuation with Increase of Marangoni Number

The variations of the flow regime and surface temperature fluctuation with an increase in the Marangoni number are presented in Fig. 15a, b, c,..h, which were first published in Ueno et al. [67]. In these figures, depicted are the top view of the liquid bridge with an aspect ratio ($A = H/D \simeq 0.32$, with $D = 5$ mm), time series of surface temperature variation, its Fourier spectrum, and reconstructed PPS (pseudo-phase space) for various Ma_H. A constant delay time for the PPS reconstruction is 0.1 s in Fig. 15b to f while 0.06 s in Fig. 15g and h.

First, the case of the smallest temperature difference is depicted in Fig. 15a. The fluctuations in the surface temperature of the liquid bridge are found very small. The flow observed in the top view of the liquid bridge is steady, with particle motion primarily restricted to the radial direction. An interesting observation is that the fine particles introduced for flow visualization, despite initially being dispersed uniformly, start to separate into particle-rich and deficient zones.

Observation indicates that the fluid and particles in the central region remain almost stagnant and is 'invaded' by the anisotropic deformation of the surrounding dynamic region. As a result, particles in the central region are gradually drawn into the particle-rich zone and tend to be trapped there, leading to a more pronounced disparity in particle concentration between the two zones.

Moving on to Figs. (b) and (c), oscillations in the surface temperature become apparent, with distinct peaks at the fundamental period and its multiples observed in the spectrum. In Fig. (b), the shape of the particle-deficient zone is rather complicated and repeats the deformation almost at the same location. Thus it is called the "Standing wave". Figure. (c), on the other hand, exhibits a clear triangular shape of the particle-deficient inner zone, which rotates in one azimuthal direction without changing its shape and is called the "Travelling wave". The mode number of these cases are $m = 3$. The mode number m can be given as $m = \pi D/\lambda_{HTW}$, where λ_{HTW} is the cirumferencial wave length of HTW. The reconstructed pseudo phase spaces (PPS) of Figs. (b) and (c) both show an almost circular shape albeit somewhat deformed.

Next, we would look at Figs. (d, e, f). The clear triangle observed in Fig. (c) once breaks down with increasing Ma, and the standing wave reappears in Fig. (d). Furthermore, as the Marangoni number is increased further, the flow transitions again into a traveling wave. The temperature waveform in Figs. (d, e, f) is slightly more complex compared to Figs. (b, c), but the periodicity is still maintained and the distinct peaks can be observed in the fluctuation spectrum. The reconstructed PPS in these Ma numbers generally remain circular, but their deformations become more pronounced.

A distinct feature in Fig. (f) is the aggregation of the fine particles into a closed loop. This topic will be treated later in the subsequent subsections.

In Figs. (g) and (h), as the temperature difference is increased further, clear peaks in the time-varying temperature spectrum diminishes and also the distinct loop shapes

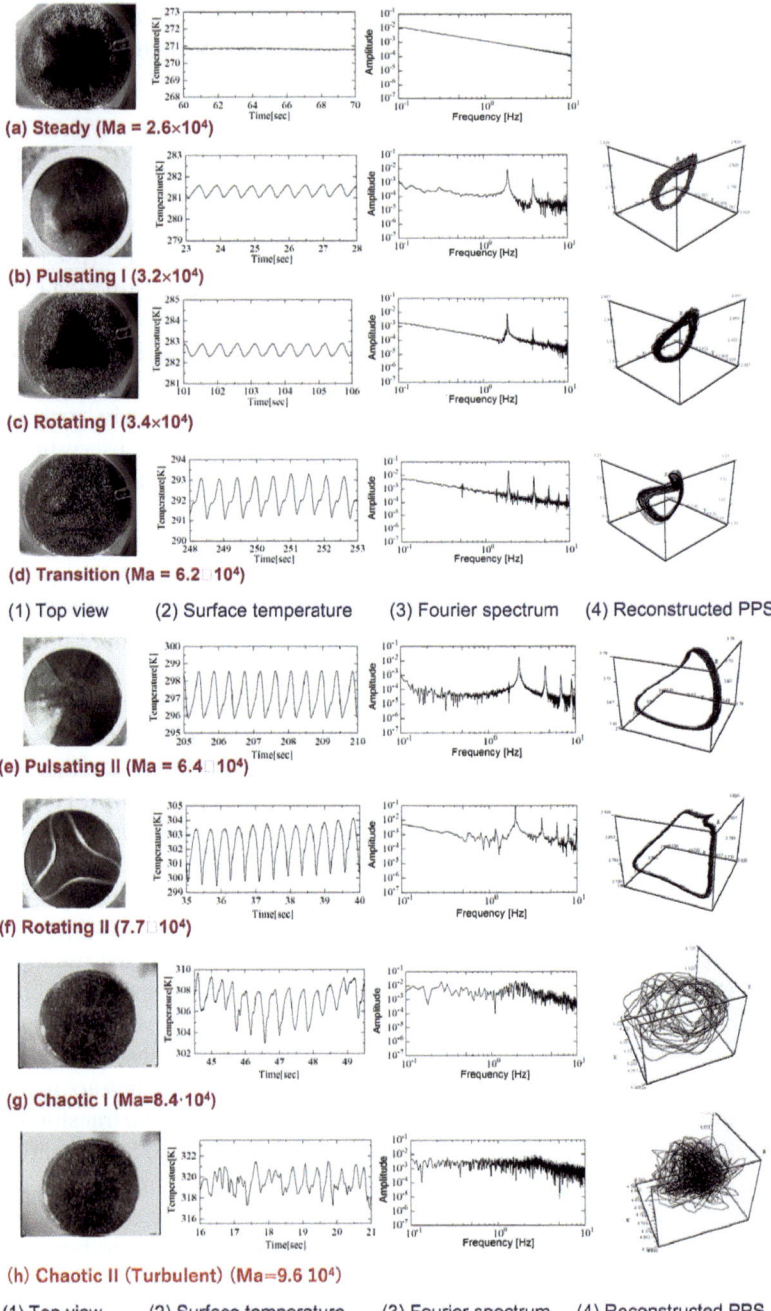

Fig. 15 a, b, c, d Evolution of the flow regime and surface temperature fluctuation with increase of Marangoni number Ma_{CH}. Reused from Ueno et al. [67] with permission of AIP. **e, f, g, h** ibid

in the reconstructed PPS have disappeared, indicating that the periodic nature is going to be lost. The flow field clearly transitions into chaotic state and the flow might be close to the turbulence. When considering the definition of turbulence, the author recalls a sentence stated in the introduction of a well-known textbook on turbulence by Tannekes and Lumley [66]: "It is very difficult to give a precise definition of turbulence. All one can do is list some of the characteristics of turbulent flows." The mentioned characteristics of the turbulence are such as irregularity, (high) diffusivity, fluctuation of vorticity and dissipation. With these in mind, when examining Figs. (g) and (h), one may notice at least the latter (h) exhibits these characteristics sufficiently well, indicating that the flow of Fig. (h) is close to a state of turbulence.

3.4 Particle Accumulation Structure (PAS) and Azimuthal Mode Numbers

Let us now revisit the distinctive feature observed in Fig. 15f, where particles are evidently gathered into a closed loop resembling a three-blade windmill. This phenomenon was first identified by Schwabe's group when observing the liquid bridge from its lateral side, employing side illumination with a sheet of light. They introduced the term Particle Accumulation Structure (PAS) for this phenomenon, Schwabe et al. [55].

Several years later, the present author's group also encountered this phenomenon independently in their laboratory experiments, Kawamura and Harada [19]. Subsequently, we elucidated the three-dimensional structure of PAS by using the transparent top rod [64].

A sketch of its three-dimensional shape is given in Fig. 16. The black closed curve loop represents a string of the PAS. The blue toroidal shape indicates the particle-rich region, which we named the "toroidal core". Since the PAS string depicted with the black line wraps around this toroidal core, the shape of PAS also rotates in the same direction as the blue arrow. Although their rotation direction was random for each individual experimental run, that of this case is counterclockwise as indicated with the blue arrow. Interestingly, the "particles" on the black string are traveling in the opposite direction as indicated by the black arrow. The motion of the particles in the toroidal core is essentially radial because they are driven downwards by the thermocapillary convection over the surface, but they are also slowly rotating azimuthally in the same direction as the envelope of the toroidal core.

Next, we would pay attention to the relationship between the azimuthal mode number m and the aspect ratio A or Γ. Figure 17 depicts top views of the PASs for a wide range of aspect ratios $A = H/D$ except for m = 1. We can find that the mode number increases as the aspect ratio decreases. In the next figure of Fig. 18, a product mA is plotted against the aspect ratio A. In determining the mode number, the top view of PAS and the particle deficient zone offer the straightforward approach.

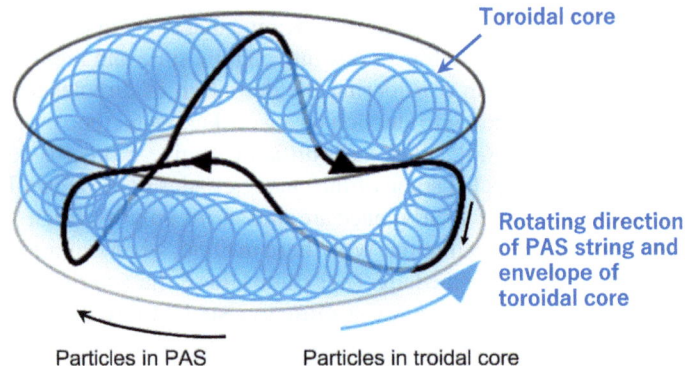

Fig. 16 Schematic illustration of the PAS (black line) and toroidal core (blue area) with m = 3. Reproduced from Author's LabDataArchive: Produced by Tanaka & Kawamura

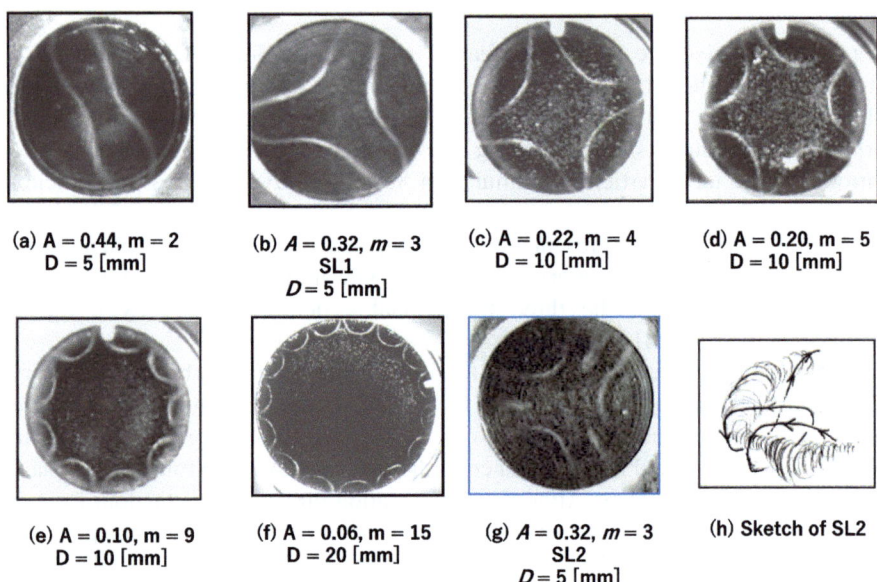

(a) A = 0.44, m = 2
D = 5 [mm]

(b) A = 0.32, m = 3
SL1
D = 5 [mm]

(c) A = 0.22, m = 4
D = 10 [mm]

(d) A = 0.20, m = 5
D = 10 [mm]

(e) A = 0.10, m = 9
D = 10 [mm]

(f) A = 0.06, m = 15
D = 20 [mm]

(g) A = 0.32, m = 3
SL2
D = 5 [mm]

(h) Sketch of SL2

Fig. 17 Top view of PAS with various mode numbers m. Reproduced from Author's LabDataArchive, Produced by Yazawa and Tanaka [73]

From these figures, one may notice that the product *mA* shows an approximately constant value relative to *A* as seen in Fig. 18. This constancy was first pointed out by Preisser et al. [34] through their experiments. They reported that the constant value was approximately $mA \simeq 1.1$. Figure 18 indicates that this product indeed falls within a range around 1.0 and decreases slightly down to $mA \simeq 0.9$ for $A \simeq 0.05$.

Preisser et al. [34] provided a discussion on a reason of this constancy. For simplicity, we will consider only the case of small aspect ratios of $H < R$. In this case,

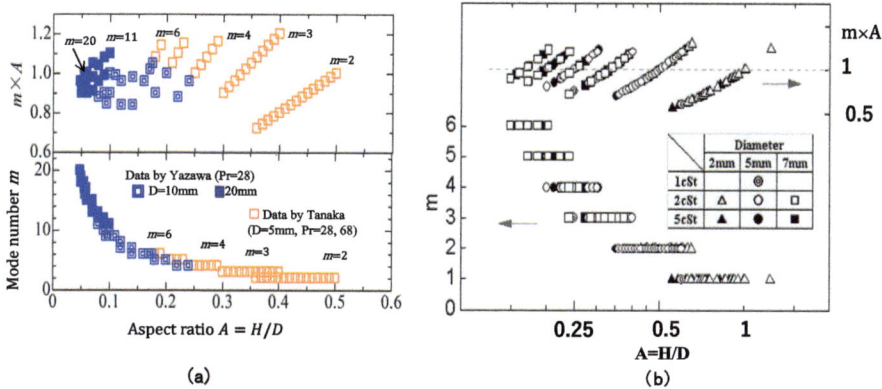

Fig. 18 a, b Emergence range of HTW; Relation between the product mA and the aspect ratio $A = H/D$. **a** Reproduced from Author's LabDataArchive,. Presented by Yazawa and Kawamura [73]. **b** Reprinted from Ueno et al. [67] with permission of AIP

as discussed already, the flow field near the surface can be expected to be dominated mainly by the height H. Then the azimuthal wavelength of the HTW, λ_{HTW}, will be able to be approximated as $\lambda_{HTW} = \xi H$, where ξ is a constant. Since $m\lambda_{HTW} = \pi D$, we find $m(H/D) = \pi/\xi$, which indicates the approximate constancy of the product mA. Since the product $m(H/D)$ was experimentally found about unity, the value of the assumed constant must be $\xi \simeq \pi$. According to Fig. 18a, b, this constancy holds fairly well down to $m = 2$. The case of $m = 1$ is rather peculiar; thus it will be discussed later separately.

Next, we would like to pay attention to the skipped figure of Fig. 17g. This is the case of $m = 3$; but one may notice that an additional turn is observed near the tips of the blades as sketched in Fig. 17h. In the author's group, the waveform of Figs. (g) and (h) has been referred to SL2 (Spiral Loop 2), while the basic case seen in Fig. 17b as SL1 (Spiral Loop 1). The SL2 occurs usually at a higher Marangoni number compared to SL1 for $m \geq 2$. The classification of the spiralling of PAS string will be discussed later again in Sect. 3.6.

At the last of this subsection on the mode number of PAS, we will describe a PAS in which two mode numbers appear to coexist in an experiment with a rod of a large-diameter ($D = 20$ mm). According to Fig. 19a by 72], twelve semicircles (m = 12) are visible on the outermost ring. In addition, strings of two connected semicircles can be seen inside the outermost ring. The periodicity of the connected pair of inner circle is $m = 6$. Accordingly, we have named this phenomenon a "Mixed-mode PAS." The spectral intensities of the temperature fluctuation is shown in Fig. 19b. In addition to the fundamental frequency f1, its subharmonics f1/2 arises at half the frequency of f1.

Detailed observations by Yazawa [72] revealed that the inner and outer semicircles are not independent of each other but are connected as illustrated schematically in

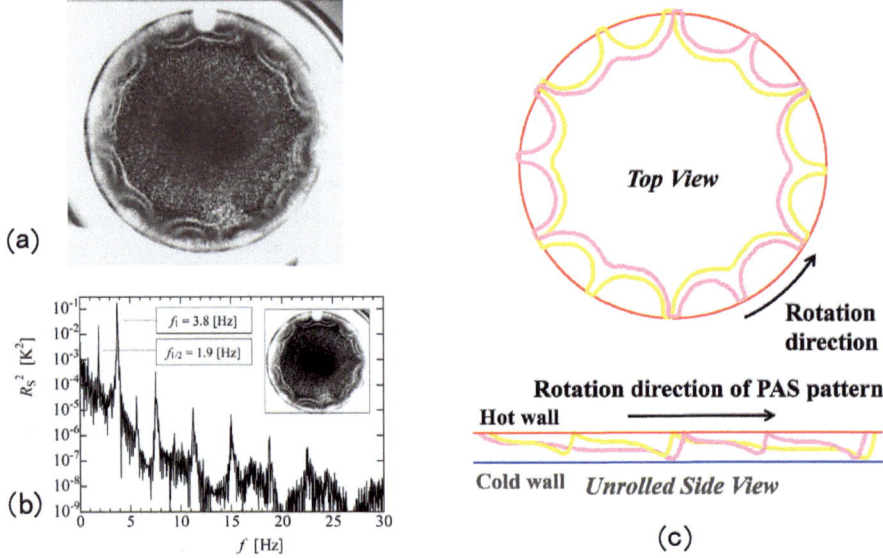

Fig. 19 Mixed mode PAS with fundamental mode number of m = 12. **a** is Top view, **b** and **d** are schematic illustrations, **c** is spectral intensities of the temperature fluctuation. Reproduced from Author's LabDataArchive, Produced by Yazawa

more detail in Fig. 19c. Referring to the purple and yellow lines, these two lines are identical and staggered with each other.

3.5 Particle Accumulation Structure (PAS) in a Microgravity Experiment

In the early phases of PAS research, it was assumed that gravity might exert a rather significant influence on the occurrence of the PAS. Indeed, ground-based experiments examining the heating position (top versus bottom) of the liquid bridge showed that PAS readily occurred if heated from the top and was rather less frequent when heated from the bottom, suggesting an effect of gravity. Therefore, under Prof. Schwabe's leadership and collaboration, we conducted an experiment on emergence and feature of the PAS under microgravity using the sounding rocket MAXUS-6. This rocket offered a microgravity duration of approximately 12 min, which is very long for rockets of this kind. The conception, preparation, course of the flight experiment, and the results obtained are already reported in Schwabe et al. [58]. The key points will be described hereafter.

For the experiments, we adopted a system that had been experienced in ground experiments, where the diameter of the liquid bridge was set at 6 mm, and its length was adjustable between the aspect ratio of $\Gamma = 0.5$ and 1.0, where $\Gamma = H/R$.

The first candidate of the test fluid was the 2 cSt silicone oil. However, during the last stage of preparatory tests, it was found rather hard to form a stable liquid bridge from silicone oil without the use of FC-antiwetting on the sapphire rod. So it was decided to use n-decane (Pr = 15) as test fluid which gives a more stable liquid bridge because of its higher surface tension compared to that of silicone oil. Only with the n-decane it was possible to form a stable liquid bridge without using the antiwetting FC-paint.

As for the tracer particles, diamond particles with an average diameter of about 15 μ were employed, because with this size of particles we got the shortest formation time of PAS in the preparatory experiments.

A distinctive feature of the experiments at ESRANGE is that the assembly and also the operation of the apparatus are both conducted by the researchers themselves. This, however, can be quite demanding, as may be seen from our experiences, which will be described in some detail to convey to readers an impression of such work under highest stress. That is, in addition to the urgent change of the test fluid mentioned above, we found that one of the most essential facilities, which was to mix the particles in the liquid before and during the rocket launch to prevent their sedimentation, has not been prepared when we arrived at the launch site. Consequently, we had to hastily develop an emergency procedure and tested it only during two nights before launch. We placed particles on the lower end rod to entrap them into the liquid bridge with the injected fluid, taking a potential risk of particle loss during launch of the rocket. This emergency procedure successfully functioned to visualize the string of the PAS, although, as depicted in Fig. 20a, the dispersion of particles into the liquid bridge was insufficient, with a significant amount remaining on the lower end rod.[2]

The rocket was launched from ESRANGE Space Center in northern Sweden in November 2004. The launch of the rocket and our experiment were performed successfully. Schwabe took charge of operating the apparatus by himself. (Schwabe did the injection and retraction of the test liquid to establish a liquid bridge with an aimed size and shape.) Shiho Tanaka (from the Science University of Tokyo) was in charge guessing the flow field for its readiness to form PAS, and for installing the correct temperature difference, if needed. After the successful emergence of PAS, we intentionally destroyed it a few times by retraction/ injection of liquid. We were thus able to repeat the experiment of PAS-formation a few times during the limited microgravity time; that is, a rare possibility in such short-time rocket experiments.

A characteristic of these rocket experiments was that the experimental apparatus can be retrieved after landing by a helicopter and returned to the researchers already that afternoon. This was enabled because the apparatus was planned to fall onto a plain field nearby, which changed from a wetland into an ice field during the wintertime, making retrieval feasible.

---

[2] Author of this chapter, D. Schwabe, would like to express his sincere appreciation to his coworker, A. Mizev from ICMM in Perm, Russia, for his essential contributions to the urgent change of the test fluid in the MAXUS-6 experiment, as well as to the preparation and testing of the particle mixing device under such pressing circumstances.

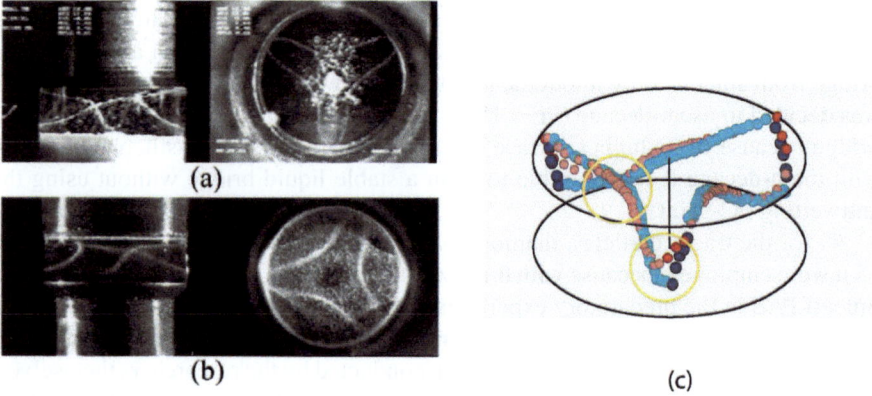

Fig. 20 PAS with m = 3 under micro and normal gravities: **a** under microgravity **b** under normal gravity. **c** illustrates an overhead view. Reused from (**a, b**) Schwabe et al. [58] Permission of Springer Nature, **c** from Tanaka [65] with permission

In the experiment, we first focused on the PAS of $m = 3$, which is most reliably observed in the on-ground experiments. Using the favored conditions known from the 1 g (the normal gravity) experiments, we were able to generate PAS with $m = 3$ in microgravity in a rather short time after the launch. That is, the PAS emerged under nearly identical conditions to those under the 1 g. A photograph of the emerged PAS is presented in Fig. 20a, where it is compared with the one under 1 g condition, Fig. 20b. The data suggests that the $m = 3$ PAS exhibits similar shape in µg and 1 g environments. Some differences in shape were noted. That is, in the top view (on the right), the PAS-string under µg extends closer to the center of the liquid bridge. Meanwhile, in the side view (on the left), the bottom-most protrusion of the PAS-string in µg is closer to the bottom surface than that in 1 g. These variances must be due to the effects of buoyancy in the on-ground experiment.

The shapes of the PAS in 1 g and µg are compared in Fig. 20c, in which the red and blue dots indicate the measured points on the PAS in 1 g and µg, respectively. Both of them show close agreement, indicating that the influence of gravity is small at this aspect ratio. Closer observation indicates, however, that some differences can be seen in the locations of the two yellow circles in (c). Specifically, in the yellow circle at the lower center of the figure, particles in the 1 g reach less closer to the bottom compared to the case of the micro-g. Since the system was heated from the top in the 1 g-experiment, the buoyancy acted in an upward direction, which tended to counteract the thermocapillary force and thus the thermocapillary effect was compensated.

After successfully capturing the PAS of $m = 3$, the aspect ratio was increased to produce $m = 2$. The search was continued with use of optimal conditions for $m = 2$; however, the standing wave predominated, hindering the transition to a travelling wave, which is the necessary condition for the emergence of PAS. Finally, as the microgravity time neared its end, the PAS of $m = 2$ was observed at last, albeit transitionally, during the procedure of returning to $m = 3$ by retracting liquid from the liquid bridge in line with the predetermined experimental plan.

While there were some unexpected developments, the experiment with use of the sounding rocket was performed successfully with emergence and observation of PAS under the microgravity. For detailed information on the experiment and a more in-depth analysis, refer to Schwabe et al. [58].

3.6 Particle Accumulation Structure (PAS) with m = 1

Up to here, we have been discussed the PAS with $m \geq 2$. We now turn our attention to PAS with mode number $m = 1$. In this case, we needed a liquid bridge of $A \gtrsim 1.0$ for the HTW of $m = 1$ to emerge. In the author's group, to form a liquid bridge of $A \gtrsim 1.0$ on ground we employed a thin liquid column of $D = 1.5$ mm. In case of this small dimension, a larger temperature difference was required to obtain an aimed Marangoni number, which increased liquid evaporation and thus caused a limitation on experimental time. Additionally, the flow velocity was increased and it made observation more challenging.

The shape of the PAS with $m = 1$ expected from photos of Fig. 17 would be a circle rotating eccentrically, because a single temperature wave form will propagate over the surface so that the circumferential symmetry will be lost. Indeed, Sasaki [42] found experimentally a deformed circular PAS for $m = 1$, as seen in top view of Fig. 21a, where white dashed lines are enhancement of the PAS patterns, because they were rather obscure due to the rapid movement of particles.

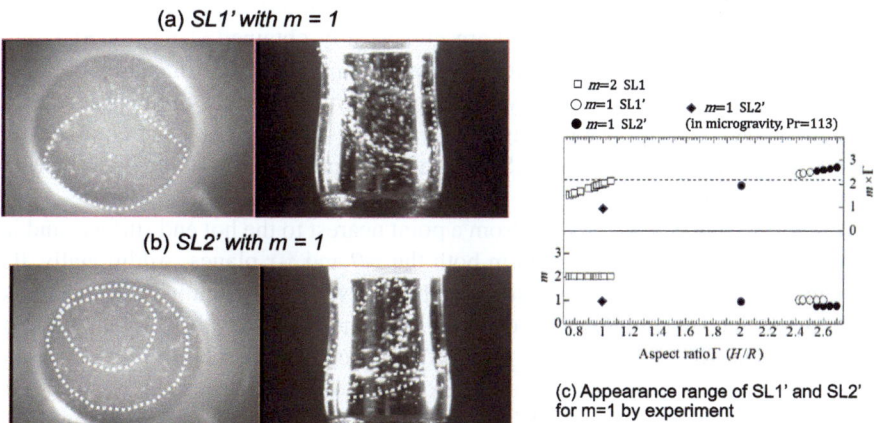

Fig. 21 Top and side views of PAS with m = 1, ($\Gamma = 2.55$, Pr = 28): **a** SL1' and **b** SL2'. **c** Plot of m × Γ versus Γ in normal g by Sasaki [42], $\Gamma = 1$ (in μg) with solid diamond by Sakata et al. [40]. Reproduced from Author's LabDataArchive, Produced by Sasaki [42]

In addition to this fundamental shape, however, a more complexed shape of PAS in HTW of $m = 1$ was observed, as shown in Fig. 21b, Sasaki et al. [41]. Between these two, (a) corresponds to the expected shape, as it forms a single loop during one cycle of PAS string, aligning with the conventional designation of SL1 for $m \geq 2$. On the other hand, (b) forms an extra loop during one cycle, thus we have referred it as SL2 following those for $m \geq 2$.

This type of PAS, called SL2 of $m = 1$, was observed also in our microgravity experiment in Kibo with a smaller aspect ratio of $\Gamma = 1.0$, which is described in section "Effect of Aspect Ratio". It was later analyzed by Ueno's group in detail [40].

Lappa's group, Capobianchi and Lappa [4], obtained a PAS quite similar to Fig. 21b through numerical calculation. (As for the numerical method on the PAS, a description will be given in the following subsection.) Their computational results are in good agreement with the characteristics of the SL2 of $m = 1$ shown in Fig. 21b. They pointed out that, while in the PAS with a larger mode number such as $m \geq 2$, the PAS string wraps around the toroidal core (see Fig. 17h), this case is distinct in that it rotates around the central axis of the liquid column, leading them to propose an alternative naming. We understand their point, and since the naming of the SL1 and SL2 have been to represent the number of turns, we would modify the notation for $m = 1$ here as SL1' and SL2' to express that the string turns around the central axis once and twice, respectively.

Sasaki [42] recognized another unique behavior of SL2' with $m = 1$; that is, in the conventional cases of SL2 with $m \geq 2$, a particle appears twice on the liquid column's surface during one period of HTW. However, in case of SL2' with $m = 1$, it requires about two periods of HTW during a particle appear once over the surface and complete its single cycle along the string.

This aspect was later confirmed in more detail with use of numerical method by Hojo [15]. Figure 22(1-a) and (2-a) are numerically obtained shapes of SL1' and SL2' with $m = 1$. Their shapes differ somewhat from those experimentally obtained ones (Fig. 21a, b). This is probably because of their differences in the aspect ratio, Pr number and also in gravity. Figure 22(1-b)–(1-f) and (2-b)-(2-f) compare the travelling of particles along the PAS string and the propagation of the hydrothermal wave (HTW) over the surface.

A reference particle is released from a point nearest to the hot end surface, and its subsequent trajectories are plotted in both the z-θ and z-r planes. Additionally, the propagation of the HTW is illustrated, with its advancing azimuthal angle denoted by $\Delta\theta_{\text{HTW}}$. Upon completing one cycle of PAS, the HTW advances by an angle of $\Delta\theta_{\text{HTW}} = 2.8\pi$ for SL1' and 3.8π for SL2', respectively, as seen in Figs. (1-f) and (2-f). Accordingly, for a particle to close one cycle of PAS string, SL1' undergoes approximately one period of HTW, whereas SL2' requires about two periods.

As for the representation of various winding shapes of PAS, Kuhlmann's group introduced several types of more detailed representations. One of them is a notation of $L_{m,n}^{l}$, where L means the line of PAS, the subscript m denotes the azimuthal mode number of the HTW, n the winding numbers of the PAS string around the vertical axis, and the superscript l is the one around the basic-flow vortex [36]. The winding

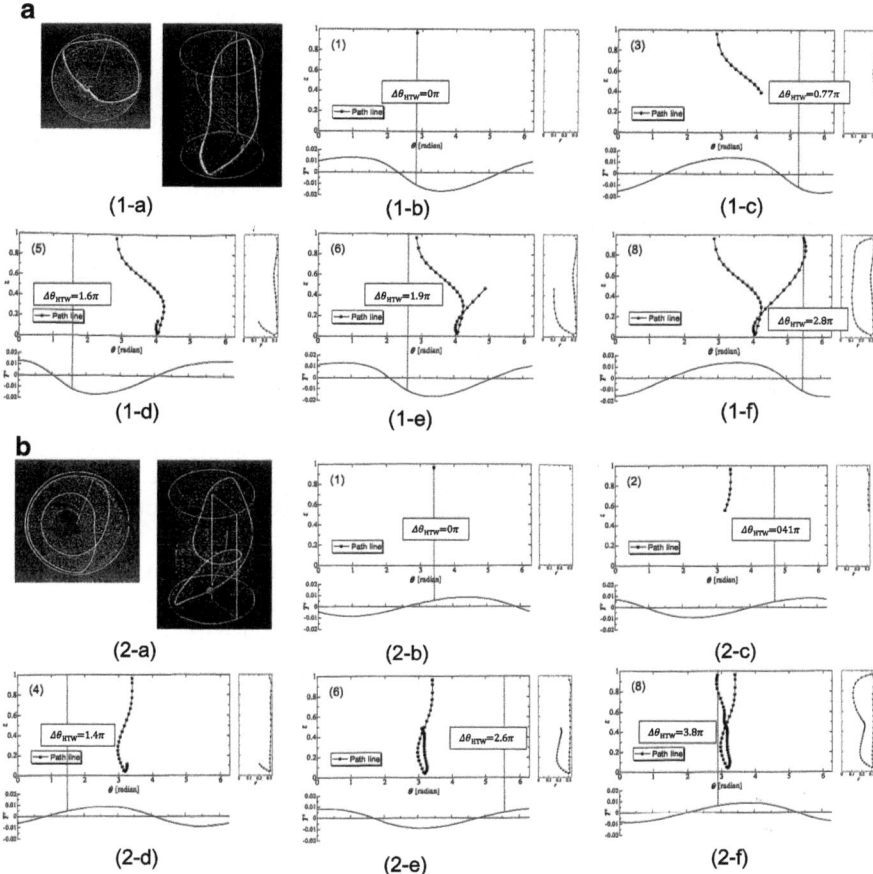

Fig. 22 Travelling of a particle along PAS and propagation of HTW in case of $m = 1$. Panel (**a**): SL1', Ma = 9,000, $\varGamma = 3.0$, Pr = 4.38), (**b**): SL2' (Ma = 7,000, $\varGamma = 3.0$, Pr = 4.38) **a**: Top and overhead views of PAS string (Note: both shots not synchronized) **b–f**: Relation between travelling of a particle and propagation of HTW. Reproduced from Hojo [15] with permission

number means the number of turns until the string closes on itself. Note that the position and the letters of suffixes here are modified from the original proposal. According to this notation, the SL1 and SL2 of PAS with $m = 3$ depicted by Fig. 17b and g are $L_{3,1}^3$ and $L_{3,1}^6$, respectively. As for $m = 1$, SL1' and SL2' in Figs. 21a and b, can be expressed respectively as $L_{1,1}^1$ and $L_{1,2}^1$.

3.7 Numerical Analyses and Formation of Particle Accumulation Structure (PAS)

When the present author (H.K.) started this series of research on the thermocapillary flows, his laboratory was conducting numerical analysis of the turbulent flow through the method known as Direct Numerical Simulation (DNS). Therefore, this method was applied to the field of thermocapillary flow also. The computational method used was the standard finite difference method, where computational meshes were established three dimensionally within a liquid bridge and also in surrounding air, if necessary.

In this type of thermocapillary flow, a significant velocity gradient occurs in the radial direction near the fluid surface. Therefore, it is crucial to create an unevenly spaced mesh with finely placed mesh points near the surface of the liquid bridge. An example of the mesh produced is shown in Fig. 23b. This mesh number of 50 thousands was rather moderate. We employed ten times or even more larger ones, if necessary. The calculation method for the fluid flow and the particle motions followed common approaches thus they are not repeated here.

First, we numerically realized thermocapillary flow with an aimed Marangoni number. Then, particles of a size equivalent to those in actual experiments were placed uniformly within the system. The flow velocity at each particle's position was determined by interpolation from the surrounding calculation points. We advanced the particles explicitly using the interpolated velocity and a time step Δt, i.e., the so-called one-way method.

At the beginning of our simulation, we encountered a peculiar phenomenon, that is, the particles dispersed in the liquid bridge vanished rapidly from the liquid region. We immediately noticed that this was because, while continuity equations enforced conservation for the fluid, such conservation principle was not applied to the particles. Accordingly, particles near the liquid surface travelled beyond the system boundary and be lost from the calculation domain. Based on this experience, some adjustments were implemented when particle's position after Δt exceeded the boundaries, as

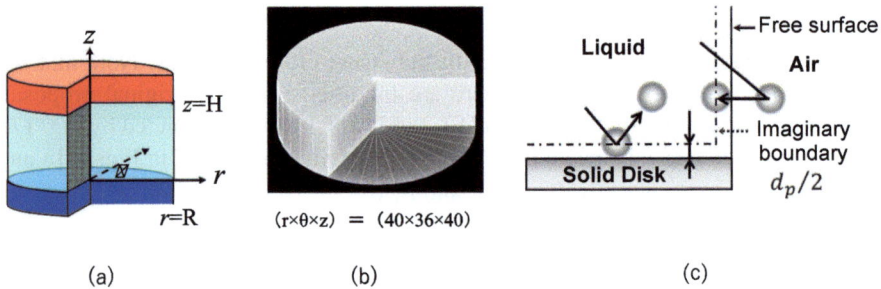

Fig. 23 Numerical analysis of the PAS. **a** co-ordinate system, **b** an example of calculation mesh, **c** treatment of particles near the free surface, Reproduced from Author's LabDataArchive, **a** Labo's own work, **b** produced by Takatsuka M & **c** by Seki T

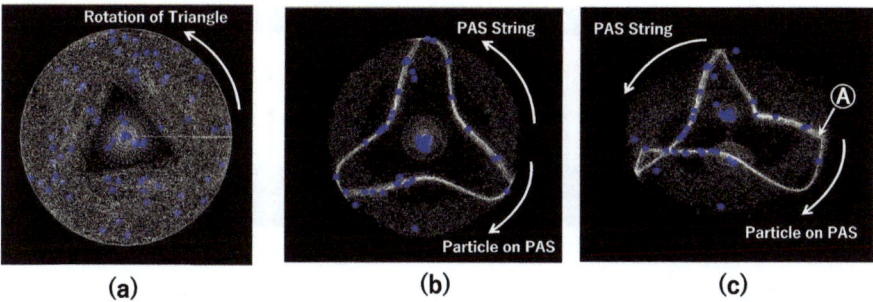

Fig. 24 Numerically reproduced particle patterns. White small dots are numerical particles and blue large ones are position markers. **a** Particle deficient region with $m = 3$, **b** Top view of PAS ($m = 3$), **c** Overhead view. Reproduced from Author's LabDataArchive, Produced by Takatsuka [62]

depicted in Fig. 23c. In case of the liquid surface, we simply relocated the particle center inwards by an amount of its radius, while for the solid walls, we assumed the elastic reflection. However, we learned gradually that the key aspect was to bring the particles back into the calculation domain depending on its size and details of its procedure were not significant.

Examples of the results obtained from this numerical simulation are shown in Fig. 24 [62]. Figure 24a represents a top view of the particle deficient zone corresponding to Fig. 15c. Figure 24b represents the PAS seen in Fig. 15f. Figure 24c is its overhead view and reproduces well the sketch of Fig. 16. Thus we found the numerical results reproduced effectively the experimentally observed features such as the particle deficient zone and the shape of PAS string.

As for the particle motion, we introduced the so-called B.B.O. (Basset–Boussinesq–Oseen) equation neglecting the Basset term [11]. However, we were not able to find noticeable effects by the additional equation at least in terms of the overall shape of the PAS and of the particle behaviors, but we have still retained the equation in the following analyses.

Next, we will discuss the reasons why such Particle Aggregation Structures (PAS) occur. The key characteristics of PAS include, firstly, particles aggregation into a relatively thin, closed string, and secondly, the stable feature of PAS string amidst the comparatively chaotic movements of particles outside the PAS.

As for the aggregation, we may simplify for the aggregation process as occurring in two directions, circumferentially and radially.

First, let us discuss the circumferential aggregation. Figure 25 illustrates (a) velocity and temperature distributions and (b) profile of particles close to the surface (unrolled). As seen in Fig. (a), the hydrothermal wave emerges and the resultant velocity and temperature profiles travel from left to right in this example. The coldest regions arise close to the bottom surface and the fluid velocity vectors are directed towards the cold regions. If Figs (a) and (b) are compared, the positions of the cold temperature regions and the accumulated particles correspond well with each other.

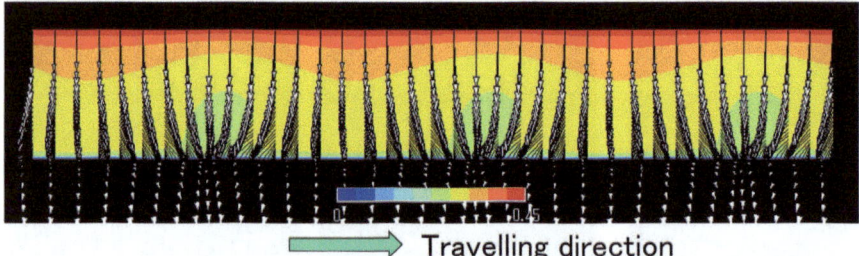

⟹ Travelling direction

(a) The temperature and velocity distributions on the free surface (unrolled) : Ma=9,000

(b) Particle profile near the free surface (unrolled).

Fig. 25 Velocity and temperature distributions and particle profile over and near the fluid surface (Pr = 4.4). Reproduced from Author's LabDataArchive. Produced by Takatsuka [63]

In addition, the cluster line of particles moves also from left to right at the same velocity as the temperature wave.

Note that the marked point Ⓐ corresponds to the Ⓐ in Fig. 24c. These observations indicate that the particles are circumferentially gathered, being accompanied by the fluid that is attracted to the colder temperature regions due to thermocapillarity. A similar discussion was presented by Schwabe et al. [59], emphasizing the importance of the gathering mechanism at the cold spot and the thin surface flow layer.

Then, we will move to the radial aggregation. Figure 26 depicts considerations on radial behavior of particles with several different sizes close to a fluid surface. Takatsuka [62, 63], Seki [60].

For simplicity, the flow is assumed to be *two-dimensional*, and its velocity field is depicted in Fig. 26b, where the upper-right corner (red elliptical circle) of Fig. (b) is enlarged. In the vicinity of the fluid/air interface, the thermocapillary effect accelerates the surface flow downwards causing a significant velocity gradient in the radial direction. Consequently, the streamlines are closely spaced there, which plays a significant role for aggregation of the particles.

Consider three streamlines (A → A', B → B', C → C') where A → A' represents the streamline that reaches closest to the surface, and C → C' the one that remains deeper in the fluid. Now, a large particle (green) and a small one (purple) are both released from the same point A. It is reasonable to assume that these particles remain within the fluid due to surface tension not protruding out of the fluid surface.

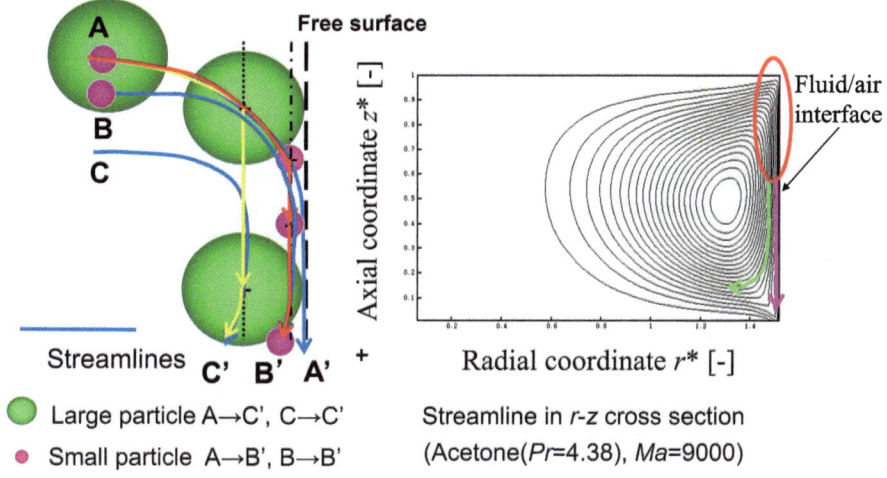

(a: Left) Schematic enlargement of encircled top-right corner of (b: Right)

Fig. 26 Particle movements near the liquid surface; particles are forced to shift to different streamlines depending on their size. Reproduced from Author's LabDataArchive, Produced by Seki [60]

In this scenario, particles with finite sizes cannot consistently follow their original streamline A-A'; instead, they must transition to different lines within the fluid depending on their own size. Specifically, the larger particle (green) shifts to the C–C' path, while the smaller one (purple) shifts to the B-B'. Consequently, particles accumulate along specific paths depending upon their own size, if their original streamlines approach closer to the surface than their radius.

A similar discussion was made by Kuhlmann's group [13], and was named the "finite-size-tracer-model". They applied it to their own numerical simulation and reproduced the PAS with $m = 3$ and subsequently various types of PAS, too.

Subsequently, this group applied this method with certain modifications to conduct a large number of calculations, discovering numerous instances of similar closed streamlines, which they term "Coherent particulate structure" e.g., Muldoon and Kuhlmann [31]. They have also examined the stability of this structure, a topic that will be discussed later in this section.

In addition to the above discussions, it is crucial that, for the PAS to retain its cyclic string stable, the particles injected into the liquid bridge from the bottom region must return at a proper point near the top surface without being dispersed inside of the liquid column.

Several research groups collaborated in the JEREMI-Project on this subject, which is overviewed in Kuhlmann, et al. [21]. With respect to the mechanism of the formation of the PAS, Shevtsova's group has been attempting an approach based on a concept of "phase-locking" between particles and hydrothermal waves. They say that "synchronization due to phase locking is ubiquitous in nature. The present modeling suggests that PAS formation in thermocapillary flows is another instance of this general phenomenon" in Pushkin et al. [35]. Further progress is found in Melnikov et al. [27].

Kuhlmann's group, on the other hand, claims that the stable structure of the PAS can be attributed to the mathematical theory known as the "KAM (Kolmogorov-Arnold-Moser) theorem". The KAM theorem is a fundamental mathematical concept, indicating that there exist quasi-periodic orbits which can remain stable in a chaotic environment provided that the perturbations are kept within certain limits. The theory itself addresses advanced mathematical concepts, which are beyond the scope of this book. Those who are interested may refer to an introductory monograph: 'The KAM Story' by Dumas [9].

If we examine the PAS with these characteristics in mind, we find that the PAS loop is periodic due to its closed nature; and additionally, the PAS remains stable within a certain range of Ma numbers, while it collapses as the flow becomes more chaotic. These observations align well with the characteristics indicated by the KAM theorem.

In addition, the stable structure described by the KAM theorem possesses distinctive features. Specifically, the stable domain is encircled by clusters of streamlines, and these clustered streamlines form closed loops by themselves. This closed hollow loop structure is commonly referred to as KAM torus or KAM tori.

With this perspective, the Kuhlmann's group has conducted numerous numerical analyses. For example, Mukin and Kuhlmann [30], and Romano and Kuhlmann [36] mainly focused on the PAS of $m = 3$, conducting numerical calculations and successfully reproducing the $m = 3$ of SL1 and SL2 numerically. Later, Barmak et al. [1] analyzed the PAS with $m = 1$ in a liquid column with an aspect ratio of $\Gamma = H/R = 1.0$. Furthermore, they identified various types of KAM tori with a larger number of spirals. Effects of several parameters such as the density ratio of particle and fluid, size of particle and the Reynolds number of the thermocapillary flow. Finally, a comprehensive overview on recent advances in research of PAS is given in this book by Ueno I. Those interested in this issue are recommended to visit section "Experimental Study on Coherent Structures by Particles Suspended in Half-Zone Thermocapillary Liquid Bridges" of this book.

At the end of these sections concerning the PAS, we would like to pay attention to the interest and research in similar phenomena in related fields. During the time when we started exploring and working with PAS (early 2000's), there were growing interests and developments in the field of microfluidics.

Referring to Davis et al. [7], the principle behind it is illustrated in Fig. 27. In this case, an array of microscale posts is arranged in a regular pattern, and small and large particles are introduced into the system as examples. Around the posts, the flow wraps around and accelerates locally, causing the spacing between streamlines to decrease. Consequently, larger particles are unable to stay on the streamline they initially followed, leading them to cross over to an outer streamline according to their size, which is based on the same principle as that of the PAS.

As a result, larger particles tend to flow downstream diagonally, as depicted. On the contrary, small particles may flow straight downstream. This enables the separation

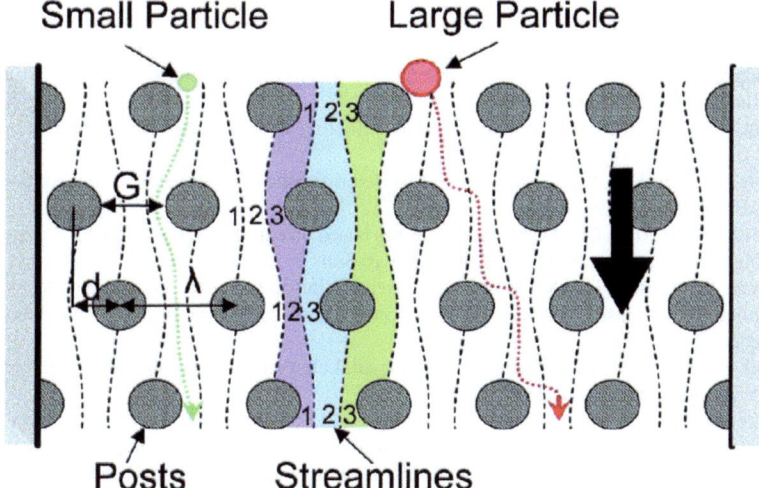

Fig. 27 Array of micro poles. Principle of the Deterministic Lateral Displacement (DLD) to separate micro particles in fluid based on their size. Reused from Davis et al. [7]: permission of NAS. Copyright (2006) National Academy of Sciences U.S.A. with permission

of particles in the fluid based on their size. A typical example of the application of this principle is the separation of red and white blood cells, Davis et al. [7]. In the field of microfluidics, this method is known as Deterministic Lateral Displacement (DLD). An experiment demonstrating high-resolution separation of microspheres with diameters of 0.8, 0.9, and 1.03 mm using a matrix of micropoles was described by Huang et al. [16]. Readers interested in a more comprehensive understanding may refer to a review article by Sajeesh and Sen [39], for example.

Acknowledgements The authors of this chapter (H. Kawamura & D. Schwabe) would like to express their sincere gratitude to BMBF and DLR (Germany) for providing the opportunity for the experiment using the sounding rocket MAXUS-6, to Astrium-Space (Germany) for their technical support, and to JAXA and MEXT (Japan) for their supports for ground experiments and travel assistances. In addition, the author of this subsection (H.K.) appreciates his leading contributions of Prof. Schwabe in the sounding rocket experiment.

References

1. Barmak I, Romano F, Kuhlman HC (2021) Finite-size coherent particle structures in high-Prandtl-number liquid bridges. Phys Rev Fluids 6:084301. https://doi.org/10.1103/PhysRevFluids.6.084301
2. Birikh RV (1966) Thermocapillary convection in a horizontal layer of liquid. J Appl Mech and Tech Phys 3:43
3. Bénard H (1900) Les tourbillons cellulaires dans une nappe liquide. Rev G en Sci, Pures et Appl 11261

4. Capobianchi P, Lappa M (2020) On the influence of gravity on particle accumulation structures in high aspect-ratio liquid bridges. J Fluid Mech 98:A29. https://doi.org/10.1017/jfm.2020.882
5. Chang CE, Wilcox WR (1975) Inhomogeneities due to thermocapillary flow in floating zone melting. J Crystal Growth 28:305–312
6. Chun CH (1980) Marangoni convection in a floating zone under reduced gravity. J Crystal Growth 48:600–610
7. Davis JA, Inglis DW, Morton KJ et al (2006) Deterministic hydrodynamics: taking blood apart. PNAS 103(40):14779–14784
8. Duffer T (ed) (2010) Crystal growth processes based on capillarity, Czochralski, floating zone, shaping and crucible techniques. John Wiley & Sons
9. Dumas HS (2014) The KAM story, A friendly introduction to the content, history, and significance of classical Kolmogorov-Arnold-Moser Theory. World Scientific Publishing, Singapore
10. Frank S, Schwabe D (1997) Temporal and spatial elements of thermocapillary convection in floating zones. Exp Fluids 23:234–251
11. Hinze JO (1975) Turbulence 2nd ed. MacGrow Hill:462–463
12. Hirata A, Nishizawa S, Sakurai M (1994) Oscillatory features of Marangoni convection in a silicone oil liquid bridge. In: 19th International symposium on space and technology science. Yokohama, Japan, ISTS 94-h-06
13. Hofmann E, Kuhlmann HC (2011) Particle accumulation on periodic orbits by repeated free surface collisions. Phys Fluids 23:072106
14. Hojo A (2007) Ground investigation for microgravity experiment of particle accumulation structure (PAS) in thermocapillary convection. Proceedings of 3rd Internnational symposium on physical sciences in space (ISPS 2007). Nara, Japan, A2–6
15. Hojo A (2008) Investigation of Marangoni convection in liquid bridge with high aspect ratio aiming on-board experiment in International Space Station. Master thesis, E603779455, Tokyo University of Science
16. Huang LR et al (2004) Continuous particle separation through deterministic lateral displacement. Science 304(5673):987–990
17. Irikura M, Arakawa Y et al (2005) Effect of ambient fluid flow upon onset of oscillatory thermocapillary convection in half-zone liquid bridge. Microgravity Sci Technol 16:176–180
18. Kawame S (2007) Numerical analysis of thermocapillary convection in liquid bridges with consideration of surrounding air. Master thesis, E603654377, Tokyo University of Science
19. Kawamura H, Harada H (1998) Observation of oscillatory Marangoni convection in a liquid column, C321. Proceedings of 35th national heat transfer symposium of Japan, pp 749–750
20. Kousaka Y, Kawamura H (2006) Numerical study on effect of heat loss upon the critical Marangoni number in a half-zone liquid bridge. Microgravity Sci Technol 18:141–145
21. Kuhlmann HC, Lappa M, Melnikov D, Shevtsova V, Ueno I, et al(2014) The JEREMI-project on thermocapillary convection in liquid bridges. Part A: overview of particle accumulation structures. FDMP 10(1):1–36
22. Landau LD, Lifshitz EM (1974) Hydrodynamik. Akademieverlag, Berlin, p 270
23. Lappa M (2010) Thermal convection: patterns, evolution and stability. John Wiley & Sons
24. Leypoldt J, Kuhlmann HC, Rath HJ (2001) Stability of hydrothermal-wave states. ZAMM 81(S3):785–786
25. Li YR, Zhang L, Wu C (2018) Experimental study on complex flow of a binary mixture in Czochralski configurations with different aspect ratios and rotation rates. Int J Heat Mass Transf 117:835–845
26. McGrath J, Jimenez M, Bridle H (2014) Deterministic lateral displacement for particle separation: a review. Lab Chip 14:4139–4158
27. Melnikov DE, Pushkin DO, Shevtsova VM (2013) Synchronization of finite-size particles by a traveling wave in a cylindrical flow. Phys Fluids 25:092108
28. Mizev AI, Schwabe D (2009) Convective instabilities in liquid layers with free upper surface under the action of an inclined temperature gradient. Phys Fluids 21:112102

29. Muehlner KA, Schatz MF, Petrov V, et al. (1997) Observation of helical travelling wave convection in a liquid bridge. Phys Fluids 9.https://doi.org/10.1063/1.869304
30. Mukin RV, Kuhlmann HC (2013) Topology of hydrothermal waves in liquid bridges and dissipative structures of transported particles. Phys Rev E 88:053016
31. Muldoon FH, Kuhlmann HC (2013) Coherent particulate structures by boundary interaction of small particles in confined periodic flows. Physica D 253:40–65
32. Nishino K et al (2015) Instability of thermocapillary convection in long liquid bridges of high Prandtl number fluids in microgravity. J Cryst Growth 420:57–63. https://doi.org/10.1016/j.jcrysgro.2015.01.039
33. Ostrach S (1979) Convection due to surface tension gradients. COSPAR Space Research 19, pp 563–570, Oxford/New York: Pergamon
34. Preisser F, Schwabe D, Scharmann A (1983) Steady and oscillatory thermocapillary convection in liquid columns with free cylindrical surface. JFM 129:545–567
35. Pushkin DO, Melnikov DE, Shevtsova VM (2011) Ordering of small particles in one-dimensional coherent structures by time-periodic flows. Phys Rev Lett. https://doi.org/10.1103/PhysRevLett.106.234501
36. Romano F, Kuhlmann HC (2018) Finite-size Lagrangian coherent structures in thermocapillary liquid bridges. Phys Rev Fluids 3:094302
37. Rudolph P (2010) Defect formation during crystal growth from the melt. In: Dhanaraj G et al (eds) Springer handbook of crystal growth. Springer, Heidelberg, pp 159–201
38. Ryzhkov II (2011) Thermocapillary instabilities in liquid bridges revisited. Phys Fluids 23:082103-1-6
39. Sajeesh P, Sen AK (2014) Particle separation and sorting in microfluidic devices: a review. Microfluid Nanofluid 17:1–52
40. Sakata T et al (2022) Coherent structures of m = 1 by low Stokes number particles suspended in a half-zone liquid bridge of high aspect ratio: Microgravity and terrestrial experiments. Phys Rev Fluids 7:014005
41. Sasaki Y, Tanaka S, Kawamura H (2005) Particle accumulation structure in thermocapillary convection of small liquid bridge. In 6th Japan/China workshop on microgravity sciences, 23A031 (Oct. 2005, Takeo, Saga, Japan)
42. Sasaki Y (2006) Particle accumulation structure in thermocapillary convection of small liquid bridge. Mater Thesis E603547894, Tokyo University of Science
43. Schwabe D (2005) Hydrodynamic instabilities under microgravity in a differentially heated bridge with aspect ratio near the Rayleigh-limit: experimental results. Adv Space Res 36:36–42
44. Schwabe D, Frank S (1999) Experiments on the transition to chaotic thermocapillary flow in floating zones under Microgravity. Adv Space Res 24(10):1391–1396
45. Schwabe D, Mizev AI (2011) Particles of different density in thermocapillary liquid bridges under the action of travelling and standing hydrothermal waves. Eur Phys J Special Topics 192:13–17
46. Schwabe D, Preisser F, Scharmann A (1982) Verification of the oscillatory state of thermocapillary convection in a floating zone under low gravity. Acta Astronaut 9:265–273
47. Schwabe D, Scharmann A (1979) Some evidence for the existence and magnitude of a critical Marangoni number for the onset of oscillatory flow in crystal growth melts. J Crystal Growth 46:125–131
48. Schwabe D, Scharmann A (1985) Messung der kritischen Marangoni-Zahl für den Übergang von stationärer zu oszillatorischer thermokapillarer Konvektion unter Microgravitation: Ergebnisse der Experimente in den ballistischen Raketen. TEXUS 5 und TEXUS 8, Z Flugwiss Weltraumforsch 9(1):21–28
49. Schwabe D, Scharmann A, Preisser F (1982) Studies of marangoni convection in floating zones. Acta Astronaut 9(3):183–186
50. Schwabe D, Scharmann A, Preisser F, Oeder R (1978) Experiments on surface tension driven flow in floating zone melting. J Crystal Growth 43:305–312. https://doi.org/10.1016/0022-0248(78)90387-1

51. Schwabe D, Velten R, Scharmann A (1990) The instability of surface tension driven flow in models for floating zones under normal and reduced gravity. J Crystal Growth 99:1258–1264
52. Schwabe D, Zebib A, Sim BC (2003) Oscillatory thermocapillary convection in open cylindrical annuli. Part 1 Experiments under microgravity. J Fluid Mech 491:239–258. https://doi.org/10.1017/S002211200300541X
53. Schwabe D, Scharmann A, Preisser F (1977) German conference on crystal growth of the German association for crystal growth (DGKK), Stuttgart
54. Schwabe D, Scharmann A (1984) Measurements of the critical Marangoni number of the laminar oscillatory state of thermocapillary convection in floating zones. Proceedings of the 5th European Symposium on Materials Science under Microgravity: Schloss Elmau 5–7 November 1983 (ESA Special Paper-222)
55. Schwabe D, Hintz P, Frank S (1996) New features of thermocapillary convection in floating zones revealed by tracer accumulation structures (PAS). Microgravity Sci Technol IX/3:163–168
56. Schwabe D (2002) Standing wave of oscillatory thermocapillary convection in floating zones under microgravity observed in the experiment. MAUS G141, Adv Space Res 29(4):651–660
57. Schwabe D (2005) Hydrothermal waves in a liquid bridge with aspect ratio near the Rayleigh limit under microgravity. Phys Fluids 17112104. https://doi.org/10.1063/1.2135805
58. Schwabe D, Tanaka S, Miezev A, Kawamura H (2006) Particle accumulation structures in time-dependent thermocapillary flow in a liquid bridge under microgravity. Microgravity Sci Techol XVIII-3/4:117–127
59. Schwabe D, Miezev I, Udhayasankar, M Tanaka S (2007) Formation of dynamic particle accumulation structures in oscillatory thermocapillary flow in liquid bridges. Phys Fluids 19:072102
60. Seki T (2006), Numerical analysis of particle accumulation phenomena in thermocapillary convection in liquid bridges. Master thesis, E603547936, Tokyo University of Science
61. Smith MK, Davis SH (1983) Instabilities of dynamic thermocapillary liquid layers. Part1. Convective Instabilities. J Fluid Mech 132
62. Takatsuka M, Tanaka S, Ueno I, Kawamura H (2002) Dynamic particle accumulation structure of Marangoni convection in liquid bridge: 2 Numerical simulation: Proc. thermal engineering. conf. JSME A234:307–308. https://doi.org/10.1299/jsmeptec.2002.0_307
63. Takatsuka M (2004), Numerical analysis of particle behaviors in Marangoni convection in a liquid bridge. Master thesis, E603308925, Tokyo University of Science
64. Tanaka S, Takatsuka M, Ueno I, Kawamura H (2002) Dynamic particle accumulation structure of Marangoni convection in liquid bridge: 1 Experimental study: Proc. thermal engineering. conf. JSME A233:305–306. https://doi.org/10.1299/jsmeptec.2002.0_305
65. Tanaka S (2007) Flow regimes and particle accumulation structures in the thermocapillary convection in a liquid bridge. Dissertation theses, Tokyo University of Science
66. Tannekes H, Lumley JL (1972) A first course in turbulence. MIT Press:1–3
67. Ueno I, Tanaka S, Kawamura H (2003) Oscillatory and chaotic thermocapillary convection in a half-zone liquid bridge. Phys Fluids 15(2):408–416. https://doi.org/10.1063/1.1531993
68. Velten D, Schwabe D, Scharmann A (1991) The periodic instability of thermocapillary convection in cylindrical liquid bridges. Phys Fluids (2):267–279
69. Xiao S, Xu S (2014) High-efficiency silicon solar cells–materials and devices physics, in critical preview in solid and materials sciences. Taylor Francis 39:277–317
70. Xu JJ, Davis SH (1984) Convective thermocapillary instabilities on long bridges. Phys Fluids 27(5):1102–1107
71. Yano T et al (2018) Report on microgravity experiments of dynamic surface deformation effects on Marangoni instability in high-Prandtl-number liquid bridges. Microgravity Sci Technol 30(5):599–610
72. Yazawa S (2011) A study on the Marangoni convection in a liquid bridge of small aspect ratios. Master thesis, Suwa Tokyo University of Science
73. Yazawa S, Kawamura H (2010) Structure of particle accumulation gathering in Marangoni convection in a short liquid bridge. Presented at 38th COSPAR Scientific Assembly, Session G (G01), Bremen, Germany

Open Access This chapter is licensed under the terms of the Creative Commons Attribution 4.0 International License (http://creativecommons.org/licenses/by/4.0/), which permits use, sharing, adaptation, distribution and reproduction in any medium or format, as long as you give appropriate credit to the original author(s) and the source, provide a link to the Creative Commons license and indicate if changes were made.

The images or other third party material in this chapter are included in the chapter's Creative Commons license, unless indicated otherwise in a credit line to the material. If material is not included in the chapter's Creative Commons license and your intended use is not permitted by statutory regulation or exceeds the permitted use, you will need to obtain permission directly from the copyright holder.

Effect of Heat Exchange, Control and Suppression of Thermocapillary Convection

Koichi Nishino, Yasuhiro Kamotani, Masaki Kudo, and Junichiro Shiomi

Abstract The effects of interfacial heat transfer on thermocapillary convection in liquid bridges of high Pr fluids placed in ground-based experiments are described in this chapter. Heat exchange due to convective and radiative heat transfer along the liquid bridge surface can sensitively affect the thermocapillary convection. Section 2 shows the effects of heating/cooling of the liquid bridge on the instability responsible for the transition from steady to oscillatory state of convection, while Sect. 3 focuses on the suppression and control of convection through the local surface heating of the liquid bridge. Section 4 deals with a full zone liquid bridge heated with a ring heater, where the thermocapillary convection which is driven by the thermal radiation from the ring heater can be controlled by the minimum heating power given by the second ring heater. The interfacial heat transfer also becomes important in microgravity as described in Sect. 6 of Chapter "Microgravity Experiments in Kibo Onboard the International Space Station."

K. Nishino (✉)
Yokohama National University, Yokohama, Japan
e-mail: nishino-koichi-fy@ynu.ac.jp

Y. Kamotani
Case Western Reserve University, Cleveland, USA
e-mail: yxk@case.edu

M. Kudo
Tokyo Metropolitan College of Industrial Technology, Tokyo, Japan
e-mail: kudo@metro-cit.ac.jp

J. Shiomi
The University of Tokyo, Tokyo, Japan
e-mail: shiomi@photon.t.u-tokyo.ac.jp

1 Introduction

Koichi Nishino

This chapter highlights the effects of interfacial heat transfer on fundamental features of thermocapillary convection in liquid bridges of high Prandtl number fluids. Thermocapillary convection is driven by the surface-tension difference along the interface between liquid and surrounding gas, where the heat exchange across the interface occurs unavoidably. This heat exchange can sensitively affect the thermocapillary convection through the alteration of the temperature distribution along the interface. Both convective and radiative heat transfer occur depending on the geometry and conditions where the liquid bridge is placed. The rate of heat exchange needed for the observation of its effects can be very small because the temperature distribution of only the thin liquid layer adjacent to the surrounding gas determines the magnitude and overall patterns of the convection. This feature is contrast to that of buoyant convection in which the volumetric heating/cooling is needed to change the driving force of convection. Such a sensitive feature of thermocapillary convection on the interfacial heat transfer has provided a unique target for the previous studies.

The effects of interfacial heat transfer in liquid bridge have been studied from basically two different viewpoints: one is the study of its effects on the instability of thermocapillary convection, and the other is the utilization of interfacial heat transfer for the control and suppression of convection. The former viewpoint is presented in Sect. 2, where the effects of heating and cooling on the transition from steady to oscillatory convection in ground-based (or 1 g) experiments is described. Heat exchange due to natural convection of surrounding gas is shown to affect the onset of transition. Such a viewpoint may bridge the gap between previous experimental results and theoretical/computation results on the instability. The latter viewpoint is focused in Sect. 3, where the localized and minimum surface heating of the liquid bridge placed in 1g is shown to be able to change, suppress and even control the instability of convection. It is demonstrated that the heating scheme called as predeterminant circumferential heating (PDCH) can reduce remarkably the temperature oscillation in the liquid bridge. Section 4 deals with a full zone liquid bridge formed in 1 g, which is heated with a ring heater placed at the middle height of the liquid bridge. Strong radiative heat transfer generates thermocapillary convection and it is shown that the PDCH using the second ring heater can control the oscillatory flow by using minimum heating power.

The effects of interfacial heat transfer also become important for the liquid bridge formed in microgravity (or μg) experiments in which the natural convection due to temperature difference diminishes. The thermocapillary convection along the interface generates the convective heat transfer between liquid and surrounding gas. Furthermore, the radiative heat transfer between liquid bridge and surrounding walls exists. Such unavoidable effects of interfacial heat transfer in μg experiments are described in Sect. 6 of Chapter "Microgravity Experiments in Kibo Onboard the International Space Station", manifesting the importance of interfacial heat transfer in the study of thermocapillary convection.

2 Effect of Interfacial Heat Loss/Gain on Stability of Thermocapillary Flow of High Prandtl Number Fluids

Yasuhiro Kamotani

This article discusses the effects of interfacial heat transfer on the critical Marangoni number of thermocapillary flow in liquid bridges of high Prandtl number fluids. In experiments, the ambient condition is varied by changing the cold wall and ambient gas temperatures. It is shown that relatively small interfacial heat loss, when the Biot number (Bi) associated with the interfacial heat transfer is less than about unity, can change the critical Marangoni number (Ma_c) several times. There is no such strong effect of Bi when it is larger than unity or when it is negative (heat gain). This phenomenon of increased Ma_c with small heat loss, which is the main topic of this article, is observed both in normal and microgravity experiments. A unique feature of this phenomenon is that the critical condition does not depend on the overall temperature difference for the liquid flow. There is no plausible explanation for this phenomenon at present.

2.1 Introduction

Thermocapillary flow is driven by a heat-induced surface tension gradient along a liquid free surface. It is usually dominated by buoyancy-induced flow in normal gravity except at a very small scale. However, it can become dominant in microgravity. For this reason, thermocapillary flow has been investigated in a small system in 1 g (on the order of a few mm) and in a larger system in microgravity. Thermocapillary flow is known to become oscillatory under certain conditions. Ever since oscillatory thermocapillary flow was discovered near the end of the 1970s [3, 38], it has been an active research subject both in 1 g and in microgravity. The most widely studied configuration is a liquid bridge configuration where a liquid column is suspended between two differentially heated rods. Therefore, the present article is mainly for this configuration.

The important dimensionless parameters associated with thermocapillary flow in a liquid bridge, in the absence of gravity, are known to be Marangoni number (Ma), Prandtl number (Pr), and bridge aspect (length-to-diameter) ratio (*A*). Recent 1 g and microgravity experiments on thermocapillary flows are conducted with high Pr fluids because oscillatory thermocapillary flows of high Pr fluids are not yet fully understood. Here, high Pr means a Pr larger than about 10 (mainly silicone oils). Therefore, thermocapillary flows of high Pr fluids are discussed in the present paper.

For a given Pr fluid in a given bridge configuration, the transition from steady to oscillatory flow should occur at a certain Ma, called the critical Ma (Ma_c). Although this is largely true, Masud et al. [28] found that Ma cannot correlate the critical experimental data under certain conditions. Masud et al. [28] discussed various additional parameters that may affect the critical condition. Among them is the heat transfer at the liquid free surface (usually heat loss). By estimating the dimensionless heat transfer rate, Biot number (Bi), in their experiments, they concluded that Bi was too small for the interfacial heat transfer to be important.

In the experiments by Kamotani et al. [12, 13] in which the cold wall temperature was varied, it was found that Ma_c changed as much as four times. It was determined that the change in the cold wall temperature altered the natural convection in the surrounding air. Since the free surface shear stress induced by the air motion was found to be very small, it was concluded that the change in the air flow caused a change in the interfacial heat transfer rate, which in turn affected Ma_c. Yano et al. [53] also showed this sensitivity of Ma_c to the interfacial heat transfer by changing the ambient temperature condition.

This sensitivity of Ma_c to the surrounding air flow was demonstrated also in the experiment in which a thin disc was placed around the bridge without touching it and moved along the columns axis [13]. It was shown that the liquid flow changes from steady to oscillatory or oscillatory to steady by simply changing the location of the disc. Several investigators showed that Ma_c increased by a few times by placing partition disks around the liquid bridge thereby changing the ambient airflow [7, 12, 15].

Some investigators studied, theoretically and numerically, the effects of the interfacial heat transfer on the instability of Marangoni convection in liquid bridges [21, 29, 50, 51]. There are several experimental and numerical studies in which the effects of co-axial ambient gas flow on the basic flow and on the instability are investigated [5, 8, 39, 52, 55]. Wang et al. [47] did an experiment with free surface heat gain and showed that the situation is quite different from that with heat loss.

In the microgravity experiment in the Kibo module of the International Space Station, called the Dynamic Surf (DS) experiment, one of its main objectives was to investigate this interfacial heat transfer effect in larger liquid bridges than those used in normal gravity tests [53]. The enhanced Ma_c for Bi less than unity was also observed in the DS experiments.

Among the various effects of the ambient conditions on the instability, perhaps the most notable one is the substantial increase in Ma_c with relatively small heat loss that was shown in Kamotani et al. [12, 13] and Yano et al. [53]. An important question is why the interfacial heat transfer, considered earlier to be negligible, can have such an appreciable effect on Ma_c. Xun et al. [51] showed in their linear stability analysis of the interfacial heat transfer effect that a sharp local maximum of Ma_c exists for Bi less than about unity for the oscillation mode of azimuthal wave number $m = 0$ instead of experimentally observed $m = 1$. Thus, this phenomenon has not yet been explained satisfactorily. There are some other unique features associated with this phenomenon. Therefore, this phenomenon is discussed in detail in this article.

2.2 Ground-Based Work on Interfacial Heat Transfer Effects

Besides aspect ratio (A) there is one more parameter characterizing the liquid bridge configuration. When the interface is straight (right circular cylinder), the volume of the liquid bridge is $V_0 = \pi D^2 H/4$, where D and H are the diameter and the length of the liquid bridge, respectively. By reducing the liquid volume, one can have a concave free surface with volume V. The volume ratio is defined as $V_R = V/V_0$. Most of the data presented in this article is for $V_R = 1$.

In a ground-based test, the orientation is vertical with the hot wall up. The cold wall temperature (T_C) is usually set nearly equal to the ambient air temperature (T_A). Due to a temperature difference between T_H and T_A, upward moving natural convection is generated in the surrounding air. In the present article, the information on the liquid flow as well as on the air motion is obtained by numerically simulating the entire flow field. The numerical model is detailed in [11, 12, 28, 47]. There is also radiation exchange between the liquid free surface and the environment. The emissivity of the fluid surface is known to be 0.9 [34]. It is noted that the heat transfer rate by radiation compared to that by natural convection is found to be about 20% or smaller.

The computed interfacial heat transfer rate, normally heat loss, is non-dimensionalized as the Biot number, Bi. The Biot number represents the ratio of free surface heat transfer rate to the conduction heat transfer rate between the hot and cold walls through the liquid. Usually the interfacial heat transfer rate is represented by the heat transfer coefficient. However, since the heat transfer rate is computed in the work, the Biot number is defined based directly on the computed heat transfer rate. The average Biot number is based on the total heat transfer rate (Q) from the free surface and defined as Bi $= Q/(2\pi H \lambda \Delta T)$ [13], where λ is the thermal condctivity of the liquid and $\Delta T = T_H - T_C$. Bi is defined to be positive for heat loss. For $V_R < 1$, the interface is self-facing so that there is a radiation exchange with itself. To calculate this radiation exchange accurately is very time-consuming because of complex computations of many view factors. Therefore, Bi is not computed for $V_R < 1$ in this article.

To show the magnitude of interfacial heat transfer in a typical ground based test, consider the conditions of Ma $= 1.8 \times 10^4$, Pr $= 52$, $D = 3$ mm, and $A = 0.67$ with $T_C = T_A = 23$ °C. Bi is computed to be 0.31. More importantly if the heat loss is compared to the overall convection heat transfer rate through the liquid, instead of conduction heat transfer rate as in Bi, the ratio is 0.04. Since this latter ratio is relatively small, it was thought that the free heat surface heat transfer would not change Ma_c substantially [28].

However, Ma_c was found to change appreciably in the tests where the cold wall temperature was varied [12, 45]. As shown in Fig. 1, when T_C was changed from 15 to 42 °C, Ma_c changed about four-fold.

For incompressible flow, the temperature difference, rather than the temperature itself, is important. Therefore, instead of T_C, the temperature difference $\Delta T_A = T_A - T_C$ is considered to affect Ma_c. To check this, tests were performed in an oven

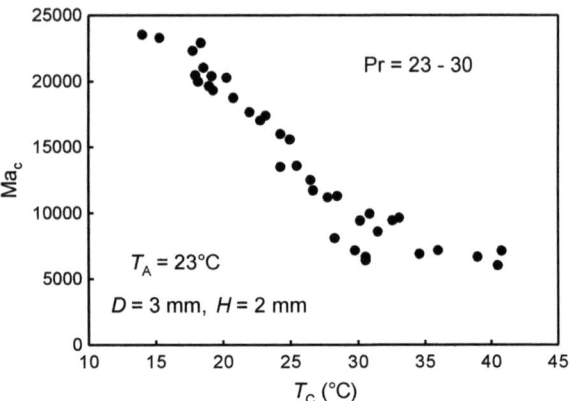

Fig. 1 Effect of cold wall temperature on Ma_c

where both T_A and T_C were varied [12, 13]. Indeed, the values of Ma_c measured under various conditions correlate well with ΔT_A as seen in Fig. 2, although the effect of ΔT_A depends on the diameter of the liquid bridge. When ΔT_A is zero or negative, the natural convection is upward everywhere along the free surface, and heat is lost from the surface. When ΔT_A is positive, the air near the cold wall moves downward, and less heat is lost from the surface. Consequently, as ΔT_A increases, the overall natural convection along the free surface becomes weaker and thus the net heat loss from the surface decreases. Thus, Fig. 2 suggests that Ma_c increases substantially as the net heat loss from the free surface becomes smaller.

To see the interfacial heat transfer effect quantitatively, the values of Bi are computed for the conditions of Fig. 2 as well as for the data taken with a smaller Pr fluid [12, 13]. They are presented in Fig. 3.

Bi is positive for all cases in Fig. 3, namely the net heat transfer is a loss from the surface. The figure shows that the critical Marangoni number, which is denoted here as Ma_{cr}, is very sensitive to Bi when Bi is less than about 0.7. This parametric

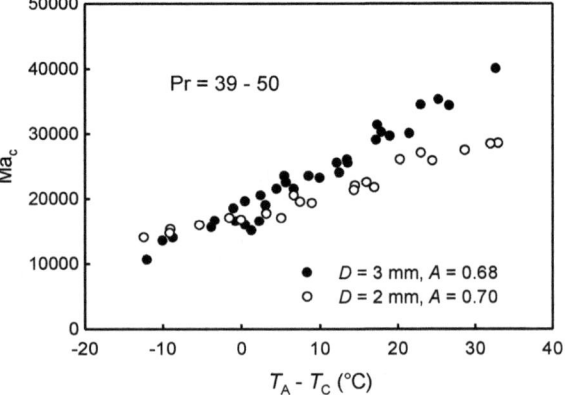

Fig. 2 Effect of temperature difference $T_A - T_C$ on Ma_c

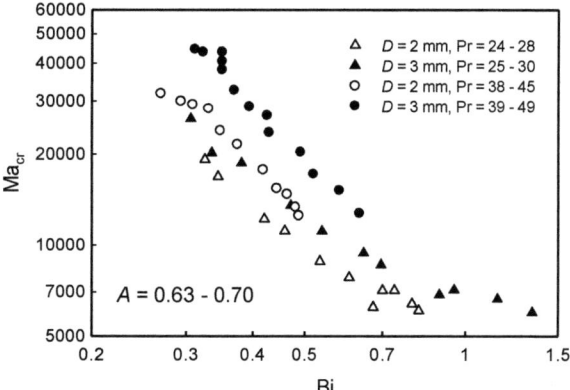

Fig. 3 Effect of Biot number on the critical Marangoni number

range is called the Bi-sensitive regime herein. As Bi increases, or as more heat is lost, Ma_{cr} decreases down to about $Ma_{cr} = 6{,}000$ when Bi is about 0.7. Ma_{cr} remains nearly constant beyond this Bi. Since Ma_{cr} increases as the heat lost is reduced, it is interesting to increase ΔT_A further (by increasing T_A experimentally) so that Bi becomes negative (net heat gain).

Figure 4 shows how Ma_{cr} changes if Bi is changed from positive to negative [47]. It is clear that there are two very different trends. Ma_{cr} is very sensitive to Bi when Bi is positive, but it is not significantly affected by Bi with net heat gain. There is a gap between the heat loss and gain data in Fig. 4 because the critical ΔT are very much different between those cases. While Ma_{cr} depends also on the bridge diameter when Bi is positive, it does not depend on the diameter when Bi is negative within the parametric range of the experiment. The smallest value of |Bi| for the heat gain tests is 0.04. It can be shown that the effect of this small |Bi| on the basic flow field is nearly negligible. It seems then that the heat gain data represent the situation where the interfacial heat transfer is negligible, so that Ma alone determines the critical condition for given Pr and A. This is true even when Bi = -0.32, where the basic flow field is somewhat modified by the free surface heat transfer, which suggests that Ma_{cr} is not sensitive to the interfacial heat transfer in the range of Bi studied herein. In this regard, the fact that Ma_{cr} remains nearly constant when the heat loss becomes large (Bi > 0.7), also suggests that Ma_{cr} is not sensitive to the surface heat loss. Therefore, the fact that Ma_{cr} becomes very sensitive to small heat loss is something unique, and it cannot be explained simply by the total free surface heat loss or Bi.

From the past studies with free surface heat loss (e.g. Masud et al. [28]) it is known that Ma_c is sensitive to the free surface shape or V_R, called the shape effect. As shown in Fig. 5 for the heat loss tests, Ma_c depends not only V_R but also on T_C (or total heat loss). Ma_c for $V_R = 0.87$ is more sensitive to the interfacial heat transfer than the $V_R = 1.0$ case. Ma_c for $V_R \leq 0.75$ is not much affected by the heat transfer.

Fig. 4 Effect of interfacial heat loss/gain on Ma_{cr}

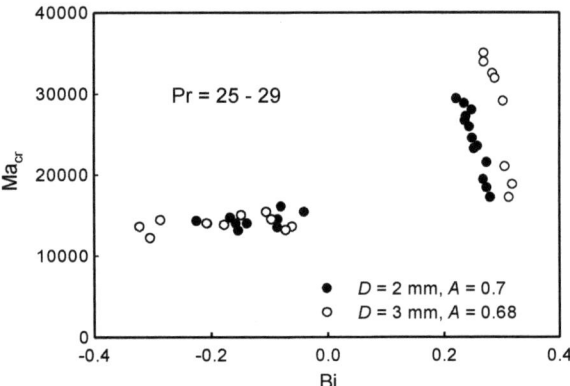

Fig. 5 Effects of cold wall temperature and volume ratio on Ma_c

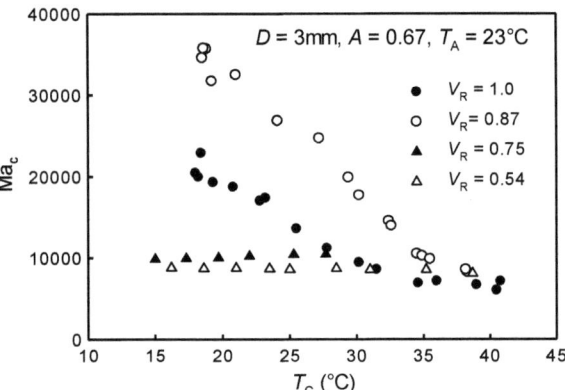

However, with increasing T_C (or increasing heat loss) the data all converge, namely the shape effect disappears.

In contrast, with heat gain Ma_c is nearly independent of V_R as shown in Fig. 6. Therefore within the experimental conditions of the above tests, Ma_c is not affected appreciably by V_R nor T_C (or interfacial heat transfer) under the heat gain condition as well as under a large heat loss condition. Thus the shape effect exists only in the Bi sensitive regime. Therefore, the sensitivity of Ma_c to small heat loss condition is something special.

The most notable thing about the critical condition for the small heat loss condition is found to be the fact that the temperature difference $T_H - T_A$ is nearly constant at the critical condition, as shown in Fig. 7 [12, 45, 46]. This fact is somewhat counter-intuitive because the critical condition does not depend on T_C or ΔT. It also shows that the critical condition does not depend on Bi. In any case, the fact that the temperature difference $T_H - T_A$ is important implies that the fluid mechanics and heat transfer near the hot wall, the so-called hot corner [11], plays an important role in the onset of oscillatory flow.

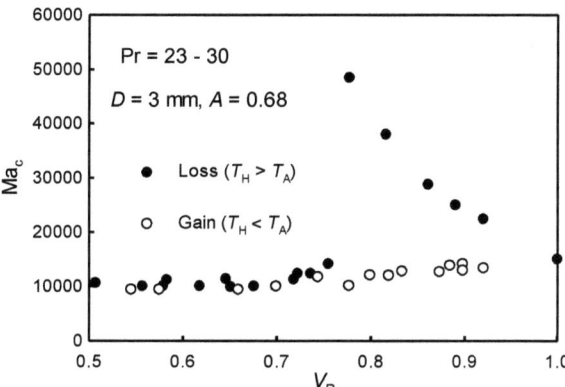

Fig. 6 Effect of heat loss/gain on shape effect

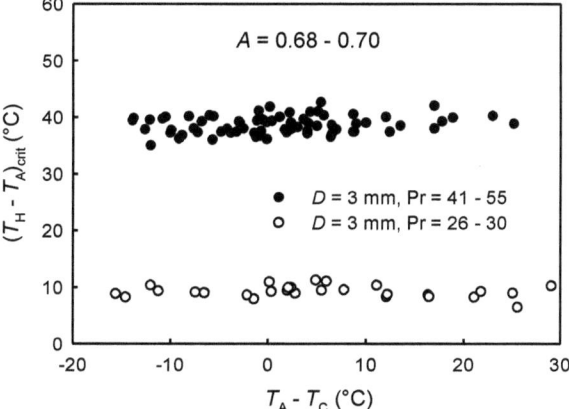

Fig. 7 Temperature difference $T_H - T_A$ at critical condition

Another important experimental fact is this. Oscillatory thermocapillary flow of high Prandtl fluids was investigated, besides the present liquid bridge configuration, in a circular dish configuration both in 1 g and in μg (e.g., Kamotani et al. [10]), where the fluid was heated by a cylindrical heater located at the center and cooled by the dish side wall. No Bi-sensitive regime was observed in this configuration.

So far available ground-based data have been discussed. Due to the Bond number constraint in 1 g, the liquid bridge size is limited so that one cannot cover a wide range of Biot number. Moreover, the interfacial heat transfer is predominant by natural convection in 1 g, while in microgravity the natural convection is suppressed, so it is possible to investigate the interfacial heat transfer effect under different conditions. Microgravity experiments have been performed recently as reported by Yano et al. [54], and the results pertinent to the interfacial heat transfer effect are discussed in Sect. 6 of Chapter "Microgravity Experiments in Kibo Onboard the International Space Station".

2.3 Summary Remarks

Effects of interfacial heat transfer on the transition to oscillatory Marangoni convection in liquid bridges of high Prandtl number fluids are discussed in this article. The past experiments, in which the cold wall temperature as well as the ambient temperature were varied systematically, revealed that the critical Marangoni number, Ma_c, becomes very sensitive to those temperature variations under certain conditions. This sensitivity was attributed to the interfacial heat transfer. The interfacial heat transfer rates were computed by numerical simulations and non-dimensionalized as the Biot number, Bi. It is shown that Ma_c increases several times as Bi is decreased below about $Bi = 0.7$, so this increased Ma_c occurs when the interfacial heat transfer is relatively small, called the Bi sensitive regime herein. However, Ma_c is not sensitive to Bi if Bi is small but negative (heat gain). Ma_c is known to be sensitive to the liquid volume ratio, V_R, which is called the shape effect. It is shown that this shape effect occurs only in the Bi sensitive regime.

A very unique feature of the Bi sensitive regime is that the temperature difference between the hot wall and the ambient, $T_H - T_A$, is found to be nearly constant at the critical condition. This means that the critical condition does not depend on ΔT, which represents the driving force for the liquid flow.

One important remaining question regarding the stability of thermocapillary flow of high Prandtl fluids is how to explain and simulate this Bi sensitive regime. More work is needed to do this. It may require additional features that were not investigated in the past.

3 Control and Suppression of Oscillatory Thermocapillary Convection in a Half-Zone Liquid Bridge

Masaki Kudo and Junichiro Shiomi

In this subsection, we review our work in which a remarkable attenuation of nonlinear oscillatory thermocapillary convection was attained in a half-zone (HZ) liquid bridge of a high Prandtl number (Pr) fluid.

3.1 Introduction

After Eyer et al. [4] revealed that detrimental striations were caused by oscillatory thermocapillary (Marangoni) convection in floating zone processing for semiconductors, research to reduce the oscillation was strongly promoted. Low-Prandtl number (Pr) fluids have electrical conductivity, therefore, the flow can be stabilized with the use of a magnetic field. In a high-Pr fluid, on the other hand, heat transfer plays an important role in the stability of the thermocapillary flow [48]. Two methods for reducing the oscillatory flow are described here. First, "control" is the method of reducing the oscillation without changing the base flow. Next, "suppression" is the method of suppressing the oscillation by modifying the base flow while maintaining the basic properties of the HZ liquid bridge.

An attempt to control the thermocapillary wave instability on a plane fluid layer was made by Benz et al. [1]. The temperature signal and phase information detected by thermocouples near the cold end of the layer were fed forward to control a laser that heated the downstream fluid surface along a line. For an annular configuration, Shiomi et al. [40, 41] applied feedback control based on a simple cancellation scheme. They modified the surface temperature locally with heating using local surface temperature signals obtained at different azimuthal locations. Using two sensor-actuator pairs, a significant attenuation of oscillation was obtained in a range of Marangoni number. In a half-zone liquid bridge, Petrov et al. [35, 36] demonstrated a reduction of oscillatory thermocapillary convection through nonlinear feedback control. They modified the surface temperature locally with cooling and heating using local surface temperature signals obtained at different azimuthal locations. They succeeded in the complete reduction of the oscillation up to 8% of the critical Marangoni number. In actual zone melting processing, highly strong nonlinear convection occurs in a melt, therefore, it is required to dampen the stronger oscillation compared to that of Petrov et al. [35, 36].

In response to this previous research, we succeeded in the complete reduction of the oscillation up to 40% of the critical Marangoni number using two kinds of simple control methods [16, 42]. Additionally, we found that an additional heater with continuous heating is able to suppress the emergence of the oscillatory flow [17]. We refer to this suppression method as PreDeterminant Circumferential Heating (PDCH). A detailed description of PDCH will be provided in the latter part of this subsection. In this subsection, we present our accomplishments in the control and suppression of oscillatory thermocapillary convection in the half-zone liquid bridge.

3.2 Control with Local Heating to Cancel a Dominant Modal Structure (Single Mode Control)

We started with a control method called "single mode control," in which we aimed to control a single mode. The system consisted of thermometers, a data acquisition and control device, and a controller [42]. We adopted the Constant Current Thermometer (CCT) to measure the temperature on the free surface. The operating principle of the CCT is the same as that of Constant Current Anemometry (CCA). The sensing part of the probe was made of thin platinum wire (diameter: 2.5 µm), which minimized the meniscus caused by the probe on the free surface. The heater was manufactured in the same manner as the sensor, and the tip was made of 10% rhodium-platinum, which has high electrical resistance among Wollaston wires. We used Lab-VIEW® as the data acquisition and control device. This control system can be applied to the actual zone melt method because the response time is fairly short compared to the period of oscillatory flow caused in the melt zone.

"Hydrothermal wave (HTW) instability" causes the flow transition from a two-dimensional steady flow to a three-dimensional oscillatory flow in high-Prandtl number fluids [43]. In HTW instability, a thermal wave propagates on the free surface at an angle. An idea was proposed that if the thermal wave is canceled by local temperature modification, the oscillatory flow would be damped. The control scheme, originally developed by Shiomi et al. [40, 41], was modified and applied to the liquid bridge system.

To reduce the oscillatory flow, temperature oscillations were damped by locally heating the lower temperature region on the free surface. The simplest control scheme would be to place the temperature sensor and heater at the same position. However, this configuration poses a difficulty because the liquid bridge is too small to place both the sensor and heater at a sufficiently close distance. Therefore, they were placed at separate azimuthal positions. The positions were determined considering the circumferential periodicity of the surface temperature distribution. In the case of oscillatory convection with $n = 2$, temperature fluctuation is in phase every π (rad) in the azimuthal direction at the same height. In that case, the sensor and the heater should be placed π (rad) apart azimuthally. Thus, the free surface was locally heated by the heater when the sensor detected a lower temperature. The control scheme is formulated as follows.

$$Q(\varphi + 2\pi/n, t) = \begin{cases} -G_1 \cdot \theta(\varphi, t) & \{\theta(\varphi, t) < 0\} \\ 0 & \{\theta(\varphi, t) \geq 0\} \end{cases} \quad (1)$$

$$Q(\varphi + \pi/n, t) = \begin{cases} 0 & \{\theta(\varphi, t) < 0\} \\ G_1 \cdot \theta(\varphi, t) & \{\theta(\varphi, t) \geq 0\} \end{cases} \quad (2)$$

Here, Q is the heating output from the heaters (W), G_1 is the control gain, φ is the azimuthal position (rad) of the heater, t is the time, and n is the wave-number of the modal structure. The non-dimensional surface temperature fluctuation is calculated as $\theta = T'/\Delta T$, where T' represents the fluctuation of the surface temperature in time, and ΔT is the temperature difference between the top and bottom rods. The gain G_1 was kept constant throughout the control process. The control scheme formulated in Eq. (1) was named the "Inverted" scheme, while Eq. (2) was named the "Non-inverted" scheme, respectively.

The target is a liquid bridge with $\Gamma = 1.0$, where the modal structure with $n = 2$ dominates after the onset of oscillatory convection. Here, the aspect ratio is defined as $\Gamma = H/R$, where R is the radius of the liquid bridge. A silicone oil with a viscosity of 5 cSt (Pr = 68) was adopted as the test fluid. When two pairs of sensors and heaters were used, the azimuthal position of the sensors and heaters was defined according to Fig. 8a. Two sensors were placed at the height of $0.5H$, with their tips inserted at a depth of approximately 100 μm into the surface. The heaters were placed approximately 250 μm away from the surface. The damping performance was examined by varying the gain G at the same Ma.

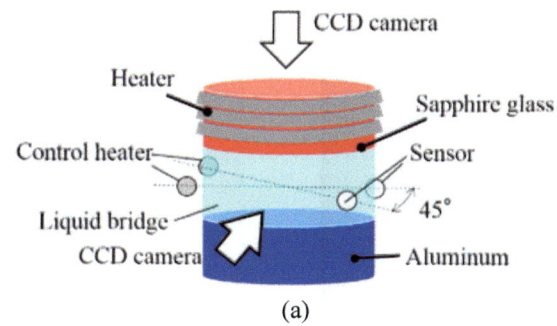

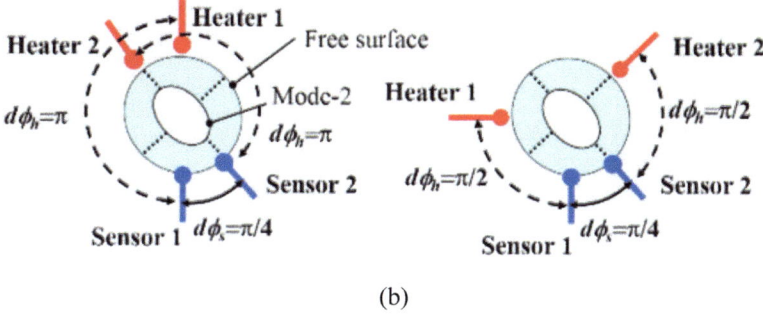

Fig. 8 Experimental set-up (**a**) and the top view of the locations with two pairs of sensors and heaters (**b**). Reprinted from Shiomi et al. [42] with permission from Cambridge University Press. All rights reserved

At first, applying only one sensor/heater pair results in a standing wave with nodes at the sensor/heater positions [35]. Therefore, two sensor/heater pairs were adopted in two configurations (Fig. 8b).

For small Ma ($Ma_c < Ma < 1.4\, Ma_c$), the top view of the visualized flow is shown in Fig. 9. The standing wave structure is destabilized by the control and eventually transitions to stationary convection. It can be observed that local heating significantly alters the entire flow field, as shown in Fig. 9d.

Figure 10 shows a time series of temperature fluctuations and the corresponding heater output during the successful application of inverted control. Both sensors recorded the same temperature changes as shown in Fig. 10. The control was initiated at $t = 50$ s and terminated at $t = 130$ s. The temperature amplitude rapidly decreases upon the initiation of control and is damped to within a few % of the initial value within 30 s. Upon cessation of control, the amplitude gradually increases. The corresponding heater output starts at approximately 1.5 mW at the beginning of control and reduces to a few μW as the temperature oscillation is diminished.

Figure 11 presents a top view of the flow field in a case where the inverted control failed due to overheating. The flow exhibited a transition from a traveling wave with $n = 2$ to a standing wave with $n = 1$ during the control. It is assumed that

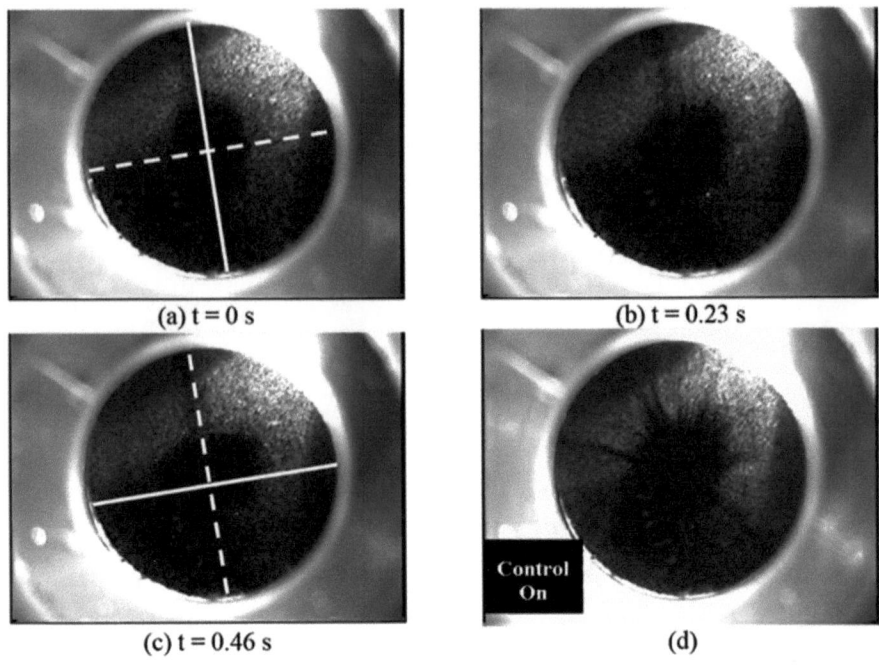

Fig. 9 Flow visualization of a standing wave with $n = 2$ in panels (**a**) and (**c**), while panel (**d**) represents the flow without control. The figure demonstrates the transition from the standing wave to a two-dimensional steady flow achieved through single mode control. $Ma = 1.18 Ma_c$. Reprinted from Shiomi et al. [42] with permission from Cambridge University Press. All rights reserved

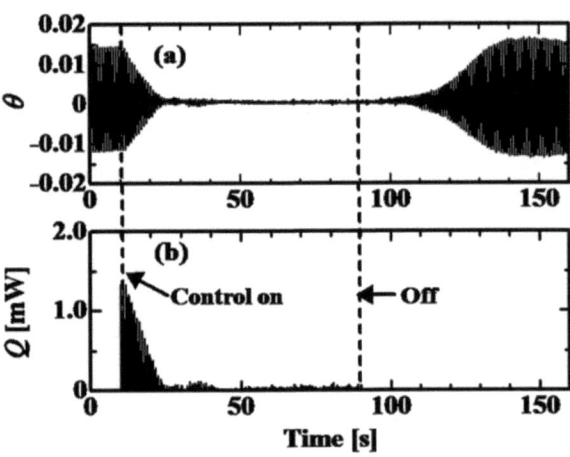

Fig. 10 a Time history of the non-dimensional temperature signal $\theta(\varphi = 0)$. **b** Simultaneously measured heater output power $Q(\varphi = \pi)$. $Ma = 1.18 Ma_c$. Reprinted from Shiomi et al. [42] with permission from Cambridge University Press. All rights reserved

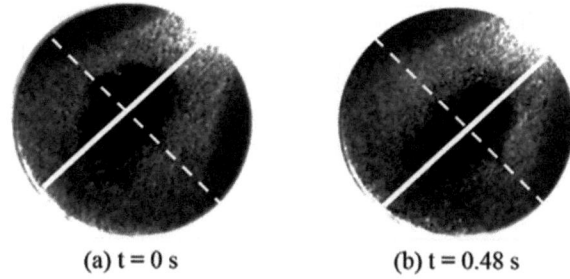

Fig. 11 Flow visualization of an excited standing wave mode with $n = 1$. Reprinted from Shiomi et al. [42] with permission from Cambridge University Press. All rights reserved

(a) t = 0 s (b) t = 0.48 s

the excitation of oscillation with $n = 1$ caused the control failure. This is because the inverted control amplifies odd-numbered mode oscillations. The problem was resolved by applying the non-inverted control scheme.

The trends of the amplitude of temperature fluctuation with respect to the control gain G_1 are shown in Fig. 12a. The damping ratio γ is defined as the ratio of the root mean squared (RMS) values of θ with and without control. In the case of the inverted control, the amplitude shows a downwardly convex shape against G_1. When G_1 exceeds the minimum θ, the amplitude increases due to the amplification of $n = 1$ oscillation. On the other hand, in the case of the non-inverted control, the amplitude does not have a global minimum against G_1. This is because the non-inverted control does not amplify $n = 1$ oscillation.

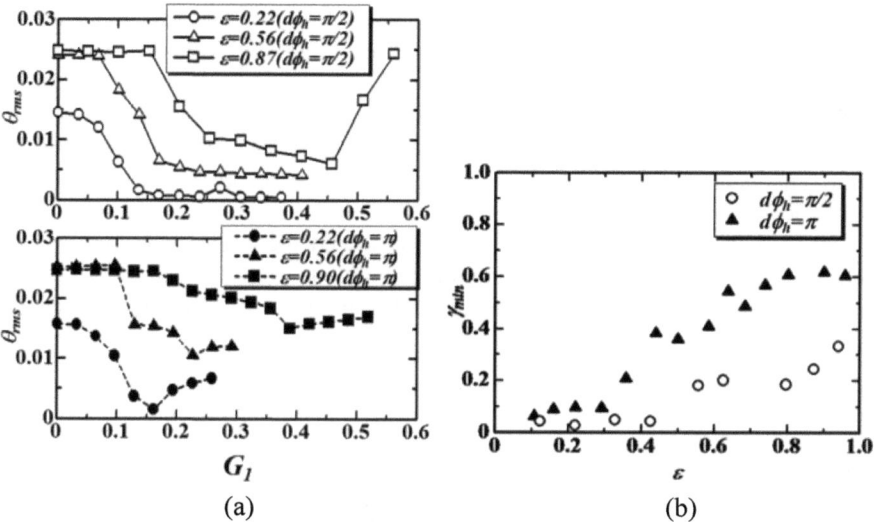

Fig. 12 a RMS of θ for various values of G_1. **b** Performance of the single mode control across a range of ε. Circles represent the damping ratio γ ($d\varphi = \pi/2$) with the optimal gain $G_{1,\text{opt}}$. Triangles represent γ ($d\varphi = \pi$) with the optimal gain $G_{1,\text{opt}}$. Reprinted from Shiomi et al. [42] with permission from Cambridge University Press. All rights reserved

The best control performance of the two control schemes is recorded for various over-critical parameters ε in Fig. 12b. This parameter, ε, is defined as $\varepsilon = (Ma - Ma_c)/Ma_c$. Complete damping of the oscillation is achieved in the range of $\varepsilon < 0.3$ for the inverted control and $\varepsilon < 0.45$ for the non-inverted control, respectively. This difference can be attributed to the tendency of the oscillation with $n = 1$ to be amplified.

3.3 Control with Local Heating to Cancel Several Modal Structures (Multimode Control)

The single mode control was designed to suppress a dominant oscillation that occurred without control. However, this method failed to reduce unintended oscillation that was excited by the control. As a result, our second control method, known as 'the multimode control' [16], considered multiple modes. This method employed a control algorithm to estimate the azimuthal distribution of surface temperature fluctuations involving various modes, aiming to address nonlinear thermocapillary oscillation.

The sensors and heaters were made in the same manner as the single mode control. Four sensors were used to measure local surface temperatures. They were placed at different azimuthal positions (Sensor 1 was at $\varphi_{S1} = 0$, Sensor 2 at $\varphi_{S2} = \pi/4$, Sensor 3 at $\varphi_{S3} = \pi$, Sensor 4 at $\varphi_{S4} = 5\pi/4$). Two heaters were employed to heat the surface locally (Heater 1 was at $\varphi_{H1} = 3\pi/2$, Heater 2 at $\varphi_{H2} = 3\pi/4$). All sensors and heaters were located at the mid-height of the bridge. The azimuthal positions of the sensors and heaters are shown in Fig. 13.

The single mode control was compared with the multimode control. Paired sensors and heaters (Sensor 1-Heater 1, Sensor 2-Heater 2) were located in the same positions as the sensors in the multimode control. Sensors (Sensor 3 and Sensor 4) monitored the surface temperature only.

Since only the oscillation with $n = 2$ was considered, the single mode control unintentionally excited oscillation with an untargeted mode ($n = 1$) when $Ma > 0.4Ma_c$. As the instability is global, the observed modal structure at the surface reflects the

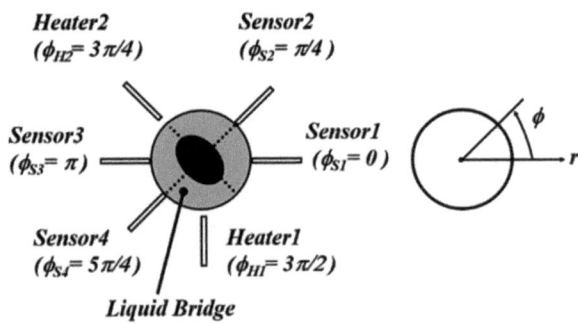

Fig. 13 Experimental set up: configuration of sensors and heaters. Reprinted from Kudo et al. [16] with permission from Elsevier. All rights reserved

overall flow field structure. In the multimode control, the spatiotemporal distribution of temperature variations was reconstructed using local surface temperature signals obtained by sensors located at different azimuthal positions at each time step.

The azimuthal distribution of surface temperature variations consists of modes with various azimuthal wave numbers. Each mode is formed by a pair of traveling waves propagating in opposite azimuthal directions [27]. The temperature distribution $\Theta(\varphi, t)$ is described as the superposition of modes from $n = 1$ to N.

$$\begin{aligned}\Theta(\varphi, t) &= \sum_{n=1}^{N} \{A_{n(c.c)}(t) \cdot \sin(n\varphi - 2\pi f_n(t) \cdot t + \eta_{n(c.c)}) \\ &\quad + A_{n(c)}(t) \cdot \sin(n\varphi + 2\pi f_n(t) \cdot t + \eta_{n(c)})\} \\ &= \sum_{n=1}^{N} \{A_n(t) \cdot \sin(n\varphi - 2\pi f_n(t) \cdot t) + B_n(t) \cdot \cos(n\varphi - 2\pi f_n(t) \cdot t) \\ &\quad + C_n(t) \cdot \sin(n\varphi + 2\pi f_n(t) \cdot t) + D_n(t) \cdot \cos(n\varphi + 2\pi f_n(t) \cdot t)\} \end{aligned} \quad (3)$$

where, $A_{n(c)}(t)$ and $A_{n(c.c)}(t)$ are the amplitudes of traveling waves that propagate in a clockwise and counter-clockwise direction with the nth mode, respectively. The angle φ represents the azimuthal location (in radians), $f_n(t)$ denotes the frequency of the nth mode, and η is the initial phase. This can be re-expressed by the amplitudes $A_n(t) - D_n(t)$, which include η.

To accurately reconstruct $\Theta(\varphi, t)$, $f_n(t)$ was calculated using the autocorrelation of the surface temperature signals $\theta(\varphi, t)$. The frequency $f_n(t)$ was estimated from the time delay at which the autocorrelation coefficient reaches its first maximum. The period of the dominant mode can be calculated using the autocorrelation as well. Finally, $A_{n(c)}(t)$ and $A_{n(c.c)}(t)$ were computed by minimizing the difference between $\Theta(\varphi, t)$ and $\theta(\varphi, t)$ at the measurement points in the spatiotemporal domain through least square fitting.

For the current Γ, the dominant mode is 2. Mode 1 was excited by excessive heating of the surface. In this study, $\Theta(\varphi, t)$ capturing only two modes ($n = 1, 2$) was achieved using four sensors due to limited space for sensor and heater placement. Since the frequency $f_2(t)$ varied considerably during the control, it was evaluated dynamically. On the other hand, as $f_1(t)$ showed minimal variations during the control, it was set to a constant value determined as $f_1(t) = 1.0$ Hz in the preliminary research. In this experiment, $A_n(t) - D_n(t)$ values were obtained using the data $\theta(\varphi, t)$ from the last 1 s and $f_2(t)$ from the last 3 s.

Heat outputs from the heaters were calculated based on $\Theta(\varphi, t)$ at the respective positions of the heaters.

$$Q(\varphi_{Hk}, t) = \begin{cases} G \cdot \Theta(\varphi_{Hk}, t + \tau^*) & \{\Theta(\varphi_{Hk}, t) < 0\} \\ 0 & \{\Theta(\varphi_{Hk}, t) \geq 0\} \end{cases} \quad (4)$$

Fig. 14 Damping performance γ of the two control methods with respect to the control gain G. Reprinted from Kudo et al. [16] with permission from Elsevier. All rights reserved

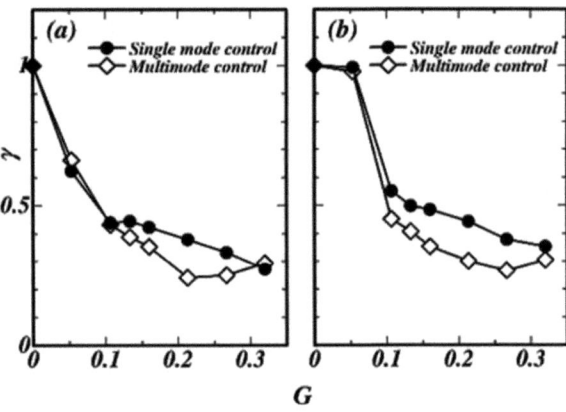

where Q represents the heating output from the heaters (W), G denotes the control gain, φ_{Hk} represents the azimuthal position (rad) of the kth heater, and τ^* indicates the timing of heat input (s). The gain G was maintained constant throughout the control process. Considering that the heater is located approximately 250 μm away from the liquid bridge, it takes about 0.25 s for the heat to reach the liquid bridge due to propagation delays. Thus, τ^* was set to 0.25 s to compensate for this time delay caused by heat propagation.

The performance of the multimode control is compared to that of the single mode control in terms of control performance. Figure 14 illustrates the damping performance in relation to G. The damping performance, denoted as γ, is defined as follows:

$$\gamma = \frac{A_{\text{with control}}}{A_{\text{without control}}} \quad (5)$$

where $A = \sqrt{\overline{\{A_{1(c)}\}^2} + \overline{\{A_{1(c,c)}\}^2} + \overline{\{A_{2(c)}\}^2} + \overline{\{A_{2(c,c)}\}^2}}$.

Figure 14a and b show the γ obtained at ε of approximately 0.5 and 0.7, respectively. The closed circles and open diamonds represent the γ obtained from single-mode control and multimode control, respectively. In both figures, significantly better damping of the oscillation is achieved by multimode control compared to single-mode control over a wide range of γ. The untargeted modes get excited when high γ are applied in both control methods. Single-mode control excites oscillations with a standing wave pattern of $n = 1$, as reported by Shiomi et al. [42]. In this case, a particle free zone with a circular shape travels in a radial direction. On the other hand, in multimode control, a pulsating particle free zone with a polygonal shape is located in the center of the liquid bridge.

Figure 15 shows the comparison of power spectra of the surface temperature oscillation for both control methods. Figure 15a, b and c demonstrate the spectra of oscillations without control, with single-mode control, and with multimode control, respectively. In Fig. 15a, the primary peak of the spectrum at 0.9 Hz ($P1$) indicates traveling wave oscillations with $n = 2$. In Fig. 15b, the spectrum peak at 1.2 Hz ($P3$) suggests standing wave oscillations with $n = 2$. In addition to $P3$, two peaks exist: $P2$ and $P4$, with frequencies lower and higher than that of $P3$, respectively. As shown by Shiomi et al. [42], the energy at the $P2$ peak belongs to oscillations with $n = 1$. In Fig. 15c, $P4$ remains, whereas $P2$ disappears. The spectral comparison demonstrates that multimode control delays the destabilization of mode-1 with respect to single-mode control. The results of spectral analyses agree with observations made from flow visualization, where the oscillations with $n = 1$ do not seem to be excited by multimode control, although no conclusions can be drawn from this. Since the visualized flow becomes complex due to the newly appearing mode $P4$, wave number $P4$ cannot be identified at this point. However, assuming that there is one temporal mode for each wave number, it is likely that $P4$ belongs to oscillations with $n \geq 3$.

The number of sensors limited the identification of modes to 2. If more sensors could be placed around the liquid bridge, the amplitude of higher modes could have been evaluated. However, the size of the liquid bridge restricted the number of sensors in the current experiment.

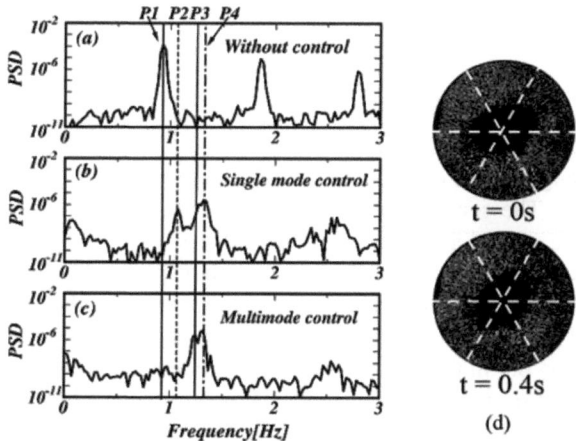

Fig. 15 **a–c** Spectral analysis of temperature signals with two control methods. $P1$, $P2$, $P3$, and $P4$ show the Fourier components of the waves with $n = 2$ (without control), $n = 1$ and 2 (with control), and $n \geq 3$, respectively (**a** $G = 0$, **b** $G = 0.32$, **c** $G = 0.32$, Ma $= 1.7$Ma$_c$). **d** Excitation of other modes by the multimode control ($G = 0.32$, Ma $= 1.7$Ma$_c$). Reprinted from Kudo et al. [16] with permission from Elsevier. All rights reserved

3.4 Oscillation Suppression by PreDeterminant Circumferential Heating (PDCH)

We initially conceived PreDeterminant Circumferential Heating (PDCH) as the control method that targets unspecific modes for further reduction. However, the PDCH was found to be a kind of suppression method through our experiment [18]. The experimental setup for the PDCH is shown in Fig. 16. The system consisted of thermometers, data acquisition and control devices, and a controller. Constant Current Thermometer (CCT) were adopted to measure the temperature on the free surface. The sensing part of the probe was made of thin platinum wire (diameter: 2.5 μm). A ring-shaped electrical heater was used for the controller, and a control input was supplied by a DC regulated power supply through an operational amplifier. The heater was made of a nichrome wire with a diameter of 200 μm and was formed into a circle with a diameter of 3 ± 0.2 mm. The heater was arranged concentrically around the liquid bridge at several different heights. Two sensors were placed at a height of $0.25 H$ and were spaced at intervals of $\pi/4$ azimuthally, slightly touching the free surface.

In the PDCH, the heat output Q was kept constant both in terms of time and azimuthally.

$$Q(\varphi, t) = \text{const.} \tag{6}$$

Damping performance was compared with the control method, the single mode control. In the single mode control, two sensor and heater pairs were placed around the liquid bridge. The sensors and heaters were spaced π (rad) apart azimuthally. Two sensors were spaced $\pi/4$ (rad) apart azimuthally.

Figure 17 shows the damping performance with respect to Ma for the single mode control and the PDCH. For the PDCH, the performance for the liquid bridge with two different Pr fluids was evaluated. While a complete reduction of the oscillation is achieved for $\varepsilon \leq 0.4$ in the case of the single mode control, a perfect reduction is achieved for $\varepsilon \leq 2$ in the PDCH. It should be noted that the heat output, Q, of the

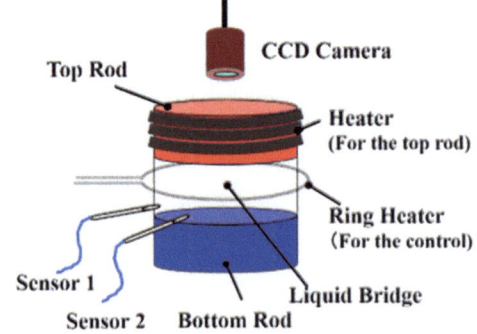

Fig. 16 Experimental set up for the PreDeterminant Circumferential Heating (PDCH). Reprinted from Kudo and Kawamura [18]. All rights reserved

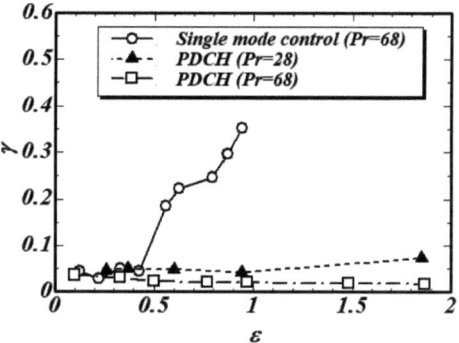

Fig. 17 Damping performance γ versus ε for the control methods and the PDCH. Reprinted from Kudo and Kawamura [18]. All rights reserved

PDCH is about 10 times higher than that of the single mode control, but it accounts for about 10% of the total heat input to the liquid bridge itself.

Figure 18 shows the damping performance at different heights of the heater in both the single mode control and the PDCH. The heater was placed at $z = (7/8)H$ (near a heated rod, 'Top'), $(1/2)H$ (middle point, 'Middle'), or $(1/8)H$ (near a cooling rod, 'Bottom'). Both the single mode control and the PDCH perform best at $(1/2)H$ and worst at $(7/8)H$. In the case of reduction at $z = (1/2)H$ by the single mode control, the traveling wave is damped to a weak standing wave. On the other hand, with the PDCH, traveling waves are completely damped out. If the heat output exceeds the global minimum, the performance decreases due to the excitation of the oscillation with $n = 1$. In the case of reduction at $z = (7/8)H$, the single mode control is unable to dampen the oscillation at all. It is observed that a standing wave oscillation is amplified and changes to a traveling wave under control. The same phenomenon is also confirmed for the PDCH. In the case of suppression at $z = (1/8)H$, the trend of damping performance is different for the single mode control and the PDCH. Although the single mode control cannot dampen the oscillation, the PDCH can achieve complete reduction by applying a higher heat output than the one applied at $z = (1/2)H$. In this case, excitation of the oscillation with $n = 1$ is not observed.

3.5 Conclusions and Recent Research Trends

We succeeded in achieving an appreciable reduction of nonlinear thermocapillary convection in a half-zone liquid bridge of a high-Pr fluid. When the control methods (single and multimode control) were applied, complete reduction of the temperature oscillation was achieved in the high non-linear region for $Ma < 1.4Ma_c$. Furthermore, a significant reduction of the oscillation was achieved for $Ma < 2.0Ma_c$.

When the suppression method (PDCH) was applied, complete reduction of the temperature oscillation was achieved up to the high nonlinear region of $Ma < 3.0Ma_c$. It should be noted that the heat output Q of the control methods was about 10% of that of the PDCH, making these control methods effective and energy-saving. For

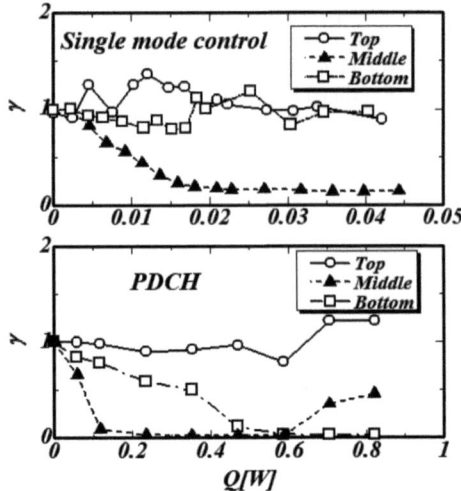

Fig. 18 Effect of the position of heater upon the damping performance (Ma = 1.77Ma$_c$). Reprinted from Kudo and Kawamura [18]. All rights reserved

future research, it would be interesting to apply the PDCH to the liquid bridge with low-Pr fluids (Pr < O (100)) and compare its performance with suppression using a magnetic field.

Recent research trends in the field include advanced control algorithms and exploration of control inputs beyond heat sources and electric fields. Muldoon [32, 33] conducted numerical simulations using an optimal control algorithm to suppress HTW instability occurring in a thin liquid film of a high-Pr fluid. Multiple heat sources were strategically positioned within a specific area on the free surface, resulting in nearly complete suppression in that area and significant reduction throughout other regions. Lappa [26] investigated the impact of periodic changes in gravitational acceleration on flow patterns in low-Pr fluid layers, specifically studying thermocapillary convection.

4 Thermocapillary Convection in a Full Zone Liquid Bridge

Masaki Kudo

Experimental and theoretical studies on thermocapillary convection in a model of the floating-zone (FZ) method, using a high Prandtl number fluid, will be described. Emphasis will be placed on the effect of buoyancy and the geometry of the liquid bridge on flow instability. A method to suppress flow instability in the FZ method is proposed, utilizing an additional ring heater with a low heat input.

4.1 Introduction

The full-zone (FZ) model more closely mimics the floating-zone method in crystal growth. The flow structure in the FZ model is more complicated in comparison with that in the half-zone (HZ) model discussed in the previous subsection. Therefore, the amount of research conducted for the FZ model was significantly less than that for the HZ model. Regarding the critical condition of flow transition in high Prandtl number (Pr) fluids, three types of effects have been experimentally studied to the best of the author's knowledge. These include the effect of heating conditions on a ring-shaped heater by Kamotani and Lee [9], as well as the effects of volume ratio and buoyancy by Sakurai et al. [37]. In terms of numerical work, Bouizi et al. [2] investigated the effect of Pr on the critical condition in the liquid bridge. Three numerical studies were conducted for low Pr fluids, focusing on the effect of the liquid bridge's shape on the flow structure by Lappa [22, 23, 24], Houchens and Walker [6], the effect of Pr on the critical condition by Bouizi et al. [2], and the solutecapillary force by Minakuchi et al. [30]. Thus, the critical condition of flow transition was investigated under limited conditions.

In response to these previous studies, we investigated the effect of geometry on the critical condition of flow transition in high Pr fluids through experiments and classified unsteady flow patterns, as conducted by Kudo et al. [19]. Furthermore, the effect of airflow around the liquid column on the critical condition was examined in a high Pr fluid through experiments carried out by Kudo et al. [20]. Additionally, the effect of buoyancy on the critical condition was investigated in a high Pr fluid through numerical simulations conducted by Motegi et al. [31]. Building upon our findings regarding the classification of unsteady flow patterns, Lappa [25] studied the time-series changes of unsteady flow patterns in a high Pr fluid using numerical simulations. In this subsection, we present our achievements regarding the effects of buoyancy and geometry of the liquid bridge on the critical conditions of flow transition.

4.2 Geometry of a Full-Zone Liquid Bridge

In the FZ liquid bridge, a liquid column is suspended between two coaxial rods with the same diameter, as shown in the left figure of Fig. 19. In the HZ method, on the other hand, half of the floating zone is simulated, as shown in the right figure of Fig. 19. The HZ model is more widely used for basic research on the floating zone method because of its simplicity. However, the FZ mimics the actual crystal growth configuration more closely. In FZ, the top and bottom rods are maintained at the same temperature using a chiller, etc. A ring heater, which is installed coaxially with the liquid bridge at half the height of the bridge, is used to add an axial temperature difference to the free surface of the liquid bridge. The aspect ratio of the liquid bridge is defined as $\Gamma_H = H/R$, where H is the half-height of the liquid bridge, and R is

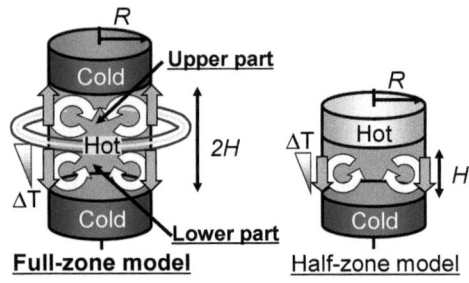

Fig. 19 Geometries of a FZ liquid bridge and a HZ liquid bridge. Reprinted from Kudo et al. [19]. All rights reserved

the radius of the rods. The Marangoni number (Ma) is a non-dimensional value that represents the strength of the thermocapillary flow, as defined by Eq. 7. In this work, Ma is calculated using the half-height of the liquid bridge (H) and the temperature difference (ΔT [K]). ΔT is defined as the difference between the temperature of the free surface nearest to the ring heater and the average temperature of the top and bottom rods. In this dimensionless number, σ_T is the temperature coefficient of the surface tension, ρ is the density of the liquid bridge, ν is the kinematic viscosity of the liquid bridge, κ is the thermal diffusivity of the liquid bridge, Re is the Reynolds number for thermocapillary flow, and Pr is the Prandtl number of the liquid bridge. The parameter ε, which represents the supercritical state, is defined by Eq. 8. Ma_c represents the critical value of Ma. The Rayleigh number (Ra) is described by Eq. 9. In this dimensionless number, β is the thermal expansion coefficient, and g is the gravitational acceleration.

$$\mathrm{Ma} = \frac{|\sigma_T|\Delta T H}{\rho \nu \kappa} = \mathrm{Re} \cdot \mathrm{Pr} \qquad (7)$$

$$\varepsilon = \frac{\mathrm{Ma} - \mathrm{Ma_c}}{\mathrm{Ma_c}} \qquad (8)$$

$$\mathrm{Ra} = \frac{g \beta H^3 \Delta T}{\nu \kappa} \qquad (9)$$

4.3 Effect of Buoyancy on Flow Instability

The effect of buoyancy on the stability of axisymmetric basic flow was investigated using the FZ model for high-Prandtl fluids. In ground-based experiments, the reflection symmetry of the flow field around the mid-plane is broken by buoyancy and the distorted geometry of the liquid-bridge.

First, regarding the effects of buoyancy, Bouizi et al. [2] investigated the transition from steady to oscillatory convection under zero-gravity conditions using linear stability analysis (LSA). The target liquid-bridge was set to have an aspect ratio (Γ_H) of 1 and a fluid's Prandtl number (Pr) of 20. The liquid volume was set to $V_o = 2\pi R^2 H$, and the contact lines were pinned to the edges of the upper and lower rods. When the buoyancy effect is absent and the Marangoni number is smaller than the critical Marangoni number Ma_c, the velocity and temperature fields are axially symmetric and have reflection symmetry around the mid-plane of the liquid-bridge. After the onset of the oscillation, all the disturbance components are in phase opposition about the mid-plane (see the left picture in Fig. 21). This leads to the symmetry breaking around the mid-plane.

We investigated the thresholds for breaking stability as a function of Ra using LSA. We focused on the two typical modes of the FZ model: the antisymmetric (reflection antisymmetric) and symmetric (reflection symmetric) modes, which will be illustrated in Fig. 21 later. Furthermore, we compared the results of the FZ and HZ models. Below the critical Marangoni Reynolds number, Re_c, the steady flow with symmetry is stable. Figure 20 shows neutral stability curves for different flow modes as a function of Ra. The most dangerous mode (the mode with the smallest Re_c) is the antisymmetric mode with $m = 1$ around Ra $= 0$. It is suggested that with increasing Ra, the most dangerous mode changes from the antisymmetric mode with $m = 1$ to the symmetric mode with $m = 1$ and then to the steady mode with $m = 1$. When temperature deviations have opposite signs in the upper and bottom halves of the cross section through the central axis of the liquid bridge, the mode is called an antisymmetric mode. The left figure in Fig. 21 shows the free surface temperature distribution of the antisymmetric mode. When the temperature deviations have the same sign in each half of the cross section, the mode is the symmetric one. The right figure in Fig. 21 shows the surface temperature distribution of the symmetric mode. If Ra $= 0$, these two modes have perfect antisymmetry or symmetry around the horizontal plane of mid-height, while they are no longer symmetric or antisymmetric if Ra is not zero.

The effect of buoyancy on thermocapillary convection in the FZ model is very complex compared with that in the HZ model. In the HZ model, the basic flow is stabilized by buoyancy regardless of the heating direction (heated from below or above), and Re_c with top heating is less than that with bottom heating [44, 49]. However, this conclusion cannot be directly applied to the FZ model. The influence of buoyancy on the symmetric and antisymmetric modes differs. For example, with respect to $m = 1$, the antisymmetric mode becomes more stable (Re_c increases) as the buoyancy increases (Ra increases), but the symmetric one becomes more unstable as the buoyancy increases (Fig. 20).

4.4 Effect of Geometry of a Liquid Bridge on Flow Instability

The influence of the geometry of the FZ liquid bridge on the flow transition was investigated through laboratory experiments. The experiments were performed on the ground using a small-sized liquid bridge to reduce the effects of buoyancy. In

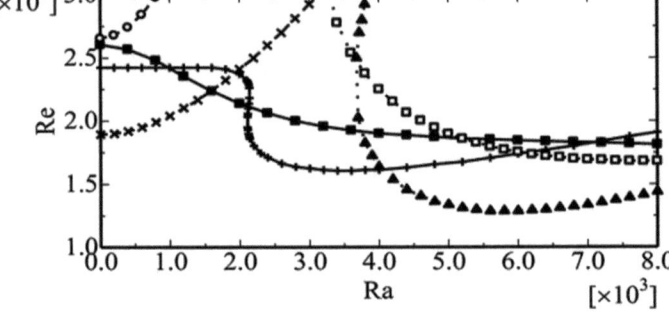

Fig. 20 Neutral stability curves for different flow modes as a function of the Rayleigh number, Ra. Reprinted from Motegi et al. [31] with permission from AIP Publishing. All rights reserved

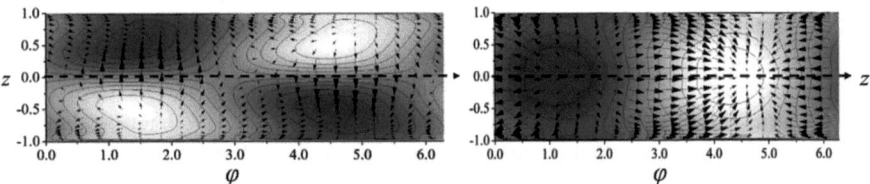

Fig. 21 Perturbation temperature and velocity distribution of oscillatory modes at the free surface. φ is the azimuthal angle of the liquid bridge, and z is the central axis passing through the liquid bridge. $z = 0$ represents the middle height of the liquid bridge. (Left: antisymmetric mode, right: symmetric mode, Ra = 0) Reprinted from Motegi et al. [31] with permission from AIP Publishing. All rights reserved

this chapter, we report the results of our research on the effect of the aspect ratio of the liquid bridge on the critical point and flow pattern in the FZ liquid bridge.

First, we briefly describe the experimental apparatus and conditions. Figure 22 illustrates the experimental apparatus. A FZ liquid bridge was suspended between two coaxial rods with equal diameters. The volume of the liquid bridge was almost the same as that of a cylinder, with a diameter of $2R$ and a height of H. The aspect ratio of the liquid bridge is defined as $\Gamma_H = H/R$. As shown in Fig. 22, the top rod was made of artificial sapphire, which has high thermal conductivity and transparency. The liquid bridge was visualized from the top through the top rod. The bottom rod was made of aluminum, which also has high thermal conductivity. These top and bottom rods were maintained at an equal temperature using a thermostatic bath. The temperature of these rods and the surface temperature of the liquid bridge at $H/2$ were measured using K-type thermocouples. Thermocapillary convection was induced by heating the

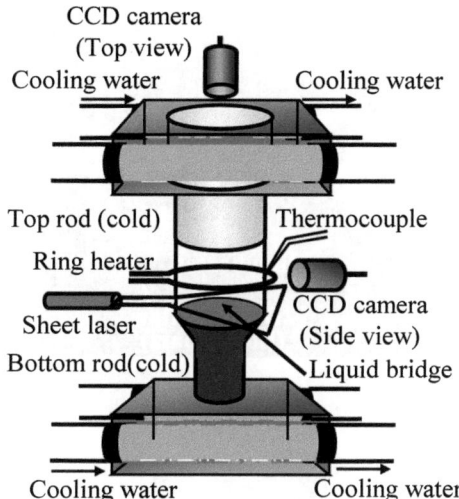

Fig. 22 Experimental apparatus. Reprinted from Kudo et al. [19]. All rights reserved

liquid bridge with a ring-shaped heater, which was located concentrically at the midheight of the bridge, as shown in Figs. 22 and 23. The ring heater was formed from a nichrome wire with a diameter of 0.1–0.2 mm. Sphere micro-particles were dispersed for flow visualization. A silicone oil (Shin-Etsu Chemical Corp, KF-96L-2cSt) was used as the test fluid.

Figure 23 shows the flow structure of a cross section and a longitudinal section under a condition of $Ma \ll Ma_c$ in the liquid bridge. The cross section is visualized with using a sheet-like red laser. White spots are the microparticles.

In the FZ liquid bridge, two-dimensional steady flow is observed below the Ma_c, similar to the HZ liquid bridge (Fig. 23a). The flow pattern differs between the upper and lower parts in relation to the horizontally placed ring heater at half the height of the liquid bridge. The vortex of the upper and lower mainstreams is depicted by solid lines in Fig. 23a. The vortex in the upper part is large and extends into the lower part. The upper vortex is observed to protrude into the central axis of the lower liquid bridge (Fig. 23a and c). This is due to the buoyant effect in a 1 g environment. Although the liquid bridge is small to minimize the effects of buoyancy, the upper liquid bridge, which is heated from below, is more influenced by buoyancy compared to the lower column, which is heated from above.

First, Fig. 24 illustrates the relationship between the shape of the liquid bridge (Γ_H) and the Ma_c. The Ma_c values range from 0.5×10^4 to 1.0×10^4, making them approximately half of the Ma_c values observed in the HZ model. One of the reasons for this discrepancy is that the length scale of Ma_c in the FZ model is determined based on the height H, which, in turn, depends on the temperature boundary.

If the length scale is based on the distance between the rigid walls ($2H$), then the Ma_c of FZ is roughly equal to that of the HZ liquid bridge. However, in the following, the Ma_c is still based on the length scale of the thermal boundary H.

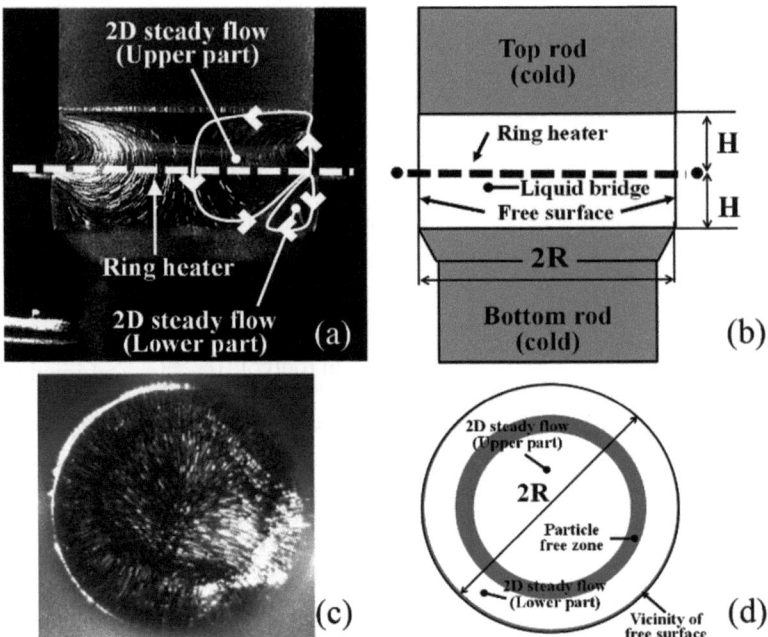

Fig. 23 Dimensions of the FZ liquid bridge: **a** photograph of a side view, **b** illustration of a side view, **c** photograph of a top view in a horizontal plane at the lower part of the liquid bridge, **d** illustration of a top view in a horizontal plane at the lower part of the liquid bridge, Ma < Ma$_c$. Reprinted from Kudo et al. [19]. All rights reserved

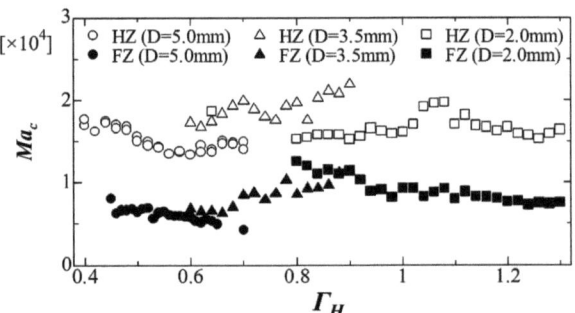

Fig. 24 The relation between critical Marangoni numbers Ma$_c$ and aspect ratio Γ_H for a FZ and a HZ liquid bridge. D represents the diameter of the liquid bridge. Reprinted from Kudo et al. [19]. All rights reserved

Next, Fig. 25 presents the relationship between the shape of the liquid bridge and the azimuthal modal numbers (m). One of the remarkable aspects in the FZ model is that multimodal structures existed across the tested aspect ratios of Γ_H. Another point of interest is that the equation $\Gamma_H \times m \approx 2$ [14, 44], which was found for the HZ model, is not simply applicable to the FZ. Regarding the first point, there is a difference between the LSA presented in Sect. 4.3 and the experiments. That is, in

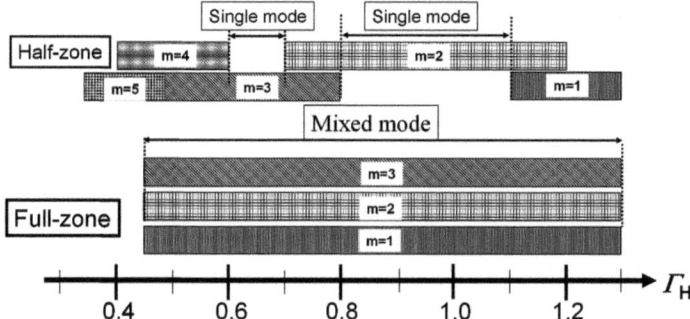

Fig. 25 The relation between dominant modal structures and aspect ratio for FZ and HZ liquid bridge. Reprinted from Kudo et al. [19]. All rights reserved

the LSA, a single mode appeared after the Ma_c, whereas multiple modes appeared from the beginning in the experiments.

Regarding the second point, the difference in the relationship between the aspect ratio Γ_H and the mode numbers m between the HZ and FZ liquid bridges is attributed to the evaluation of Γ_H based on temperature boundaries. Considering that Mode 1 is dominant when Γ_H equals 1 in the results of LSA [2, 31], if we apply another aspect ratio Γ_H' ($= 2H/R = 2\Gamma_H$) based on the boundaries of the rigid walls, then Γ_H' × m ≈ 2, which is nearly the same as that of the HZ liquid bridge.

To investigate the behavior of the mixed modes in this case, we examined the time series variation of each mode. Using the HZ method, the dominant mode could be easily determined visually as the geometry of the particle-free zone was clearly observed from the upper rod end face. However, it was challenging to do so using the FZ method, as depicted in Fig. 23c.

In the FZ method, the liquid column below the ring heater was horizontally irradiated with slit light and photographed from the upper rod edge (Fig. 22). The shape of the particle-free zone was fitted to the waveform (y), which is a superposition of modes with circumferential wavenumbers 1 to m, using the least-squares method. Then, the intensity (A_m) and spatial phase (θ) of each mode were estimated. The greater the intensity of the oscillatory flow, the larger the displacement of the distance from the center of the liquid bridge to the edge of the particle-free zone. Specifically, the movies of the visualized flow were cut into time-series still images with equal time intervals. The distance ($y(\Phi)$) from the center of the liquid bridge to the edge of the particle-free zone was then measured from $\Phi = 0$ to 2π, with an increment of $\Delta\Phi = \pi/18$ for each image. The angle Φ represents the circumferential angle (rad), with counterclockwise being positive.

The superimposed waveforms were approximated as the superposition of the circumferential wavenumbers $m = 1$ to n, expressed by Eq. 10. They were then fitted using the least squares method to obtain $A1_m$ and $A2_m$.

$$y(\Phi) = \sum_{m=1}^{n} A_m \sin(m\Phi + \theta)$$

$$= \sum_{m=1}^{n} \{A1_m \sin(m\Phi) + A2_m \cos(m\Phi)\} \quad (10)$$

where mode numbers are summed up to $m = 5$ in the HZ method, while it goes up to $m = 10$ in the FZ method.

As shown in Figs. 26 and 27, the ratio of the amplitude of each mode to the total modes, referred to as the magnitude ratio $M_R(m)$ here, and the phase inclination $\theta(m)$ of mode m are calculated using the following equations.

$$M_R(m) = \sqrt{A1_m^2 + A2_m^2} / \sqrt{\sum_{m=1}^{n}(A1_m^2 + A2_m^2)} \quad (11)$$

$$\theta(m) = \tan^{-1}(A2_m/A1_m) \quad (12)$$

The calculated results for $\Gamma_H = 0.50$ are shown in Fig. 28. One may recognize from Fig. 28a that the dominant mode structure cannot be estimated from inspection

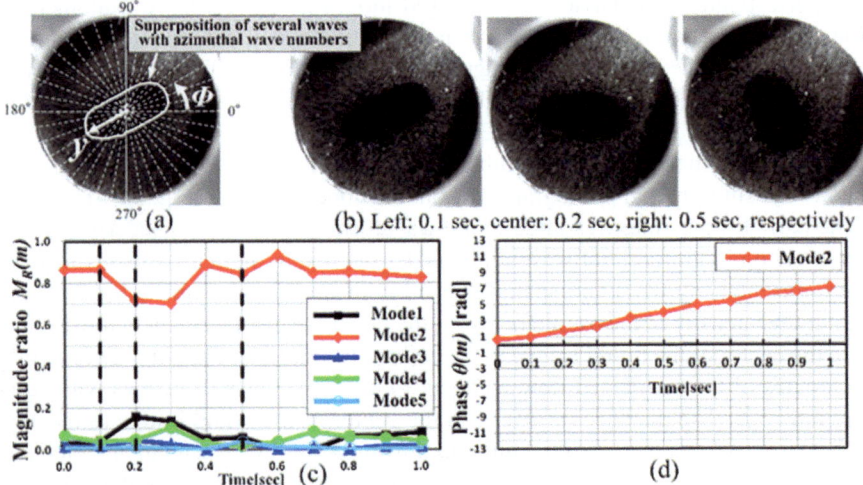

Fig. 26 Behavior of each mode in the traveling wave-type oscillatory flow in the HZ model ($\Gamma_H = 1.0$, $\varepsilon = 0.6$). The traveling wave-type oscillatory flow is characterized by the constant increase (clockwise traveling (CW)) or decrease (counterclockwise (CCW) traveling) of phase. **a**: Example of the least square fitting between a particle-free area and the superposition of several waves with azimuthal wave number (at 0.1 s), **b**: top views of the oscillatory flow ($m = 2$, CW), **c**: time series of magnitude ratio $M_R(m)$, Eq. 11, **d**: time series of phase $\theta(m)$ of mode with $m = 2$, Eq. 12. Reprinted from Kudo et al. [19]. All rights reserved

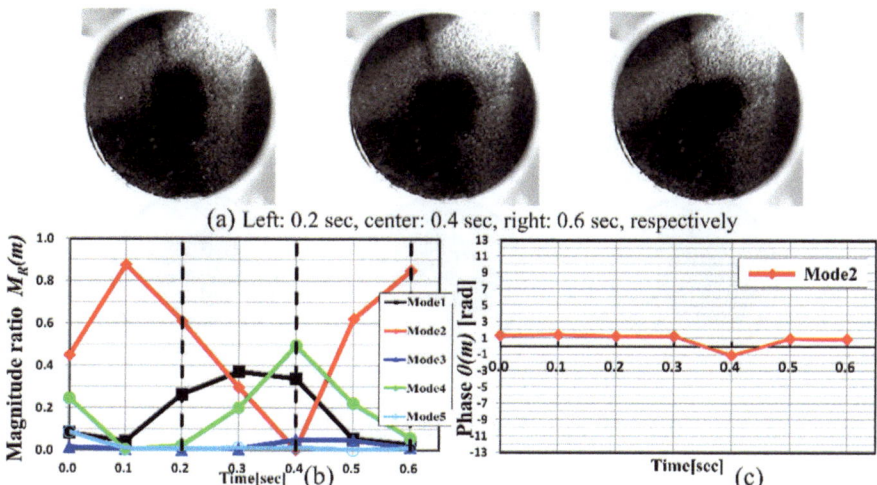

Fig. 27 Behavior of each mode in the standing wave-type oscillatory flow in the HZ model ($\Gamma_H = 1.0$, $\varepsilon = 0.13$). The standing wave-type oscillatory flow is characterized by the constant phase: **a** top views of the oscillatory flow ($m = 2$), **b** time series of the magnitude of each mode to the total modes, **c** time series of the phase of the mode with $m = 2$. Reprinted from Kudo et al. [19]. All rights reserved

due to the complex shape of the particle-free zone. Figure 28b shows the time series of the magnitude ratio of each mode to the total. Modes 6 and above are excluded from the figure because their magnitude ratios are too small. The dominant mode is not fixed, and the magnitude ratios of modes 1 to 3 change from time to time. Next, some of the characteristic structures are shown. At 0.7 s, the particle-free zone has a complicated shape, and Fig. 28b shows that it is a superposition of modes 1 to 3. At 1.1 s, the particle-free zone is nearly elliptical in shape. In Fig. 28b, Mode 2 becomes dominant. At 1.3 s, the particle-free zone appears to be pentagonal, and Fig. 28b shows that Mode 3 is dominant. Looking at the time series data as a whole, a trade-off among the modes can be observed; for example, when modes 1 and 3 are weak, mode 2 is dominant, and in other periods, they are vice versa.

In this case with $\Gamma_H = 0.5$, the shape of the particle-free zone is easily recognizable, and the time series data can be arranged in relation to the oscillation period. We investigated the mode shapes based on the time history of the phase (Fig. 28c). Specifically, Modes 2 and 3 are dominant, and both exhibit traveling waves. Mode 2 travels clockwise, while Mode 3 travels counterclockwise for the majority of the examined period. Mode 1 is a less dominant component, and its traveling direction is reversed in a short period.

The results of the modal analysis for several typical aspect ratios Γ_H show that, unlike the HZ method, multiple modes tend to appear, and their occurrence rates vary depending on the aspect ratio Γ_H. Figure 29 shows the occurrence ratios for approximately 1 s. It is interesting to note that there exists a dominant mode in two of these tested cases, and fewer dominant modes are mixed in the rest of the cases.

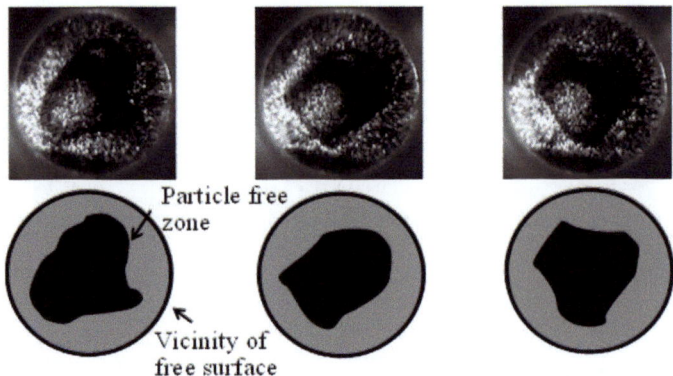

(a) Left: 0.7 sec, center: 1.1 sec, right: 1.3 sec, respectively

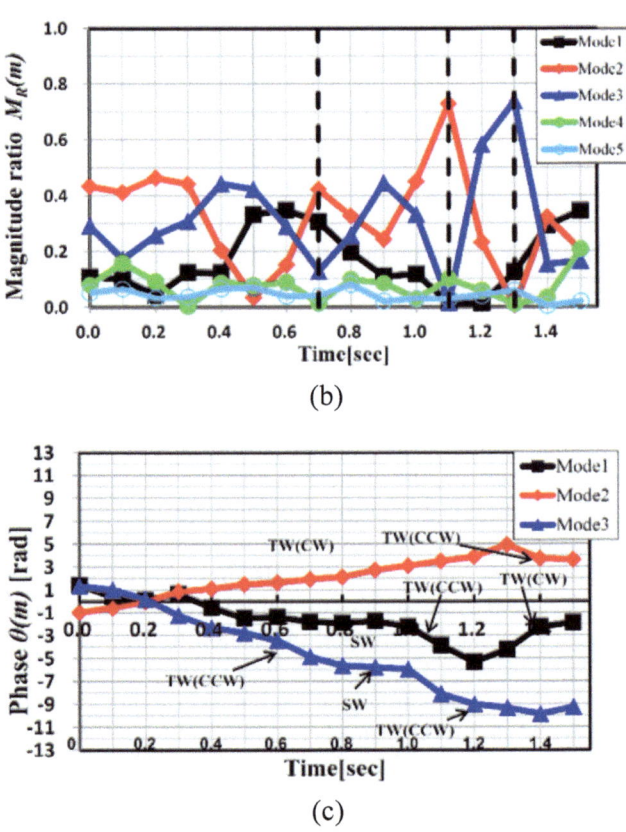

Fig. 28 Behavior of each mode in the oscillatory flow in the FZ model ($\Gamma_H = 0.50$, $\varepsilon = 1.0$). The dominant modal structures are either standing or traveling waves and change irregularly from one to another: **a** upper (top views of the oscillatory flow), lower (outline drawings of the particle-free zone shapes), **b** time series of magnitude ratios, **c** time series of phases of modes with $m = 1, 2, 3$. Reprinted from Kudo et al. [19]. All rights reserved

Specifically, $m = 2$ is dominant for $\varGamma_H = 0.63$, and $m = 1$ is dominant for $\varGamma_H = 0.88$. Thus, the product $m·\varGamma_H$ is 1.26 and 0.88, respectively, and these values are in good agreement with the relation $m·\varGamma_H \simeq 1$ found in this study as the preferred mode number for the FZ. With a further decrease in the aspect ratio ($\varGamma_H = 0.50$), higher mode numbers occupy a larger proportion. On the other hand, with an increase in the aspect ratio ($\varGamma_H = 1.2$), higher mode numbers emerge again. The reason for this latter tendency is not well understood. Generally, in the case of the HZ, the flow field becomes more complex in a tall liquid bridge. Additionally, the increased buoyancy effect may also make the flow field more complex. These complex flow fields might be recognized as components with higher mode numbers.

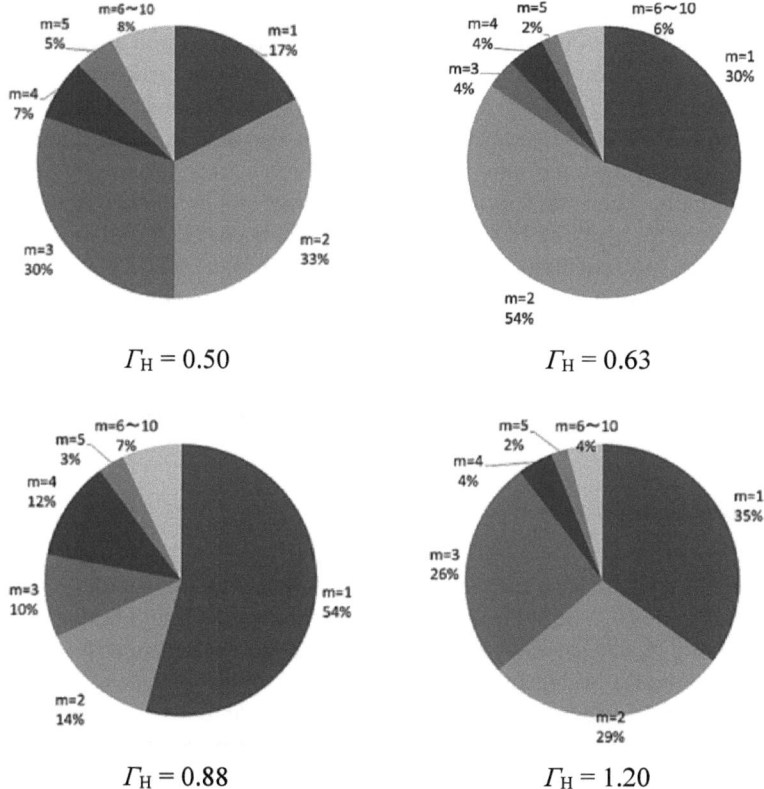

Fig. 29 Frequency of appearance of each modal structure ($m = 1 \sim 10$) in the FZ model. Reprinted from Kudo et al. [19]. All rights reserved

4.5 Oscillation Suppression by Predeterminant Circumferential Heating (PDCH)

Based on the findings mentioned in Sect. 3 of Chapter "Thermocapillary Convection in Liquid Bridges of Finite Length", we propose the application of the PreDeterminant Circumferential Heating (PDCH) to the floating zone method as well. Kudo [17] pointed out that a stable mode structure does not appear for oscillatory flow in an FZ liquid column, making PDCH a suitable suppression method for the floating zone method. Therefore, we conducted a preliminary experiment on PDCH for FZ liquid columns. Figure 30 illustrates the experimental results. The experimental apparatus is the same as shown in Fig. 22. In addition to the ring heater (1), which provides a temperature difference to the liquid column, a ring heater (2) was used for PDCH. Heater (2) is a 50-micron diameter nichrome wire processed into a ring shape with a radius of 3 ± 0.2 mm. Initially, the oscillatory flow near the critical point ($\varepsilon \approx 0.3$) alternates between Fig. 30(a-1) and (a-2). When the vortex above heater (1) widens towards the lower rod, the vortex below the heater also widens radially. Conversely, when the upper vortex narrows away from the lower rod, the lower vortex also narrows radially. This characteristic behavior represents oscillatory flow in an FZ liquid column. By applying PDCH, when heater (2) is placed at $(3/4)H$, the oscillatory flow tends to strengthen. On the other hand, when heater (2) is placed at $(1/4)H$, the oscillatory flow tends to be dampened. The heat generated by ring heater (2) was 0.15W at $(1/4)H$. When the PDCH reduced the oscillatory flow, a flow structure was observed in which the vortex on the upper side of heater (1) spread to the lower side, while the vortex on the lower side narrowed radially.

4.6 Conclusions

Regarding the effect of buoyancy on the flow transition of thermocapillary convection in a floating zone (FZ) liquid bridge, the steady flow in the liquid bridge above and below the ring heater becomes anti-symmetric (reflection asymmetric) at the mid-height of the liquid bridge in the case of Ra = 0. With increasing Ra, the most dangerous mode changes from the anti-symmetric mode to the symmetric one.

Regarding the effect of the geometry of the liquid bridge on the flow transition, Ma_c is approximately half that of the half-zone (HZ) method when evaluated with the aspect ratio Γ_H based on the distance of the temperature boundaries. Moreover, in addition to the mixing of oscillatory modes, the relationship between Γ_H and the circumferential wavenumber of oscillatory modes m is different from that of the HZ method. If the height $2H$ and the aspect ratio Γ_H' based on the velocity boundary are adopted, Ma_c and the relationship between Γ_H' and m are almost the same as those of the HZ liquid bridge. The dominant mode number changes from time to time, and the type of oscillation, either traveling or standing, also changes temporally.

Fig. 30 PDCH for a FZ liquid bridge: **a**-1 and **a**-2 without heat input, and flow is oscillating, **b** with heat input, and flow is steady. Reprinted from Kudo [17]. All rights reserved

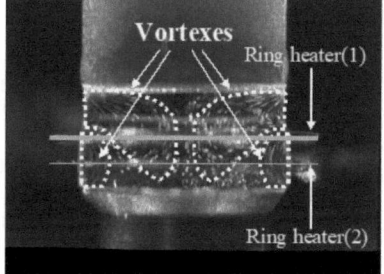

(a-1) 0s Heater (2)-off

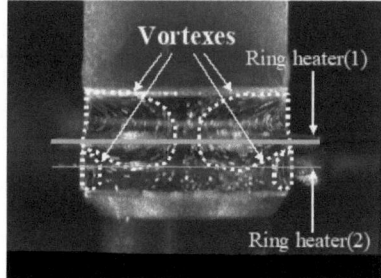

(a-2) 0.4s Heater (2)-off

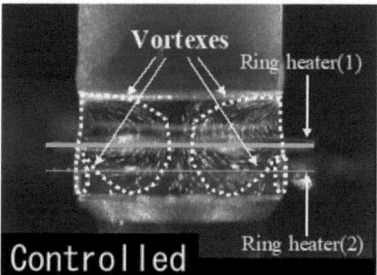

(b) Controlled- Heater (2)-on

As an application of the PreDeterminant Circumferential Heating (PDCH) to the FZ method, an additional ring heater was placed at $(3/4)H$ from the bottom. Then, the oscillatory flow tended to be enhanced, while when placed at $(1/4)H$, the oscillatory flow was damped.

References

1. Benz S, Hintz P, Riley RJ, Neitzel GP (1998) Instability of thermocapillary-buoyancy convection in shallow layers. Part 2. Suppression of hydrothermal waves. J Fluid Mech 359:165–180
2. Bouizi O, Delcarte C, Kasperski C (2007) Stability study of the floating zone with respect to the Prandtl number value. Phys. Fluids 19(114102):1–14
3. Chun CH, Wuest W (1979) Experiments on the transition from steady to the oscillatory Marangoni convection of a floating zone under reduced gravity effect. Acta Astronaut 6:1073–1082
4. Eyer A, Leiste H, Nitsche R (1985) Floating zone growth of silicon under microgravity in a sounding rocket. J Cryst Growth 71:173–182
5. Herrada MA, Lopez-Herrada JM, Vega EJ, Montanero JM (2011) Numerical simulation of a liquid bridge in a coaxial gas flow. Phys Fluids 23:1–11
6. Houchens BC, Walker JS (2005) Modeling the floating zone: instabilities in the half zone and full zone. J Thermophys Heat Transf 19(2):186–198
7. Irikura M, Arakawa Y, Ueno I, Kawamura H (2005) Effect of ambient flow upon onset of oscillatory thermocapillary convection in half-zone liquid bridge. Microgravity Sci Technol 16:176–180
8. Jayakrishnan R, Tiwari S (2018) Effect of ambient conditions on flow and heat transfer in a liquid bridge. J Thermophys and Heat Transf 32(2):380–391
9. Kamotani Y, Lee KJ (1989) Oscillatory thermocapillary flow in a liquid column heated by a ring heater. Physicochem Hydrodyn 11:729–736
10. Kamotani Y, Lee JH, Ostrach S, Pline A (1992) An experimental study of oscillatory thermocapillary convection in cylindrical containers. Phys Fluids A 4:955–962
11. Kamotani Y, Ostrach S (1998) Theoretical analysis of thermocapillary flow in cylindrical columns of high Prandtl number fluids. J Heat Transf 120(6):758–764
12. Kamotani Y, Wang L, Hatta S, Selver L, Yoda S (2001) Effect of free surface heat transfer on onset of oscillatory thermocapillary flow of high Prandtl number fluid. J Jpn Soc Microgravity Appl 18(4):281–288
13. Kamotani Y, Wang L, Hatta S, Wang A, Yoda S (2003) Free surface heat loss effect on oscillatory thermocapillary flow in liquid bridges of high Prandtl number fluids. Int J Heat Mass Transf 46:3211–3220
14. Kawamura H, Ono Y, Ueno I (2001) Transition and modal structure of oscillatory Marangoni convection in liquid bridge. Trans Jpn Soc Mech Eng Ser B 67(658):1466–1473 (in Japanese)
15. Kousaka Y, Kawamura H (2006) Numerical study on the effect of heat loss upon the critical Marangoni number in a half-zone liquid bridge. Microgravity Sci Technol 18:141–145
16. Kudo M, Shiomi J, Ueno I, Amberg G, Kawamura H (2005) Experiment on multimode feedback control of non-linear thermocapillary convection in a half-zone liquid bridge. Adv Space Res 36:57–63
17. Kudo M (2006) Control of oscillatory flows in thermocapillary convection in a liquid bridge. Tokyo University of Science, Doctoral thesis, 77–82 (in Japanese)
18. Kudo M, Kawamura H (2006) Control of oscillatory thermocapillary convection in a half-zone liquid bridge by circumferential round heating. Therm Sci Eng 14(2):1–7 (in Japanese)
19. Kudo M, Ueno I, Kawamura H (2014) Transition of thermocapillary convection in a full-zone liquid bridge. Trans JSME 80 TEP0095 (in Japanese)
20. Kudo M, Akiyama Y, Takei S, Motegi K, Ueno I (2015) Effect of ambient air flow on thermocapillary convection in a full-zone liquid bridge. Interfacial Phenom Heat Transf 3(3):231–242
21. Kuhlmann HC, Rath HJ (1993) Hydrodynamic instabilities in cylindrical thermocapillary liquid bridges. J Fluid Mech 247:247–274
22. Lappa M (2003) Three-dimensional numerical simulation of Marangoni flow instabilities in floating zones laterally heated by an equatorial ring. Phys Fluids 15:776–789

23. Lappa M (2004) Combined effect of volume and gravity on three-dimensional flow instability in noncylindrical floating zones laterally heated by an equatorial ring. Phys Fluids 16:331–343
24. Lappa M (2005) Analysis of flow instabilities in convex and concave floating zones heated by an equatorial ring under microgravity condition. Comput Fluids 34:743–770
25. Lappa M (2016) On the onset of multi-wave patterns in laterally heated floating zones for slightly supercritical conditions. Phys Fluids 28(124105):1–23
26. Lappa M (2016) Control of convection patterning and intensity in shallow cavities by harmonic vibrations. Microgravity Sci Technol 28:29–39
27. Leypoldt J, Kuhlmann HC, Rath HJ (2000) Three-dimensional numerical simulation of thermocapillary flows in cylindrical liquid bridge. J Fluid Mech 414:285–314
28. Masud J, Kamotani Y, Ostrach S (1997) Oscillatory thermocapillary flow in cylindrical columns of high Prandtl number fluids. J Thermophys Heat Transfer 11(1):105–111
29. Melnikov DE, Shevtsova V (2014) The effect of ambient temperature on the stability of thermocapillary flow in liquid column. Int J Heat Mass Transfer 74:185–195
30. Minakuchi H, Okano Y, Dost S (2004) A three-dimensional numerical simulation study of the Marangoni convection occurring in the crystal growth of $SixGe1-x$ by the floating zone technique in zero gravity. J Cryst Growth 266:140–144
31. Motegi K, Kudo M, Ueno I (2017) Linear stability of buoyant thermocapillary convection for a high-Prandtl number fluid in a laterally heated liquid bridge. Phys Fluids 29(044106):1–11
32. Muldoon F (2013) Control of hydrothermal waves in a thermocapillary flow using a gradient-based control strategy. Int J Num Methods Fluids 72(1):90–118
33. Muldoon F (2018) Numerical study of hydrothermal wave suppression in thermocapillary flow using a predictive control method. Comput Math Math Phys 58(4):493–507
34. Pline A, Jacobson TP, Wanhainen JS, Petrarca DA (1990) Hardware development for the surface tension driven convection experiment. J Spacecr Rockets 27:312–317
35. Petrov V, Schatz MF, Muehlner KA, Vanhook SJ, McComick WD, Swift JB, Swinney HL (1996) Nonlinear control of remote unstable states in a liquid bridge convection experiment. Phys Rev Lett 77:3779–3782
36. Petrov V, Muehlner KA, Vanhook SJ, Swinney HL (1998) Model-independent nonlinear control algorithm with application to a liquid bridge experiment. Phys Rev E 58:427–433
37. Sakurai M, Tamura A, Kinoshita A, Hirata A (1998) Marangoni convection in a liquid bridge that is heated by a ring heater in a full zone model. J Jpn Soc Microgravity Appl 15:419–424
38. Schwabe D, Scharmann A (1979) Some evidence for the existence and magnitude of a critical Marangoni number for the onset of oscillatory flow in crystal growth melts. J Cryst Growth 46:125–131
39. Shevtsova V, Gaponenko YA, Nepomnyashchy A (2013) Thermocapillary flow regimes and instability caused by a gas stream along the interface. J Fluid Mech 714:644–670
40. Shiomi J, Amberg G, Alfredsson H (2001) Active control of oscillatory thermocapillary convection. Phys Rev E 64:031205–031211
41. Shiomi J, Amberg G (2002) Active control of a global thermocapillary instability. Phys Fluids 14:3039–3045
42. Shiomi J, Kudo M, Ueno I, Kawamura H, Amberg G (2003) Feedback control of oscillatory thermocapillary convection in a half-zone liquid bridge. J Fluid Mech 496:193–211
43. Smith MK, Davis SH (1983) Instabilities of dynamic thermocapillary liquid layers. Part 1: convective instabilities. J Fluid Mech 132:119–144
44. Velten R, Schwabe D, Scharmann A (1991) The periodic instability of thermocapillary convection in cylindrical liquid bridges. Phys Fluids A3(3):267–279
45. Wang L (2003) Effects of free surface deformation and heat transfer on thermocapillary flows of high Prandtl fluids. PhD dissertation, Case Western Reserve University
46. Wang A (2005) Effects of free surface heat transfer and shape on thermocapillary flow of high Prandtl fluids. PhD dissertation, Case Western Reserve University
47. Wang A, Kamotani Y, Yoda S (2007) Oscillatory thermocapillary flow in liquid bridges of high Prandtl number fluid with free surface heat gain. Int J Heat Mass Transf 50:4195–4205

48. Wanschura M, Shevtsova VM, Kuhlmann HC, Rath HJ (1995) Convective instability mechanism in thermocapillary liquid bridges. Phys Fluids 7:912–925
49. Wanschura M, Kuhlmann HC, Rath HJ (1997) Linear stability of two-dimensional combined buoyant thermocapillary flow in cylindrical liquid bridges. Phys Rev E 55:7036
50. Xun B, Li K, Chen PG, Hu WR (2009) Effect of interfacial heat transfer on the onset of oscillatory convection in liquid bridge. Int J Heat Mass Transf 52:4211–4220
51. Xun B, Li K, Hu WR, Imaishi N (2011) Effect of interfacial heat exchange on thermocapillary flow in a cylindrical liquid bridge in microgravity. Int J Heat Mass Transf 54:1698–1705
52. Yano T, Maruyama K, Matsunaga T, Nishino K (2016) Effect of ambient gas flow on the instability of Marangoni convection in liquid bridges of various volume ratios. Int J Heat Mass Transf 99:182–191
53. Yano T, Hirotani M, Nishino K (2018) Effect of interfacial heat transfer on basic flow and instability in a high-Prandtl-number thermocapillary liquid bridge. Int J Heat Mass Transf 125:1121–1130
54. Yano T, Nishino K, Matsumoto S, Ueno I, Komiya A, Kamotani Y, Imaishi N (2018) Report on microgravity experiments of dynamic deformation effects on Marangoni instability in high-Prandtl-number liquid bridges. Microgravity Sci Technol 30:599–610
55. Yasnou V, Gaponenko Y, Mialdun A, Shevtsova V (2018) Influence of a coaxial gas flow on the evolution of oscillatory states in a liquid bridge. Int J Heat Mass Transf 123:747–759

Open Access This chapter is licensed under the terms of the Creative Commons Attribution 4.0 International License (http://creativecommons.org/licenses/by/4.0/), which permits use, sharing, adaptation, distribution and reproduction in any medium or format, as long as you give appropriate credit to the original author(s) and the source, provide a link to the Creative Commons license and indicate if changes were made.

The images or other third party material in this chapter are included in the chapter's Creative Commons license, unless indicated otherwise in a credit line to the material. If material is not included in the chapter's Creative Commons license and your intended use is not permitted by statutory regulation or exceeds the permitted use, you will need to obtain permission directly from the copyright holder.

Microgravity Experiments in Kibo Onboard the International Space Station

Taishi Yano⊙, Koichi Nishino⊙, Satoshi Matsumoto⊙, Ichiro Ueno⊙, Hiroshi Kawamura, and Yasuhiro Kamotani

Abstract This chapter reviews a series of microgravity experiments on the thermocapillary convection in a liquid bridge of high-Prandtl-number fluid (i.e., MEIS, Marangoni-UVP, and Dynamic Surf projects), which has been conducted in the Japanese Experiment Module—Kibo—onboard the International Space Station. The silicone oils with the Prandtl number of $Pr = 67\text{--}207$ were used as the test liquid, and the liquid bridge was formed between the disks with diameter of 10 mm to 50 mm. The driving force of convection was generated by imposing the temperature difference between these disks, and the resultant thermocapillary convection was observed with various measurement techniques. The first experiment started in 2008 and continued until 2020, during which period many findings and important experimental data were obtained. Starting with general information about the project, the experimental equipment, procedures, and typical results are introduced in this chapter. It contains the onset conditions of thermocapillary-convection instability for a variety of experimental conditions, the measurement results of velocity and temperature fields, the dynamic behavior of the liquid-bridge free surface, the transition process

T. Yano (✉)
Kanagawa University, Yokohama, Japan
e-mail: t-yano@kanagawa-u.ac.jp; ft102124oh@jindai.jp

K. Nishino
Yokohama National University, Yokohama, Japan
e-mail: nishino-koichi-fy@ynu.ac.jp

S. Matsumoto
Japan Aerospace Exploration Agency, Tsukuba, Japan
e-mail: matsumoto.satoshi@jaxa.jp

I. Ueno
Tokyo University of Science, Noda, Japan
e-mail: ich@rs.tus.ac.jp

H. Kawamura
Tokyo University of Science & Suwa University of Science, Professor Emeritus, Chino, Japan
e-mail: kawa@rs.sus.ac.jp

Y. Kamotani
Case Western Reserve University, Cleveland, OH, USA
e-mail: yxk@case.edu

to the chaotic or turbulent states, and so on. In addition to the experimental results, the related numerical simulations and theoretical analyses are also introduced.

1 Outline of Microgravity Experiment

Taishi Yano, Koichi Nishino, and Satoshi Matsumoto

The project of microgravity (μg, hereinafter) research called Marangoni Experiment in Space (Marangoni Exp/MEIS)—the official name of the project is *"Chaos, Turbulence and its Transition Process in Marangoni Convection"*—started in 2008 as the first thermocapillary-convection experiment in the International Space Station (ISS) and as the first science experiment in the Kibo Japanese Experiment Module onboard the ISS [1]. After that, two projects called Marangoni UVP (also known as MaranGoniat) and Dynamic Surf—their official names are *"Spatio-temporal Flow Structure in Marangoni Convection"* and *"Experimental Assessment of Dynamic Surface Deformation Effects in Transition to Oscillatory Thermocapillary Flow in Liquid Bridge of High Prandtl Number Fluid"*, respectively—have been conducted so far. The outlines of these projects are summarized in Table 1. More details on the projects can be found in Data Archives and Transmission System (DARTS)[1] maintained by Center for Science-satellite Operation and Data Archive (C-SODA) at Institute of Space and Astronautical Science (ISAS)/Japan Aerospace Exploration Agency (JAXA). In this chapter, these projects are referred to as MEIS, UVP, and DS, respectively. As shown in Table 1, each project consisted of multiple missions, depending on the research targets (e.g., effects of diameter D, Prandtl number Pr, and so on). A total of more than 300 days of experiments were conducted through these projects, and a lot of scientific knowledge were achieved.

1.1 Fluid Physics Experiment Facility (FPEF)

All experiments were performed in the RYUTAI rack (Fig. 1a) mounted on the Kibo module, which is a payload rack specialized in fluid dynamics research. This rack was equipped with multiple facilities, and the Fluid Physics Experiment Facility (FPEF, Fig. 1b) and Image Processing Unit (IPU, Fig. 1c) were mainly used for the thermocapillary-convection experiments. FPEF was designed to observe thermocapillary convection in μg and consisted of the power supply system, exchangeable experiment cell, measurement hardware, and so on [2]. A cassette packed with test liquid and disks for the liquid-bridge formation system were installed in the exchangeable experiment cell, which was filled with argon or neon at the absolute pressure of 88.2–101.3 kPa and the room temperature of 18.3–26.7 °C [3]. Figure 2

---

[1] https://www.darts.isas.jaxa.jp/iss/kibo/experiment/marangoni.

Table 1 Outline of μg experiments on thermocapillary convection in a liquid bridge of high-Prandtl-number fluid conducted aboard the International Space Station (ISS)

	Marangoni Experiment in Space				
Short title	MEIS-1	MEIS-2	MEIS-3	MEIS-4	MEIS-5
Start date[*1]	Aug 19, 2008	Jul 9, 2009	Sep 20, 2011	Aug 28, 2010	Jun 26, 2012
End date[*1]	Oct 17, 2008	Aug 25, 2009	Feb 8, 2012	Dec 22, 2010	Feb 25, 2013
Experiment days	31	18	29	27	32
PI(s)[*2,3]	Prof. Hiroshi Kawamura (Tokyo University of Science, Japan)[*4]				
	Prof. Koichi Nishino (Yokohama National University, Japan)[*5]				
Coordinator	Dr. Satoshi Matsumoto (JAXA, Japan)				
Experiment cell	MS30	MS30	MS30	MI50	MI50
Liquid	KF-96L-5cs	KF-96L-5cs[*6]	KF-96-20cs[*6]	KF-96-20cs	KF-96-10cs
Gas	Argon	Argon	Argon	Argon	Argon
D (mm)	30	30	30	50	50
Pr	67	67	207	207	112

	Marangoni UVP		Dynamic Surf		
Short title	UVP-1	UVP-2	DS-1	DS-2	DS-3
Start date[*1]	Jan 29, 2010	Jan 27, 2011	Sep 13, 2013	Dec 1, 2014	Nov 19, 2015
End date[*1]	Jul 16, 2010	Feb 27, 2020	Feb 27, 2014	Apr 15, 2015	Nov 10, 2016
Experiment days	23	56	33	31	30
PI(s)[*2,3]	Prof. Shin-ichi Yoda (JAXA, Japan)		Prof. Yasuhiro Kamotani (Case Western Reserve University, USA)		
Coordinator	Dr. Satoshi Matsumoto (JAXA, Japan)		Dr. Satoshi Matsumoto (JAXA, Japan)		
Experiment cell	MI50	MI50	MD30	MD10	MD30
Liquid	KF-96L-5cs	KF-96-10cs	KF-96L-5cs	KF-96L-5cs	KF-96-10cs
Gas	Argon	Argon	Argon	Argon/Neon	Argon
D (mm)	50	50	30	10	30
Pr	67	112	67	67	112

[*1]Japan Standard Time (JST), [*2]Principal Investigator, [*3]Affiliation is the one at that time, [*4]MEIS-1, [*5]MEIS-2 to 5, [*6]Containing photochromic dye

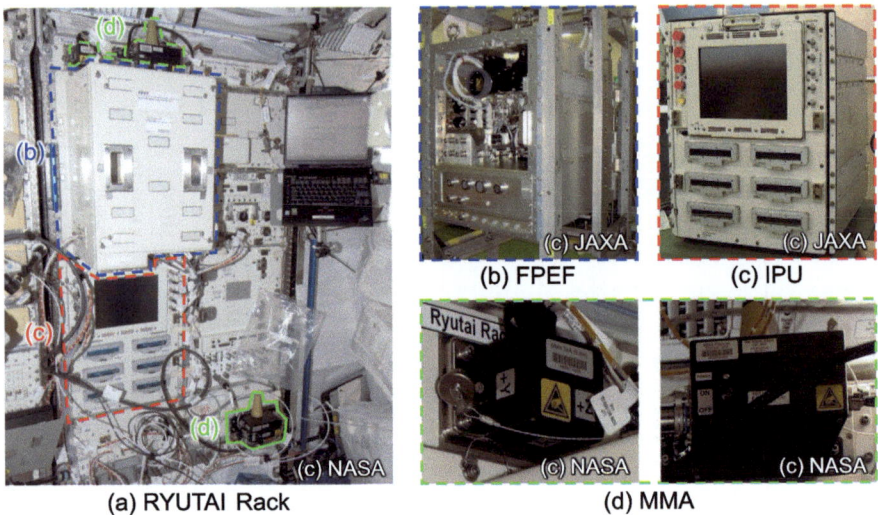

Fig. 1 Photographs of **a** RYUTAI rack, **b** Fluid Physics Experiment Facility (FPEF), **c** Image Processing Unit (IPU), and **d** Microgravity Measurement Apparatus (MMA) installed in the Kibo module (courtesy of NASA/JAXA). (©2018 The Jpn. Soc. Microgravity Appl., adapted with permission from Yano et al. [4]. All rights reserved)

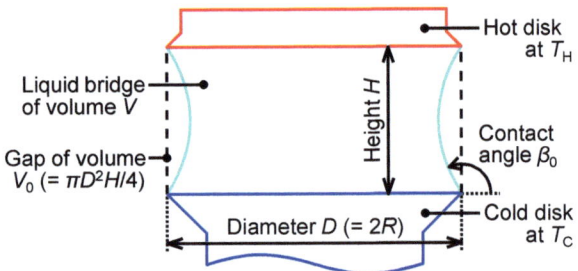

Fig. 2 Geometry of the liquid bridge with height H suspended in a gap between cold and hot disks with diameter D (and radius R)

shows the schematic diagram of the target geometry. The liquid bridge of height H was formed in a gap between a pair of concentric disks with equal diameter D (and radius R (= $D/2$)). One disk was controlled at lower temperature and the other at higher temperature to generate a thermocapillary convection; therefore, the former is referred to as the cold disk and the latter as the hot disk in this chapter. The temperatures of these disks are, respectively, denoted by T_C and T_H. Figure 3 shows pictures of liquid bridges with three different diameters: (1) $D = 10$ mm, (2) $D = 30$ mm, and (3) $D = 50$ mm, wherein the diameter-based aspect ratio A_r (= H/D) is unity in all cases. Due to the existence of gravity, the typical diameter of the liquid bridge in terrestrial experiments is 2–5 mm. It is difficult (or impossible) to stably hold a liquid bridge with a diameter greater than 10 mm for a long time (say, minutes to hours) except under μg environment on the space station at the time of writing this document.

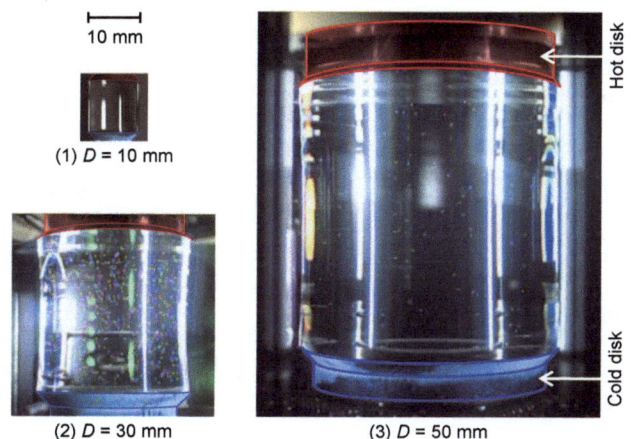

Fig. 3 Liquid bridges with $A_r = 1$, wherein disk diameters are (1) $D = 10$ mm, (2) $D = 30$ mm, and (3) $D = 50$ mm. Photographs are shown in almost actual scale. (Photo courtesy of JAXA. All rights reserved)

1.2 Test Liquid and Tracer Particles

In all projects, the test liquid was silicone oil (Silicone Fluid KF-96, Shin-Etsu Chemical Co., Ltd., Japan), but the physical properties were different depending on the research targets. The products used in each mission are shown in Table 1. Silicone Fluids KF-96L-5cs, KF-96-10cs, and KF-96-20cs are referred to hereinafter as 5-, 10-, and 20-cSt (centistokes) silicone oils, respectively. Their physical properties at 25 °C are listed in Table 2. The values of density ρ, kinematic viscosity ν, thermal conductivity λ, specific heat at constant pressure c_p, volumetric thermal expansion coefficient β, and surface tension σ are based on the catalog data [5], and the values of dynamic viscosity $\mu\ (= \rho\nu)$, thermal diffusivity $\kappa\ (= \lambda/(\rho c_p))$, and Prandtl number Pr $(= \nu/\kappa)$ are calculated from them. We note that the values of c_p in the catalog are in cal/(g · K), and they are converted to J/(kg · K) by the International Steam Table calorie (i.e., 1 cal$_{IT}$ = 4.1868 J). The values of temperature coefficient of surface tension $\sigma_T\ (= \partial\sigma/\partial T)$ are measurement data [6, 7]. Two values of σ_T are given for 5- and 20-cSt silicone oils: one for pure silicone oil and the other for photochromic-dye-containing silicone oil. The solution of the dye was 1,3,3-trimethyl-6'-nitrospiro[indoline-2,2'-chromene] (TNSB, prepared by Prof. Masahiro Kawaji, University of Toronto, Canada), and it was blended at 0.05 % by weight [8, 9]. The effect of photochromic dye on other physical properties were neglected here.

The physical properties of silicone oils were assumed to be constant except for σ and ν. The value of σ was a linear function of temperature T (°C) as

$$\sigma(T) = \sigma_0 + \sigma_T(T - 25), \tag{1}$$

Table 2 Physical properties of silicone oils used in μg experiments at 25 °C

	KF-96L-5cs	KF-96-10cs	KF-96-20cs
Prandtl number Pr ($= \nu/\kappa$)	67	112	207
Density ρ (kg/m^3)	915	935	950
Kinematic viscosity ν (m^2/s)	5.0×10^{-6}	10.0×10^{-6}	20.0×10^{-6}
Dynamic viscosity μ (Pa · s)	4.58×10^{-3}	9.35×10^{-3}	1.90×10^{-2}
Thermal conductivity λ (W/(m · K))	0.12	0.14	0.15
Specific heat at constant pressure c_p (cal/(g · K)) or (J/(kg · K))	1.76×10^3 or 0.42	1.68×10^3 or 0.40	1.63×10^{-3} or 0.39
Thermal diffusivity κ (m^2/s)	7.46×10^{-8}	8.94×10^{-8}	9.67×10^{-8}
Volumetric thermal expansion coefficient β (1/K)	1.09×10^{-3}	1.06×10^{-3}	1.04×10^{-3}
Surface tension σ (N/m)	1.97×10^{-2}	2.01×10^{-2}	2.06×10^{-2}
Temperature coefficient of surface tension σ_T (N/(m · K))	-6.58×10^{-5} [*1] -6.26×10^{-5} [*2]	-6.12×10^{-5} [*1]	-6.24×10^{-5} [*1] -5.85×10^{-5} [*2]

[*1] Pure silicone oil, [*2] Photochromic-dye-containing silicone oil

where σ_0 is σ at the reference temperature T_0 (= 25 °C). According to the manufacturer's data [5], the value of ν depends on T (but in the range from -25 °C to 250 °C) as

$$\nu(T) = \nu_0 \exp\left(\frac{25 - T}{273.15 + T}\right), \tag{2}$$

where ν_0 is ν at T_0. In this chapter, the representative kinematic viscosity $\bar{\nu}$ is evaluated as the arithmetic mean of ν at T_C and T_H. Figure 4 shows temperature dependences of $\bar{\nu}$ and Pr (= $\bar{\nu}/\kappa$) as a function of the temperature difference between supporting disks ΔT (= $T_H - T_C$), wherein T_C is constant at 20 °C. As shown in this figure, both $\bar{\nu}$ and Pr decrease monotonically with ΔT. Although the values of Pr strongly depend on ΔT, the representative Pr for 5-, 10-, and 20-cSt silicone oils are denoted by Pr = 67, 112, and 207, respectively, in this chapter.

The silicone oils contained spherical acrylic tracer particles for the flow visualization (custom-made particle of MX series, Soken Chemical & Engineering Co., Ltd., Japan). The surface of each particle was coated with a gold-nickel alloy, which was

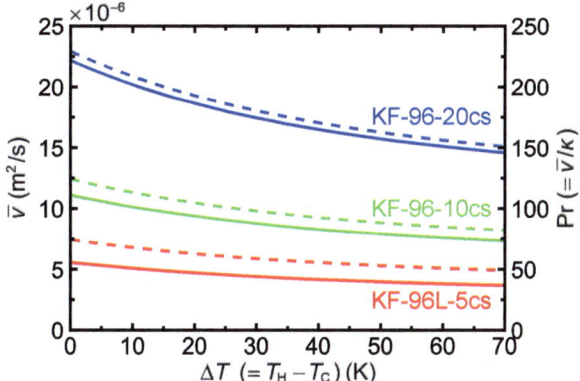

Fig. 4 Temperature dependences of kinematic viscosity $\bar{\nu}$ (*solid lines*) and Prandtl number Pr (*bashed lines*) of silicone oils, wherein the cold-disk temperature T_C is assumed to be 20 °C

important for providing good visibility and for keeping dispersibility in the long-term μg experiments. The tracer particles with average diameters of 30, 180, 200 μm or a combination thereof were mixed into the silicone oil before being launched to the ISS. Their average densities were 1486, 1364, and 1296 kg/m³, respectively. The average diameter and total number (for 180 and 200 μm-diameter particles) or total weight (for 30 μm-diameter particle) of tracer particles in each mission are listed in Table 3.

Table 3 Average diameter, density, and total number/mass of tracer particles in each mission

	MEIS-1	MEIS-2	MEIS-3	MEIS-4	MEIS-5
Diameter (μm)	30/180	180	180	180	180
Density (kg/m³)	1486/1364	1364	1364	1364	1364
Total number[*1] or mass[*2]	18.6 mg/15	500	2000	500	10000
	UVP-1	UVP-2	DS-1	DS-2	DS-3
Diameter (μm)	30	30	200	30	200
Density (kg/m³)	1486	1486	1296	1486	1296
Total number[*1] or mass[*2]	140 mg	140 mg	2000	0.0585 mg	2000

[*1] Without the unit, [*2] With the unit of (mg)

2 Experiment Cell and Measurement Apparatus

Taishi Yano, Koichi Nishino, and Satoshi Matsumoto

The experiment cell with $D = 10$ mm was named MD10. Both cold and hot disks were made of aluminum alloy covered with a black anodic oxide layer, and three K-type thermocouple sensors were embedded in each disk to measure T_C and T_H. For $D = 30$ mm, there were two types of experiment cell: MS30 for MEIS project and MD30 for DS project. Their basic functions were the same, but measurement apparatuses and performance were somewhat different. For $D = 50$ mm, the same experiment cell MI50 was used in MEIS and UVP projects. The cold disks in MS30, MD30, and MI50 were made of aluminum alloy with a black anodic oxide layer, of which temperatures were measured by three flush-mounted K-type thermocouple sensors. The hot disks in these cells were made of synthetic sapphire. The present sapphire was transparent and had a relatively high thermal conductivity of $\lambda_{Al_2O_3} = 42$ W/(m · K) (e.g., typical thermal conductivity of glass is 1 W/(m · K)). Two indium tin oxide (ITO) film sensors were mounted flush with the front side of the hot disk (i.e., surface in contact with the silicone oil) and three platinum resistance thermometer sensors (Pt sensors, hereinafter) were mounted on the back side.

All sensors were calibrated on the ground before being launched to the ISS; however, those for T_H measurement had considerable residual bias. In MD10 used in DS-2, outputs of thermocouple sensors embedded in cold and hot disks showed a slight difference of 0.04 K (Table 4), even though the disks were closed and the temperature inside the experiment cell was nearly uniform. This temperature difference was added to the thermocouple sensor outputs to correct T_H. On the other hand, the bias of ITO sensor was corrected on the basis of the one-dimensional steady heat conduction through the fluid (i.e., silicone oil or argon) and sapphire as reported by Nishino et al. [6]. In the simple model shown in Fig. 5, the heat flux densities through the fluid and sapphire can be evaluated as $q_{fluid} = -\lambda_{fluid}(T_H - T_C)/H$ and $q_{Al_2O_3} = -\lambda_{Al_2O_3}(T_{Pt} - T_H)/H_{HD}$, respectively, by Fourier's law, where λ_{fluid} is the thermal conductivity of the fluid (i.e., λ in Table 2 for silicone oil and 0.018 W/(m · K) for argon), T_{Pt} is the temperature on the back side of the hot disk from Pt sensors, and H_{HD} (= 12.5 mm) is the height of the hot disk. The temperatures T_C and T_{Pt} were measured for sufficiently short H (i.e., 1 mm or less) under nearly isothermal condition so that there was no (or extremely weak) convection in the fluid. The value of T_H was evaluated using these temperatures by assuming that the heat flux densities through the fluid and sapphire were equivalent (i.e., $q_{fluid} = q_{Al_2O_3}$). As the results, there were discrepancies of -0.2 K to 2.4 K between the evaluated T_H and ITO sensor outputs, and such temperature differences were adopted as the correction values of the ITO sensors (Table 4). This temperature correction of T_H was performed once in each mission.

The cold-disk temperature was controlled by the Peltier elements through heat exchange with the coolant water with the temperature of 16–23 °C. The controlling capability of T_C in MS30 and MI50 was roughly 5–20 °C. On the other hand, that in MD10 and MD30 was roughly 5–40 °C because both cooling and heating of the cold

Table 4 Materials of disks and correction values for T_H in each mission

	MEIS-1	MEIS-2	MEIS-3	MEIS-4	MEIS-5
Cold disk	Aluminum				
Hot disk	Sapphire				
Correction value (K)	2.0	2.4	1.2	1.7	0.9
	UVP-1	UVP-2	DS-1	DS-2	DS-3
Cold disk	Aluminum		Aluminum		
Hot disk	Sapphire		Sapphire	Aluminum	Sapphire
Correction value (K)	1.29	0.33	0.1	0.04	−0.2

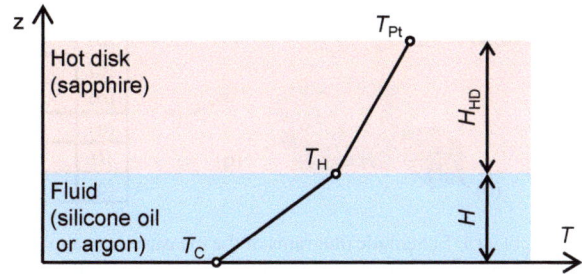

Fig. 5 One-dimensional steady heat conduction through the fluid (i.e., silicone oil or argon) and sapphire

disk was possible by changing the direction of electric current passing through the Peltier elements. The hot-disk temperature in MD10 was controlled by embedded polyimide heaters, while that in other experiment cells was controlled by the ITO film heater formed on the back side of the disk. The temperature T_H could be raised up to 90 °C. However, since there was no cooling function, it could not be lowered below T_C; therefore, the relation $\Delta T (= T_H - T_C) > 0$ had always been established. The temperature uniformity on the surface in contact with silicone oil was confirmed to be less than ±0.1 K for both cold and hot disks [6].

All disks had a 45° sharp pinning edge, as shown in Fig. 3, to prevent the liquid creeping over the disk edge. Additionally, the side surface of each disk was coated with an anti-wetting chemical (Fluoro Surf, FS-1010TH, FluoroTechnology Co.,LTD., Japan). These measures for preventing the liquid leakage were very important in the on-orbit experiment aboard the ISS because each mission had to be continued for several weeks to several months or more than a year without receiving any additional test liquids. However, these measures were not perfect, and the amount of silicone oil was gradually lost during the mission due to the liquid dripping from the disk edge and evaporation. Therefore, the science team members and operation staffs had always checked the amount of test liquid and had carried out the experiment under feasible condition with remaining silicone oil.

Schematic diagrams of the measurement system are shown in Fig. 6(1)–(3). As shown in these figures, the common measurement apparatuses for all experiment cells

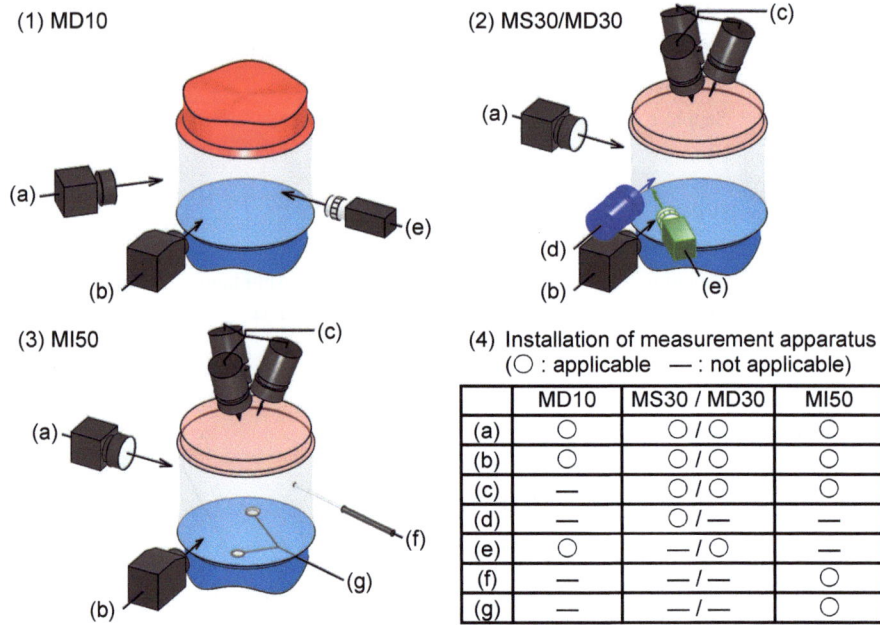

Fig. 6 (1)–(3) Schematic diagrams of the measurement apparatuses installed in 10, 30, and 50 mm-diameter experiment cells. (4) Installation status of each apparatus (i.e., **a** side-view camera, **b** IR camera, **c** top-view cameras, **d** camera for photochromic dye activation technique, **e** MIDM, **f** insertion thermocouple sensor, and **g** ultrasonic velocity profiler)

were (a) a side-view camera and (b) a black and white (B/W, hereinafter) infrared (IR) camera. The side-view camera was mainly used to observe the liquid-bridge shape and the overall flow pattern. The images in MD10 were gray scale, while those in other experiment cells were color scale. The IR camera, of which the wavelength sensitivity was 8–14 μm, was used to measure the temperature distribution on the liquid-bridge free surface. The model of IR camera was changed in 2013, and the new one was used from DS project. Catalog values for the minimum temperature resolutions before and after the model change were less than 0.1 K and 0.05 K, respectively.

Three finger-sized B/W cameras (i.e., (c) top-view cameras) were mounted near the hot disk in MS30, MD30, and MI50. These cameras were arranged every 120° in circumferential direction. Since the hot disk made of sapphire, ITO film sensors, and ITO film heater were all transparent and the Pt sensors were placed so as not to obstruct the view, the top-view cameras could observe the flow inside a liquid bridge without being affected by the curvature on the free surface. The images captured by top-view cameras were mainly used to observe the qualitative flow pattern and to measure the quantitative velocity field as discussed in Sect. 5.

The unique measurement apparatus in MS30 was (d) a camera for photochromic dye activation technique [8, 9]. A pulsed gaseous nitrogen (GN_2) laser beam (with the wavelength of 337.1 nm, pulse energy of 300 mJ, and grade of Class 3B) was split into two, and two spots on the liquid-bridge free surface at the same height but at different azimuthal positions were excited. The camera for photochromic dye activation technique observed the excited dye traces so that the free-surface velocity of thermocapillary convection could be measured. A diode laser illuminated the liquid bridge to improve the contrast between excited and nonexcited liquids, and the photochromic dye traces appeared as a dark pink in a light pink background. We note that this method was not performed in MEIS-1, because it was not the research target in this mission.

The dynamic behavior of the liquid-bridge free surface was the main research target of DS project. To measure the delicate motion of the free surface, (e) a micro-imaging displacement meter (MIDM) [10] was installed in MD10 and MD30. The present MIDM consisted of a microscope objective (i.e., 20×magnification for MD10 and 5×magnification for MD30) and a B/W camera equipped with an imaging lens (i.e., 1×magnification for MD10 and 2×magnification for MD30), consequently, the total optical magnifications were 20× and 10× for MD10 and MD30, respectively. All data obtained by cameras mentioned above were digitized to 720(W)×480(H) pixels (px) 8-bit images by the IPU (Fig. 1c). An image in this size was landscape and resizing the image to 640(W)×480(H) px results in a pixel aspect ratio of approximately 1:1. The data were downlinked to the ground at 29.97 fps (frames per second) with a time lag of a few seconds; therefore, the experiments were performed almost in real time.

Other measurement apparatuses equipped in MI50 were (f) an insertion thermocouple sensor and (g) an ultrasonic (or ultrasound) velocity profiler [11]. The insertion thermocouple sensor, with K-type element, was sheathed with an Inconel. The outer diameter of this sensor was 0.25 mm, and the thermal conductivity of sheath was 14.9 W/(m · K) at 20 °C. The extension of the thermocouple sensor and the axis of the liquid bridge intersected at right angle to each other. The present sensor was traversable in a radial direction (i.e., the direction of approaching or leaving the free surface) and an axial direction (i.e., the height direction of the liquid bridge). The moving range in radial direction was 9.8 mm inside and 12.0 mm outside from the disk edge, while that in axial direction was 44.6 mm below the front side of the hot disk. The insertion thermocouple sensor was mainly used to measure the gas temperature in the vicinity of the liquid bridge but was also used to measure the interior temperature of the liquid bridge. There were two transducers for the ultrasonic velocity profiler, which were embedded in the cold disk: one measuring line was parallel to the axis of the liquid bridge and the other was inclined 30° with respect to it. They were used to clarify the spatiotemporal structure of thermocapillary convection. The basic frequency of the pulse emission from transducers was 8 MHz. The ultrasonic velocity profiler was used only in UVP project and not in MEIS-4 and 5. The data obtained from the insertion thermocouple sensor and the ultrasonic velocity profiler were downlinked to the ground in almost real time with time resolutions of 10 Hz and 5.44 Hz, respectively.

Figure 6(4) summarizes the installation of measurement apparatuses in each experiment cell, and Fig. 7 shows the proper locations and measurement directions of

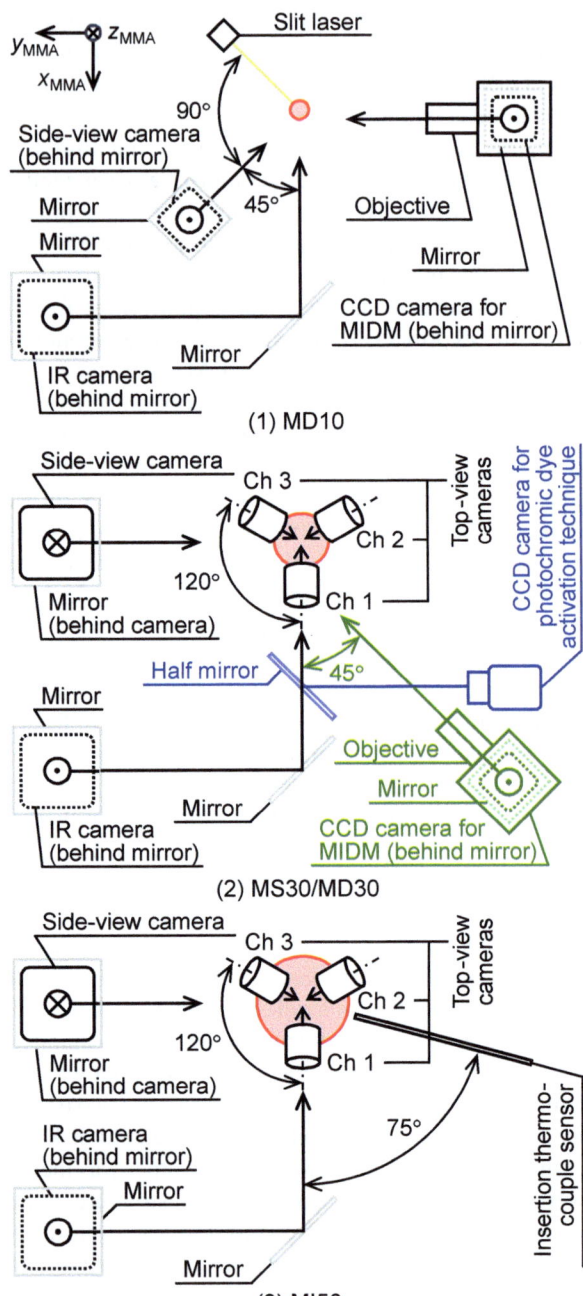

Fig. 7 Layout of measurement system observed from the hot-disk side for (1) MD10, (2) MS30/MD30, and (3) MI50. Hot disks are indicated by *red circles*. Equipment for photochromic dye activation technique (*blue*) and MIDM (*green*) in (2) are unique in MS30 and MD30, respectively. Axes (x_{MMA}, y_{MMA}, z_{MMA}) illustrated in the upper left represent the coordinate system for the microgravity measurement apparatus (MMA)

apparatuses except the ultrasonic velocity profiler. Due to the size limitation of the FPEF, cameras other than the top-viewing ones used mirrors to observe the liquid bridge. In the FPEF, the maximum number of concurrently accessible cameras was five, and the camera channel was switched according to the research targets in MS30 and MD30, which have a total of six cameras. More details on the measurement apparatuses will be given in corresponding sections.

3 Experimental Procedure

Taishi Yano, Koichi Nishino, Satoshi Matsumoto, and Hiroshi Kawamura

The liquid bridge in the present μg experiments possessed free surface and was not firmly restrained except the contact surface with the supporting disks. Since the surface tension of the silicone oil—the force to maintain the cylindrical shape—was not very strong, the liquid bridge was susceptible to the residual acceleration of gravity on the ISS called g-jitter (more detail on g-jitter will be presented in Sect. 7). The experiments were, therefore, basically conducted at nighttime on the ISS (i.e., from 21:00 to 6:00 in Greenwich Mean Time—GMT). This period corresponded to the sleeping time of the astronauts (i.e., GMT 21:30–6:00) so that the human-induced disturbances were expected to be minimal. All experiments were performed remotely from Tsukuba Space Center (TKSC) of JAXA in Japan by monitoring telemetry data and images sent from the FPEF and IPU. The time difference between ISS and TKSC is nine hours and above experiment time corresponds to 6:00–15:00 in Japan Standard Time (JST).

Two examples of typical experimental procedures are shown in Fig. 8, which indicates the time series of liquid-bridge height H (*black line*) and temperatures (*blue*, *red*, and *green lines* for T_C, T_H, and ΔT, respectively). The presence or absence of shading indicates the time zone during which the communication between the ISS and TKSC is active or inactive. These time zones are called AOS (acronym of acquisition of signal) and LOS (acronym of loss of signal), respectively. All telemetry data (exclude a few parts) were stored in the FPEF, and the images from each camera could be recorded on hard disk drives installed in the IPU; therefore, most data during LOS were available. Basically, the procedures of each experimental run consisted of three steps: (I) liquid-bridge formation step, (II) observation and measurement step, and (III) liquid-bridge recovery step.

Firstly, the procedure of step (I) is briefly introduced. Pictures of the front side of the cold disk are shown in Fig. 9. We note that the picture for $D = 10$ mm (Fig. 9(1)) was taken on the ground before μg experiments; therefore, there is neither liquid bridge, hot disk, nor O-ring. Other pictures for $D = 30$ mm and 50 mm (Figs. 9(2) and (3)) were taken by the top-view camera of Ch 1 during the μg experiment for $A_r = 0.33$ (see Fig. 7 for the camera channel). As shown in Fig. 9, one or two injection holes were drilled near the center of the cold disk. The bellow-type tank filled with silicone oil was located behind these holes. When the cold disk moved

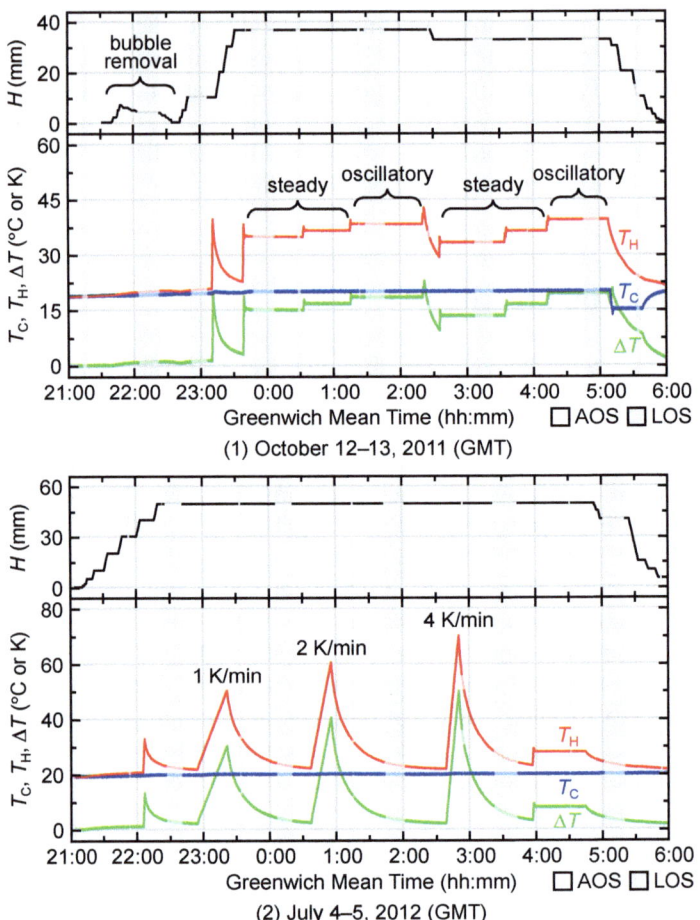

Fig. 8 Examples of telemetry data: liquid-bridge height (upper) and temperatures (lower). Absence/presence of shading indicates the time zone during which the communication between the ISS and ground is active/inactive (AOS/LOS)

and the gap between the supporting disks widened, the test liquid was automatically supplied through the injection holes, and the liquid bridge was formed between the disks. The test liquid was sucked into the tank when the gap closed. This system was designed to transport the silicone oil with the volume of $\pi D^2 \Delta H/4$ for the displacement of the cold disk of ΔH. In MS30, MD30, and MI50, the movement of the cold disk was always accompanied by the transportation of the test liquid. On the other hand, MD10 could move the cold disk independently of it. Another tank was installed behind the main tank. Such an additional tank could supply the test liquid without moving the cold disk. The minimum amount of silicone oil that could be supplied/sucked using this tank was approximately ±0.01 cc; therefore, the

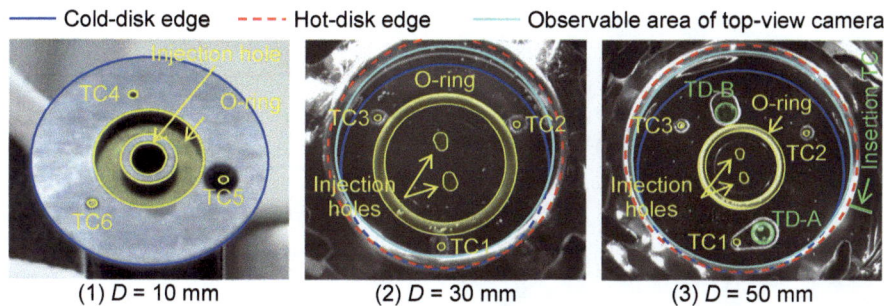

Fig. 9 End-view pictures of the cold disk taken (1) by ordinary camera on the ground and (2)–(3) by top-view camera (Ch 1) during the µg experiments. Disk diameters are 10, 30, and 50 mm for (1)–(3), respectively. *Blue*, *red*, and *light blue lines* indicate the cold-disk edge, hot-disk edge, and observable region of the top-view camera. Abbreviations TC and TD mean thermocouple and transducer, respectively. (Photo courtesy of JAXA. All rights reserved)

volume of the liquid bridge was precisely adjusted in the present µg experiments (e.g., the volume of the straight liquid bridge with $D = 30$ mm and $H = 15$ mm is approximately 10.6 cc). The motion of the cold disk and the liquid supply from the additional tank were controlled by stepping motors.

During the liquid-bridge formation step, small bubbles occasionally appeared in the silicone oil as shown in Fig. 10. Especially in the first week after the start of this series of experiment (i.e., MEIS-1), we were seriously bothered with the appearance of bubbles. It was considered to be caused by the bubbling of dissolved gas in the silicone oil or contamination from the ambient gas, but the correct reason remains unknown. Since the presence of bubbles, even if extremely small in size, would destroy the symmetry of the flow and temperature fields inside the liquid bridge, the removal of bubbles was necessary for the thermocapillary-convection experiment. In the ground-based experiments as well, we often encountered bubbles; however, it was rather easy to remove them with a needle or by rebuilding a new bubble-free liquid bridge. However, in the present µg experiments, bubbles had to be removed remotely. A new bubble removal method was developed in the first week of MEIS-1 through rather tough trial and errors. Telemetry data for GMT 21:40–22:40 in Fig. 8(1) correspond to the bubble removal process. The procedure for bubble removal is as follows: (1) open the gap between disks to some extent to allow bubbles to move; (2) induce the flow by increasing ΔT, and move bubbles outward from the O-ring; (3) close the disks, and pin bubbles with an O-ring so as not to penetrate inside (Fig. 10b); (4) close the disks further and/or suck the silicone oil until bubbles burst. Figure 10c shows the time series of bubble removal process. The bubble pinned by an O-ring is pressed as the disk moves ($t < 0$) and bursts between $t = 0$–0.2 s. Among these processes, the application of the temperature difference ΔT was crucial. It induced thermocapillary flow in the thin liquid film depicted in Fig. 10b. Subsequently, the induced flow thinned the film, enhancing the rupture of the bubbles. Conversely, using only mechanical methods, such as closing the disks or suctioning the liquid, we

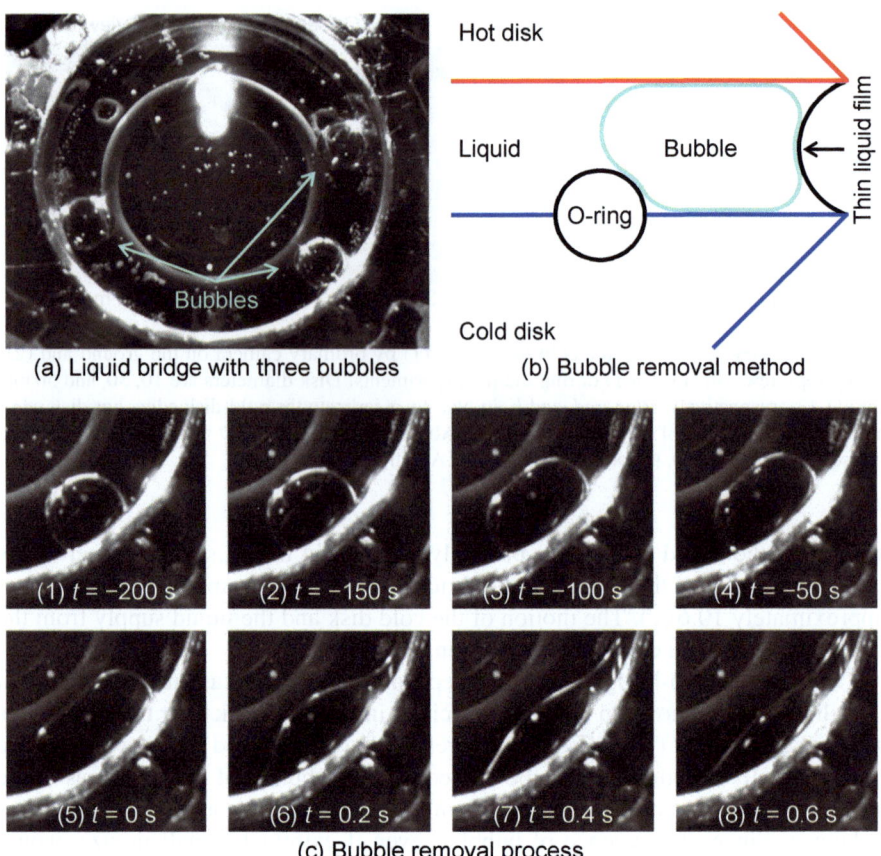

Fig. 10 **a** Liquid bridge with three bubbles, **b** schematic of the bubble removal method, and **c** time series of removing the lower left bubble in (**a**). (Photo courtesy of JAXA. All rights reserved)

were almost unable to rupture the bubbles. Using this method, bubbles in a silicone oil could be certainly removed under μg environment without any assistance by astronauts, although we had to spend some of the invaluable μg time for this process. After bubble removal, a liquid bridge with target height and volume was formed.

The step (II), observation and measurement step, was the main part of the present μg experiments. In this step, cold and hot disks were cooled/heated to the designated temperatures, and the resultant thermocapillary convection was observed with multiple measurement apparatuses mentioned in Sect. 2. As shown in Fig. 8(1) from 23:40 to 5:00 in GMT, the disk temperatures (and the resultant ΔT) were usually changed in a stepwise manner. After T_C and T_H reached the target values, a sufficiently long waiting period was given to each temperature step to attain thermal and hydrodynamic equilibrium. In the present μg experiments, the thermal diffusion time was evaluated by considering the heat conduction as

$$\tau_\kappa = \frac{L^2}{\kappa}, \tag{3}$$

where L is the characteristic length scale defined by the liquid-bridge height H and the disk radius R as

$$L = \sqrt{\left(\frac{1}{H^2} + \frac{1}{R^2}\right)^{-1}}. \tag{4}$$

The specific values of τ_κ for the straight liquid bridges of $Pr = 67$, $A_r = 1.0$, and $D = 10, 30,$ and 50 mm are $\tau_\kappa \approx 4, 40,$ and 112 minutes, respectively. The value of τ_κ for $D = 5$ mm, which is the typical diameter in terrestrial experiments, is $\tau_\kappa \approx 1$ minutes. As pointed out by Kawamura and Ueno [12], a large-scale liquid bridge, as in the present μg experiments, requires much longer thermal diffusion time to achieve a fully developed thermocapillary convection compared to a small-scale one, as in terrestrial experiments. In the present μg experiments, a waiting period longer than τ_κ has been given for each temperature step except in special cases.

After the waiting period, the states of flow (i.e., steady or oscillatory) at each ΔT step were judged to determine the onset conditions of oscillatory thermocapillary convection. The flow states were judged by multiple ways: the temperature signals of thermocouple and ITO film sensors, the motion of tracer particles observed by top- and side-view cameras, the temperature distribution observed by an IR camera, and the motion of excited dye traces in the photochromic dye activation technique. When the flow was judged to be an oscillatory state, the oscillation frequency, oscillation type (i.e., standing-wave type or rotating-wave type), and azimuthal mode number were measured. The critical ΔT for the onset of oscillatory thermocapillary convection (i.e., critical temperature difference ΔT_c) were considered to exist between the largest ΔT for steady state and the smallest ΔT for oscillatory state, and the value of ΔT_c were evaluated as the average of them. For example, ΔT_c for $A_r = 1.25$ in MEIS-3 was measured with three temperature steps as shown in GMT 23:40–2:20 in Fig. 8(1). The flow was steady in the first and second temperature steps at $\Delta T = 15.2$ K and 16.9 K, and it transitioned to an oscillatory state at the third temperature step at $\Delta T = 18.6$ K; therefore, the critical condition was evaluated to be $\Delta T = 17.7 \pm 0.9$ K. In the present μg experiments, the temperature step was set so as to bring the uncertainty less than $\pm 10\%$ with respect to ΔT_c. The experiments were performed not only by increasing ΔT but also by decreasing ΔT. As reported by Nishino et al. [6], the effect of temperature hysteresis on the determination of ΔT_c could be ignored if a sufficiently long waiting period was given.

As shown in Fig. 8(2), ΔT was changed with constant speed (i.e., 0.1–4.0 K/min) in some experimental runs to understand the influence of changing rate of ΔT on the detection of thermocapillary-convection instability, and its details will be presented in Sect. 4.

4 Instability of Thermocapillary Convection—Effects of Geometrical Conditions, Prandtl Number, and Temperature Control Method

Taishi Yano and Koichi Nishino

The main objective of the present μg research was to understand the instability of thermocapillary convection in a high-Pr liquid bridge. To accomplish this goal, the instability thresholds for the transition from a steady thermocapillary convection to an oscillatory one as well as the critical oscillation frequencies and the critical instability modes have been measured for various conditions through MEIS, UVP, and DS projects. The important experimental parameters considered in these projects were the ones related to the geometry of the liquid bridge, the physical property of the test liquid, the heat transfer condition, and the external disturbance. In this section, the effects of former two factors (i.e., geometrical condition and physical property) will be discussed. In addition to them, the effect of changing rate of disk temperatures on the development of thermocapillary-convection instability will be presented in this section.

4.1 Effect of Aspect Ratio

One of the most famous parameters that affects the instability of thermocapillary convection is the aspect ratio of the liquid bridge. In the present projects, especially in MEIS, numerous efforts have been undertaken for studying its effects. The results are presented in Fig. 11, which shows the plots of (1) the critical temperature difference ΔT_c and (2) the critical oscillation frequency f_c as a function of diameter-based aspect ratio A_r $(= H/D)$. These data were obtained under the conditions with a fixed cold-disk temperature of $T_C = 20\,°C$ and volume ratios of $0.95 \leq V_r \leq 1.00$, wherein the ambient gas was argon. The volume ratio V_r $(= V/V_0)$ is the ratio of the liquid volume V to the gap volume between the supporting disks V_0 $(= \pi D^2 H/4)$; therefore, liquid bridges were cylindrical or slightly concave in shape (Fig. 2). The original data of these plots have been reported previously [6, 13–16]. The measurement method and the evaluation method of error bars for ΔT_c are described in Sect. 3. The oscillation frequency f at each ΔT step was obtained by analyzing either the temperature signals from thermocouple or ITO sensors, the motion of tracer particles, the temperature waves traveling on the free surface (i.e., hydrothermal wave), or the motion of excited photochromic dye traces. The values of f_c plotted in Fig. 11b are f obtained for oscillatory flow with smallest ΔT (in other words, ΔT slightly larger than ΔT_c). Therefore, they are not strictly critical values, but this is not an important issue. We note that the results for $A_r = 2.5$ and 2.8, which are indicated by *stars*, should be interpreted with caution. The liquid bridges

Fig. 11 Plots of **a** critical temperature difference ΔT_c and **b** critical oscillation frequency f_c as a function of aspect ratio A_r, wherein the cold-disk temperature is fixed at $T_C = 20\,°\text{C}$. The colors indicate the Prandtl number (i.e., *red*, *green*, and *blue* for Pr = 67, 112, and 207, respectively) and the symbols indicate the disk diameter (i.e., *delta*, *circle*, and *diamond* for $D = 10, 30$, and 50 mm, respectively). The plots with *star* symbols are reference data because the liquid bridge was contaminated with bubbles during experiments

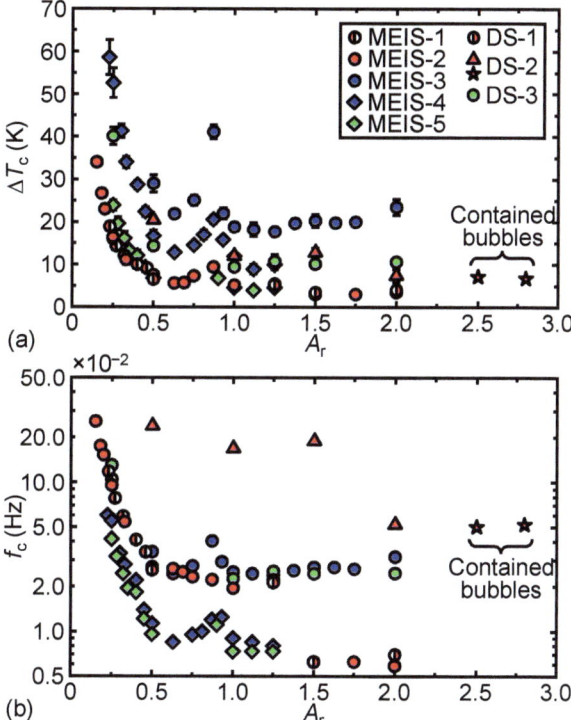

contained small bubbles when these data were acquired because there was no time to remove bubbles.

For shorter liquid bridges (say, $A_r < 1$), one can recognize a simple dependency of ΔT_c on Pr and D from Fig. 11a that the larger the Pr or smaller the D, the larger the ΔT_c. On the other hand, the values of f_c for $A_r < 1$ in Fig. 11b increase with decreasing D, but are almost independent of Pr. For shorter liquid bridges, say $A_r < 0.5$, it is confirmed that the instability mode of oscillatory thermocapillary convection strongly depends on A_r. The characteristics of oscillatory thermocapillary convection are usually represented by the azimuthal mode structure. The convection at the oscillatory state involves pairs of modal structures (e.g., a pair of hot and cold regions) in the azimuthal direction, and the number of such structures is the azimuthal mode (or wave) number m, which takes an integer value. The values of m for $A_r < 0.5$ in the present µg experiments tend to decrease with increasing A_r (Fig. 12), as in the previous ground-based studies [17–19]. However, the values of m in the present µg experiments were apparently smaller than those under normal-gravity condition. Preisser et al. [17] reported the empirical relation for $A_r \leq 0.5$ and $V_r \approx 1$ that the product of m and A_r takes an approximately constant value as

$$m A_r \approx c. \tag{5}$$

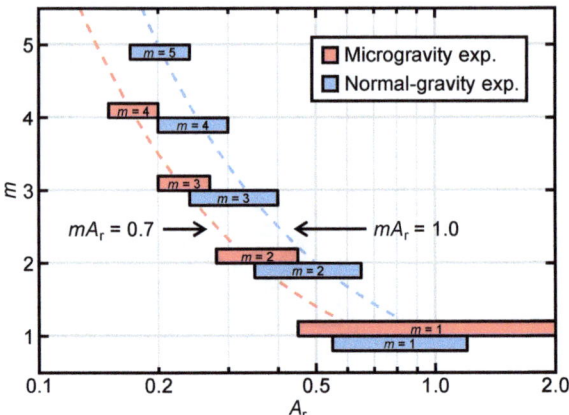

Fig. 12 The dependence of critical azimuthal mode number m on the aspect ratio A_r. Original data were reported by Nishino et al. [6] for μg experiment and Kawamura et al. [20] and Ueno et al. [19] for normal-gravity experiment. (Reprinted from Nishino et al. [6], Copyright 2015, with permission from Elsevier. Modifications have been made in this figure with the permission from the publisher. All rights reserved)

Presser et al. [17] and Ueno et al. [19] derived $c \approx 1.0$–1.1 from their terrestrial experiments, and Leypoldt et al. [18] derived $c \approx 1.2$ from their numerical simulations under normal-gravity condition. We note that their Pr were different from the present study, and liquid bridges were heated from above with respect to the gravity direction. The value of c in the present μg experiments was $c \approx 0.7$. The possible reasons for this difference are effects of buoyancy, Prandtl number, shape of the liquid bridge, heat transfer on the free surface, but the primary one is considered to be the effect of buoyancy.

Lappa et al. [21] numerically studied the effect of buoyancy by changing the heating direction. Their values of c when the liquid bridge was heated from above and was heated from below are $c \approx 0.85$ and 0.55, respectively. Figure 13 shows the flow and temperature fields of steady thermocapillary convection in a liquid bridge of 2-cSt silicone oil (Pr = 28 at 25 °C) with different gravity conditions obtained from the numerical simulations [7]: (a) $g = 9.8 \, \text{m/s}^2$, (b) $g = 0$, and (c) $g = -9.8 \, \text{m/s}^2$. These gravity conditions correspond to the ones heated from above under normal gravity, under zero gravity, and heated from below under normal gravity, respectively. We note that other conditions are $D = 5$ mm, $A_r = 0.5$, $V_r = 1.0$, $T_C = 20\,°\text{C}$, and $T_H = 38\,°\text{C}$, and the deformation due to gravity is neglected. When the liquid bridge is heated from above (Fig. 13a), as in the studies by Presser et al. [17], Leypoldt et al. [18], Ueno et al. [19], etc., an additional circulation rotating in the opposite direction with respect to the primary convection roll appears near the center of the cold disk due to the buoyancy. This additional circulation prevents the thermocapillary flow from penetrating the interior bulk region. On the other hand, when the liquid bridge is heated from below (Fig. 13c), the direction of the buoyancy-driven flow is

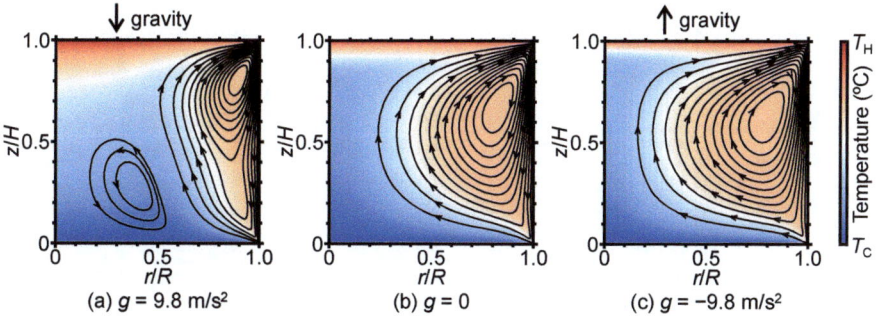

Fig. 13 Flow and temperature fields obtained by numerical simulations under different gravitational accelerations: **a** $g = 9.8\,\text{m/s}^2$, **b** $g = 0$, and **c** $g = -9.8\,\text{m/s}^2$, wherein the negative/positive g corresponds to the heating from above/below on the ground. Other conditions are $Pr = 28$, $D = 5\,\text{mm}$, $A_r = 0.5$, $V_r = 1.0$, $T_C = 20\,°C$, and $T_H = 38\,°C$. (Adapted with permission from Yano [7]. All rights reserved)

same with that of the thermocapillarity-driven surface flow, and the flow can further penetrate into the interior bulk region. The flow field under zero-gravity condition is in between them (Fig. 13b). The expansion of the convection reduces the number of flow structure that can be accommodated in the circumferential direction; therefore, the value of m (and the resultant mA_r) in μg tends to be smaller than that in normal gravity as shown in Fig. 12.

Another important characteristic of oscillatory thermocapillary flow is the oscillation type. When the modal structures swing radially or rotate azimuthally, the oscillation type is categorized as the standing-wave (or pulsating-wave) type or rotating-wave (or traveling-wave) type, respectively. They are, hereinafter, denoted by SW type and RW type, respectively. In high-Pr thermocapillary liquid bridges, a SW-type oscillatory flow tends to be preferred near the instability threshold, and it often transitions to a RW-type oscillatory flow with an increase in driving force [19, 22]. Such trend was also confirmed in the present μg experiments.

To return to the measurement technique, the azimuthal mode number and the oscillation type were basically determined by the distribution of tracer particles viewed from the hot-disk side. Since paths of tracer particles were affected by the particle-free-surface interaction [23] and the centrifugal force due to the density difference between tracer particles and test liquids [24], the region where tracer particles cannot (or seldom) enter appeared near the central part of the liquid bridge. This region is called the particle-depletion zone [23], particle-free zone [25], or void region [26]. As shown in Fig. 14, the azimuthal mode and the oscillation type (i.e., (a) SW or (b) RW) can be determined by the shape and behavior of the particle-depletion zone, wherein the data were obtained in MEIS-5 (i.e., $D = 50\,\text{mm}$, $Pr = 112$), and the aspect ratios are $A_r = 0.50$, 0.32, and 0.25 for (i) $m = 1$, (ii) $m = 2$, and (iii) $m = 3$, respectively. We note that ΔT for these results are much higher than the critical values (i.e., $\Delta T/\Delta T_c = 1.1$–$2.7$) because the easy-to-understand data are selected. It is also noted that the particle images shown in Fig. 14 are phase-averaged

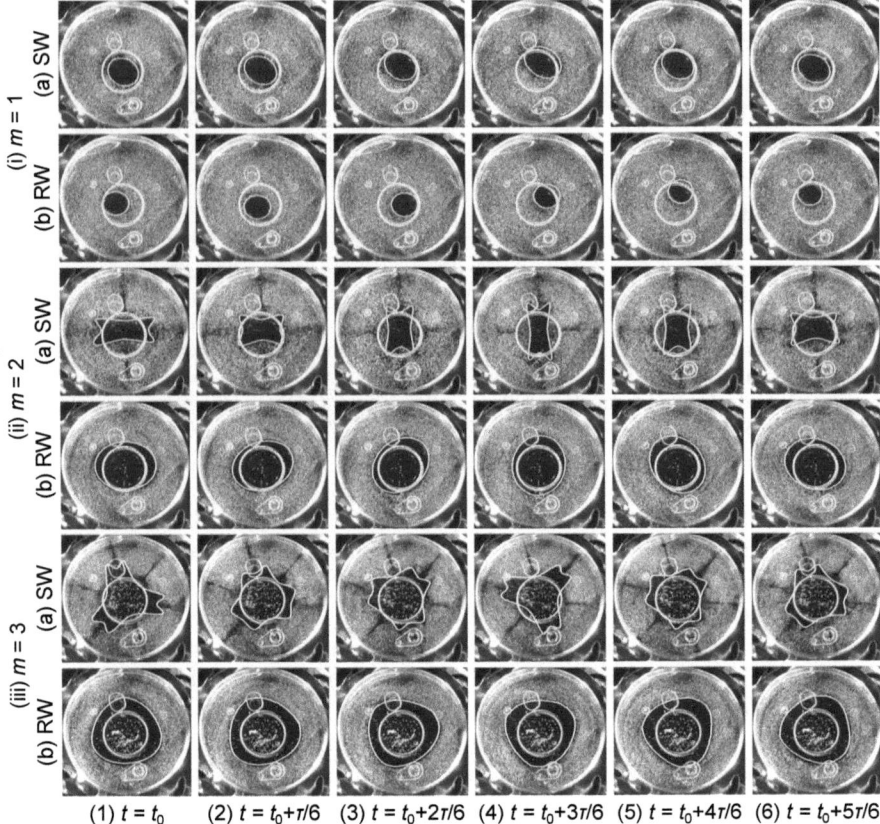

Fig. 14 Time series of particle-depletion zones for every 1/6 oscillation period τ. Azimuthal mode numbers are (i) $m = 1$, (ii) $m = 2$, and (iii) $m = 3$, and oscillation types are **a** standing-wave (SW) type and **b** rotating-wave (RW) type

ones for providing good visibility. In the case of SW-type oscillation, the particle-depletion zone stretches and contracts along a single or multiple radial-axial planes and is stationary in the circumferential direction. The symmetry planes are located every $2\pi m$ in the circumferential direction, and the number of them corresponds to m [27]. In the case of RW-type oscillation, the particle-depletion zone rotates at a constant angular velocity (i.e., $2\pi f/m$) with maintaining its shape. The value of m can be determined from the shape (or number of vertices) of the particle-depletion zone: circle, ellipse, triangle, square, pentagon, ... correspond to $m = 1, 2, 3, 4, 5,$ ..., respectively.

Under certain conditions, tracer particles seeded in a liquid bridge follow a single or multiple closed paths. Such a phenomenon of tracer particles has attracted attention since the mid-1990s [28, 29] and is now widely known as the particle accumulation structure (PAS). The MAXUS-6 sounding-rocket experiment conducted by Dietrich Schwabe (Justus Liebig University Giessen, Germany), Hiroshi Kawamura (Tokyo University of Science, Japan), and others marked the first observation of PAS under the μg environment (see Sect. 3 in Chapter "Thermocapillary Convection in Liquid Bridges of Finite Length"), and they revealed the effects of gravity, but not deterministic, on PAS. One of the important objectives through a series of μg experiments aboard the ISS is to observe the PAS in a large-scale liquid bridge, and it was successively accomplished. Figures 15(i) and (ii) show the time series of PASs for $m = 1$ and 2, respectively. The former was observed in DS-3 (i.e., $D = 30$ mm, Pr $= 112$) under the conditions of $A_r = 0.50$, $V_r = 0.95$, and $\Delta T = 8.2$ K ($T_C = 40.0\,°C$ and $T_H = 48.2\,°C$), and the later was in MEIS-3 (i.e., $D = 30$ mm, Pr $= 207$) under the conditions of $A_r = 0.35$, $V_r = 0.80$, and $\Delta T = 54.0$ K ($T_C = 20.0\,°C$ and $T_H = 74.0\,°C$). In these results, particle images taken by a top-view camera of Ch 2 and a side-view camera are phase averaged over 10 oscillation periods (i.e., $\tau = (1/f) = 32.8$ s and 15.4 s for Figs. 15(i) and (ii), respectively), and contrasts are adjusted. We note that the side-view images are oriented as looking at the top-view images from the 7 o'clock direction (see Fig. 7 for the camera arrangement) and are treated with background subtraction for providing better recognition of PAS. From these results, one can see that many tracer particles are accumulated on the spiral and elliptical paths for $m = 1$ and 2, respectively. According to the previous terrestrial experiments [19], the PAS is likely to form at ΔT several times larger than ΔT_c. The PASs observed in the present μg experiments were, however, formed near the instability threshold, and the PAS could not be observed at $\Delta T \gg \Delta T_c$. In addition to the above-mentioned conditions, PASs were also observed in MEIS-3 for $A_r = 0.35$ and $V_r = 0.76$–0.80 and in MEIS-5 for $A_r = 0.28$ and $V_r = 0.95$, both of which appeared at ΔT slightly larger than ΔT_c. The discrepancy between the terrestrial and μg experiments is considered to be due to the effects of gravity, liquid-bridge size or shape, and so on, but the exact reason remains unclear, and further study is needed. More detailed discussions on the PAS observed in the present μg experiments can be found in Refs. [30, 31] and Sect. 1 in Chapter "Surface-Tension Related Flows in Microgravity and Microscale: Liquid Bridges" of this book.

The IR imaging technique is another method that can determine the instability mode of thermocapillary convection. Figure 16 shows examples of data analysis for IR images corresponding to Fig. 14. The time series of image intensities approximately 5 mm below the hot disk (i.e., marked region in Fig. 16(1)) were extracted, and the following postprocessing were applied: (1) longitudinally averaging over 11 px height; (2) phase averaging over 4–6 oscillation periods; and (3) subtracting the time-averaged data. As a result, the temperature fluctuations on the free surface shown in Fig. 16(2) were obtained. It is easy to distinguish the SW-type oscillation and the RW-type oscillation from Fig. 16(2). Since a pair of hydrothermal waves with the same angular speed and the opposite circumferential direction propagate in the SW-type oscillatory thermocapillary convection, a mirror symmetry temperature pattern with 'X'-shaped colder regions appears in this oscillation type [32]. On the other hand, a rotational symmetry temperature pattern with inclined colder regions appears in the RW-type oscillatory thermocapillary convection because one hydrothermal wave occupies the counterpropagating one.

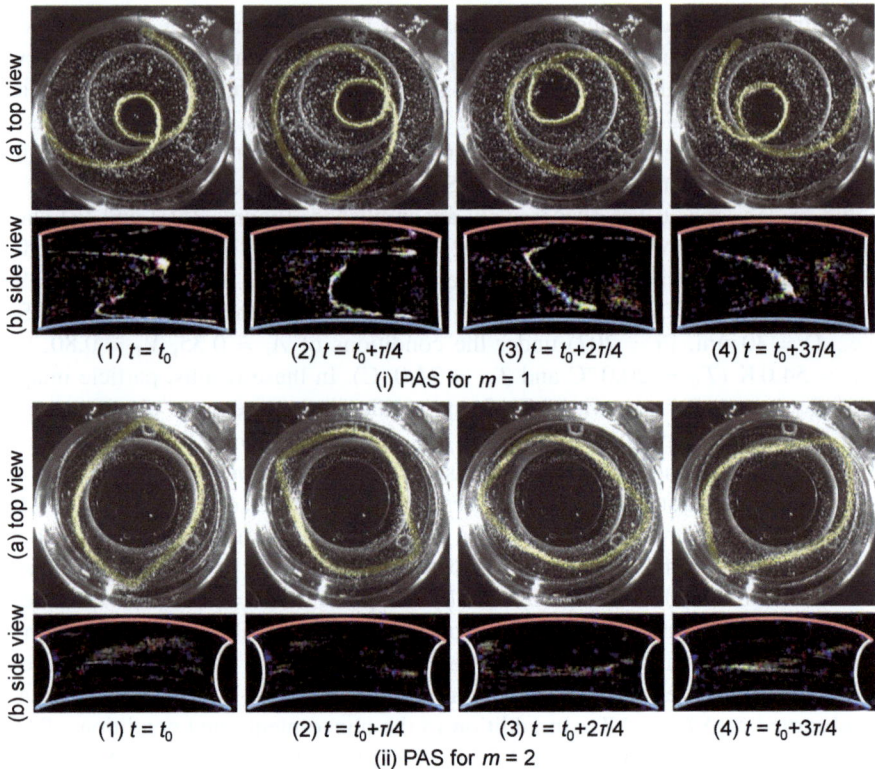

Fig. 15 Time series of particle accumulation structures (PAS) for every 1/4 oscillation period τ: **a** the top-view image and **b** the side-view image. Azimuthal mode numbers are (i) $m = 1$ and (ii) $m = 2$

The azimuthal mode number can be determined from the difference in circumferential positions of the adjacent colder and hotter regions (i.e., $\Delta\theta = |\theta_{\text{cold}} - \theta_{\text{hot}}|$) because the relation $m = \pi/\Delta\theta$ holds. For example, at $(t - t_0)/\tau = 1/6$ in Fig. 16(2)-(iii-a), the centers of hotter and colder regions are located at $\theta_{\text{hot}} \approx -37°$ ($\sin\theta_{\text{hot}} \approx -0.6$) and $\theta_{\text{hot}} \approx 24°$ ($\sin\theta_{\text{cold}} \approx 0.4$), respectively, and $\Delta\theta \approx 61°$. This means that three hotter regions and three colder regions exist in the circumferential direction, and the resultant azimuthal mode number is evaluated to be $m = 3$.

The particle images captured by top-view cameras were mainly used to determine the instability mode of oscillatory thermocapillary convection except for DS-2 whose experiment cell (MD10) had no top-view cameras (Fig. 6(1)). The instability modes in DS-2 were determined using the above-mentioned IR imaging technique unless the IR camera could not properly observe the hydrothermal wave due to the limitation of the field of view and the inadequate setting of detectable temperature range.

The instability of thermocapillary convection in a liquid bridge is normally organized using following dimensionless numbers:

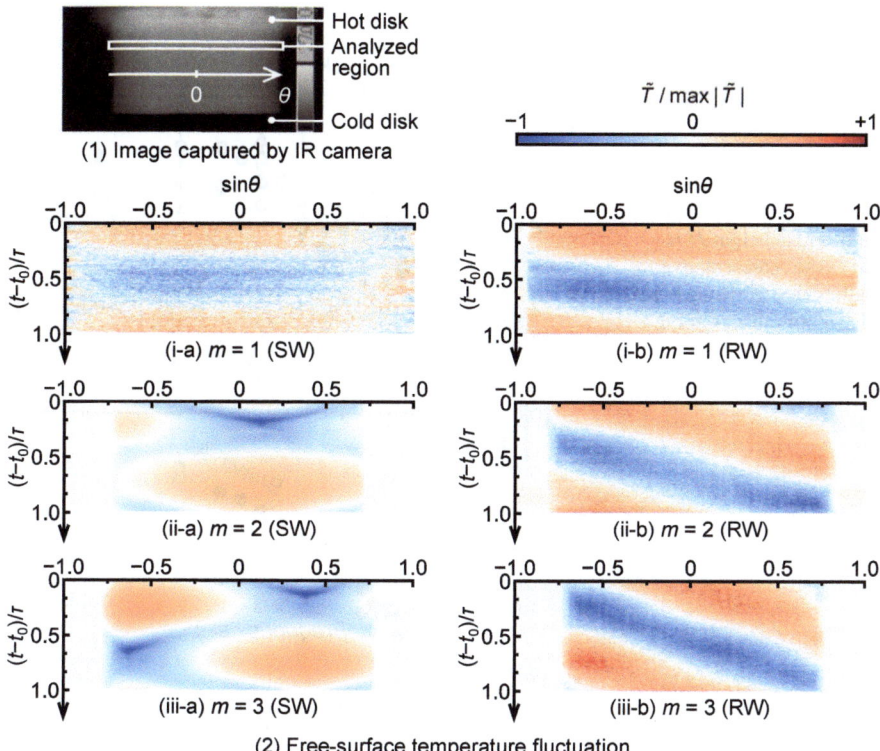

Fig. 16 Traveling of hydrothermal waves: (1) the image captured by IR camera and (2) the time series of free-surface temperature fluctuation. Azimuthal mode numbers are (i) $m = 1$, (ii) $m = 2$, and (iii) $m = 3$, and oscillation types are **a** standing-wave (SW) type and **b** rotating-wave (RW) type. (Photo in (1) courtesy of JAXA. All rights reserved)

$$\mathrm{Ma} = \frac{|\sigma_T|\Delta T L}{\rho \bar{\nu} \kappa}, \tag{6}$$

$$F = \frac{L^2}{\kappa \sqrt{\mathrm{Ma}}} f, \tag{7}$$

where the former is the Marangoni number, the latter is the dimensionless oscillation frequency [17], and L is the representative length scale. The first natural choice for L is the liquid-bridge height H as in many previous studies [19, 33–35]. The critical Marangoni number $\mathrm{Ma_c}$ and the critical dimensionless oscillation frequency F_c, which are Ma and F for $\Delta T = \Delta T_c$ and $f = f_c$ in Fig. 11, are plotted as a function of A_r in Fig. 17, wherein the representative length scale is H (i.e., $L = H$). For $A_r < 1.00$, the curves connecting $\mathrm{Ma_c}$ (i.e., the neutral stability curves) show common tendencies: (1) decrease for $A_r < 0.50$, (2) increase for $A_r = 0.50$–0.87,

Fig. 17 Plots of **a** critical Marangoni number Ma_c and **b** critical dimensionless oscillation frequency F_c as a function of aspect ratio A_r, wherein the cold-disk temperature is fixed at $T_C = 20\,°C$ and the liquid-bridge height H is used as the representative length scale. The same color and symbol rules as in Fig. 11 are applied

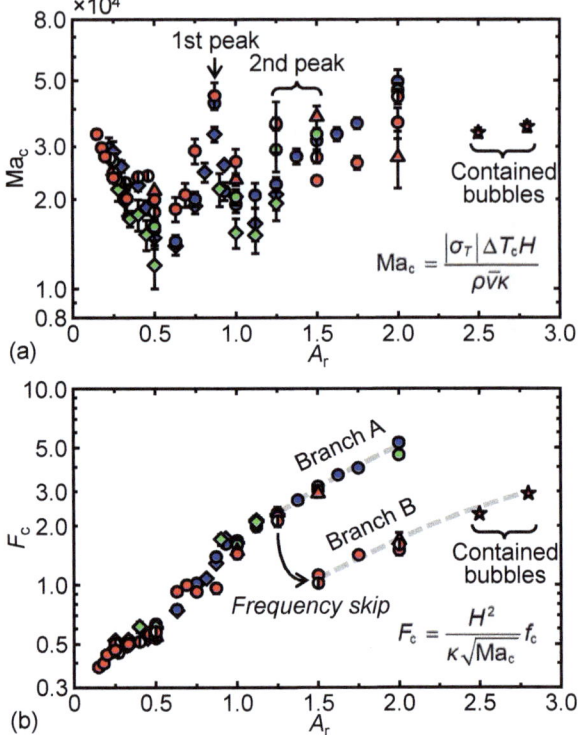

and (3) decrease again for $A_r = 0.87$–1.00; therefore, the neutral stability curves exhibit the local minimum at $A_r \approx 0.50$ and the local maximum at $A_r \approx 0.87$. On the other hand, the plots of F_c monotonously increase with A_r for $A_r < 1.00$. The values of Ma_c and F_c with different Pr and D show similar magnitude for the same A_r. Additionally, the present Ma_c and F_c for $A_r < 1.00$ are reasonably consistent with those in the ground-based studies by Preisser et al. [17], Velten et al. [34], and many other researchers as demonstrated by Nishino et al. [6]. Hence, these dimensionless quantities are good parameters for understanding the instability of thermocapillary convection in relatively shorter liquid bridges not only under normal gravity but also under μg. We note that some previous ground-based studies show different A_r dependence of Ma_c (see Sect. 3 in Chapter "Thermocapillary Convection in Liquid Bridges of Finite Length" for the details of ground-based studies), especially for small-A_r conditions, but the reason for such discrepancy is not clarified yet.

In contrast to the results for $A_r < 1.00$, the neutral stability curves for $A_r > 1.00$ show somewhat complex trends and are dependent on Pr and D. For Pr = 67, the neutral stability curves show a local maximum at $A_r = 1.25$–1.50, at which F_c discontinuously changes to a lower value. Such a discontinuous change in oscillation frequency is sometimes referred to as the *frequency skip* [36–38]. The values of A_r exhibiting a local maximum of Ma_c and a *frequency skip* are $A_r = 1.25$ and 1.50 for

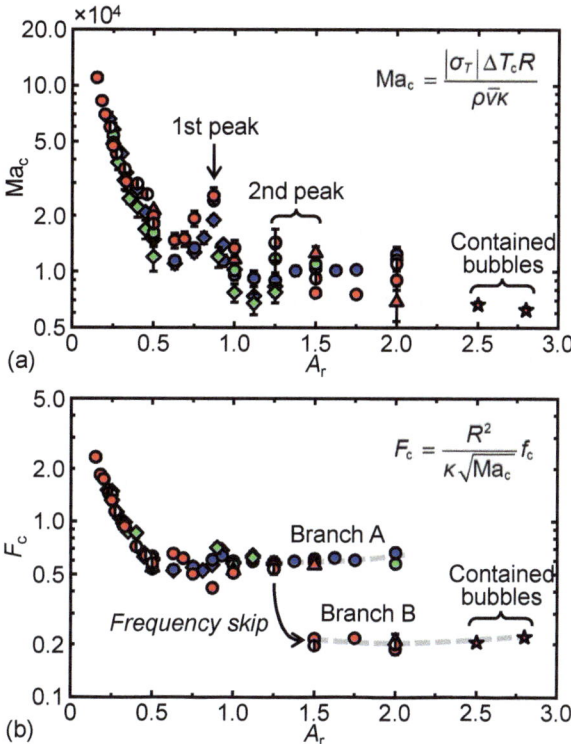

Fig. 18 Plots of **a** critical Marangoni number Ma_c and **b** critical dimensionless oscillation frequency F_c as a function of aspect ratio A_r, wherein the cold-disk temperature is fixed at $T_C = 20\,°C$ and the disk radius R is used as the representative length scale. The same color and symbol rules as in Fig. 11 are applied

$D = 10$ and $30\,mm$, respectively. For other Pr, both a local maximum of Ma_c and a *frequency skip* do not appear, and Ma_c and F_c increase monotonically with A_r. As a result, the connecting curves of F_c show two branches (i.e., branch A and branch B in Fig. 17b). On the whole, the values of Ma_c and F_c increase with increasing A_r; therefore, the thermocapillary convection seems to be stabilized (in other words, it becomes difficult to transition to the oscillatory state) and the oscillation frequency of convection seems to increase. However, these results should be interpreted with caution because the increases in Ma_c and F_c for $A_r > 1.00$ are intrinsically due to the increase in a liquid-bridge height H. For this reason, Nishino et al. [6] proposed to use the disk radius R as the representative length scale (i.e., $L = R$) for longer liquid bridges, and Ma_c vs. A_r and F_c vs. A_r plots using R as the representative length scale are shown in Fig. 18. The big change in Ma_c shown in Fig. 17a are not recognized in Fig. 18a, and the values of Ma_c converge to $Ma_c = 8 \times 10^3$–12×10^3. We note that the value of Ma_c for $A_r = 2.5$ and $Pr = 28$ reported by Schwabe [39, 40], who carried out the μg experiment using a sounding rocket MAXUS 4, is in accord with the present results [6]. The values of F_c that does not experience the *frequency skip* (i.e., branch A) converge to $F_c \approx 0.6$, while those experience the *frequency skip* (i.e., branch B) converge to $F_c \approx 0.2$. The difference in flow and temperature fields between branches A and B will be discussed in Sects. 5 and 6. The results of present μg experiments suggest that the liquid-bridge height H is much suitable for the representative length scale for shorter liquid bridges, while the disk radius R is much suitable for longer liquid bridges. In the remainder of this chapter, H is used as the representative length scale for $A_r < 1$ and R is for $A_r \geq 1$, unless otherwise noted.

4.2 Effect of Volume Ratio

Another important geometrical parameter in the thermocapillary liquid bridge is the volume ratio $V_r (= V/V_0)$, which is the ratio of the liquid volume V to the gap volume between supporting disks $V_0 (= \pi D^2 H/4)$ as defined previously. The present μg experiments were carried out with $V_r = 0.95$–1.00 in many cases; however, in MEIS and DS projects, several experiments were performed by varying V_r significantly to investigate its effect on the thermocapillary-convection instability. Figure 19 shows a variety of liquid-bridge shapes: (a) $A_r = 0.35$, (b) $A_r = 0.50$, and (c) $A_r = 1.00$. As shown in these snapshots, the liquid-bridge shape changes from cylindrical to concave one with decreasing V_r. The target V_r was mainly less than unity to minimize liquid

Fig. 19 Snapshots of liquid bridges with various shapes: **a** $A_r = 0.35$, **b** $A_r = 0.50$, and **c** $A_r = 1.00$ (Photo courtesy of JAXA. All rights reserved)

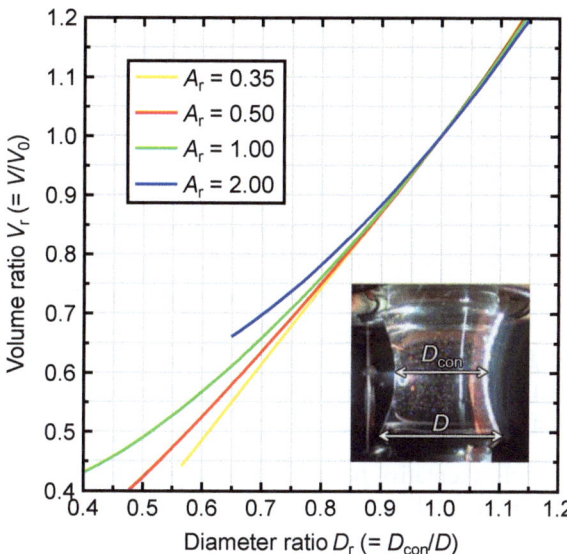

Fig. 20 Relation between volume ratio V_r and diameter ratio D_r under zero-gravity condition for $A_r = 0.35, 0.50, 1.00,$ and 2.00. (Photo courtesy of JAXA. All rights reserved)

leakage from the disk edges, and experiments with convex liquid bridge (i.e., $V_r > 1$) were very limited in the present projects.

Since the direct measurement of liquid volume was difficult in the present experimental setup, the value of V_r was evaluated from the diameter ratio D_r ($= D_{con}/D$), which is the ratio of the diameter at the most concave or convex region (sometimes called liquid-bridge neck) D_{con} to the disk diameter D [41, 42]. During the μg experiments, the value of D_r was obtained by measuring neck and disk diameters on the image sent from a side-view camera. As shown in Fig. 20, V_r under zero-gravity condition is in clear one-to-one correspondence with D_r; therefore, the value of V_r can be determined in situ by measuring D_r if the relationship between them is known in advance. Incidentally, the liquid bridge under normal-gravity condition can have both concave and convex regions so that the definition of D_r and its relation to V_r is somewhat ambiguous.

In the present μg experiments, the shapes of liquid bridge and the resultant V_r-D_r relationships were obtained by solving the following Young-Laplace equation [43, 44]:

$$\frac{d^2r}{ds^2} = -\frac{dz}{ds}\left(-\frac{Bo}{A_r^2}\frac{z}{D^2} + \frac{C}{\sigma} - \frac{1}{r}\frac{dz}{ds}\right), \tag{8}$$

$$\frac{d^2z}{ds^2} = \frac{dr}{ds}\left(-\frac{Bo}{A_r^2}\frac{z}{D^2} + \frac{C}{\sigma} - \frac{1}{r}\frac{dz}{ds}\right), \tag{9}$$

where s is the curve arc length, r and z are, respectively, the radial and axial positions on the free surface as a function of s, Bo is the (static) Bond number, and C is the constant derived from the pressure difference between liquid and gas. The Bond number represents the ratio of hydrostatic pressure to the capillary pressure, and is defined as

$$\text{Bo} = \frac{\Delta \rho g H^2}{\sigma_0}, \tag{10}$$

where $\Delta\rho$ is the density difference between liquid and gas. Since Bo ≈ 0 in the μg environment on the ISS, Eqs. 8 and 9 become simplified. The problem was solved numerically using the shooting method [44]. The boundary (or initial) conditions at $s = 0$, which is on the cold disk, are as follows: (1) $r = R$, (2) $z = 0$, (3) $dr/ds = \cos \beta_0$, and (4) $dz/ds = \sin \beta_0$, where β_0 is the (macroscopic) contact angle on the cold disk (Fig. 2). The shape of liquid bridge with designated V_r was obtained by the following procedures:

(1) select tentative β_0 and C and integrate Eqs. 8 and 9 by using the trapezoidal rule or the explicit Euler method [45] with a sufficiently small interval of s until z reaches H,

(2) adjust C and repeat procedure (1) if condition $r = R$ at $z = H$ is not fulfilled,

(3) evaluate V_r by integrating the obtained r-z curve, and

(4) adjust β_0 and repeat procedures (1)–(3) if the evaluated V_r is not consistent with the designated one.

In procedures (2) and (4), the tolerances for R and V_r were set to $D \times 10^{-6}$ and 10^{-8}, respectively. The above calculation assumes an isothermal liquid bridge; therefore, σ is constant and no flow exists. The variation of σ and the internal flow affect the liquid-bridge shape significantly only when the imposed ΔT is extremely large as demonstrated in Sect. 2 of Chapter "Surface-tension Related Flows in Microgravity and Microscale: Liquid Bridges".

The shapes of liquid bridge for various V_r and constant A_r ($= 0.50$) calculated from the Young-Laplace equation and those extracted from the experimental data are compared in Fig. 21. In the experimental data, the image captured from a side-view camera was processed so that the liquid bridge appeared black against a white background (right images in Fig. 21), and the free surface was detected with the silhouette method [46]. The computational and experimental results are in reasonable agreement, and such agreement suggests the favorable accuracy of adjustment of V_r in the present μg experiments.

The effect of V_r on the instability of thermocapillary convection was studied for typical A_r. Figure 22 shows plots of (a) Ma$_c$ and (b) F_c as a function of V_r for relatively shorter liquid bridges, wherein (1) $A_r = 0.35$, (2) $A_r = 0.40$, and (3) $A_r = 0.50$. The results include data with various D (*triangle*: 10 mm, *circle*: 30 mm, *diamond*: 50 mm) and Pr (*red*: 67, *green*: 112, *blue*: 207). The azimuthal mode number and oscillation type can be recognized from the symbol (*open*: $m = 1$, *double*: $m = 2$) and the color shading (*dark*: SW, *light*: RW), respectively. The

Fig. 21 (Left) Comparisons of liquid-bridge shapes for $A_r = 0.50$ between those obtained from experimental data (*open circles*) and the Young-Laplace equation (*dashed lines*) for various volume ratios: $V_r = 0.61$, 0.70, 0.81, 0.90, and 0.99. (Right) Binarized side-view images used for shape detection

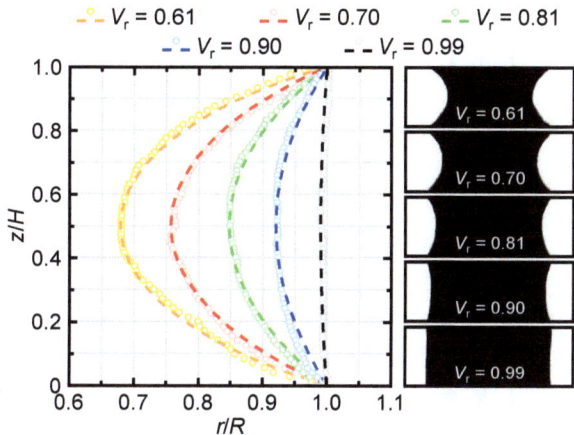

results achieved in the MEIS project were reported previously [7, 47], but those in the DS project are the first appearance. For $A_r = 0.35$, the volume ratio affects the onset conditions of oscillatory thermocapillary convection, and the value of $\mathrm{Ma_c}$ increases with decreasing V_r, while it has little effect on F_c (Fig. 22(1)). In this A_r, the critical azimuthal mode is always $m = 2$ and SW-type oscillation is preferable except for the minimum V_r. The opposite tendency is seen for $A_r = 0.40$ that $\mathrm{Ma_c}$ is insensitive to V_r while F_c has positive linear relationship with V_r (Fig. 22(2)). The instability mode is influenced by the liquid-bridge shape and the decrease in V_r causes a switching of m from 2 to 1 at $V_r = 0.80$–0.83, as with the increase in A_r (Fig. 12). The switching of m with the change of V_r was observed in previous ground-based studies [48–51]. For high-Pr cases, such switching of m is often accompanied by local maximization of $\mathrm{Ma_c}$ in contrast to the present μg experiments. Both $\mathrm{Ma_c}$, F_c, and the oscillation type (but not m) are more responsive to V_r for $A_r = 0.50$ (Fig. 22(3)). The value of $\mathrm{Ma_c}$ increases with decreasing V_r, suddenly drops at $V_r = 0.55$–0.65, and converges to an almost constant value for much smaller V_r. The value of F_c decreases in the $\mathrm{Ma_c}$-increasing region, shows local minimum at $V_r = 0.55$–0.65, and increases sharply in the $\mathrm{Ma_c}$-constant region. The interesting feature observed in the results for $A_r = 0.50$ is that the SW-type instability is dominant for larger V_r while the RW-type one is dominant for smaller V_r, and the switching of oscillation type occurs at which $\mathrm{Ma_c}/F_c$ becomes a local maximum/minimum.

The effect of V_r was also studied for longer liquid bridges. The V_r dependence of (a) $\mathrm{Ma_c}$ and (b) F_c for $A_r = 1.00$, 1.25, and 1.50 are shown in Fig. 23. For all results, $D = 30$ mm and $m = 1$, and the aspect ratio can be distinguished from the symbol (*circle*: $A_r = 1.00$, *square*: $A_r = 1.25$, *diamond*: $A_r = 1.50$). The same color rules as in Fig. 22 are applied. The values of $\mathrm{Ma_c}$ and F_c gradually increase with decreasing V_r, and both drop rapidly at $V_r \approx 0.7$. For $V_r < 0.7$, $\mathrm{Ma_c}$ converges to almost constant value while F_c increases again (but more sharply). The rapid decrease in F_c at $V_r \approx 0.7$ is reminiscent of the *frequency skip* observed in F_c–A_r plots for nearly cylindrical liquid bridges (Figs. 17 and 18). The interesting difference between

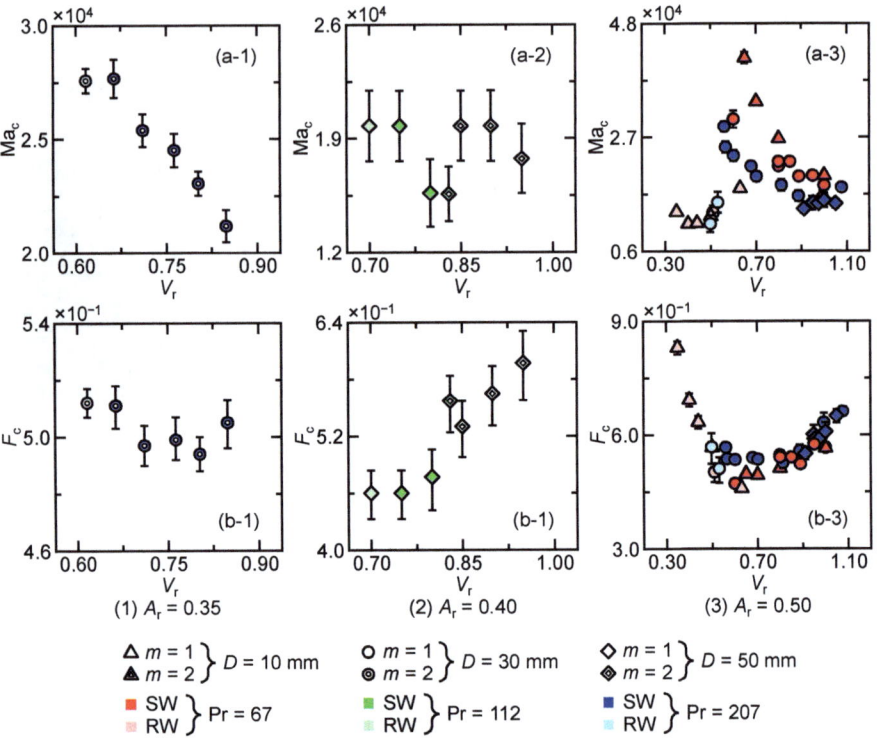

Fig. 22 Plots of **a** critical Marangoni number Ma_c and **b** critical dimensionless oscillation frequency F_c as a function of volume ratio V_r for shorter liquid bridges: (1) $A_r = 0.35$, (2) $A_r = 0.40$, and (3) $A_r = 0.50$. The cold-disk temperature is fixed at $T_C = 20\,^\circ\text{C}$

the *frequency skip* in F_c–A_r plot and that in F_c–V_r plot is the oscillation type. The SW-type oscillation mode is the most critical for a wide range of A_r and the RW-type one is unusual when $V_r = 0.95$–1.00. On the other hand, it is obvious that the RW-type oscillatory flow is the primary for small V_r, especially on the branch with lower F_c when $A_r = 1.00$–1.50. Since the preference of SW- or RW-type instabilities is the nonlinear feature [22] and is not completely understood yet, the effect of V_r on the oscillation type remains unclear.

In 2016, the Chinese research group has conducted μg experiments in the Tiangong-2 space laboratory [52, 53]. One of their main research targets was the shape effect on the instability of thermocapillary convection in a liquid bridge with $Pr = 67$. They carried out the experiments by changing V_r and A_r very finely and drew instability maps in A_r-V_r space. Kang et al. [52] found two instability modes: one has a high oscillation frequency and is preferable for larger V_r and smaller A_r and the other has a low oscillation frequency and is preferable for smaller V_r and larger A_r. They defined a geometric factor (i.e., $\{A_r - 3.2V_r + 1.4\}$) and suggested that the high-/low-frequency mode appears when this value is negative/positive. When the

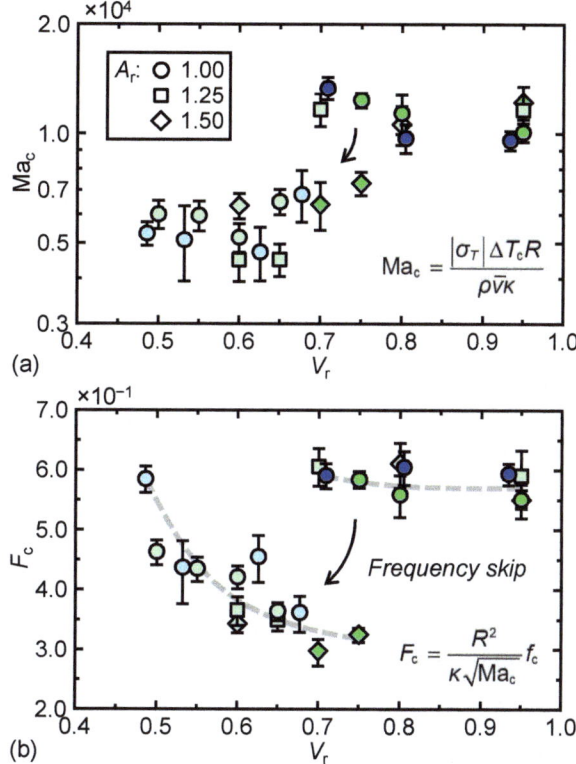

Fig. 23 Plots of **a** critical Marangoni number Ma_c and **b** critical dimensionless oscillation frequency F_c as a function of volume ratio V_r for longer liquid bridges: $A_r = 1.00$ (*circle*), 1.25 (*square*), and 1.50 (*diamond*). The cold-disk temperature is fixed at $T_C = 20\,°C$

geometrical factor is evaluated for the data in Fig. 23, it becomes positive after the *frequency skip* and vice versa except for very close conditions at which the *frequency skip* occurs. Kang et al. [53] reported that the RW (traveling wave in their notation)-type instability is dominant in the low-frequency mode while the SW-type instability is dominant in the high-frequency mode. The present results are also consistent with their results, suggesting that the instability mode after-/before the *frequency skip* is corresponding to the low-/high-frequency mode reported by Kang et al. [52, 53].

4.3 Effect of Prandtl Number

The Prandtl number $Pr \,(= \nu/\kappa)$, which gives the relative importance of the momentum diffusion to the thermal diffusion, is crucially important parameter for the instability of thermocapillary convection. Table 5 lists Prandtl numbers of typical liquids used in the study on thermocapillary convection. It is well known that the instability mechanism of thermocapillary convection in a liquid bridge is essentially different between low- and high-Pr fluids [54, 55]. In low-Pr liquid bridges (say, $Pr \ll 1$),

Table 5 Typical Prandtl numbers of various fluids used in the study of thermocapillary convection

Fluid with Pr < 1	Pr	Fluid with Pr > 1	Pr
Silver (Ag)	0.0063[*1]	Potassium chloride (KCl)	1.025[*6]
Silicon (Si)	0.009[*1]	Acetone (C_3H_6O)	4.3[*7]
Tin (Sn)	0.016[*1]	Sodium nitrate ($NaNO_3$)	7.025[*6]
Molybdenum (Mo)	0.025[*2]	n-decane ($C_{10}H_{22}$)	13.5[*8]
Mercury (Hg)	0.0258[*3]	Hexadecane ($C_{16}H_{34}$)	42[*9]
Gallium antimonide (GaSb)	0.04[*4]	Tetracosane ($C_{24}H_{50}$)	49.194[*6]
Gallium arsenide (GaAs)	0.068[*5]	Fluorinert (FC-43)	64[*9]

[*1]Hibiya et al. [59], [*2]Jurisch and Löser [60], [*3]Jurisch et al. [61], [*4]Cröll et al. [62], [*5]Rupp et al. [56], [*6]Schwabe et al. [63], [*7]Simic-Stefani et al. [64], [*8]Watanabe et al. [65], [*9]Kamotani et al. [66]

the primary instability of thermocapillary convection is the hydrodynamical one. The flow transitions from an axisymmetric steady state to a nonaxisymmetric steady state, and then transitions to an oscillatory state with further increase in Ma [56]. In contrast, the thermocapillary convection transitions from an axisymmetric steady state to an oscillatory state directly in high-Pr liquid bridges due to the hydrothermal-wave instability [54, 55]. According to the studies by Chen et al. [57] and Levenstam et al. [58], the threshold at which such instability mechanism changes is Pr $\approx$ 0.057 for $A_r = 0.5$.

Figure 24 shows plots of Ma_c as a function of Pr for $A_r = 0.5$ and $V_r = 0.95-1.00$ measured in the present μg experiments (*colored circles*). Also included are literature data for the same A_r and V_r obtained in the terrestrial experiment (*open white symbols*), the linear stability analysis (LSA, hereinafter; *cross symbols or line plots*), and the direct numerical simulation (DNS, hereinafter; *double symbols*). Some data with $A_r \neq 0.5$ obtained experimentally under normal-gravity condition are also plotted here for reference (*open gray symbols*: $A_r = 0.6$ in Shevtsova et al. [67]; $A_r = 0.48$ in Hayashida et al. [68]; $A_r = 0.43$ in Watanabe et al. [65]). The temperature dependence of Pr is taken into account for the present results (Fig. 4), while that is not for the literature and their original Pr are used. Preisser et al. [17], Kamotani et al. [66], and Ueno et al. [19] measured Ma_c for multiple D, and the error bars indicate its upper and lower limits. As described above, the thermocapillary convection in a low-Pr liquid bridge becomes oscillatory via two-step transition. The primary transition to the nonaxisymmetric steady state is represented by *light colors* and the secondary transition to the oscillatory state is represented by *dark colors*. Wanschura et al. [55] reported Ma_c (critical Reynolds number Re_c (= Ma_c/Pr) in their original manuscript) for many more values of Pr, but some data are thinned out in Fig. 24 for better readability.

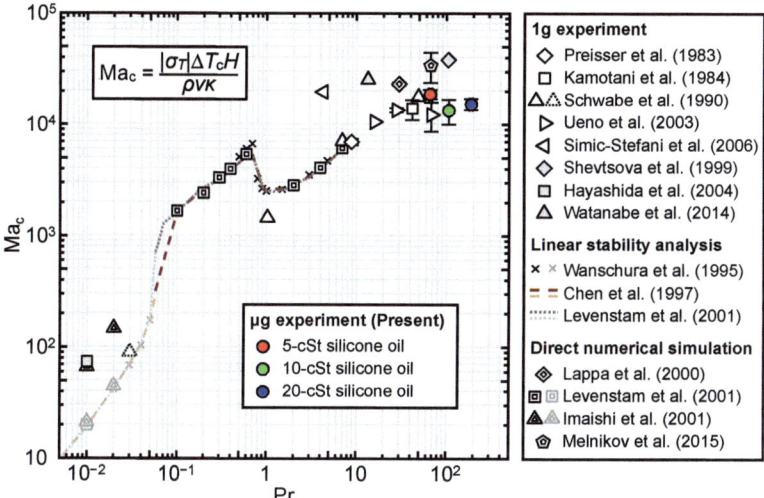

Fig. 24 Effect of Prandtl number Pr on the critical Marangoni number Ma_c for the liquid bridge of $A_r = 0.5$ measured in MEIS and DS projects (*colored circles*). Also included are literature data measured under normal-gravity environment for $A_r = 0.5$ (*white symbols*) and $A_r \neq 0.5$ (*gray symbols*), those obtained by the linear stability analysis (*cross symbols* or *line plots*), and those obtained by the direct numerical simulation (*double symbols*)

For $0.01 < Pr < 0.057$, Ma_c increases monotonically with Pr. The value of Ma_c for the primary transition is of the order of $O(10–10^2)$, and the secondary transition occurs when Ma increases several times. Schwabe et al. [63] predicted the instability threshold of the secondary transition for $Pr \approx 0.03$, which is the typical Prandtl number of liquid Si, as $Ma_c \approx 90$ (*white delta with dotted outline*) from the linear extrapolation of Ma_c obtained for KCl, $NaNO_3$, and $C_{24}H_{50}$ ($Pr \approx 1$, 7, and 49, respectively; *white delta with solid outline*). This predicted value is consistent with the measurement result of Cröll et al. [62], who reported $Ma_c = 140–200$ for liquid Si. However, the behavior of the neutral stability curve for $0.057 < Pr < 10$ is actually more complex as shown in Fig. 24. The value of Ma_c increases up to $Pr \approx 0.7$, decreases sharply at $0.7 < Pr < 1$, and then increases again. In this Pr range, the critical azimuthal mode number and the critical oscillation type change frequently depending on Pr [22, 58].

Since few samples are available for the experiments with low- and intermediate-Pr liquid bridges, theoretical and computational studies are influential for $Pr < 10$. On the other hand, the number of theoretical and computational results for $Pr > 10$ is limited due to the difficulty in treating high-Pr fluids, which stems, e.g., from a very thin thermal boundary layer in the vicinity of the solid disk. Therefore, experimental studies tend to be carried out preferentially to understand the thermocapillary-convection instability in high-Pr liquid bridges. As shown in Fig. 24, the values of Ma_c for $Pr > 10$ converge to 1×10^4–2×10^4, and no significant effect of Pr on Ma_c was observed. Kang et al. [52] measured $Ma_c = 2.1 \times 10^4$–2.2×10^4 for $Pr = 67$,

$A_r = 0.50$, and $V_r = 0.90$–0.98 in their μg experiment described in Sect. 4.2. Their results are also in accordance with the trend in Fig. 24.

As with LSA and DNS, there are difficulties in measuring Ma_c in the experiment with high-Pr fluids. It follows from Eq. 6 that it requires a large temperature difference ΔT or a large length scale L to increase Ma when the viscosity (and the resultant Pr) of the test liquid is large. To the best of authors' knowledge, the thermal diffusivity of high-Pr liquids does not change much compared to the kinematic viscosity, and $\kappa = O(10^{-8}$–$10^{-7})$ holds for the typical liquids with Pr > 1 in Table 5 as well as the silicone oils used in the present μg experiments. In the terrestrial experiments, the height of liquid bridge is limited to a few millimeters; therefore, an effective way to increase Ma is to increase ΔT. However, for example, it requires $\Delta T = 236$ K to realize Ma $= 2 \times 10^4$ in a liquid bridge with $D = 5.0$ mm and $H = 2.5$ mm (i.e., $A_r = 0.5$) when the test liquid is 20-cSt silicone oil (Pr $= 207$). Such a large ΔT causes a variety of experimental problems—e.g., increase in heat loss, change in physical properties, evaporation of test liquid, etc. These problems can be broken through in the absence of gravity, and almost pure effect of Prandtl number on the thermocapillary-convection instability was studied for Pr $= 67$–207 in the present μg experiments. We emphasize that the measurement of Ma_c for Pr $\gg 100$ without above-mentioned problems is the first achievement in the world. Recent computational studies provide excellent agreement of Ma_c with experimental data even under high-Prandtl-number condition of Pr ≈ 67, due to progress in analytical techniques and improvements in computer performance [6, 69]. In the near future, numerical simulations and stability analyses for much larger Pr are expected to be performed; therefore, the data obtained in the present μg experiments are highly valuable.

4.4 Effect of Changing Rate of Disk Temperatures

Prior to the present μg experiments aboard the ISS, several researchers have carried out μg experiments with large-scale thermocapillary liquid bridges utilizing the sounding rocket or the Space Shuttle, e.g., Monti et al. [70], Monti and Fortezza [71], Hirata et al. [72], Kawamura et al. [73], Carotenuto et al. [74], etc. In these previous μg experiments, the convection remained steady even at Ma higher than the expected Ma_c from terrestrial experiments, theoretical studies, or numerical simulations. For example, the research team in Italy has conducted a series of μg experiments in the TEXUS sounding rockets, and they reported $Ma_c = 5 \times 10^4$–9×10^4 for 5-cSt silicone oil (Pr $= 74$) and $Ma_c = 10 \times 10^4$–12×10^4 for 2-cSt silicone oil (Pr $= 30$), wherein the size of liquid bridge was $D = 18$ mm and $H = 18$ mm [71]. We note that they originally reported ΔT_c, but it was converted to Ma_c using physical properties in Ref. [75]. The Japanese research teams have also conducted a series of TR-IA sounding-rocket experiments with the liquid bridge of $D = 15$ mm or 50 mm, and they observed steady thermocapillary convection at relatively high-Marangoni-number conditions, say, Ma $\gg 2 \times 10^4$ [72, 73]. Such increase in Ma_c—in other words, stabilization of steady thermocapillary convection—is, as pointed out by researchers themselves, considered to be due to the short waiting period for the thermal equilibrium, because the achievable μg duration in sounding rockets is less than 15 minutes at most [76].

The Space Shuttle could provide a much longer μg duration; however, the values of Ma_c for liquid bridges with $D = 30, 45$, and 60 mm measured during the Spacelab D2 (or STS-55) mission, which was the scientific research program in the laboratory mounted onboard the Space Shuttle Columbia, were still beyond expectations [74, 77]. The reason of this discrepancy is considered to be twofold: (1) the matter of short waiting period for thermal equilibrium, and (2) the possibility of other dominant parameters besides Marangoni number [78]. These topics were examined in MEIS and DS projects. In this section, the result of former examination is introduced, and the latter will be discussed in Sect. 7.

In the real situation, thermocapillary convection does not transition to an oscillatory state as soon as Ma exceeds instability threshold, and it requires a certain time for the observer to recognize the unsteady motion. Therefore, too rapid change in ΔT or insufficient waiting period after the change of ΔT can affect the measurement accuracy of the onset condition of oscillatory thermocapillary convection. Kawamura and Ueno [12] were concerned about this problem and conducted a series of terrestrial experiments with various changing rates of disk temperature difference (i.e., $D_t \Delta T$, where D_t denotes the time derivative in Euler's notion). They showed a notable increase in Ma_c with an increase in the dimensionless changing rate of disk temperature difference $dT^*/dt^* = D_t \Delta T \{L^2/(\Delta T_c \kappa)\}$, where L is the characteristic length scale given in Eq. 4, and they suggested that dT^*/dt^* should be less than 0.1.

The effect of changing rate of ΔT on the development (or detection) of oscillatory thermocapillary convection was also investigated in MEIS-3, 4, and 5 [6]. As shown in Fig. 8(2), it was done by increasing T_H linearly while keeping T_C constant at 20 °C. The examined values of $D_t \Delta T|_{T_C=20°C}$ were 0.1–1.0 K/min in MEIS-3, 0.1–0.4 K/min in MEIS-4, and 1.0–4.0 K/min in MEIS-5, and the aspect ratios are $A_r = 0.50$ in MEIS-4 and $A_r = 1.00$ in MEIS-3 and 5. Figure 25a shows time series of ΔT (*dark lines*, see the left vertical axis) and variations of free-surface temperature T_{surf} (*light lines*, see the right vertical axis) obtained in MEIS-4, wherein T_{surf} were measured with an insertion thermocouple sensor placed near the free surface without touching it. The horizontal dashed line displays ΔT_c (= 16.7 K) measured under the condition of $D_t \Delta T = 0$ (i.e., ΔT_c in Fig. 11), and $t = 0$ corresponds to the time at which each ΔT reaches this temperature difference. The values of T_{surf} are offset to be zero at $t = 0$. There are some gaps in the plots of T_{surf} because the measurement range of the present thermocouple sensor was ±2.5 K, and the dynamic range must be adjusted manually. It follows from Fig. 25a that T_{surf} increases linearly with time and starts oscillation at ΔT above $\Delta T_c|_{D_t \Delta T=0}$ due to the thermocapillary-convection instability.

The onset of oscillation can be seen more clearly in Fig. 25b, which shows temperature fluctuations from linear regression lines δT_{surf}. For $D_t \Delta T = 0.1$ K/min, δT_{surf} starts oscillation shortly after ΔT exceeds $\Delta T_c|_{D_t \Delta T=0}$. On the other hand, for $D_t \Delta T = 0.2$ K/min and 0.4 K/min, it is obvious that the oscillation of δT_{surf}

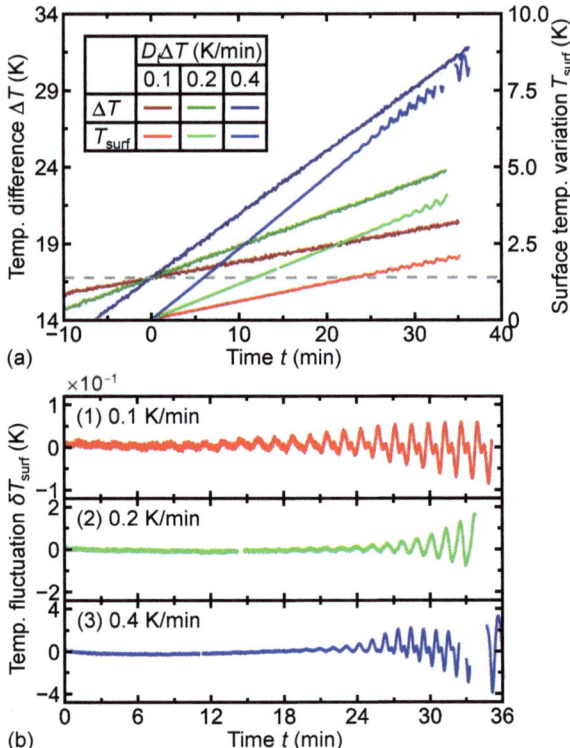

Fig. 25 Effect of changing rate of disk temperature $D_t \Delta T$ on the development of oscillatory thermocapillary convection. **a** Time series of temperature difference ΔT and surface temperature variation T_{surf}. **b** Temperature fluctuations near the free surface δT_{surf} with changing rates of (1) 0.1 K/min, (2) 0.2 K/min, and (3) 0.4 K/min

starts much later, and ΔT becomes much larger than $\Delta T_c|_{D_t \Delta T=0}$ when the oscillation can be recognized. The amplitudes of δT_{surf} grow with time for all cases. In MEIS-4 and 5, which used MI50 having an insertion thermocouple sensor, the onset condition of instability $\Delta T_c|_{D_t \Delta T \neq 0}$ was determined by the extrapolation of the δT_{surf} amplitude to the zero value. In MEIS-3, it was determined qualitatively by the experimenter with the data sent from observation cameras because there was no insertion thermocouple sensor in MS30 as described in Sect. 2.

Since it does not make much sense to discuss the effect of $D_t \Delta T$ in dimensional form when the size of liquid bridge and the physical properties of test liquid are not constant, Nishino et al. [6] proposed a new dimensionless quantity based on the idea derived from the following energy equation:

$$\frac{\partial T}{\partial t} + (\boldsymbol{u} \cdot \nabla)T = \kappa \nabla^2 T, \quad (11)$$

where $\boldsymbol{u}$ is the velocity vector. When Eq. 11 is made dimensionless with the physical scales in the first row of Table 6 (i.e., U^*, ΔT, H, and H/U^*), one can find the Marangoni number (Eq. 6) with H being the representative length scale, where U^*

Table 6 Scales for making energy equation dimensionless

	Velocity	Temperature	Length	Time
Usual way (e.g., Kuhlmann [35])	U^* $(= \|\sigma_T\|\Delta T/\mu)$	ΔT	H	H/U^*
Nishino et al. [6]	U_{ref} $(= \|\sigma_T\|T_{\text{ref}}/\mu)$	T_{ref} $(= (D_t\Delta T)H^2/\kappa)$	H	H/U_{ref}

$(= |\sigma_T|\Delta T/\mu)$ is the so-called characteristic thermocapillary velocity. In contrast, Nishino et al. [6] proposed to replace ΔT with T_{ref} in the physical scales, where

$$T_{\text{ref}} = \frac{(D_t\Delta T)H^2}{\kappa}. \tag{12}$$

Accordingly, the physical scales for velocity and time change into $U_{\text{ref}} (= |\sigma_T|T_{\text{ref}}/\mu)$ and H/U_{ref}, respectively, as shown in the second row of Table 6. As a result, the energy equation becomes dimensionless as follows:

$$\frac{\partial \tilde{T}}{\partial \tilde{t}} + (\tilde{u} \cdot \tilde{\nabla})\tilde{T} = \frac{1}{(D_t\Delta T^*)^2}\tilde{\nabla}^2\tilde{T}, \tag{13}$$

where the upper tilde ($\tilde{\ }$) denotes dimensionless variables. $D_t\Delta T^*$ in Eq. 13 is the dimensionless changing rate of disk temperature difference and is defined as

$$D_t\Delta T^* = \frac{H}{\kappa}\sqrt{\frac{|\sigma_T|(D_t\Delta T)H}{\rho\nu}}. \tag{14}$$

Figure 26 shows plots of Ma_c as a function of $D_t\Delta T^*$ measured in MEIS [6, 7]. Also included are those reported by Carotenuto et al. [74]. Their test liquid was 5-cSt silicone oil with $\text{Pr} = 74$, and they measured Ma_c for various D and A_r during the Spacelab D2 mission. We note that the physical properties in Savino and Monti [75], who were in the same research group with Carotenuto et al. [74], are used to evaluate their $D_t\Delta T^*$. A significant increase in Ma_c with $D_t\Delta T^*$ can be recognized in Fig. 26, revealing the noteworthy effect of the temperature-control (changing) method on the measurement of Ma_c. The present result suggests that the changing rate of disk-temperature difference should be $D_t\Delta T^* < 100$ to suppress the measurement uncertainty lower than 10%. Carotenuto et al. [74] has conducted experiments with $D_t\Delta T = 1$ K/min. This value is typical in terrestrial experiments and results in $D_t\Delta T^* \approx 25$ for $D = 5$ mm; however, it results in $D_t\Delta T^* \approx 1141$ for $D = 30$ mm when the test liquid is 5-cSt silicone oil and $A_r = 1$. It can be concluded that the considerably large Ma_c measured in the previous Space-Shuttle mission is very likely

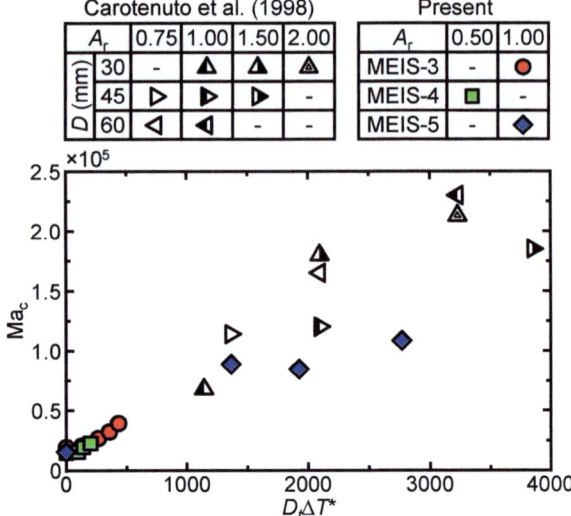

Fig. 26 Comparison of critical Marangoni number Ma_c measured under various conditions of dimensionless changing rate of disk temperature difference $D_t \Delta T^*$. Data for $A_r = 1.00$ were originally reported by Nishino et al. [6]. Also included are literature data reported by Carotenuto et al. [74], who measured Ma_c with a changing rate at 1 K/min during the Spacelab D2 mission

due to the effect of too large changing rate of ΔT, as has been pointed out before [6].

5 Flow Measurement with Contactless Techniques

Taishi Yano and Koichi Nishino

As described in Sect. 2, FPEF was equipped with a variety of measurement techniques. This section mainly focuses on the three-dimensional particle tracking velocimetry (hereinafter, 3-D PTV) and the photochromic dye activation technique. The former is a powerful technique for measuring the spatial structure of internal bulk flow, and the latter is effective for measuring the surface flow, both of which are very important for understanding the thermocapillary convection in a liquid bridge.

5.1 Inner Flow Measurement with Three-Dimensional Particle Tracking Velocimetry (3-D PTV)

Contactless flow measurement using fine tracer particles is useful way to understand the structure of thermocapillary convection which is susceptible to disturbances. As described in Sect. 1, the liquid bridges formed in the present μg experiments contained tracer particles for flow visualization, and their motions were observed by multiple cameras. Except for DS-2, the hot disk was made of transparent sapphire,

Fig. 27 Schematic diagram of coordinate systems: physical coordinate system (x, y, z) drawn with black axes; image coordinate systems (X, Y) drawn with white axes; camera coordinate systems (x_c, y_c, z_c) drawn with colored axes (i.e., *red*, *green*, and *blue* for x_c, y_c, and z_c, respectively). Origins of camera coordinate systems are located at each perspective point (x_0, y_0, z_0) in the physical coordinate system

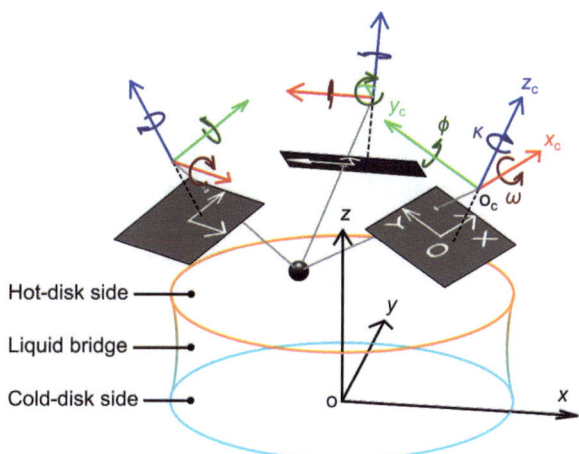

and three top-view cameras could observe inside the liquid bridge from different orientations simultaneously, as illustrated in Figs. 6, 7, and 27. In the 3-D PTV, the positional information of tracer particles taken by each camera—i.e., positions on a (two-dimensional) image coordinate—are reconstructed into three-dimensional ones by means of the principal of stereoscopic vision. The 3-D PTV used in this study was originally developed by Nishino et al. [79] and was customized by Yano et al. [80] to accommodate to the present experimental facility. The details have already been reported elsewhere [81]; therefore, only a brief introduction is given here.

When a reference point at (x, y, z) in the (three-dimensional) physical coordinate is projected on (X, Y) in the (two-dimensional) image coordinate of a certain camera, their relationship under the collinearity condition can be written as

$$X = -c \frac{a_{11}(x - x_0) + a_{12}(y - y_0) + a_{13}(z - z_0)}{a_{31}(x - x_0) + a_{32}(y - y_0) + a_{33}(z - z_0)} + \Delta X, \quad (15)$$

$$Y = -c \frac{a_{21}(x - x_0) + a_{22}(y - y_0) + a_{23}(z - z_0)}{a_{31}(x - x_0) + a_{32}(y - y_0) + a_{33}(z - z_0)} + \Delta Y, \quad (16)$$

where a_{ij} is a component of the following rotation matrix (Fig. 27):

$$\mathbf{A} = \begin{bmatrix} 1 & 0 & 0 \\ 0 & \cos \omega & -\sin \omega \\ 0 & \sin \omega & \cos \omega \end{bmatrix} \begin{bmatrix} \cos \phi & 0 & \sin \phi \\ 0 & 1 & 0 \\ -\sin \phi & 0 & \cos \phi \end{bmatrix} \begin{bmatrix} \cos \kappa & -\sin \kappa & 0 \\ \sin \kappa & \cos \kappa & 0 \\ 0 & 0 & 1 \end{bmatrix}. \quad (17)$$

ΔX and ΔY in Eqs. 15 and 16 are correction terms for aberration. In this study, the influence of lens distortion is considered as

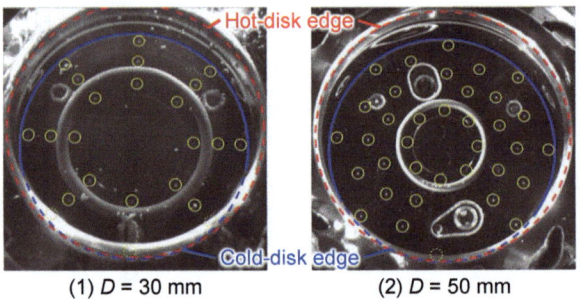

(1) $D = 30$ mm (2) $D = 50$ mm

Fig. 28 Pictures of calibration dots marked on the cold-disk surface for (1) $D = 30$ mm and (2) $D = 50$ mm. Colored lines with *yellow*, *blue*, and *red* indicate calibration dot, cold-disk edge, and hot-disk edge, respectively. Calibration dots in the blind spots are surrounded by dotted lines

$$\Delta X = X_0 + (X - X_0)(k_1 r^2 + k_2 r^4), \tag{18}$$

$$\Delta Y = Y_0 + (Y - Y_0)(k_1 r^2 + k_2 r^4), \tag{19}$$

where $r^2 = \{(X - X_0)^2 + (Y - Y_0)^2\}/c^2$. These equations include 11 camera parameters: (x_0, y_0, z_0) for the physical coordinates of the perspective point (or the camera center); (ω, ϕ, κ) for the rotation angles of the camera; c for the principal distance in the pixel dimension; (X_0, Y_0) for the shifts of the principal point in the image coordinate; (k_1, k_2) for the lens distortion coefficients. We note that X needs to be replaced by X/A_{px} when the pixel aspect ratio A_{px} (i.e., Y to X) is not unity. Additionally, in some situations where further accuracy is required, the correction terms for the decentering distortion are added to Eqs. 18 and 19 [81]. However, the present facility provided sufficient accuracy without them, so that most subsequent results of 3-D PTV do not consider the decentering distortion.

As shown in Fig. 28, small dots for the camera calibration, whose x- and y-positions in the physical coordinate system are known in advance, were marked on the surface of the cold disk. The number of dots for 30 mm-diameter disks (MS30 and MD30) was 23 and that for 50 mm-diameter disks (MI50) was 36, although some of them were hidden in the blind spots of the camera. In each series of MEIS and DS projects, the cold disk was precisely traversed in the z-direction in a stepwise manner, and images of calibration dots were captured at each step. Consequently, hundreds of pairs of known (x, y, z) and (X, Y) were obtained, and the camera parameters were determined using the Levenberg-Marquardt method [82]—one of the least-squares solutions. The resultant calibration errors were evaluated to be less than 1 px for all top-view cameras, which is acceptable for the 3-D PTV.

Once all the camera parameters (including A_{px} if necessary) are determined, and the position of the tracer particle in the image coordinate system (X, Y) is obtained, three unknowns remain in Eqs. 15 and 16 (i.e., x, y, and z). Since these equations hold for all top-view cameras (although camera parameters and (X, Y) are specific for each), the three-dimensional position (x, y, z) of the target tracer particle can be determined as long as two or more cameras observe the same tracer particle.

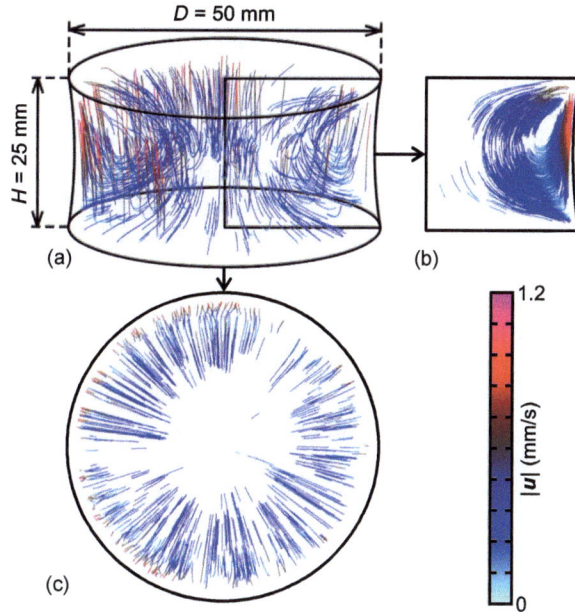

Fig. 29 Particle trajectories of steady thermocapillary convection measured by 3-D PTV for Pr = 207, $A_r = 0.50$, $V_r = 0.95$, $T_C = 20.0\,°C$, $T_H = 35.4\,°C$, and $\Delta T = 15.4$ K: **a** bird's-eye view, **b** r-z cross-sectional view, and **c** r-θ cross-sectional view

Actually, due to calibration errors and other uncertainties, it is almost impossible to solve the problem exactly, and some processes are required to obtain (x, y, z), but the details are not covered here (see, e.g., Yano and Nishino [81]). In principle, the minimum number of cameras required for stereoscopic vision is two; however, three cameras were used in this study wherever possible because a large number of tracer particles on the image would lead to the incorrect reconstruction and the appearance of ghost particles (i.e., tracer particles reconstructed in positions not actually existing). The present 3-D PTV first tracks the motions of individual tracer particles on each image coordinate and later reconstructs their trajectories into the three-dimensional space. Such procedures can reduce the number of ghost particles, but again, the details are not provided here (see, Yano [7]). A typical example of the 3-D PTV measurement is shown in Fig. 29: (a) bird's-eye view, (b) r-z cross-sectional view, and (c) r-θ cross-sectional view of particle trajectories for 450 s. The original data were obtained in MEIS-4 ($D = 50$ mm and Pr = 207), and the experimental conditions are $A_r = 0.50$, $V_r = 0.95$, $T_C = 20\,°C$, and $T_H = 35.4\,°C$. The given ΔT (= 15.4 K) was set lower than the critical value (i.e., $\Delta T_c = 16.7$ K); therefore, the flow should be axisymmetric and steady. The particle trajectories exhibit an axisymmetric and steady toroidal flow pattern, and this result qualitatively indicates the validity of the present 3-D PTV. The three-dimensional velocity measurements of thermocapillary convection in a liquid bridge under μg environment have been performed in the previous sounding-rocket experiments (e.g., TR-IA-4 by Kawamura et al. [73] and TR-IA-6 by Nishino et al. [83]), but there is no result of obtaining velocity data with a spatial resolution as high as that in the present μg experiments.

A variety of basic flow patterns of thermocapillary convection with ΔT slightly lower than each instability threshold (i.e., $\Delta T/\Delta T_c = 0.6$–$1.0$) for various A_r and for (a) $Pr = 67$ and (b) $Pr = 207$ are shown in Figs. 30a and b, respectively. The data were obtained in MEIS-2 and 3 so that $D = 30$ mm and $T_C = 20.0\,°C$. When the Prandtl number is small (i.e., $Pr = 67$), the convection roll elongates longitudinally with increasing A_r, and an additional recirculation embedded in the global convection roll arises in the cooler side of the liquid bridge at $A_r \geq 1.25$. Such a detachment of the convection roll is often observed for the thermocapillary flow in large-A_r liquid bridges and in liquid layers [84, 85]. The basic flows for $A_r = 1.25$ and 1.50 both have two roll structures, but their balance is different depending on the liquid-bridge height: for $A_r = 1.25$, rolls are comparable in size (Fig. 30a-4); for $A_r = 1.50$, smaller and larger rolls appear on the warmer and cooler sides, respectively (Fig. 30a-5). It should be pointed out that the instabilities on branches A and B are preferable for $A_r = 1.25$ and 1.50, respectively, as shown in Figs. 17 and 18. The development sequences of the basic flow pattern with $\Delta T/\Delta T_c$ for $A_r = 1.25$ and $Pr = 67$ is shown in Fig. 31. When the given temperature difference is small (i.e., $\Delta T/\Delta T_c = 0.54$), the basic flow pattern is similar to that for $(A_r, \Delta T/\Delta T_c) = (1.50, 0.99)$ (see Figs. 30(a-5) and 31(1)). The convection roll on the warmer side grows with increasing $\Delta T/\Delta T_c$, while that in cooler side shrinks, and eventually their sizes become comparable at $\Delta T/\Delta T_c = 0.83$.

The change in basic flow pattern with A_r for $Pr = 207$ also follows a similar process, whereas an additional recirculation appears at $A_r = 1.50$. The interesting aspect to be pointed out is that the basic flow pattern immediately before the onset of instability for $(Pr, A_r) = (207, 1.50)$ is significantly different with that for $(Pr, A_r) = (67, 1.50)$. When the Prandtl number is large (i.e., $Pr = 207$), the basic flow has a larger convection roll near the hot corner and a weaker recirculation in the lower half of the liquid bridge, whereas the size relationship of convection rolls for $Pr = 67$ is opposite. These results suggest the possibility that the instability on branch A/B is preferable when the basic flow has larger convection roll near the hot/cold disk. More details on the relationship between the thermocapillary-convection instability in large-A_r liquid bridges and the flow field will be discussed later in this section.

The appearance of the additional recirculation is related not only to A_r and Pr, but also the liquid volume. The effect of V_r on the basic flow pattern is shown in Fig. 32, which shows basic flow patterns for $Pr = 207$, $A_r = 1.00$, $V_r = 0.49$–0.93, and $\Delta T/\Delta T_c = 0.69$–$0.92$. When the shape of the liquid bridge is close to the cylindrical one, a single global convection roll occupies the interior of the liquid bridge, but it splits into two between $V_r = 0.68$ (Fig. 32(2)) and $V_r = 0.81$ (Fig. 32(3)) as the liquid bridge becomes more concave. Such a splitting of the convection roll due to the decrease in V_r has been observed in previous ground-based studies, e.g., Masud et al. [42]. It is interesting to point out that the *frequency skip* caused by the change in V_r (Fig. 23) seems to have a correlation with the change in the basic flow pattern (Fig. 32), but further studies are required for more details.

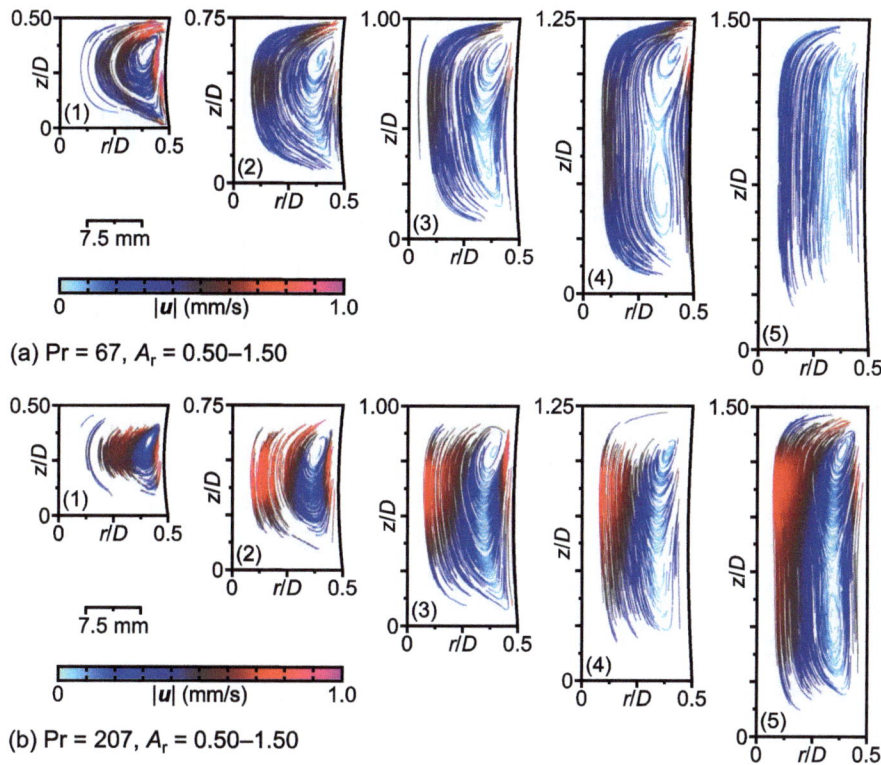

Fig. 30 Particle trajectories in a steady thermocapillary liquid bridge measured with 3-D PTV for various A_r: **a** $Pr = 67$ and **b** $Pr = 207$. (Adapted with permission from Yano [7]. All rights reserved)

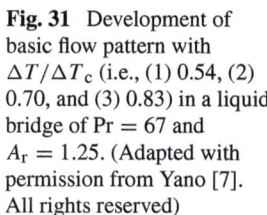

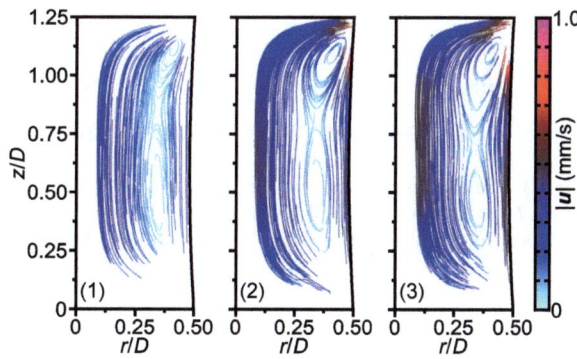

Fig. 31 Development of basic flow pattern with $\Delta T/\Delta T_c$ (i.e., (1) 0.54, (2) 0.70, and (3) 0.83) in a liquid bridge of $Pr = 67$ and $A_r = 1.25$. (Adapted with permission from Yano [7]. All rights reserved)

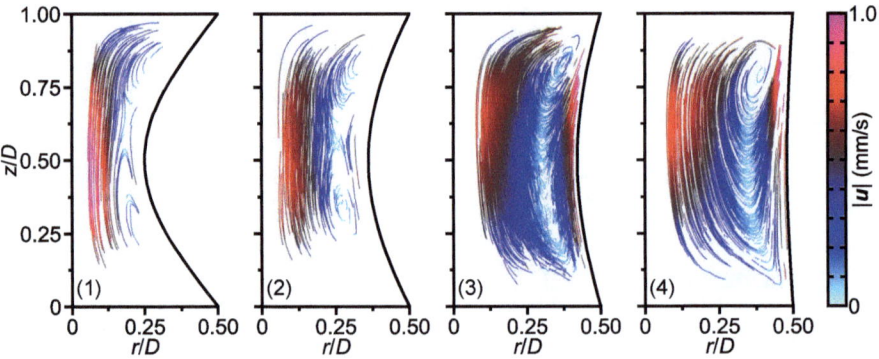

Fig. 32 Basic flow patterns in liquid bridges with various shapes for Pr = 207 and $A_r = 1.00$: (1) $V_r = 0.49$, (2) $V_r = 0.68$, (3) $V_r = 0.81$, and (4) $V_r = 0.93$. (Adapted with permission from Yano [7]. All rights reserved)

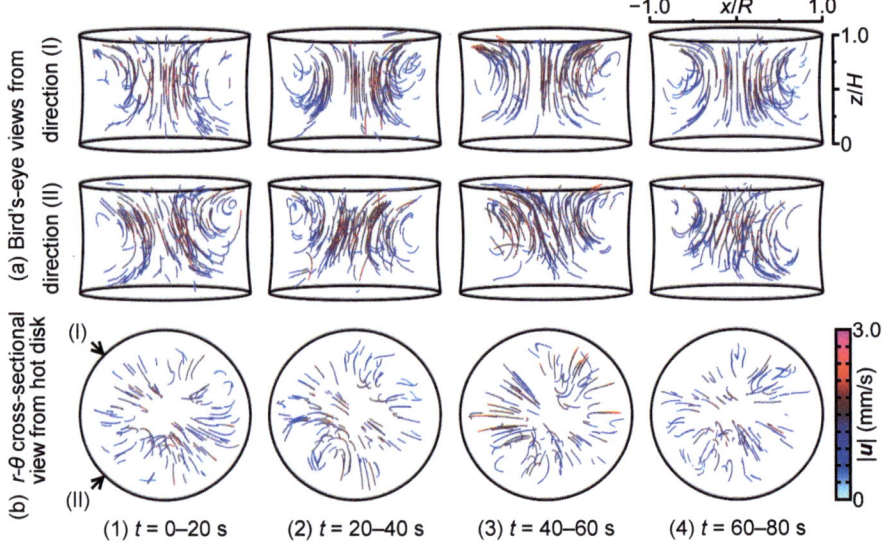

Fig. 33 Time series of particle trajectories in every 20-s interval for Pr = 67, $A_r = 0.75$, $V_r = 0.95$, and $\Delta T_c = 7.9$ K ($T_C = 20.0\,°C$ and $T_H = 27.9\,°C$). (©2014 The Jpn. Soc. Microgravity Appl., adapted with permission from Matsumoto et al. [30]. All rights reserved)

The velocity fields of thermocapillary convection in an oscillatory flow regime are also visualized by 3-D PTV. Figure 33 shows time series of particle trajectories for 20 s measured in MEIS-2, wherein Pr = 67, $A_r = 0.75$, and $V_r = 0.95$ [30]. The original particle images were acquired with slightly higher ΔT (or Ma) than the critical value (i.e., both $\Delta T / \Delta T_c$ and Ma/Ma$_c$ are approximately 1.1). The reconstructed flow fields exhibit diagonal back-and-force motion in a fixed r-z plane, because the oscillation mode in this condition is SW-type one with $m = 1$. The

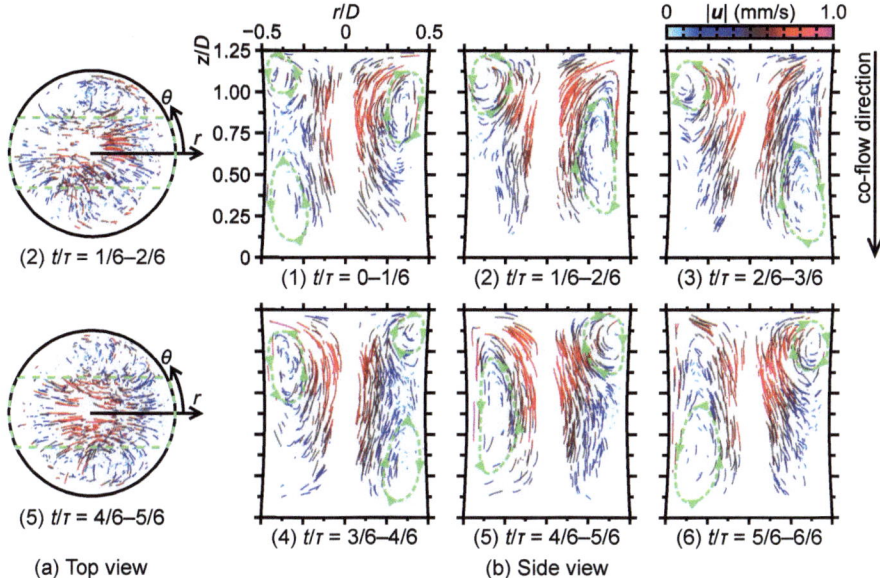

Fig. 34 Time series of particle trajectories of oscillatory thermocapillary convection for Pr = 67, $A_r = 1.25$, $V_r = 0.95$, and Ma = 1.72×10^4 ($\approx 1.22 \text{Ma}_c$) measured in MEIS-2. Each time duration for superposition is 1/6 oscillation period. (Reprinted from Yano et al. [86], with the permission of AIP Publishing. Modifications have been made in this figure with the permission from the publisher. All rights reserved)

presence of a single pair of convection rolls, one in the left side and the other in the right side of the liquid bridge, can be recognized in a bird's-eye view, and these convection rolls alternately move up and down. Such oscillatory flow pattern is reminiscent of that observed in the previous terrestrial experiments [17].

The flow fields of oscillatory thermocapillary convection in a large-A_r liquid bridge have also been measured with 3-D PTV in the present μg experiments, and several new findings have been obtained. Before the MEIS project, experimental flow measurement of thermocapillary convection in a large-A_r liquid bridge was very rare [39, 40]. In particular, to the best of authors' knowledge, the 3-D PTV in MEIS was the first quantitative measurement of three-dimensional velocity field of oscillatory thermocapillary convection in a liquid bridge with $A_r > 1$. The time series of flow fields of oscillatory thermocapillary convection for Pr = 67, $A_r = 1.25$, and $V_r = 0.95$ are shown in Fig. 34. The particle trajectories within the region marked with a *dashed line* in the top-view image are displayed in side-view images, wherein particle trajectories located in left (i.e., $90° < |\theta| < 180°$) and right (i.e., $|\theta| < 90°$) halves in the top-view image have negative and positive r values, respectively. The disk temperatures are set at slightly supercritical condition, i.e., Ma = 1.72×10^4 ($\approx 1.22 \text{Ma}_c$). The instability mode for this condition is SW type with $m = 1$, as shown in Fig. 34a. The oscillation period is measured to be $\tau = 45.5$ s, of which the

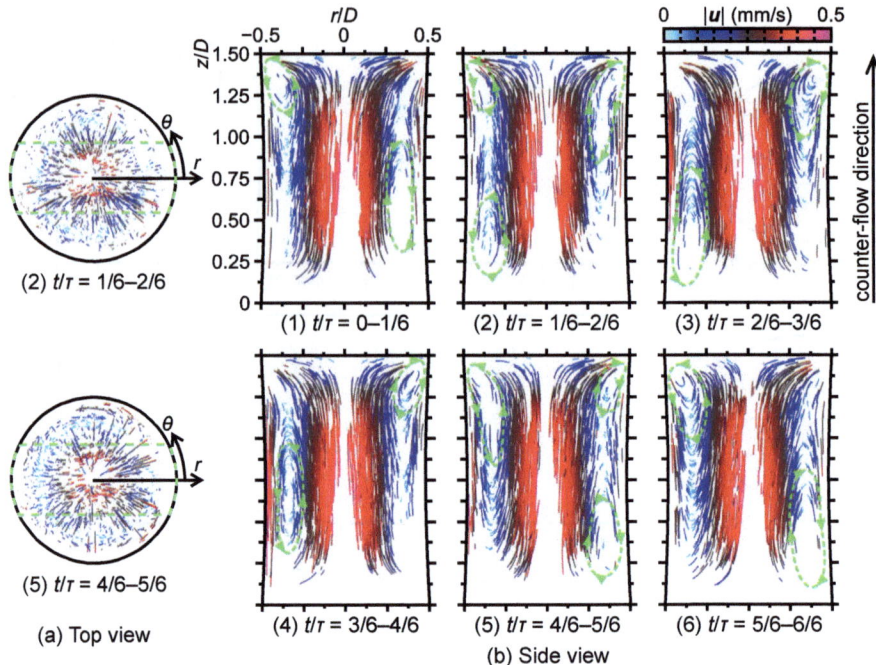

Fig. 35 Time series of particle trajectories of oscillatory thermocapillary convection for Pr = 67, $A_r = 1.50$, $V_r = 0.95$, and Ma = 7.95×10^3 ($\approx 1.03\text{Ma}_c$) measured in MEIS-2. Each time duration for superposition is 1/6 oscillation period. (Reprinted from Yano et al. [86], with the permission of AIP Publishing. Modifications have been made in this figure with the permission from the publisher. All rights reserved)

resultant dimensionless oscillation frequency F (= 0.51) lies on branch A in Fig. 18. The presence of roll structures, highlighted with *dashed green circles*, and their traveling in the longitudinal direction can be recognized in Fig. 34b. The process from the appearance to the disappearance of the upper-left convection roll in Fig. 34b-1 are as follows: (1) a new convection roll appears near the hot corner at $t/\tau = 0$–1/6; (2) it travels in the negative z-direction with growing in size and occupies entire left side for $t/\tau = 1/6$–5/6; (3) it shrinks with approaching the cold disk for $t/\tau = 5/6$–8/6 and finally disappears at $t/\tau = 8/6$–9/6 (we note that see the lower-left convection roll in 34b-1 to b-3 for $t/\tau > 1$). The same evolution of roll structures with the same time sequence offset by a half period is seen in the right side of the liquid bridge. Schwabe [39, 40] has observed the traveling of cellular structures of convection, called drifting Bénard-Marangoni cells, from the warmer side toward the cooler side in the liquid bridge of $A_r = 2.5$ in his μg experiment. Since the convection rolls in Fig. 34 have similar characteristics (e.g., traveling direction, flow pattern, oscillation frequency, etc., see Yano et al. [86] for detailed comparison) with the drifting Bénard-Marangoni cells reported by Schwabe [39, 40], they are believed

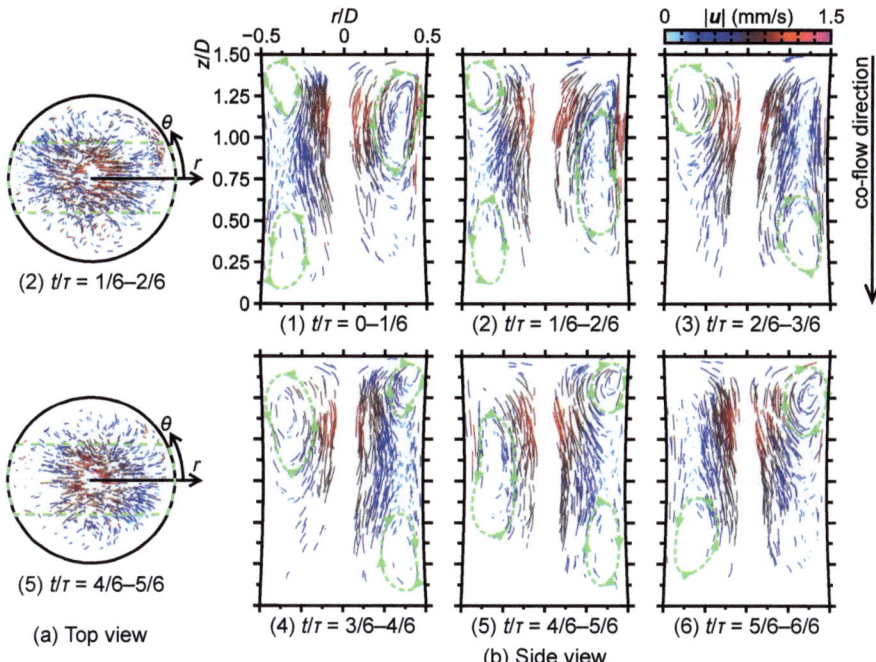

Fig. 36 Time series of particle trajectories of oscillatory thermocapillary convection for Pr = 207, $A_r = 1.50$, $V_r = 0.95$, and Ma = 1.13×10^4 ($\approx 1.09\text{Ma}_c$) measured in MEIS-3. Each time duration for superposition is 1/6 oscillation period. Note that some velocity vectors in **a** top views are not displayed in **b** side views for better visibility. (Adapted with permission from Yano [7]. All rights reserved)

to be identical. Hereinafter, the direction same as the surface flow (i.e., from the hot-disk side toward the cold-disk side) is referred to as the co-flow direction.

As described in Sect. 4.1, the oscillation frequency for Pr = 67 changes discontinuously to a lower value—or switches from branch A to branch B—at $A_r = 1.25\text{--}1.50$ (Fig. 18). The flow structure measured with 3-D PTV in MEIS-2 under the condition very close to branch B is shown in Fig. 35, wherein $A_r = 1.50$, $V_r = 0.95$, Ma = 7.95×10^3 ($\approx 1.03\text{Ma}_c$), and $F = 0.21$. The presence and the traveling of roll structures can be recognized in this result, as in Fig. 34. However, the interesting difference to be pointed out is that the convection rolls travel in the opposite direction to the surface flow (i.e., from the cold-disk side toward the hot-disk side), which is referred to hereinafter as the counter-flow direction. The traveling direction of the roll structure depends not only on A_r but also on other experimental conditions. For example, as shown in Fig. 36, the oscillatory thermocapillary convection for Pr = 207 has co-flow-direction-traveling roll structures even at $A_r = 1.50$. Such a difference in the traveling direction of roll structures is considered to be due to the effect of interfacial heat transfer rather than the effect of Pr as described in Sect. 6.2. Furthermore, the traveling direction of roll structures also changes depending on the supercritical-

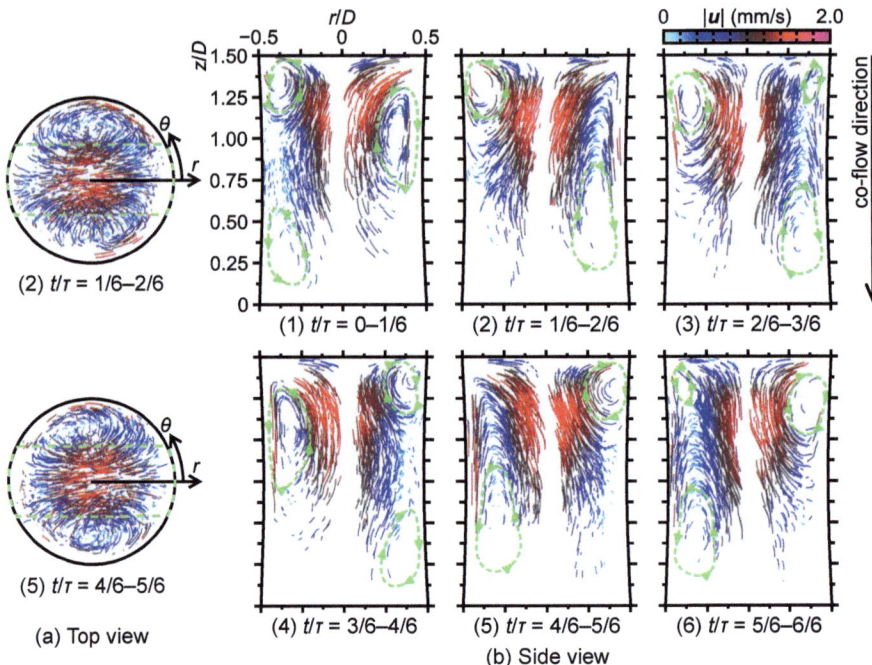

Fig. 37 Time series of particle trajectories of oscillatory thermocapillary convection at the state far beyond the instability threshold for Pr = 67, $A_r = 1.50$, $V_r = 0.95$, and Ma = 3.12×10^4 ($\approx 4.06 \text{Ma}_c$) measured in MEIS-2. Each time duration for superposition is 1/6 oscillation period. (Reprinted from Yano et al. [86], with the permission of AIP Publishing. Modifications have been made in this figure with the permission from the the publisher. All rights reserved)

ity of the temperature condition. Figure 37 shows a time series of flow fields with Ma far beyond the critical condition (i.e., Ma = 3.12×10^4 and Ma/Ma$_c \approx 4.06$) for (Pr, A_r) = (67, 1.50). The corresponding dimensionless oscillation frequency is $F = 0.58$, which lies on branch A. The result indicates the traveling of roll structures in the co-flow direction; therefore, the reversal of the traveling direction of roll structures occurred between Ma/Ma$_c$ = 1.03–4.06.

Not shown here, as far as confirmed by 3-D PTV, every roll structure in oscillatory thermocapillary convection on branch B travels in the counter-flow direction, while that on branch A travels in the co-flow direction. It is, therefore, evident that the change of critical branch characterized by the *frequency skip* shown in Fig. 18 is accompanied by the change of the traveling direction of roll structures. The results of flow measurement for large-A_r conditions suggest that the oscillatory thermocapillary convections on branches A and B possess different types (or modes) of hydrothermal-wave instability because the traveling direction of hydrothermal wave and that of roll structure are closely related to each other. More details on the relation between the hydrothermal wave and the roll structure will be presented in Sect. 6.2.

5.2 Surface Flow Measurement with Photochromic Dye Activation (PDA) Technique

Since the thermocapillary convection originates from the flow on the liquid-bridge free surface, measuring its velocity is important to understand this phenomenon. Although 3-D PTV can measure the velocity in the vicinity of the free surface (e.g., see Fig. 30), the photochromic dye activation (PDA) technique, which is a kind of molecular tagging velocimetry, is more effective for the measurement of surface velocity in the strict sense. In this technique, the excitation light, such as a pulse laser of ultraviolet light, activates the photochromic dye material dissolved in a test liquid, and the flow velocity is measured by the subsequent behavior of the discolored region. The PDA technique can measure the velocity without disturbing the flow (or with no slip between dye and fluid) and is particularly effective in measuring the flow close to a wall or a liquid-gas interface. This technique was originally developed by Popovich and Hummel [87] and has been used in the previous thermocapillary-convection experiments. For example, Simic-Stefani et at. [64] measured surface velocity of steady and oscillatory thermocapillary convections in an acetone liquid bridge in their terrestrial experiment by exciting the photochromic dye with the pulsed GN_2 laser. The surface velocity measurement by using the PDA technique has also been performed in the µg experiment in the TR-IA-6 sounding rocket by Nishino et al. [83], who reported surface velocities of 10.0 mm/s and 0.4 mm/s at 3.5 mm and 6.7 mm below the hot disk, respectively, in a 2-cSt silicone-oil liquid bridge with $D = 28$ mm, $H = 20$ mm ($A_r \approx 0.71$), and $\Delta T = 49$ K, although their liquid bridge was considerably concaved in shape due to the unexpected liquid leakage [88]. They excited photochromic dye with a GN_2 pulse laser having a cross-shaped beam.

As described in Sect. 1.2, the photochromic dye called TNSB [8, 9]—1,3,3-trimethyl-6'-nitrospiro[indoline-2,2'-chromene]—was blended into 5- and 20-cSt silicone oils used in MEIS-2 and 3 at 0.05 % by weight. This photochromic dye has several synonyms, such as 1',3',3'-trimethylindoline-6-nitrobenzospiropyran, and so on (see CAS No. 1498-88-0). Figure 38 shows the coloring of the photochromic dye observed in MEIS-2. A GN_2 pulse laser simultaneously irradiates two spots on the circumferentially shifted liquid-bridge free surface 6 mm below the hot-disk edge at $f_{GN_2} = 0.33$ Hz, where f_{GN_2} is the irradiation frequency of the GN_2 pulse laser. The laser beams enter the liquid bridge obliquely via a prism as shown in Fig. 38c. The distance between irradiation points is designed to be 5.2 mm, resulting in the approximately 20° shift in the circumferential direction. As shown in Figs. 38a and b, the excited liquid changes its color from light pink to purple (or dark pink), in fact these colors are due to illumination and unexcited silicone oils are colorless. The discolored region moves downstream as the color returns, and one can measure both the magnitude and direction of the surface velocity from dye traces. As shown in the instantaneous image in Fig. 38a, there exist tracer particles with colors similar to the excited liquid, which interfere with detecting the activated dye. However, it can be removed by postprocessing as shown in Fig. 38b, in which phase

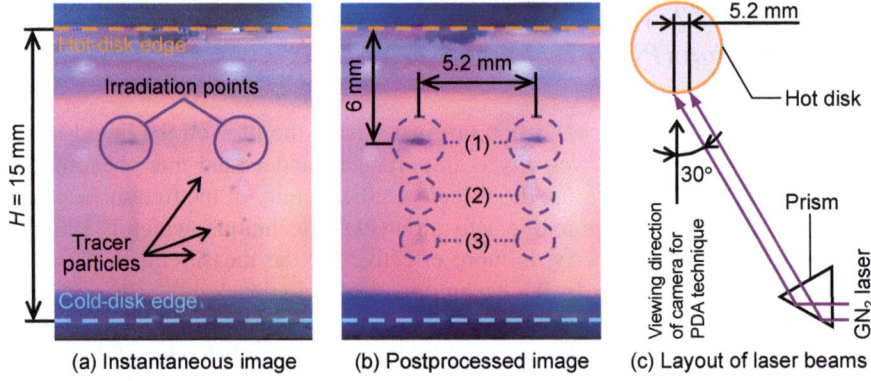

Fig. 38 Photochromic dye trace images obtained in MEIS-2 for $A_r = 0.5$: **a** instantaneous image, **b** postprocessed image, and **c** layout of laser beams. Measurement positions are 6 mm below the hot disk, and a pulsed GN$_2$ laser is irradiated at 0.33 Hz. (Photo in (**a**) courtesy of JAXA. All rights reserved)

averaging and brightness and contrast adjustments are applied to the time series of photochromic dye trace images. In this postprocessed image, *dashed circles* (1)–(3) depict dye traces activated at $t = t_0$, $t = t_0 - \tau_{GN_2}$, and $t = t_0 - 2\tau_{GN_2}$, respectively, where τ_{GN_2} is the time interval between laser irradiations ($= 1/f_{GN_2}$). The laser irradiation interval is $\tau_{GN_2} \approx 3.3$ s, so that the photochromic dye trace vanishes in approximately 10 s in this case. The duration until the color vanishes depends on the experimental conditions. The GN$_2$ pulse laser used in the present µg experiments can switch the irradiation frequency between $f_{GN_2} = 4.57 \times 10^{-4}$–10 Hz ($\pm 1\,\%$) and have sufficient continuous irradiation capability, and experimenters adjusted irradiation condition according to the situation. Hereinafter, the results of surface velocity measurements by the PDA technique in MEIS-2 are shown; therefore, the Prandtl number and volume ratio are $Pr = 67$ and $V_r = 0.95$, respectively, unless otherwise stated.

Figure 39 shows photochromic dye traces in (a) steady and (b) oscillatory thermocapillary convections for $A_r = 2.00$, wherein the values of $(\Delta T, \Delta T / \Delta T_c)$ are (a) (2.9 K, 0.83) and (b) (5.4 K, 1.54), respectively. The measurement position is 12 mm below the hot disk, and each figures shows streaks of photochromic dye for a duration of 10 s obtained by superposing the photochromic dye images. The tracer particles are removed by the image processing. The streaks move straight from the warmer side toward the cooler side and is time independent when the flow is in a steady state (Fig. 39a). On the other hand, the streaks change its moving direction periodically in an oscillatory flow state (Fig. 39b). The photochromic dye trace clearly reflects the difference between the steadiness and unsteadiness (or the symmetry and asymmetry) of thermocapillary convection; therefore, the PDA technique worked effectively not only for measuring the surface flow velocity but also for detecting the flow transition. Under conditions in which the number of tracer particles are few or the oscillation period exceeds a hundred seconds (e.g., high-A_r condition on branch

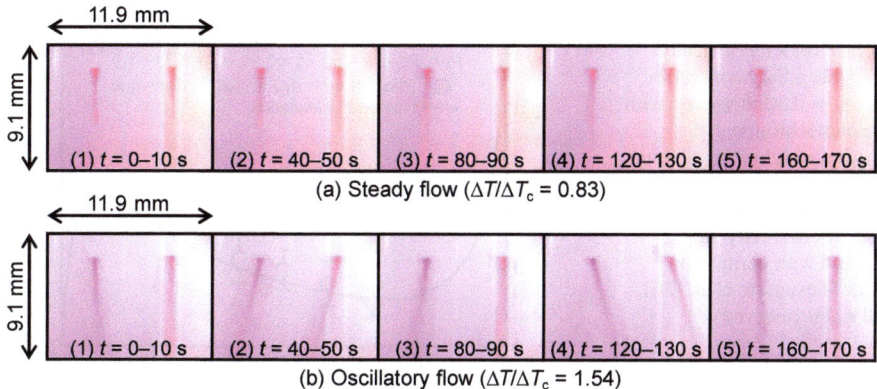

Fig. 39 Visualization of surface velocity by means of photochromic dye activation technique for Pr = 67 and $A_r = 2.00$: **a** steady flow with $\Delta T = 2.9$ K ($\Delta T/\Delta T_c = 0.83$) and **b** oscillatory flow with $\Delta T = 5.4$ K ($\Delta T/\Delta T_c = 1.54$). (©2014 The Jpn. Soc. Microgravity Appl., adapted with permission from Matsumoto et al. [30]. All rights reserved)

B in Figs. 17 and 18), it becomes difficult or takes a long time to determine whether the flow is steady or oscillatory from the motion of tracer particles. However, the PDA technique can detect the onset of instability in a shorter time under such conditions, and hence this technique was appreciated in MEIS-2 and 3.

The axial velocity component of surface flow u_z measured with the PDA technique for $A_r = 0.50$ and $\Delta T = 5.7$ K ($\Delta T/\Delta T_c = 0.76$) is compared with that obtained by the numerical simulation in Fig. 40 [30, 89]. The right vertical axis indicates the dimensionless axial velocity scaled with the (characteristic) thermocapillary velocity U^* (= 74.5 mm/s, see Sect. 4.4 for the definition). Although the numerical simulation was performed for $V_r = 1$ unlike the experiment in which a slightly slender liquid bridge with $V_r = 0.95$ was considered, the effect of this difference is considered to be trivial. Unfortunately, for safety reasons and due to the extremely thin velocity boundary layers near the disks, the surface velocity close to the cold and hot disks could not be measured by the PDA technique in the present experimental setup. However, the surface velocity around the middle height of the liquid bridge was measured well by the PDA technique, and the obtained u_z shows reasonable agreement with the result of numerical simulation. This agreement ensures the validity of the surface velocity measurement by the present PDA technique.

The PDA technique was also applied to measure the surface velocity of oscillatory thermocapillary convection in MEIS-2. Figure 41 shows the time series of the axial component u_z (top), tangential component u_θ (middle), and magnitude U_{surf} (= $(u_z^2 + u_\theta^2)^{0.5}$) (bottom) of the surface velocity of oscillatory thermocapillary convection for $A_r = 2.00$ and $\Delta T = 11.5$ K ($\Delta T/\Delta T_c = 3.29$). The measurement position is 12 mm below the hot disk and this temperature condition is far beyond the instability threshold (i.e., $\Delta T_c = 3.5$ K). The oscillation period is $\tau = 28.3$ s,

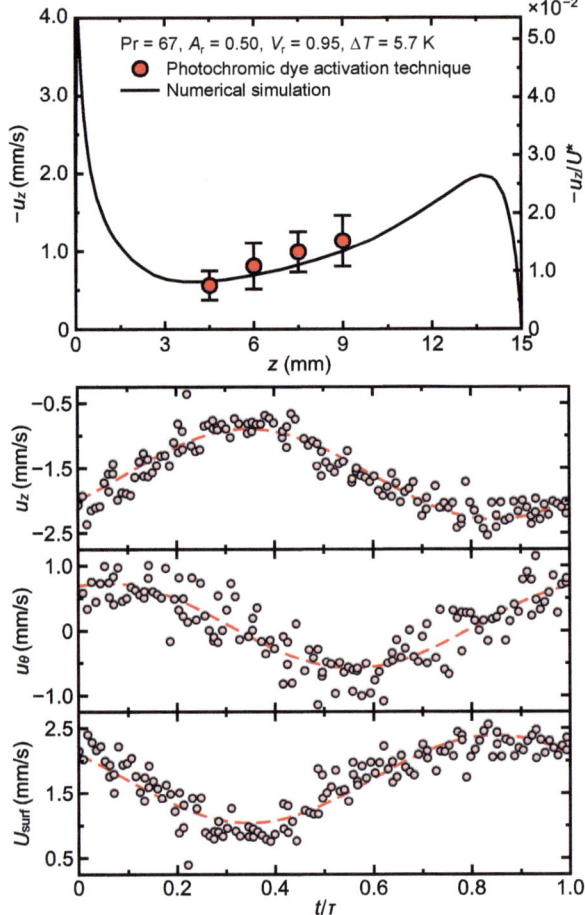

Fig. 40 Comparison of surface velocity of steady thermocapillary convection for $A_r = 0.50$ measured with the photochromic dye activation technique and that obtained from the numerical simulation. (©2014 The Jpn. Soc. Microgravity Appl., adapted with permission from Matsumoto et al. [30]. All rights reserved)

Fig. 41 Time series of axial velocity component u_z (top), tangential velocity component u_θ (middle), and magnitude of surface velocity U_{surf} (bottom) of surface flow of oscillatory thermocapillary convection for $Pr = 67$, $A_r = 2.00$, $V_r = 0.95$, and $\Delta T = 11.5$ K. The measurement position is 12 mm below the hot disk. *Red dashed lines* are sinusoidal fitting curves. Original data were reported by Suzuki [89]

and both velocity components and surface flow speed fluctuate sinusoidally with this period.

The surface velocity measurements of oscillatory thermocapillary convection were performed with a wide range of A_r in MEIS-2, and the average speeds of surface flow $\overline{U}_{surf}$ for various A_r are plotted as a function of the axial position z in Fig. 42 [89]. Each value of $\overline{U}_{surf}$ are evaluated as the mean of maximum and minimum U_{surf}. The given temperature differences (supercriticality) are $\Delta T = 7.7, 7.9, 5.6, 11.2,$ and 11.5 K ($\Delta T/\Delta T_c = 1.03, 1.08, 1.10, 3.73,$ and 3.29) for $A_r = 0.50, 0.75, 1.00, 1.50,$ and 2.00, respectively. The temperature condition is slightly supercritical for $A_r = 0.50$–1.00 while that is far beyond the instability threshold for $A_r = 1.50$ and 2.00. It follows from Fig. 42 that the average speed of surface flow generally decreases with increasing A_r in the case immediately after the transition to the oscillatory state. This is because of the decreases in the axial temper-

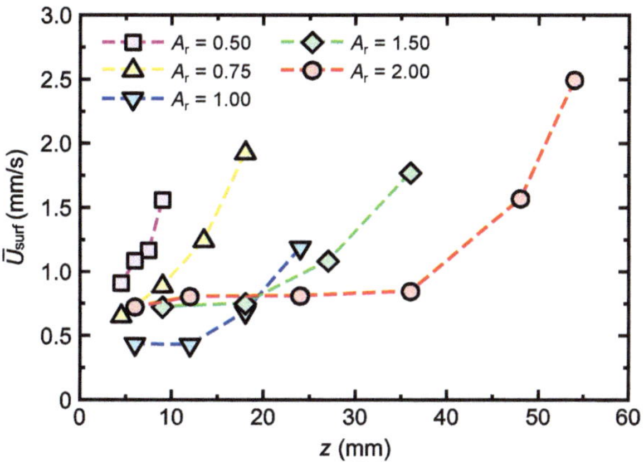

Fig. 42 Axial profiles of average speed of surface flow $\overline{U}_{surf}$ measured with the photochromic dye activation technique for various A_r. The temperature conditions are slightly beyond the instability threshold for $A_r = 0.50$–1.00 while they are far beyond it for $A_r = 1.50$ and 2.00. Original data were reported by Suzuki [89]

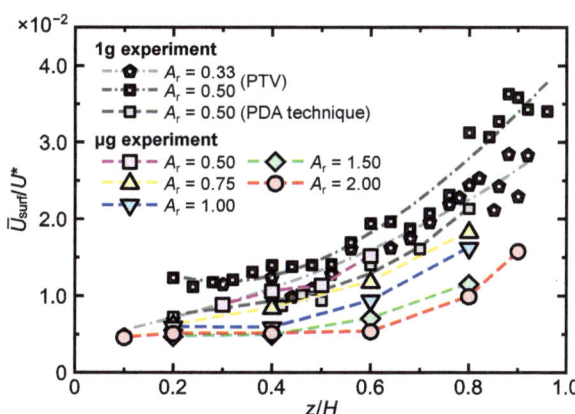

Fig. 43 Plots of dimensionless average speed of surface flow $\overline{U}_{surf}/U^*$ as a function of relative axial position to the liquid-bridge height z/H. Experimental results obtained in the previous terrestrial studies for small-scale liquid bridges are also included. Original data were reported by Suzuki [89]

ature gradient $\Delta T/H$ and the resultant driving force of thermocapillary convection due to the increase in the liquid-bridge height. In the case where ΔT is much larger than ΔT_c, the thermocapillarity is also larger, and therefore, the values of $\overline{U}_{surf}$ for $A_r = 1.50$ and 2.00 tends to be larger. To omit the effects of supercriticality and to compare the results more fairly, $\overline{U}_{surf}$ is dimensionalized with U^*, and the resultant $\overline{U}_{surf}/U^*$ are plotted as a function of the relative axial position to the liquid-bridge height (i.e., z/H) in Fig. 43 [89]. The values of U^* that are used for scaling are $U^* = 102.5, 105.3, 73.1, 153.7,$ and 158.2 mm/s for $A_r = 0.50$–2.00, respectively. This figure includes $\overline{U}_{surf}/U^*$ obtained in terrestrial experiments for liquid bridges with $D = 5$ mm and $A_r = 0.33$ and 0.50, which were measured with the PDA technique or by tracking the motion of tracer particles in the vicinity of the liquid-bridge free surface (i.e., PTV). All the data included in Fig. 43 show a similar curve in which $\overline{U}_{surf}/U^*$ gradually increases with approaching the hot disk. Additionally, $\overline{U}_{surf}/U^*$ tends to decrease with A_r, regardless of the supercriticality of the convection. In the present μg experiments, the characteristics of surface velocity in an oscillatory thermocapillary convection (e.g., the relationship with the temperature field on the free surface) are also investigated, but the details are not covered here.

6 Heat Transfer Through the Liquid-Gas Interface

Taishi Yano, Koichi Nishino, and Yasuhiro Kamotani

Since the thermocapillary convection is originated from the temperature gradient along the liquid-bridge free surface, the instability as well as the flow and temperature patterns of thermocapillary convection are naturally affected by heat transfer through the liquid-gas interface (referred to as interfacial heat transfer in this section). This effect has attracted much interest from researchers working on this phenomenon, and many studies focusing on the interfacial heat transfer have been conducted through terrestrial experiments [90, 91], theoretical analyses [92, 93], and numerical simulations [94, 95]. However, the experimental study under reduced-gravity condition were limited before the present μg experiments, especially before DS project. This section provides the results on the effect of interfacial heat transfer obtained in the DS project. In this section, it is defined as the heat-loss or heat-gain condition when the net heat transfer rate on the liquid-bridge free surface is positive or negative (in other words, the liquid bridge loses or receives heat as a whole), respectively.

6.1 Effects of Cold-Disk Temperature and Ambient Gas Properties

There are several methods to actively vary the heat-transfer condition on the liquid-bridge free surface: (1) change the relative temperature of the liquid bridge to the ambient gas, (2) change the physical properties—particularly, the thermal conductivity—of the ambient gas, (3) impose a forced gas flow around the liquid bridge, (4) change the temperature difference between the liquid bridge and the chamber wall, and so on. In general, heat is transferred via several ways, and the former two are mainly concerned with conduction, the third with convection, and the last with radiation. An additional type of heat transfer related to the thermocapillary convection is evaporation. Simic-Stefani et al. [64], who carried out experiments with acetone liquid bridges, demonstrated a significant effect of evaporative cooling on the instability of thermocapillary convection. However, amounts of evaporated liquid and the resultant latent heat were evaluated to be so trivial in the present μg experiments that the effect of evaporative heat transfer is not considered here. In DS project, the effects of interfacial heat transfer were investigated by changing the cold-disk temperature T_C and by changing the ambient gas properties. These changes mainly lead to the change in conductive heat transfer, but they also lead incidentally to the changes in convective and radiative ones.

Fig. 44 Plots of **a** critical Marangoni number Ma_c and **b** critical dimensionless oscillation frequency F_c as a function of cold-disk temperature T_C for $A_r = 0.50$, wherein the volume ratio is $V_r = 0.9$–1.0

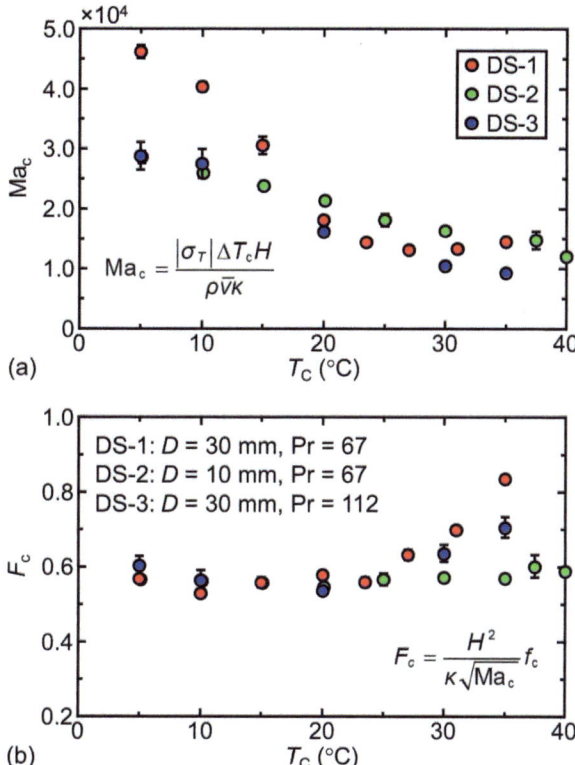

Figure 44 shows plots of (a) Ma_c and (b) F_c as a function of T_C for $A_r = 0.50$ and $V_r = 0.9$–1.0. The ambient gas is argon, and the average Prandtl number ($= \bar{\nu}/\kappa$) varies in the range $54 < \text{Pr} < 85$ in DS-1, $47 < \text{Pr} < 77$ in DS-2, and $87 < \text{Pr} < 131$ in DS-3. The plots of Ma_c show similar trend with the previous terrestrial experiments (e.g., Kamotani et al. [90]) that Ma_c decreases with increasing T_C up to a certain value (i.e., approximately 30 °C in this case). It was confirmed by numerical simulations that all cases in Fig. 44 are in heat-loss condition, even for $T_C = 5$ °C, and the amount of heat lost from the liquid bridge through the free surface increases with increasing T_C and vice versa. It follows from this result that the increase in the interfacial cooling facilitates the onset of oscillatory thermocapillary convection. For $T_C = 5$–20 °C, the slope of Ma_c-T_C curve depends on Pr and D, and larger Pr or smaller D makes the slope more gradual. As shown in Fig. 44b, the oscillation frequency shows different trend. The value of F_c is less sensitive to the interfacial heat transfer for $T_C = 5$–20 °C, while it increases with T_C for $T_C > 20$ °C. The slope of F_c-T_C curve for $T_C > 20$ °C also depends on Pr and D, and larger Pr or smaller D makes the slope more gradual.

The effects of T_C on Ma_c and F_c for longer liquid bridges, i.e., $A_r = 1.00$–2.00 and $V_r = 0.9$–1.0, are shown in Figs. 45a and b, respectively. The trends of Ma_c-T_C

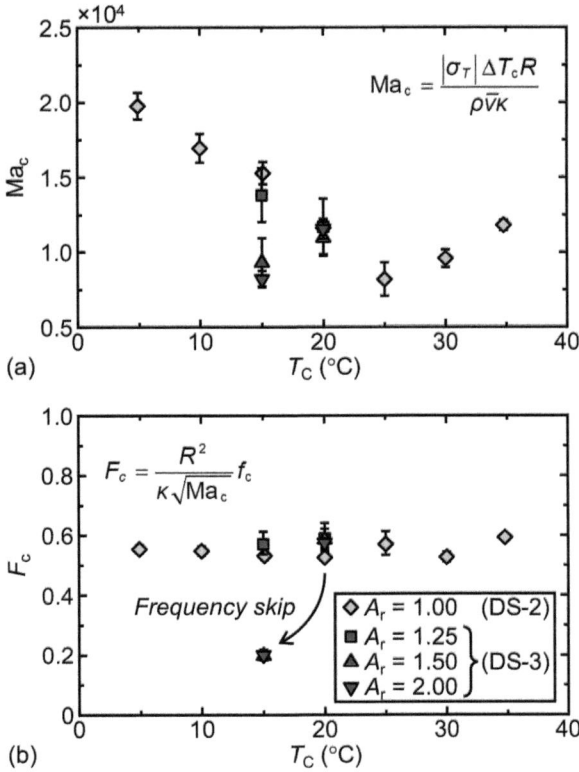

Fig. 45 Plots of **a** critical Marangoni number Ma_c and **b** critical dimensionless oscillation frequency F_c as a function of cold-disk temperature T_C for $A_r \geq 1.00$, wherein the volume ratio is $V_r = 0.9$–1.0

and F_c-T_C curves are similar to those in Fig. 44 for many cases; however, interesting feature is observed at $T_C = 15$–$20\,°C$ for $A_r \geq 1.50$. As T_C decreases from $20\,°C$ to $15\,°C$, both Ma_c and F_c decrease suddenly, and the neutral stability curve splits into two branches. This result is reminiscent of bifurcations caused by changing A_r and V_r (Figs. 18 and 23). It is found that two branches with higher or lower Ma_c and F_c are also related to the change in the characteristics of roll structures as discussed in Sect. 5.1 [96]. More details of the effect of interfacial heat transfer on the hydrothermal-wave instability will be given in Sect. 6.2.

Based on the experimental data for $D = 3$ mm, $Pr = 26$–52, and $A_r = 0.65$–0.7 obtained under normal-gravity condition, Kamotani et al. [97] demonstrated that the onset of oscillatory flow takes place at a nearly constant temperature difference between the hot disk and the ambient gas (i.e., $T_H - T_{amb}$, where T_{amb} is the temperature of the ambient gas) regardless of T_C (or $T_C - T_{amb}$) when the amount of heat lost from the liquid bridge is not so large and D_r is close to one. This fact implies the importance of heat transfer in the vicinity of the hot disk—often called the hot corner—to the instability of thermocapillary convection under weak heat-loss condition. A similar trend with Kamotani et al. [97] was recognized in the present μg experiments for $A_r = 0.50$–1.00 and $T_C \leq 20\,°C$ as shown in Fig. 46, in

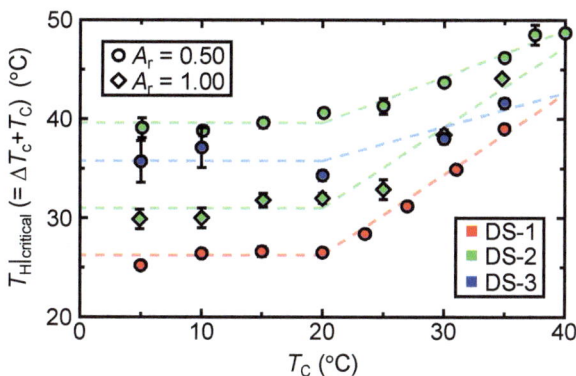

Fig. 46 Plot of hot-disk temperature T_H at critical condition (i.e., $T_H|_{\text{critical}} = \Delta T_c + T_C$) as a function of cold-disk temperature T_C for $A_r = 0.50$ (*circles*) and 1.00 (*diamonds*). *Dashed lines* are the 0th- and 1st-order fitting lines for $T_C \leq 20\,°C$ and $T_C \geq 20\,°C$, respectively, for eye guiding

which T_H at critical condition (i.e., $T_H|_{\text{critical}} = \Delta T_c + T_C$) is plotted as a function of T_C. We note that it was difficult to specify the strict value of T_{amb} in the present experimental setup. However, since the average temperature of the ambient gas was insensitive to T_C (because the spatial occupancy of the cold disk to the volume inside a chamber is small), the trend of T_H can be regarded as that of $T_H - T_{\text{amb}}$. The values of $T_H|_{\text{critical}}$ for $T_C \leq 20\,°C$ converge to each certain constant, while those for $T_C \geq 20\,°C$ increase linearly with T_C. It follows from this result that the heat transfer near the hot corner is particularly important under weak heat-loss condition, even in μg.

The effects of T_C on the instability of thermocapillary convection was also investigated for slender liquid bridges with $V_r = 0.60$, and measured Ma_c and F_c are plotted as a function of T_C in Fig. 47. In the results obtained in DS-1, Ma_c-T_C and F_c-T_C curves for $T_C \geq 20\,°C$ shows similar trend with those in Fig. 44 that Ma_c increases and F_c decreases with decreasing T_C (or decreasing in heat loss). For $T_C \leq 20\,°C$, Ma_c decreases with decreasing T_C while F_c hardly changes, and the Ma_c-T_C curve shows local maximum at $T_C = 20\,°C$, which is never seen for straight liquid bridges. The results obtained in DS-3 are similar to some of those in DS-1, but the local maximum of Ma_c is not observed because the experimental condition was limited to $T_C \geq 20\,°C$. On the other hand, both Ma_c and F_c are insensitive to the interfacial heat transfer in DS-2, as in the terrestrial experiments by Kamotani et al. [90, 97] and Wang et al. [98].

As shown in Table 1, the ambient gas was argon or neon in DS-2. Neon gas was used in the first 10 experiments in DS-2, and it was then replaced by argon gas in other experiments [16]. The purpose of this replacement is to change the heat conduction in the ambient gas. Table 7 lists physical properties of argon and neon gases at $25\,°C$ and 94 kPa, wherein the values of ρ, μ, λ, and c_p are derived from the material database of the commercial computational fluid dynamics (CFD) software STAR-CCM+ ver. 12.02 (Siemens Digital Industries Software, Texas, USA) and the others are calculated from them. As shown in this table, neon gas has roughly three-fold larger thermal conductivity compared to the argon gas (i.e., $\lambda_{\text{Ar}} = 0.018\,\text{W}/(\text{m} \cdot \text{K})$ vs. $\lambda_{\text{Ne}} = 0.049\,\text{W}/(\text{m} \cdot \text{K})$, where subscripts Ar and Ne denote argon and neon,

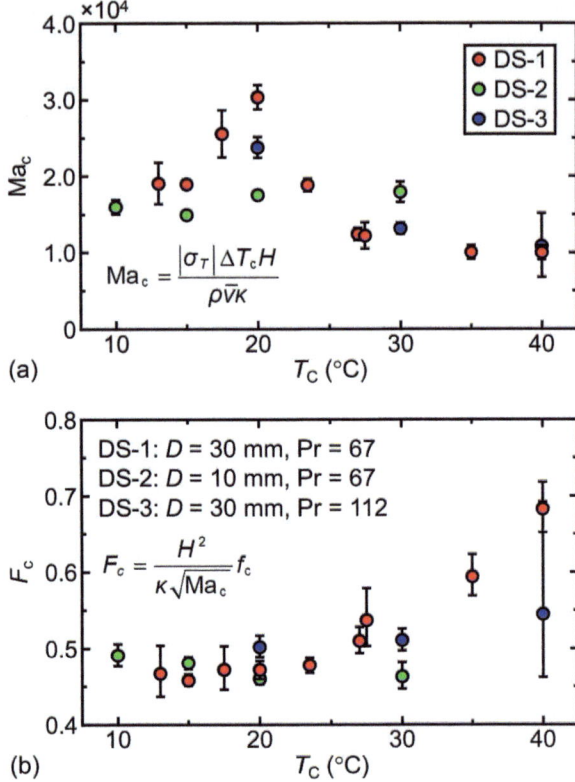

Fig. 47 Plots of **a** critical Marangoni number Ma_c and **b** critical dimensionless oscillation frequency F_c as a function of cold-disk temperature T_C for $A_r = 0.50$ and $V_r = 0.60$

Table 7 Physical properties of argon and neon gases at 25 °C and 94 kPa

Property	Argon	Neon
Prandtl number Pr ($= \nu/\kappa$)	0.7	0.7
Density ρ (kg/m^3)	1.51	0.77
Kinematic viscosity ν (m^2/s)	1.5×10^{-5}	4.1×10^{-5}
Dynamic viscosity μ (Pa · s)	2.28×10^{-5}	3.17×10^{-5}
Thermal conductivity λ (W/(m · K))	0.018	0.049
Specific heat at constant pressure c_p (J/(kg · K))	522	1030
Thermal diffusivity κ (m^2/s)	2.3×10^{-5}	6.2×10^{-5}

respectively); therefore, the interfacial heat transfer due to conduction was expected to be encouraged by changing the ambient gas from argon to neon.

Figure 48 shows plots of (a) ΔT_c and (b) f_c as a function of T_C for (1) $A_r = 0.50$ and (2) $A_r = 1.00$, wherein the *delta* and *nabla* symbols indicate the results measured under argon- and neon-ambient conditions, respectively [16]. We note that all cases

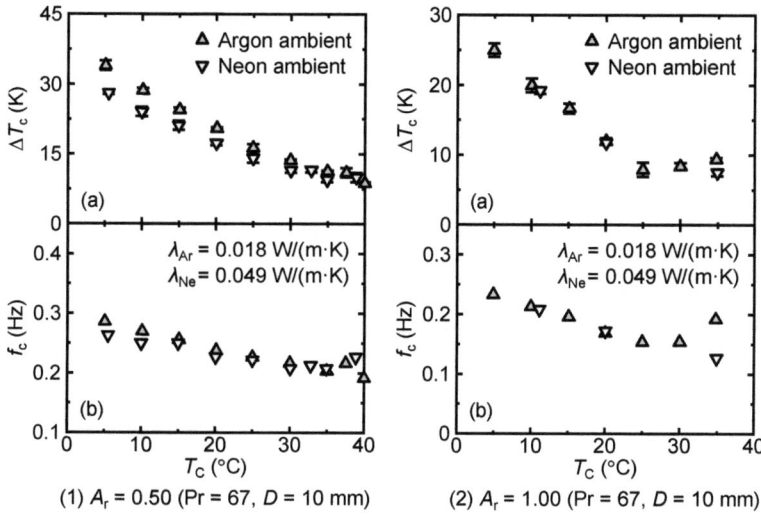

Fig. 48 Plots of **a** critical temperature difference ΔT_c and **b** critical oscillation frequency f_c as a function of cold-disk temperature T_C for (1) $A_r = 0.50$ and (2) $A_r = 1.00$. The ambient gas is argon (*delta*) or neon (*nabla*). (Reprinted by permission from Springer Nature: Springer Nature Yano et al. [16], Copyright Springer Science+Business Media B.V., part of Springer Nature 2018. Modifications have been made in this figure with the permission from the publisher. All rights reserved)

in Fig. 48 are evaluated to be in heat-loss condition, even when the ambient gas is neon. The effect of ambient-gas property is clearly seen in the results for $A_r = 0.50$ that the values of ΔT_c decrease with increasing λ of ambient gas. This feature is consistent with the discussion on Fig. 44 that the increase in interfacial heat loss destabilizes the steady thermocapillary convection. As a result of decrease in ΔT_c, the values of f_c also decrease when the ambient gas is changed from argon to neon. In contrast, such features are not recognized for $A_r = 1.00$, and both ΔT_c and f_c are independent on the kind of ambient gas. In general, the contribution of (surface-to-surface) radiation, which is irrelevant to the ambient-gas property, to the total amount of interfacial heat transfer becomes greater as A_r increases [99, 100]. Therefore, the effect of conductive cooling (and also that of convective one) seems to be weakened, and the change in ambient gas no longer affects the instability of thermocapillary convection for $A_r = 1.00$. Further discussion on the effects of radiative heat transfer will be given in Sect. 6.3.

6.2 Sensitivity of Hydrothermal Waves in Long Liquid Bridges to the Interfacial Heat Transfer

As described in Sect. 4.3, the hydrothermal-wave instability triggers the oscillatory thermocapillary convection in high-Pr liquid bridges. At the oscillatory flow state, the band-shaped region with relatively low temperature—the so-called hydrothermal wave (referred to hereinafter as HTW in this section)—propagates periodically on the liquid-bridge free surface. Figure 49 shows time series of propagating HTWs observed in MEIS-2 for $A_r = 2.00$, $V_r = 0.95$, and Pr = 67, wherein the time t is made dimensionless by the oscillation period τ $(= 1/f)$. In these results, images captured by an IR camera were phase averaged and the contrasts were adjusted. The values of ΔT are 11.4 K and 14.6 K for Figs. 49a and b, respectively. The critical temperature difference for this condition is measured to be $\Delta T_c = 3.5 \pm 0.4$ K; therefore, ΔT are much larger than ΔT_c. In both cases, one can see inclined HTWs traveling in the co-flow direction (i.e., from the warmer side toward the cooler side, see Sect. 5). However, the HTW not always propagates in that direction, and the axial traveling direction is one of the important features of HTW, especially in long liquid bridges. The present μg experiments reveal significant effects of interfacial heat transfer on the characteristics of HTW.

The color contours in Figs. 50a and b show free-surface temperature fluctuations $\tilde{T}_{surf}$ measured in DS-3 for $A_r = 1.50$, and $V_r = 0.95$, wherein the cold-disk temperatures are (a) $T_C = 15\,°C$ and (b) $T_C = 20\,°C$, respectively [96]. The values of ΔT are slightly above the instability threshold (i.e., $\Delta T/\Delta T_c = 1.1$–1.2). In these results, the following postprocessing was applied to images captured by an IR camera: (1) adjusting the time variation of overall image brightness due to the peculiarity of the IR camera; (2) phase averaging over 10 oscillation periods; (3) laterally averaging in the region marked with a *dashed rectangle* in each left figure. Through these operations, the image intensity along the axial direction (i.e., the z-direction in Fig. 50) in a given phase can be obtained with a favorable signal-to-noise ratio. The measurable temperature ranges of present IR camera were set at (a) $21.0\,°C$ to $25.0\,°C$ and (b) $25.7\,°C$ to $28.7\,°C$ for the image intensity of 0 (black) to 255 (white), and the image intensities after above postprocessing were converted to the free-surface temperatures T_{surf}. The time-averaged free-surface temperature $\overline{T}_{surf}$ was subtracted from them (i.e., $\tilde{T}_{surf} = T_{surf} - \overline{T}_{surf}$), and the time series of $\tilde{T}_{surf}$ along the z-direction are displayed horizontally in each right figure in Fig. 50, of which the vertical axis is the axial position normalized by the liquid-bridge height (i.e., z/H) and the horizontal axis is the time normalized by the oscillation period (i.e., t/τ). The present results can be regarded as $\tilde{T}_{surf}$ unfolded circumferentially at some instant because the relationship of $t/\tau = \theta/2\pi$ holds for $m = 1$, where θ is the circumferential position in the frame of reference rotating with the HTW. We note that the results are repeated for two oscillation period (or two laps) in Fig. 50, and $\tilde{T}_{surf}$ for $t/\tau = 0$–1 and that for $t/\tau = 1$–2 are identical.

The distributions of $\tilde{T}_{surf}$ in Fig. 50 show significant effects of the cold-disk temperature on the characteristics of HTW. As shown in Fig. 50a, regions with

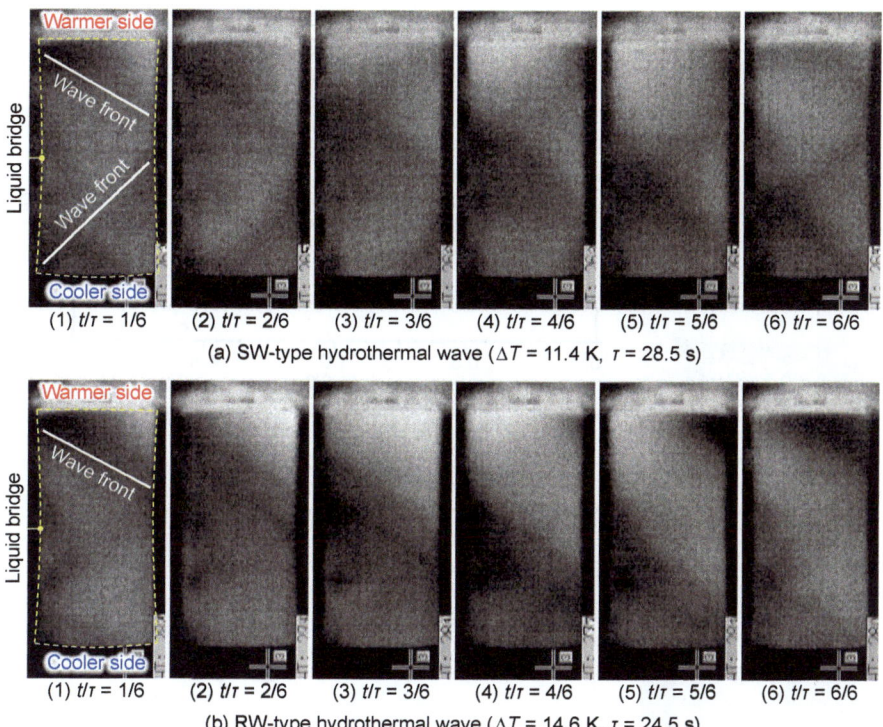

Fig. 49 Time series of free-surface temperature distribution for $A_r = 2.00$, $V_r = 0.95$, and $\Pr = 67$ measured by IR camera in MEIS-2: **a** SW-type oscillatory flow for $\Delta T = 11.4$ K and $\tau = 28.5$ s, and **b** RW-type oscillatory flow for $\Delta T = 16.6$ K and $\tau = 24.5$ s

positive and negative $\tilde{T}_{\text{surf}}$—i.e., hot and cold spots—appear alternately in the vicinity of the hot disk ($z/H > 0.75$), and inclined hot and cold band-shaped regions appear near the cold disk ($z/H < 0.75$). The former hot and cold spots are pinned to the hot disk, while the hot and cold regions on the inclined band shift upward with time. Therefore, the traveling direction of HTW for this condition is from the cooler side toward the warmer side (i.e., the counter-flow direction). The present pattern of $\tilde{T}_{\text{surf}}$ is similar to the results of numerical simulations reported by Ueno et al. [101] for liquid bridges of $\Pr = 28.1$ and $A_r \geq 1.25$ and those of numerical simulation and terrestrial experiment reported by Kawamura et al. [102] for a thin liquid layer of $\Pr = 10.3$. We note that their numerical simulations for a liquid bridge and a liquid layer were conducted with an adiabatic free surface.

On the other hand, as shown in Fig. 50b, the temperature fluctuation for $T_C = 20\,°C$ shows much simpler feature that the inclined hot and cold band-shaped regions occupy the entire free surface. The inclination angle of such band-shaped regions is turned upside down from that in Fig. 50a. The hot and cold regions shift downward with time; therefore, HTWs travel in co-flow direction. In order to evaluate the net

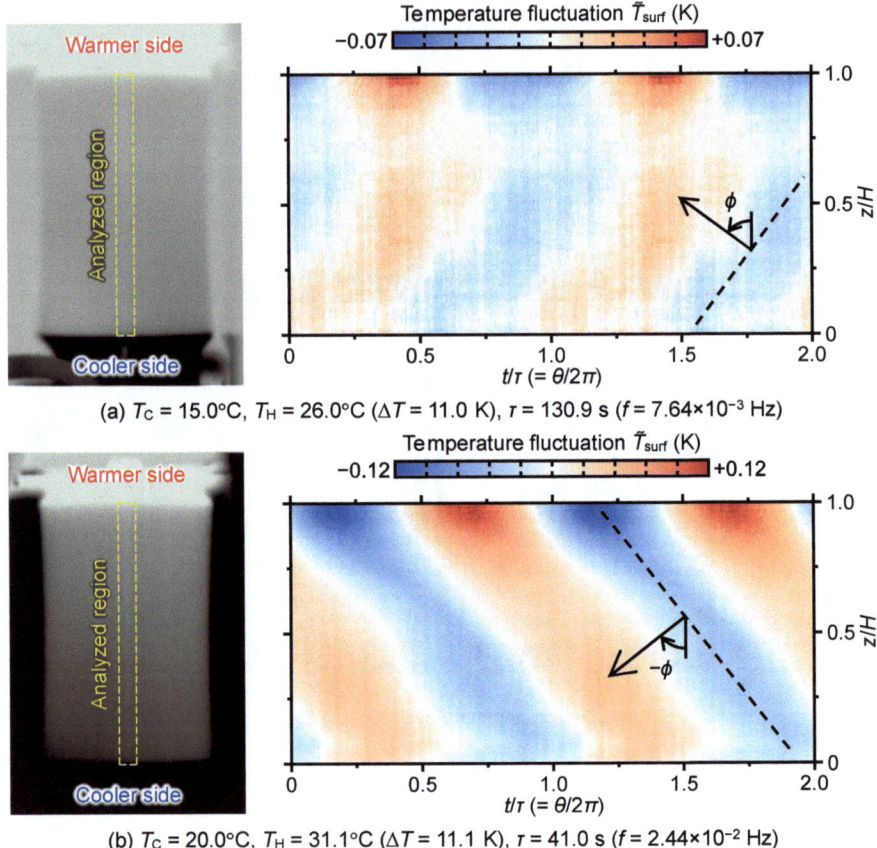

Fig. 50 Surface temperature fluctuations visualized by IR camera for Pr = 112, $A_r = 1.50$, $V_r = 0.95$, and **a** $T_C = 15\,°C$ and **b** $T_C = 20\,°C$. (Reprinted from Yano et al. [96], with the permission of AIP Publishing. Modifications have been made in this figure with the permission from the publisher. All rights reserved)

heat transfer rate on the free surface of the liquid bridge, Yano et al. [96] performed numerical simulations for steady thermocapillary convection that is expected to be seen immediately before the onset of instability. They revealed that the liquid bridges in Figs. 50a and b are in heat-gain and heat-loss conditions, respectively. It follows from these results that the HTW traveling in counter-flow/co-flow direction is preferable in a heat-gain/heat-loss condition in large-A_r liquid bridges, and the bifurcation of neutral stability curve and the *frequency skip* recognized in Fig. 45 may be caused by the change in the traveling direction of HTWs.

The flow fields corresponding to Figs. 50a and b were measured with 3-D PTV and shown in Figs. 51a and b, respectively [96]. In these results, the velocity vectors in a 12.7-mm thick r-z slice were phase superposed over several oscillation periods. Additionally, by considering a mirror symmetry of the flow field, the velocity vectors

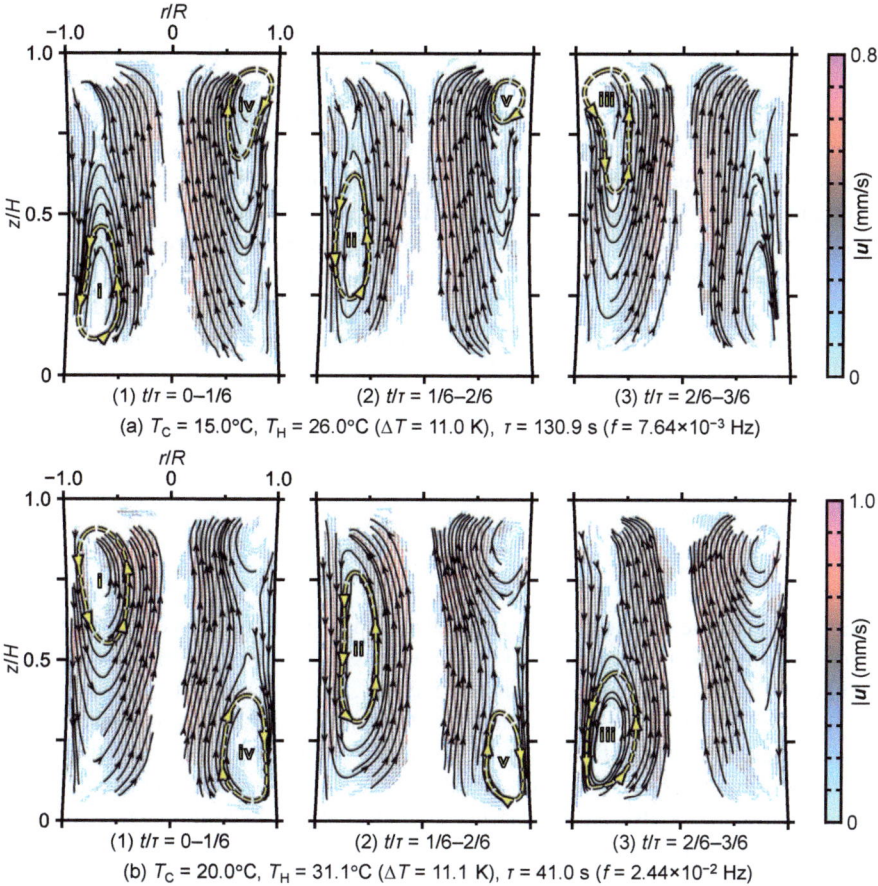

Fig. 51 Time series of velocity fields over half-oscillation periods for $Pr = 112$, $A_r = 1.50$, $V_r = 0.95$, and **a** $T_C = 15\,°C$ and **b** $T_C = 20\,°C$. (Reprinted from Yano et al. [96], with the permission of AIP Publishing. Modifications have been made in this figure with the permission from the publisher. All rights reserved)

shifted by half a period were also superposed after flipping left to right. These velocity data were interpolated into the fixed grid points to obtain velocity fields $\boldsymbol{u}(r, z)$ (*colored arrows*) and streamlines (*black lines with arrowhead*). The loops with *yellow dashed line* are eye guides for the roll structure, and loops i–v display the time evolution of a certain roll structure (by considering a mirror symmetry of the flow field, one can see that the right halves at $t/\tau = 0$–$3/6$ correspond to the left halves at $t/\tau = 3/3$–$6/6$). As shown in Fig. 51, the roll structure moves in a counter-flow/co-flow direction and shrinks with approaching the hot/cold disk when the cold-disk temperature is $T_C = 15/20\,°C$ and the heat-transfer condition is in heat gain/loss. The traveling speed as well as the traveling direction of the roll structure agree well with those of the HTW, and the flow field and the temperature field show strong correlation. This is very natural because the temperature field in high-Pr fluids is mainly controlled by convection rather than conduction. From these facts, the traveling direction of HTWs for Pr = 67, $T_C = 20\,°C$, $A_r \geq 1.50$, and $V_r = 0.95$ in Fig. 35, which could not be directory observed by an IR camera due to the insufficient temperature resolution, is expected be the counter-flow direction because roll structures travel in that direction. The bifurcation of neutral stability curve and the associated *frequency skip* at $A_r = 1.25$–1.50 shown in Figs. 17 and 18 are also expected to be related to the change in the axial traveling direction of HTWs.

Typical characteristics of HTWs for $A_r = 1.50$ and 2.00 observed in the present μg experiments are summarized in Table 8. In this table, the following Marangoni number Ma*, which is defined differently from Eq. 6, and dimensionless angular oscillation frequency ω^* are included:

$$\mathrm{Ma}^* = \frac{|\sigma_T|\partial_z T_{\mathrm{surf}} R^2}{\rho \bar{\nu} \kappa}, \tag{20}$$

$$\omega^* = \frac{2\pi \rho \bar{\nu}}{|\sigma_T|\partial_z T_{\mathrm{surf}}} f, \tag{21}$$

where $\partial_z T_{\mathrm{surf}}$ is the axial temperature gradient along the free surface. These dimensionless quantities are introduced to compare the characteristics of HTW with those obtained by LSA for an infinite liquid bridge (i.e., $A_r = \infty$) by Xu and Davis [103] and Ryzhkov [104]. In the LSA for an infinite liquid bridge, $\partial_z T_{\mathrm{surf}}$ is constant. In a high-Pr liquid bridge with finite A_r, it is, however, hardly constant because there exist thin thermal boundary layers near solid disks. According to the experimental facts, the value of $\partial_z T_{\mathrm{surf}}$ outside these thermal boundary layers becomes nearly constant, and the temperature change in this region accounts for roughly 10 % of the total temperature change. Therefore, the empirical relationship of $\partial_z T_{\mathrm{surf}} \approx 0.1 \Delta T/H$ is used here. Also included in Table 8, except for the results obtained in MEIS-2, is the traveling direction of HTW with respect to the positive z-axis (i.e., ϕ, see Fig. 50).

As shown in Table 8, the values of Ma* under the heat-loss condition tend to be somewhat larger than those under the heat-gain condition; however, no significant difference is observed. On the other hand, ω^* and ϕ clearly depend on the interfacial heat transfer. In spite of the comparable Ma*, the values of ω^* for the heat-loss condition is two- to four-fold larger than those for the heat-gain condition. The magnitude of ϕ is insensitive to the interfacial heat transfer, but its sign is not. Under the heat-gain/heat-loss condition, HTWs travel in the counter-flow/co-flow direction, and the sign of ϕ becomes positive/negative.

This feature has been well predicted by the LSA for an infinite liquid bridge with the heat exchange by Kaoru Fujimura of Sect. 4 in Chapter "Thermocapillary Convection in an Infinite Liquid Layer and in an Infinite Liquid Column". While the LSAs hitherto of this respect were on an adiabatic free surface as for the mean temperature. He examined the effects of the net heat gain/loss on the stability of the HTW in an infinitely long liquid bridge. His result is given in Table 2 of Sect. 4 of

Table 8 Characteristics of hydrothermal wave for large-A_r liquid bridges

Series	MEIS-2		DS-3				MEIS-3	
Test liquid	5-cSt silicone oil		10-cSt silicone oil				20-cSt silicone oil	
$Pr (= \bar{\nu}/\kappa)$	72	71	112	112	123	126	190	185
A_r	1.50	2.00	1.50	2.00	1.50	2.00	1.50	2.00
$T_C\ °C$	20.0		15.0		20.0		20.0	
$T_H\ °C$	23.1	23.9	26.0	24.0	31.1	31.1	41.8	45.4
$\Delta T/\Delta T_c$	1.02	1.08	1.16	1.19	1.10	1.05	1.08	1.08
Heat transfer condition[*1]	Heat gain		Heat gain		Heat loss		Heat loss	
Ma*	261	245	365	247	407	306	379	339
ω^*	0.454	0.455	0.331	0.464	0.947	1.270	1.038	1.368
$\phi\ (°)$	NA	NA	32	30	−34	−31	−27	−25
Axial traveling direction	Counter-flow[*2]		Counter-flow		Co-flow		Co-flow	

[*1]Evaluated by numerical simulations [96], [*2]Predected from flow field

Chapter "Thermocapillary Convection in an Infinite Liquid Layer and in an Infinite Liquid Column", where P-I and P-II give the results for an infinite liquid layer and infinite cylinder, respectively. The dimensionless temperature difference $\delta\Theta < 0$ corresponds to the heat gain and $\delta\Theta > 0$ to the heat loss. The direction of the HTW is given in the bottom column, which indicates that under the heat-gain ($\delta\Theta < 0$) and heat-loss ($\delta\Theta > 0$) conditions, the HTW travels in counter- and co-flow directions, respectively. This result predicts qualitatively well the flow directions obtained in the present μg experiment mentioned above. Note here that the critical Marangoni number obtained by LSA is generally much smaller for an infinite cylinder than for a finite one. As discussed above in relation to the Ma*, this difference can be attributed to the non-uniform surface temperature gradient in the finite bridge, in which the temperature gradient is large at both end regions and is much smaller in the central one, and also to the constraining effect of both end plates.

The characteristics of HTW observed in the present μg experiments are compared with those in an infinite liquid bridge with adiabatic free surface reported by Xu and Davis [103] and Ryzhkov [104] in Fig. 52. In this figure, (a) Ma*, (b) $|\omega^*|$, and (c) ϕ are plotted as a function of Pr, wherein their original data are displayed as *small squares*, with linear interpolation between them. The LSA distinguishes positive and negative ω^* but experiments cannot; therefore, the absolute value of ω^* is shown in Fig. 52b. It is important to note that the most critical azimuthal mode number in the LSA by Xu and Davis [103] is $m = 0$ for $Pr \geq 62.2$ but the neutral stability curves for $m = 1$ are shown here because the oscillatory flow with $m = 1$ was always preferable for $A_r \geq 1.50$ in the present μg experiments. Ryzhkov [104] found two different instability mode of $m = 1$ and distinguished them. Respecting his definition, results of Ryzhkov [104] correspond to $m = 1$(a) and those of Xu and Davis [103] correspond to $m = 1$(b).

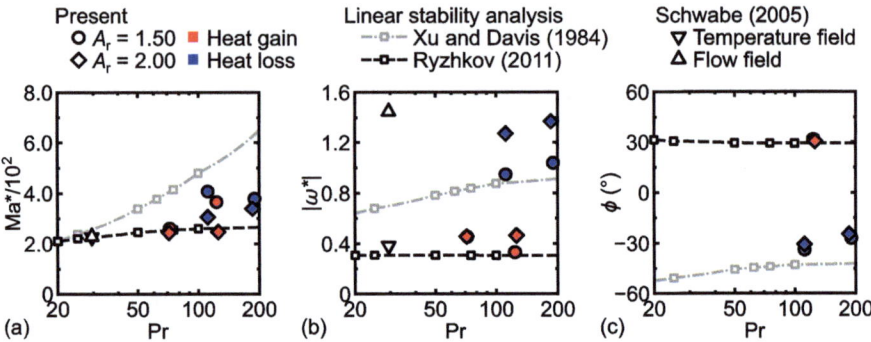

Fig. 52 Plots of **a** Ma*, **b** $|\omega^*|$, and **c** ϕ as a function of Pr for $A_r = 1.50$ (*circle*) and $A_r = 2.00$ (*diamond*). Symbols filled with *blue* or *red* are in heat-loss or heat-gain conditions, respectively. Also included are results of linear stability analysis for an infinite liquid bridge by Xu and Davis [103] and Ryzhkov [104], and results of μg experiment for $A_r = 2.5$ by Schwabe [39, 40]

As shown in Fig. 52a, Ma* for $m = 1(b)$ is always greater than that for $m = 1(a)$; therefore, the results of Xu and Davis [103] represent the characteristics of HTW beyond the instability threshold. This, however, would not be a serious problem for $|\omega^*|$ and ϕ for the following reasons. By substituting Eq. 20 to Eq. 21, the dimensionless angular oscillation frequency can be rewritten as $\omega^* = 2\pi R^2 f/(\kappa \text{Ma}^*)$. When the same instability mode is maintained, the oscillation frequency would increase linearly with ΔT [16, 39], hence one can expect that f takes a form close to the linear function of Ma*. Therefore, ω^* may be weakly dependent (or independent) on how far Ma* is from the critical value. Based on the empirical evidence, the magnitude of ϕ is also weakly dependent on Ma*. For example, the traveling direction of HTW for Figs. 49a and b are evaluated to be $\phi = -28°$ and $-32°$, respectively, which are close to ϕ near the instability threshold.

As shown in Fig. 52, the present results obtained under the heat-gain condition show excellent agreement with the theoretical results of Ryzhkov [104]. The results obtained under the heat-loss condition also agree well with the theoretical results of Xu and Davis [103] except for Ma*. It follows from these agreements that the thermocapillary convection in a long liquid bridge tends to prefer the HTW with instability modes of $m = 1(a)$ and $1(b)$ under the heat-gain and heat-loss conditions, respectively.

In Fig. 52, the experimental result for $A_r = 2.5$ and Pr = 28–29 measured by Schwabe [39, 40] in his μg experiment are included. He simultaneously observed axially traveling HTWs and roll structures (i.e., drifting Bénard-Marangoni cells in his original paper) having quite different characteristics, wherein the former travels in the counter-flow direction with smaller $|\omega^*|$ while the latter travels in the co-flow direction with larger $|\omega^*|$. As shown in Fig. 52a, the neutral stability curves of Xu and Davis [103] and Ryzhkov [104] approach each other at Pr = 28–29; therefore, the instabilities of $m = 1(a)$ and $1(b)$ may have coexisted.

Xu and Davis [103] and Ryzhkov [104] considered the effect of interfacial heat transfer in their LSA for an infinite liquid bridge; however, such consideration was applied only to the perturbation field and not to the basic flow field (their velocity of the basic flow is a function of the r-position only). To understand the instability of thermocapillary convection in more detail, it is necessary to consider the effect of interfacial heat transfer on the basic flow field. Recently, thanks to scientific advances and improvements in the performance of analysis devices, the LSA of thermocapillary convection in finite liquid bridges of high-Pr fluids has been successfully achieved [105]. In the analysis of single-phase system, the thermal boundary condition at the liquid-bridge free surface in the dimensionless form is usually given as

$$-\boldsymbol{n} \cdot \nabla \tilde{T} = \mathrm{Bi}(\tilde{T} - \tilde{T}_0), \tag{22}$$

where, $\tilde{T}$ is the dimensionless temperature, $\boldsymbol{n}$ is the unit normal vector on the free surface, and Bi is the Biot number. The temperature far beyond the liquid bridge, the cold-disk temperature, etc., are often used as the reference temperature $\tilde{T}_0$. The Biot number gives relative importance of convective heat transfer on the free surface to the conductive one in the liquid, and is defined as

$$\mathrm{Bi} = \frac{hR}{\lambda}, \tag{23}$$

where h is the heat transfer coefficient.

Figure 53 compares the results of LSA for Pr $= 67$ and for Bi $= 0$ (*orange dashed-dotted line with small circles* [14]) and Bi > 0 (*red dashed line with small circles* [6]), together with the experimental data obtained in MEIS-1, -2, and DS-1 (*large pink circles*). The reference temperature was set at T_C in the LSA. The disk radius R is used as the representative length scale here, and (a) Ma_c and (b) F_c calculated by Eqs. 6 and 7 are plotted as a function of A_r. As shown in this figure, the results of LSA for Bi $= 0$ show apparently larger Ma_c than the experimental one for $A_r < 1.25$, and the results for $A_r > 1.25$ also agree less with one another. Although the F_c-A_r curve for Bi $= 0$ seems to be consistence with the experiment at a first glance, the value of A_r where the *frequency skip* occurs is different. On the other hand, both Ma_c-A_r and F_c-A_r curves for Bi > 0 show excellent agreement with the experimental results because the values of Bi were chosen to match with the experiment, i.e., Bi $= 0.30$ for $A_r < 1$ and Bi $= 0.15$ for $A_r \geq 1$. The LSA with appropriate Bi predicts the bifurcation of neutral stability curve and the *frequency skip* at $A_r = 1.25$–1.50 accurately, and the LSA for Bi $= 0.15$ reveals that the type of HTW changes at $A_r \approx 1.4$ as in Fig. 50 [106]. It is interesting to point out that the value of A_r at which the bifurcation of neutral stability curve occurs increase with increasing Bi (and the resultant interfacial heat lost from the liquid-bridge free surface), and this trend is similar to the present μg experiments. It follows from these results that the interfacial heat transfer (and the choice of Bi) is of great importance for the HTW instability in high-Pr thermocapillary liquid bridges.

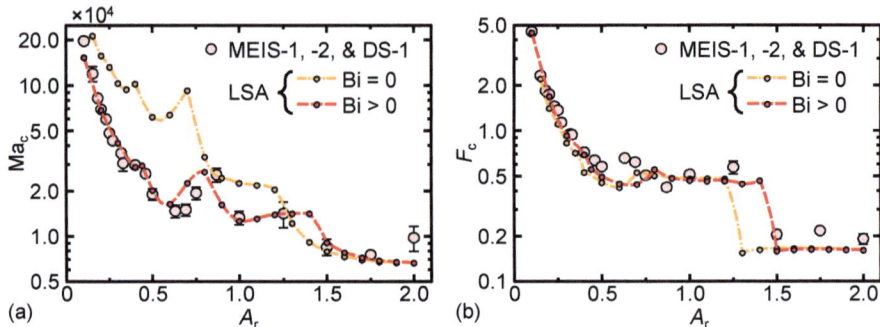

Fig. 53 Plots of **a** $\mathrm{Ma_c}$ and **b** F_c as a function of A_r for $\mathrm{Pr} = 67$: (*large pink circles*) experimental results obtained in MEIS-1, -2, and DS-1; (*orange dashed-dotted line with small circles*) results of LSA for $\mathrm{Bi} = 0$; (*red dashed line with small circles*) results of LSA for $\mathrm{Bi} > 0$

6.3 Contribution of Radiative Heat Transfer

Before the present μg experiments, the contribution of radiative heat transfer to the thermocapillary convection in a high-Pr liquid bridge had been considered minor. However, MEIS and DS projects revealed that the radiation can be the leading factor of interfacial heat transfer and can have a significant effect on the thermocapillary convection under the reduced-gravity environment even at the room temperature in contrast to the normal-gravity environment where the convective heat transfer is usually dominant. The early evaluation of the amount of heat transferred via radiation for the present μg experiments was performed by Melnikov et al. [69] and Yano et al. [96]. They determined the average temperature of the liquid-bridge free surface from the empirical relationship or the numerical simulation without considering radiation and post-evaluated the net radiative heat transfer according to the Stefan-Boltzmann law. Since there was no buoyancy and the motion of ambient gas was extremely slow in the μg environment, their evaluated radiative heat transfer sometimes overtakes convective one, especially in long liquid bridges.

Shitomi et al. [99] evaluated the local radiative heat transfer and showed significant effects of radiation on the flow and temperature fields of thermocapillary convection for 30 mm-diameter liquid bridges through CFD analyses in a two-phase system (i.e., liquid and gas). An example of the simulation domain and grid system they used are shown in Fig. 54, although the grid is made rough for better visibility. The computational grids for liquid and gas phases are depicted in gray and green, respectively. A silicone-oil liquid bridge, together with cold and hot disks, is placed inside a cylindrical chamber 90 mm in diameter filled with argon gas. The following time independent continuity, momentum, and energy equations for the liquid and gas are solved simultaneously using the STAR-CCM+ ver. 12.02:

$$\nabla \cdot \boldsymbol{u} = 0, \tag{24}$$

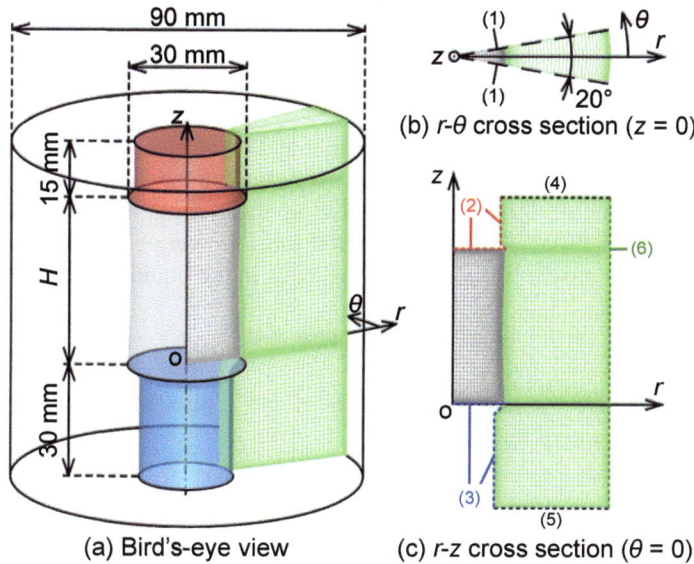

Fig. 54 Simulation domain and grid system for CFD analysis: **a** bird's-eye view, **b** r-θ cross sectional view at $z = 0$, and **c** r-z cross sectional view at $\theta = 0$. Surfaces/interfaces on which the boundary conditions are applied are (1) side surfaces of cylindrical sector, (2) hot-disk surface, (3) cold-disk surface, (4) chamber top wall, (5) chamber bottom wall, (6) chamber side wall, and liquid-gas interface (i.e., contact interface of gray and green grids)

$$(\boldsymbol{u} \cdot \nabla)\boldsymbol{u} = -\frac{1}{\rho}\nabla p + \nu \nabla^2 \boldsymbol{u}, \tag{25}$$

$$(\boldsymbol{u} \cdot \nabla)T = -\kappa \nabla^2 T, \tag{26}$$

where p is the pressure.

A three-dimensional grid system is required for taking radiative heat transfer into account in STAR-CCM+, even if the target flow is a two-dimensional one. To reduce the computational cost, the simulation domain is limited to a cylindrical sector with a central angle of 20° and is divided into two in the circumferential direction as illustrated in Fig. 54a. The periodic boundary conditions are imposed on the side surfaces (i.e., r-z cross sections at $\theta = \pm 10°$), which are indicated by *long-dashed lines* in Fig. 54b. A no-slip velocity condition ($\boldsymbol{u} = 0$) is applied at all solid-fluid interfaces indicated by *short-dashed lines* in Fig. 54c. The temperature boundary conditions at these interfaces are as follows: constant temperatures at T_H, T_C, and T_W on the hot-disk surface, cold-disk surface, and chamber side wall, respectively; adiabatic conditions on the chamber top and bottom walls. Additionally, the tangential stress balance of viscous force and thermocapillarity is imposed as the boundary condition at the liquid-gas interface as follows [35, 107, 108]:

$$\mu^{(l)} \left.\frac{\partial u_\eta}{\partial \xi}\right|_{(l)} - \frac{\partial \sigma}{\partial \eta} = \mu^{(g)} \left.\frac{\partial u_\eta}{\partial \xi}\right|_{(g)}, \tag{27}$$

where η and ξ are, respectively, the tangential and the normal directions to the free surface, and the suffixes (l) and (g), respectively, denote the quantities for the liquid and gas. Shitomi et al. [99] ignored the viscous shear stress from the gas (i.e., the right-hand side of Eq. 27) and implemented the following equivalent body force $\boldsymbol{F} = (F_r, F_\theta, F_z)$ into the radially outermost cells in the liquid bridge:

$$\boldsymbol{F} = (\boldsymbol{e}_r \cdot \nabla F_\eta, 0, \boldsymbol{e}_z \cdot \nabla F_\eta), \tag{28}$$

where $\boldsymbol{e}_r$ and $\boldsymbol{e}_z$ are the unit vectors in the r and z directions, respectively. The tangential component of the body force F_η is given as

$$F_\eta = \frac{\sigma_T}{\Delta \xi} \frac{\partial T_{\text{surf}}}{\partial \eta}, \tag{29}$$

where $\Delta \xi$ is the thickness of the radially outermost cell in a liquid bridge that depends on the axial position when $V_r \neq 1$.

For the thermal boundary condition, the local heat flux density at the free surface q_{surf} is undertaken by the convective heat flux density q_c and the radiative heat flux density q_r, and the relationship $q_{\text{surf}} = q_c + q_r$ holds. In the numerical simulations of Shitomi et al. [99], the surface-to-surface radiation in a closed system filled with gas was considered. When the gaseous region is discretized into $N_r \times N_\theta \times N_z$ cells, the number of faces on which the radiative heat exchange occurs is $N = 2N_\theta(N_r + N_z)$, and the radiative heat flux density at the i-th face can be expressed as

$$q_{r(i)} = \varepsilon_{(i)} \sigma_{\text{SB}} T^4 - \varepsilon_{(i)} \sum_{j=1}^{N} F_{ij} J_{(j)}, \tag{30}$$

where subscript (i) represents the face number, ε is the emissivity, σ_{SB} is the Stefan-Boltzmann constant ($\approx 5.67 \times 10^{-8}$ W/(m$^2 \cdot$ K^4)), F_{ij} is the view factor from the i-th face to the j-th face, and $J_{(j)}$ is the radiosity at the j-th face expressed as

$$J_{(j)} = \varepsilon_{(j)} \sigma_{\text{SB}} T^4 + (1 - \varepsilon_{(j)}) \sum_{k=1}^{N} F_{jk} J_{(k)}. \tag{31}$$

Here, q_r and T in Eq. 30 or J and T in Eq. 31 are quantities on the same face; therefore, subscripts (i) or (j) are omitted from T for convenience.

The convective heat flux density on the i-th face can be expressed as

$$q_{c(i)} = -\lambda^{(g)} \left.\frac{\partial T}{\partial \xi}\right|_{(g)}. \tag{32}$$

In the absence of gravity, the gas is driven only by the viscous shear stress from the liquid (i.e., the first term on the left-hand side of Eq. 27). Since this viscous shear stress is comparatively small, and the gas is nearly stationary in the present μg experiments, q_c is rather conductive heat transfer. In this section, it is, however, referred to as the convective heat transfer as before. As a result, the balance of heat flux density on the liquid-gas interface can be rewritten as

$$-\lambda^{(l)} \left.\frac{\partial T}{\partial \xi}\right|_{(l)} = -\lambda^{(g)} \left.\frac{\partial T}{\partial \xi}\right|_{(g)} + \varepsilon_{(i)} \left(\sigma_{SB} T^4 - \sum_{j=1}^{N} F_{ij} J_{(j)} \right). \quad (33)$$

In the numerical simulations of Shitomi et al. [99], the values of F_{ij} are estimated by the ray tracing method implemented in the STAR-CCM+, and the values of ε are assumed to be 0.5 for hot-disk surface and 0.9 for all other surfaces. In addition to the boundary conditions mentioned above, velocity continuity $\boldsymbol{u}^{(l)} = \boldsymbol{u}^{(g)}$ and temperature continuity $T^{(l)} = T^{(g)}$ are also given as boundary conditions at the liquid-gas interface. To satisfy the governing equations and boundary conditions, $\boldsymbol{u}$, T, p, and each component of q are solved iteratively based on a finite volume method.

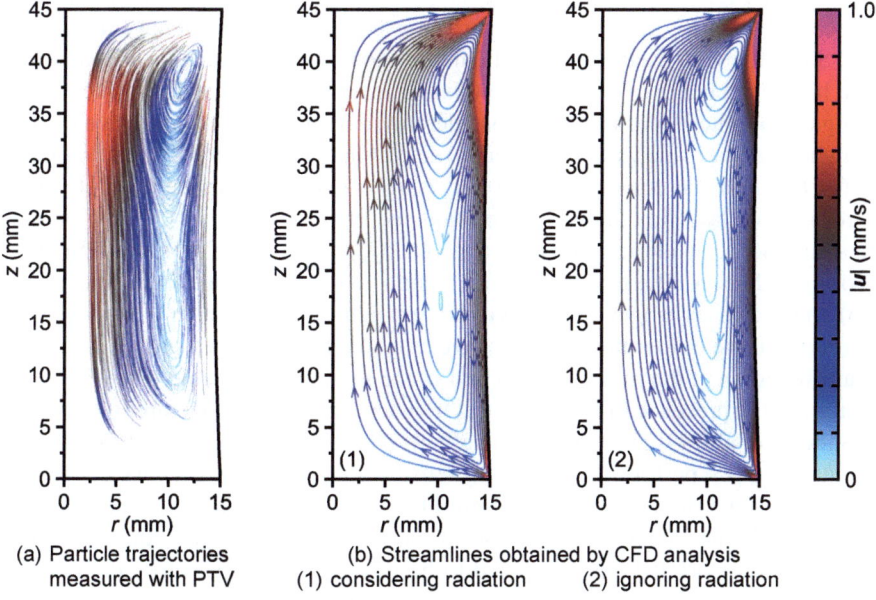

Fig. 55 Comparison of flow field for Pr = 207, A_r = 1.50, V_r = 0.95, T_C = 20.0 °C, T_H = 35.8 °C, and T_W = 23 °C: **a** particle trajectories measured with 3-D PTV and **b** streamlines obtained by CFD analyses (1) considering and (2) ignoring radiation. (Reprinted from Shitomi et al. [99], Copyright 2018, with permission from Elsevier. Modifications have been made in this figure with the permission from the publisher. All rights reserved)

A notable example of the effect of radiation on the thermocapillary convection is shown in Fig. 55, which displays flow fields for $Pr = 207$, $A_r = 1.50$, $V_r = 0.95$, $T_C = 20.0\,°C$, $T_H = 35.8\,°C$, and $T_W = 23\,°C$, where T_W is the temperature of the chamber side wall. This case corresponds to Fig. 30b-5. The value of $\Delta T\ (= 15.8\,K)$ is slightly lower than the instability threshold; therefore, the flow exhibits axisymmetric steady pattern. Figure 55a shows an experimental result obtained in MEIS-3, and the particle trajectories measured with 3-D PTV are projected on the r-z cross section. Other results show streamlines obtained by CFD analyses wherein the radiation is considered in Fig. 55b and is ignored in Fig. 55c. It is obvious that the simulation result considering radiation agrees much better with the experimental result from the perspectives of the overall flow pattern, flow speed, and center locations of convection rolls. The flow velocity is compared more quantitatively in Fig. 56a, which shows axial profiles of velocity components (u_r, u_z) at (1) $r = 7$ mm and (2) $r = 14$ mm. We note that the experimental data (i.e., *circles* and *diamonds*) are averages within a square region of 1 mm × 1 mm, and the error bars represent standard deviations. Both u_r and u_z agree better with experimental results in the numerical simulation considering radiation (*red solid lines*) than that ignoring radiation (*blue dashed lines*) for either (1) return flow at $r = 7$ mm and (2) surface flow at $r = 14$ mm. Additionally, the flow speed $|u|$ increases when the radiation exists. As shown in Fig. 56b, the free-surface temperature T_{surf} is higher than the side wall temperature $T_W\ (= 23\,°C)$ on most surface except in the very vicinity of the cold disk. Since the view factor from the liquid bridge to the chamber side wall accounts for a high percentage of the total view factor from the liquid bridge in the present configuration, the liquid bridge is considerably influenced by the radiative cooling from the chamber side wall. Specifically, the net heat transfer rates due to radiation $Q_r\ (= \int_S q_r dS)$ and convection $Q_c\ (= \int_S q_c dS)$, which are surface integrals of the corresponding heat flux densities, are evaluated to be $(Q_r, Q_c) = (151, 45)$ mW in the simulation considering radiation and $(Q_r, Q_c) = (0, 51)$ mW in the simulation ignoring radiation. The thermal radiation cools the liquid-bridge free surface (Fig. 56b) and increases tangential temperature gradient $\partial_\eta T_{\text{surf}}$, which is equivalent to the driving force of thermocapillary convection, on the entire free surface except for a narrow region at the middle height and in the vicinity of the cold disk (Fig. 56c). Therefore, the presence of radiative heat transfer increases the overall flow speed.

Shitomi et al. [99] also studied the effects of radiative heat transfer for other Pr conditions. They performed CFD analyses of a steady thermocapillary convection for a wide range of A_r in DS-3, wherein $Pr = 112$, $V_r = 0.95$, $T_C = 20.0\,°C$, and $T_W = 23\,°C$. The values of T_H were set at each instability threshold; therefore, their simulated flows are the ones expected to be seen immediately before the onset of oscillation. Figure 57 shows the heat transfer ratios $(Q_c + Q_r)/Q_{HD}$ for various A_r, where Q_{HD} is the amount of heat transferred from the hot disk to the liquid bridge per unit time. In this figure, the contributions of convective and radiative components to the heat transfer ratio can be distinguished by the color of bars. The heat transfer conditions in Fig. 57 are all heat loss because the values of $(Q_c + Q_r)$ are positive in all cases (see *circle plots* and the left vertical axis). It follows from this result that approximately 10–30 % of the heat entering to the liquid bridge through the hot-disk

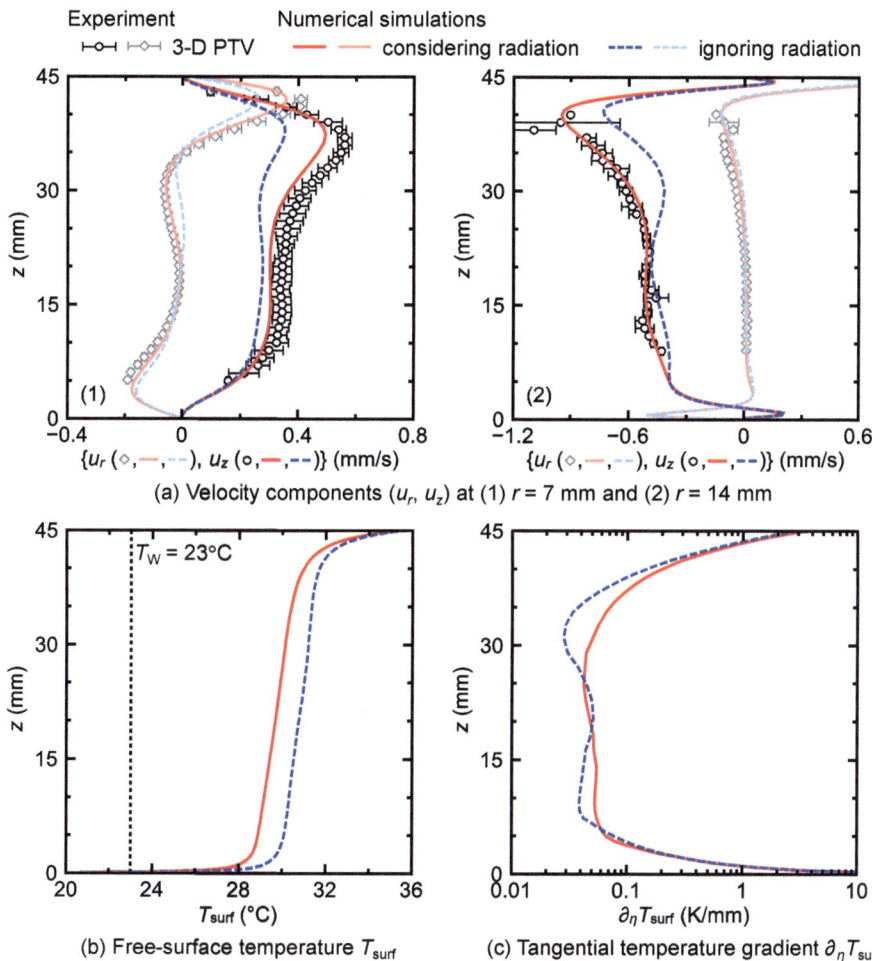

Fig. 56 Axial profiles of **a** velocity components (u_r, u_z) at (1) $r = 7$ mm and (2) $r = 14$ mm, **b** free-surface temperature T_{surf}, and **c** tangential temperature gradient $\partial_\eta T_{\text{surf}}$ for $\text{Pr} = 207$, $A_r = 1.50$, $V_r = 0.95$, $T_C = 20.0\,°\text{C}$, $T_H = 35.8\,°\text{C}$, and $T_W = 23\,°\text{C}$

surface is dissipated to the ambient gas, and the remaining heat is transferred to the cold disk. We note that the heat balance in the liquid bridge $Q_{\text{HD}} - Q_{\text{CD}} = Q_c + Q_r$ must be satisfied, where Q_{CD} is the net heat transfer rate from the liquid bridge toward the cold disk. It also can be seen from Fig. 57 that approximately 65–75 % of the heat dissipated to the ambient gas is transferred via radiation. A similar trend was confirmed in other conditions except for some circumstances (e.g., the case in which T_{surf} approaches T_W), especially the contribution of radiation tends to increase in heat-gain conditions [99].

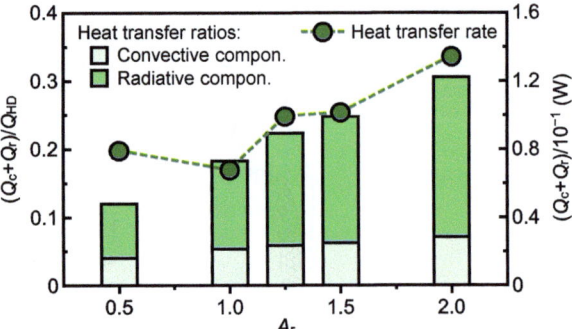

Fig. 57 Heat transfer ratio $(Q_c + Q_r)/Q_{HD}$ (*bar graph*, see the left vertical axis) and heat transfer rate from the liquid-bridge free surface to the ambient $(Q_c + Q_r)$ (*circle plots*, see the right vertical axis) for various A_r, wherein Pr = 112, $T_W = 23\,°C$, $T_C = 20.0\,°C$, and $T_H = T_C + \Delta T_c$

The most used dimensionless number to represent the interfacial heat transfer is the Biot number as described in Sect. 6.2. In the present geometry, the Biot number at the liquid-gas interface can be written as

$$\mathrm{Bi} = \frac{\{(Q_c + Q_r)/(2\pi R H)\} R}{\delta T \lambda^{(l)}}, \tag{34}$$

where δT is the characteristic temperature difference that defines the heat transfer coefficient, and the surface area of the liquid bridge is approximated as $2\pi R H$ because V_r is close to unity in many cases. The first natural choice of δT in the present configuration is $\delta T = \overline{T}_{\mathrm{surf}} - T_W$; however, it would be inappropriate in a situation where $\overline{T}_{\mathrm{surf}} \approx T_W$. For this reason, Yano et al. [96] proposed the heat transfer ratio. The Nusselt number on the hot-disk surface can be written as

$$\mathrm{Nu} = \frac{\{Q_{HD}/(\pi R^2)\} R}{\Delta T \lambda^{(l)}}. \tag{35}$$

From Eqs. 34 and 35, the heat transfer ratio can be expanded as

$$\frac{Q_c + Q_r}{Q_{HD}} = 4 \frac{\mathrm{Bi}}{\mathrm{Nu}} \frac{\delta T}{\Delta T} A_r, \tag{36}$$

which suggests that the thermocapillary convection is affected by Bi, Nu, $\delta T/\Delta T$, and A_r. Hence the discussion on the effect of interfacial heat transfer solely with Bi is insufficient. In the present μg experiments, Yano et al. [96] well organized the characteristics of HTW instability (e.g., the traveling direction) for various Pr, A_r, and temperature conditions by using the heat transfer ratio.

When the temperature difference between T_{surf} and T_W is extremely large, the effect of radiative heat transfer becomes more pronounced. Figure 58a shows the surface temperature distribution (left) and the particle trajectories (right) for Pr = 67 and $A_r = 2.00$ measured by an IR camera and a side-view camera, respectively [99]. This data was obtained in DS-1, and the temperature conditions were set at $T_C = 27.5\,°C$, $T_H = 28.9\,°C$, and $T_W = 25\,°C$. The images captured by an IR camera

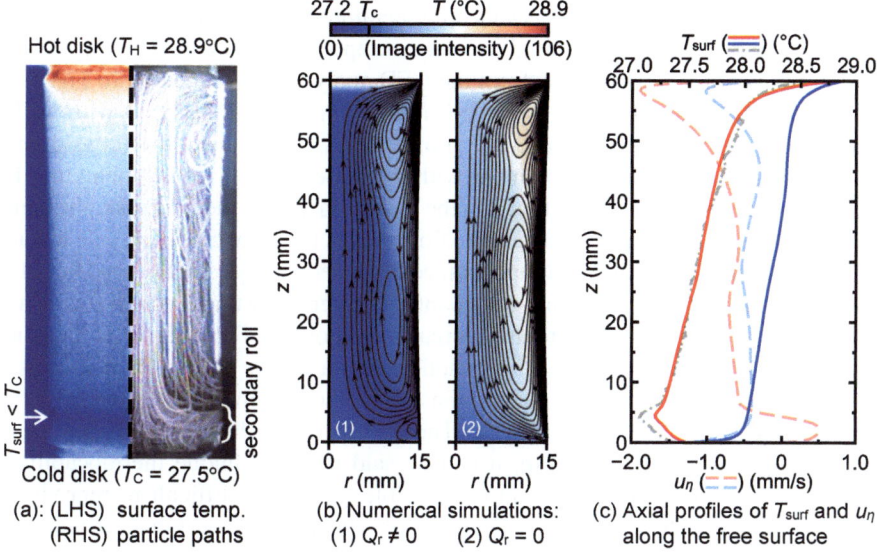

Fig. 58 Effect of radiative heat transfer for $Pr = 67$, $A_r = 2.00$, $V_r = 0.96$, $T_C = 27.5\,°C$, and $T_H = 28.9\,°C$: **a** surface temperature distribution (left-hand side) and particle paths (right-hand side) observed experimentally, **b** temperature and flow fields obtained by numerical simulations (1) considering radiation (i.e., $Q_r \neq 0$) and (2) ignoring radiation (i.e., $Q_r = 0$), and **c** axial profiles of free-surface temperature T_{surf} (*solid lines* for simulation results; *dashed-dotted line* for experimental result) and tangential velocity component u_η (*dashed lines*), wherein *red* and *blue* denote results of numerical simulations considering and ignoring radiation, respectively. (Reprinted from Shitomi et al. [99], Copyright 2018, with permission from Elsevier. Modifications have been made in this figure with the permission from the publisher. All rights reserved)

and a side-view camera were time averaged and time superposed over six minutes, respectively. We note that a color contour of the image intensity is displayed in the left-hand side of Fig. 58a (the higher the image intensity is, the higher is the temperature, and vice versa) because the present IR camera included unacceptable measurement uncertainty (i.e., approximately 1 °C). The interesting features of Fig. 58a are twofold: (1) a band-like region with $T_{surf} < T_C$ appears at $z \approx 5$ mm, and (2) a small secondary roll rotating in the opposite direction to the main roll appears at $z < 5$ mm.

The numerical simulations were performed for this condition with considering the radiation ($Q_r \neq 0$) or ignoring the radiation ($Q_r = 0$). These simulations assume steady flow and temperature fields, although the actual ones are in an oscillatory state. The results are shown in Figs. 58b-1 and b-2, respectively, wherein the streamlines (*black lines with arrowhead*) and temperature color contour on the *r-z* cross section are displayed in a superposed manner. A secondary roll near the cold disk exists in the simulation results considering radiation whereas such unique feature is not reproduced in the simulation ignoring the radiation. The temperature fields are also very different weather the radiation is taken into account or not.

The axial profiles of the free-surface temperature T_{surf} (*solid lines*, see the top horizontal axis) and the tangential velocity component u_η (*dashed lines*, see the bottom horizontal axis) obtained from the CFD analyses are compared in Fig. 58c. Also included in this figure is the free-surface temperature measured with an IR camera (*gray dashed-dotted line*). The profile of T_{surf} simulated with considering radiation shows reasonable agreement with the experimental result, including the local temperature drop at $z \approx 5$ mm. On the free surface near the cold disk, the signs of $\partial T_{\text{surf}}/\partial \eta$ and u_η are reversed, and the flow moves from the cold-disk side toward the hot-disks side, resulting in the appearance of the secondary roll. The cold liquid at $z \approx 5$ mm is transported from the surface into the interior by the action of secondary roll, and the overall bulk temperature decreases as shown in Fig. 58b-1. On the other hand, the simulation result ignoring radiation give a profile of T_{surf} that increases monotonically with the z-position, and the surface flow is always directed from the hot-disk side toward the cold-disk side. These features are distinctly different from the experimental result; therefore, it can be said that the unique thermocapillary convection shown in Fig. 58a is unreproducible (or highly difficult to reproduce) without radiation. The results obtained in the present study suggest that the radiative heat transfer can be a very important factor for the thermocapillary convection under the μg environment.

7 Dynamic Free-Surface Deformation

Taishi Yano and Koichi Nishino

As described in Sect. 1 of Chapter "Thermocapillary Convection in an Infinite Liquid Layer and in an Infinite Liquid Column", the (thermocapillary) Reynolds number Re, the Prandtl number Pr, and their product, the Marangoni number Ma ($=$ RePr) are the only parameters appear in the dimensionless governing equations of thermocapillary convection in weightlessness (i.e., the following continuity, momentum, and energy equations):

$$\nabla \cdot \boldsymbol{u} = 0, \tag{37}$$

$$\frac{\partial \boldsymbol{u}}{\partial t} + (\boldsymbol{u} \cdot \nabla)\boldsymbol{u} = -\nabla p + \frac{1}{\text{Re}}\nabla^2 \boldsymbol{u}, \tag{38}$$

$$\frac{\partial T}{\partial t} + (\boldsymbol{u} \cdot \nabla)T = \frac{1}{\text{Ma}}\nabla^2 T, \tag{39}$$

where t, $\boldsymbol{u}$, p, and T are the time, vector, pressure, and temperature with no dimension, respectively (the nabla operator ∇ is also dimensionless). For this reason, the Marangoni number is recognized as the most important parameter for understanding of the instability of thermocapillary convection, and the values of Ma_c for similar-shaped liquid bridges were expected to be comparable for any zone dimension. However, beyond this expectation, a several-fold or an order of magnitude increase in Ma_c with the increase in zone dimension was observed in the Space Shuttle and sounding-rocket missions [42, 73, 77]. To investigate this issue, Kamotani and Ostrach [78] paid special attention to the dynamic behavior of the liquid-bridge free surface. Based on the theoretical scaling analysis, they derived that the dynamic deformation (i.e., bulging and sinking) of the free surface near the hot corner, in which the surface flow mainly driven, should play an important role in the transition from a steady thermocapillary convection to an oscillatory one. In fact, the main reason for the unexpectedly large Ma_c in the previous µg experiments has been revealed to be due to the too rapid change in ΔT (or insufficient waiting period for the thermal equilibrium), as described in Sect. 4.4. However, whether the effect of dynamic deformation of the liquid-bridge free surface—the so-called dynamic surface (or free-surface) deformation DSD—on the instability of thermocapillary convection is negligible or not remains unclear despite many studies [109–112]. For this reason, the DS project has been conducted to clarify the effect of DSD on the thermocapillary-convection instability in large-scale liquid bridges.

As described in Sect. 2, the experimental cells MD10 and MD30 were equipped with the micro-imaging displacement meter (MIDM) [10] to measure the DSD. The present MIDM consisted of a microscope objective, a B/W CCD camera with an imaging lens, and a background illumination system, of which the total optical magnification and depth of field are, respectively, 20× and ±3.5 µm for MD10 and 10× and ±14.0 µm for MD30. The fields of view in the radial×axial directions are 242 µm × 321 µm for MD10 and 635 µm × 475 µm for MD30. The resultant spatial resolutions are approximately 0.5 µm/px for MD10 and 1.0 µm/px for MD30 when the image size is 640 × 480 px. We note that MIDMs in MD10 and MD30 had a different camera orientation by 90°, and the long side of the image was aligned with the z-direction in MD10, while that was aligned with the r-direction in MD30. The present MIDM system could traverse in axial, lateral, and focal directions with respect to the liquid-gas interface; therefore, the DSD along the entire liquid bridge could be measured (but not simultaneously).

The dynamic motion of the liquid-bridge free surface caused by the oscillatory thermocapillary convection has been measured in the previous terrestrial experiments by several researchers [10, 110, 113, 114]; however, the DS project is the first and thus far the only experiment in measuring the DSD in a large-scale liquid bridge and in the µg environment. Hereinafter, the results of DSD measurement in DS-2 are presented; therefore, $Pr = 67$ and $D = 10$ mm throughout this chapter unless otherwise noted. In the DS project, the optical imaging technique originally developed by Nishino et al. [46], but adequately customized for the present MIDM, was used to measure the DSD. Figure 59a shows a part of the MIDM image with the size of 200 µm × 100 µm for a liquid bridge of $A_r = 0.50$ and $V_r = 1.00$. Since the liquid bridge is illuminated by a light-emitting diode (LED) array from the backside, the liquid phase appears as a shaded area in a brighter gas phase. Figure 59b shows the horizontal distribution

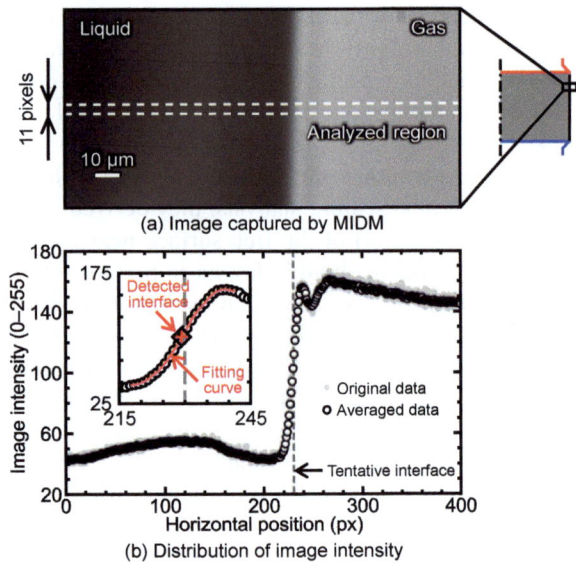

Fig. 59 Example of DSD measurement: **a** image captured by MIDM and **b** horizontal distribution of image intensity. (©2018 The Jpn. Soc. Microgravity Appl., modified with permission form Yano et al. [4]. All rights reserved)

of image intensity (0 for black and 255 for white) at the middle height in the MIDM image. The procedures for detecting the liquid-gas interface are as follows [4, 16]:

(1) average the image intensity surrounded by *dashed lines* in Fig. 59a in the vertical direction (i.e., the direction parallel to the liquid-gas interface) to improve the signal-to-noise ratio,
(2) calculate the numerical gradient of the averaged image intensity along the horizontal direction (i.e., the direction normal to the liquid-gas interface), and find the position at which the gradient becomes maximum to set it as the tentative liquid-gas interface,
(3) obtain the third-order polynomial fitting curve of the averaged image intensity in the region of a few dozen pixels centered at the tentative liquid-gas interface by the least squares method (*red line* in an enlarged view in Fig. 59b,
(4) calculate the inflection point of the fitting curve obtained in the previous step, which is regarded as the conclusive position of the liquid-gas interface, and
(5) repeat steps (1)–(4) for the entire vertical direction.

In the data analysis by Yano et al. [4, 16], the image intensity was averaged over 11 px in step (1), thus improving the signal-to-noise ratio by a factor of $\sqrt{11} \approx 3.3$, and the size of the region in step (3) was ± 12 px. Since the capillary number Ca ($= |\sigma_T| \Delta T / \sigma$), which represents the ratio of the viscous normal stress to the surface tension at the liquid-gas interface, is relatively small in large-scale thermocapillary liquid bridges (say, Ca $\ll 1$), the amount of deformation due to the internal flow is expected to be small. High-spatial-resolution measurement is, therefore, required to detect the DSD in the present µg experiments. The optical imaging technique

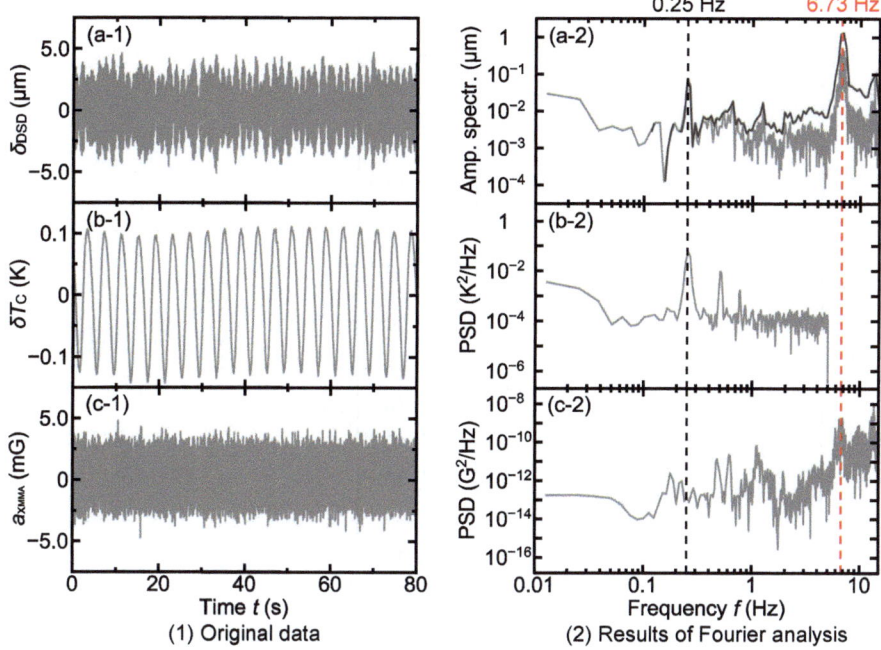

Fig. 60 (1) Time series of **a** liquid-bridge free surface displacement δ_{DSD}, **b** temperature fluctuation δT_C, and **c** g-jitter in the x_{MMA}-direction $a_{x_{MMA}}$ for $Pr = 67$, $A_r = 0.50$, $V_r = 1.00$, and $\Delta T = 20.8$ K. (2) Corresponding **a** amplitude spectrum and **b**, **c** power spectral densities (PSD). The line with *dark gray* in (a-2) is the 1/12 octave-band filtered amplitude spectrum. (Reprinted by permission from Springer Nature: Springer Nature Yano et al. [16], Copyright Springer Science+Business Media B.V., part of Springer Nature 2018. Modifications have been made in this figure with the permission from the publisher. All rights reserved)

mentioned above allows ones to measure the DSD with submicron-order accuracy, as demonstrated by Yano et al. [4, 16].

Figure 60(1) shows (a) the time series of the displacement of liquid-bridge free surface δ_{DSD} at 1.1 mm below the hot disk, together with (b) the temperature fluctuation on the cold disk δT_C and (c) the residual acceleration of gravity in the x_{MMA}-direction $a_{x_{MMA}}$, where G denotes the acceleration of gravity normalized by the earth gravity (≈ 9.8 m/s^2). These data were measured with the MIDM, the thermocouple sensor embedded in the cold disk, and the MMA (see Fig. 7 for the coordinate system of MMA), respectively. The experimental conditions are $A_r = 0.50$, $V_r = 1.00$, $\Delta T = 20.8$ K ($T_C = 20.0\,°\text{C}$ and $T_H = 40.8\,°\text{C}$), and Ma $= 2.2 \times 10^4$, and the ambient gas is neon (see Yano et al. [16] for the original results). The instability threshold for this condition is $\Delta T_c = 17.3$ K (and Ma$_c = 1.8 \times 10^4$), and the oscillation mode is the RW type with $m = 1$. In contrast to the terrestrial experiment by Ferrera et al. [114], who showed the good correlation between the behaviors of free-surface deformation and free-surface temperature fluctuation, the present

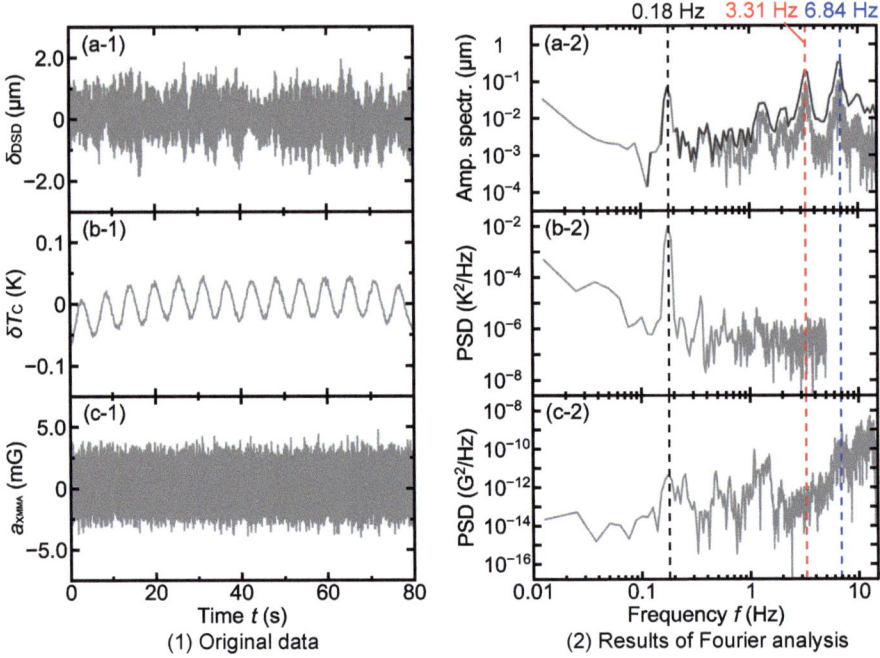

Fig. 61 (1) Time series of **a** liquid-bridge free surface displacement δ_{DSD}, **b** temperature fluctuation δT_C, and **c** g-jitter in the x_{MMA}-direction $a_{x_{MMA}}$ for $Pr = 67$, $A_r = 1.00$, $V_r = 1.00$, and $\Delta T = 12.4$ K. (2) Corresponding **a** amplitude spectrum and **b**, **c** power spectral densities (PSD). The line with *dark gray* in (a-2) is the 1/12 octave-band filtered amplitude spectrum. (Reprinted by permission from Springer Nature: Springer Nature Yano et al. [16], Copyright Springer Science+Business Media B.V., part of Springer Nature 2018. Modifications have been made in this figure with the permission from the publisher. All rights reserved)

results in Figs. 60a-1 and b-1 seem to exhibit quite different patterns. To understand the details of these results, discrete Fourier analysis was applied to the data presented in Fig. 60(1), and the resultant amplitude spectrum or power spectral density (PSD, hereinafter) are shown in Fig. 60(2). We note that results with *light gray* are the original Fourier analyzed data and that with *dark gray* in Fig. 60a-2 is the 1/12 octave-band filtered data [115]. The amplitude spectrum of δ_{DSD} in Fig. 60a-2 shows two clear peaks at $f = 0.25$ and 6.73 Hz, and the PSD of δT_C in Fig. 60b-2 shows a clear peak at $f = 0.25$ Hz, together with its harmonics at 0.50 and 0.75 Hz. Both δ_{DSD} and δT_C oscillate at 0.25 Hz, indicating that the DSD reflects the oscillatory motion of the thermocapillary convection inside the liquid bridge. As discussed later, the peak at 6.73 K in Fig. 60a-2 is caused by the external disturbances.

Figure 61 shows the similar results for $A_r = 1.00$ and $V_r = 1.00$, wherein the temperature conditions are $\Delta T = 12.4$ K ($T_C = 20.0\,°C$ and $T_H = 32.4\,°C$) and $Ma = 1.2 \times 10^4$. The measurement position of DSD is 1.3 mm below the hot disk, and the ambient gas is neon. The instability threshold for this condition is $\Delta T_c =$

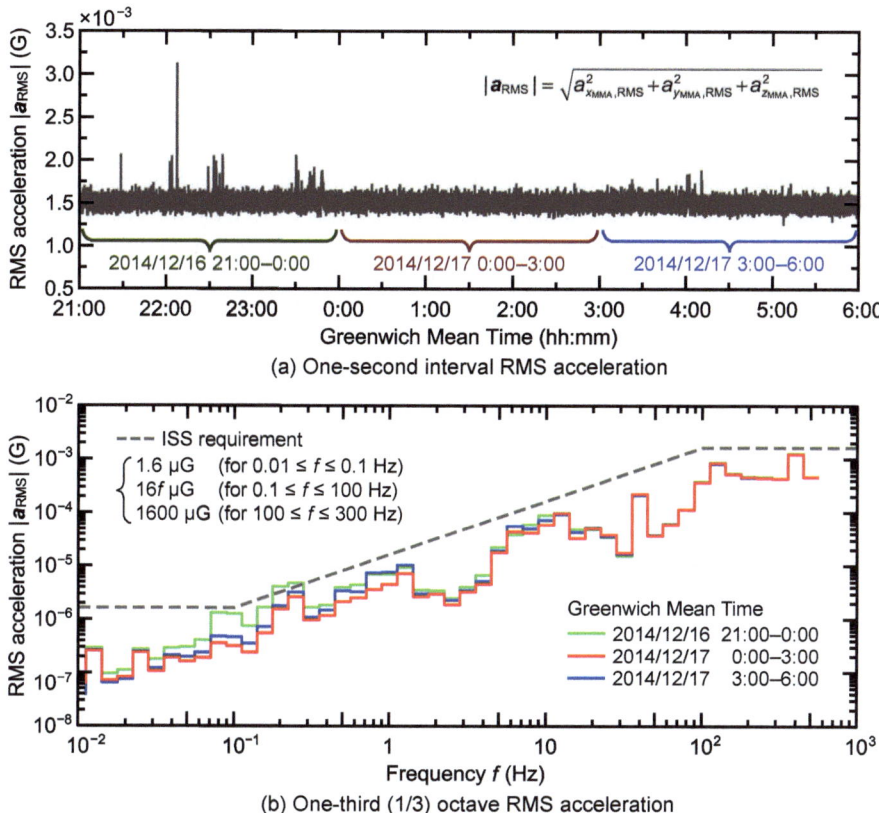

Fig. 62 G-jitter on the Kibo module in the period from 2014/12/16 21:00 to 2014/12/17 6:00 in Greenwich Mean Time (GMT): **a** one-second interval RMS acceleration and **b** one-third (1/3) octave-band filtered RMS acceleration. (©2018 The Jpn. Soc. Microgravity Appl., modified with permission from Yano et al. [4]. All rights reserved)

11.7 K (and $Ma_c = 1.1 \times 10^4$), and the oscillation mode is the RW type with $m = 1$. As shown in Fig. 61a-2, the amplitude spectrum of δ_{DSD} exhibits three peaks at $f = 0.18$, 3.31, and 6.84 Hz. Since the peak frequency at 0.18 Hz is in excellent agreement with that of PSD of δT_C in Fig. 61b-2, the oscillation at this frequency is caused by the internal thermocapillary convection. As with the results for $A_r = 0.50$, the DSD with higher frequency is due to the external disturbances.

As described in Sect. 2, there exist a residual acceleration of gravity called g-jitter on the ISS, which is caused by the crew activities, thruster firing, mechanical vibrations, docking/undocking with spacecrafts, and so on. Figure 62 shows an example of g-jitter: (a) the time series of the magnitude of one-second interval root-mean-square (RMS) acceleration $|a_{RMS}|$ measured with the MMA and (b) the three-hours interval one-third (1/3) octave-band filtered RMS acceleration [4]. The measurement date and time are given in the figure. Also plotted in Fig. 62b is the vibratory requirement of the ISS: (1) $\leq 1.6\,\mu\text{G}$ for $f \leq 0.1\,\text{Hz}$; (2) $\leq 16f\,\mu\text{G}$ for $0.1 \leq f \leq 100\,\text{Hz}$; (3) $\leq 1600\,\mu\text{G}$ for $f \geq 100\,\text{Hz}$ [116]. It is obvious that the g-jitter condition before 0:00 o'clock is rather severe than that after 0:00 o'clock. Particularly, the difference in 1/3 octave-band filtered RMS acceleration is remarkable for $f = 0.03$–$2\,\text{Hz}$. As described in Sect. 3, the period between GMT 21:30–6:00 corresponds to the sleeping time of the astronauts. There is, however, a fact that the ISS astronauts sleep less on average than they are scheduled to [117], and g-jitter caused by crew activities, of which the frequency band is mainly $f = 0.06$–$4\,\text{Hz}$ [118], is not completely eliminated even during this time slot. Thanks to the cooperation of the ISS crews, the g-jitter condition after 0:00 o'clock was quite good and very helpful to the success of the projects.

The liquid bridge can be strongly affected by g-jitter due to resonance, even if the RMS acceleration satisfies the ISS requirements. Figure 63 shows plots of dimensionless pulsation ω_R^* (i.e., dimensionless resonance angular frequency, see left vertical axis) and resonance frequency f_R (see the right vertical axis) as a function of A_r for straight liquid bridges of 5-cSt silicone oil with $D = 10\,\text{mm}$, where

$$\omega_R^* = 2\pi f_R \sqrt{\frac{\rho R^3}{\sigma}}. \tag{40}$$

These results were originally reported by (a) Sanz and Diez [119] and (b) Ichikawa et al. [120]: the former derived theoretically the resonance behavior of the liquid bridge for various oscillation mode with the Plateau technique, and the latter derived that for a single oscillation mode by considering the mass-spring-damper system. Sanz and Diez [119] studied various combinations of resonance oscillation modes in the azimuthal and axial directions (i.e., m_{LB} and N_{LB}, respectively); however, only the results for $(m_{\text{LB}}, N_{\text{LB}}) = (1, 1), (0, 2)$, and $(1, 2)$ are presented in Fig. 63a because other oscillation modes cannot be captured by the present MIDM. As described in Sect. 2, the time resolution of the CCD camera is 29.97 fps; therefore, the detectable free-surface oscillation is limited to $f < 15\,\text{Hz}$. Additionally, since the azimuthal measurement position of the MIDM is fixed, it cannot strictly distinguish m_{LB}. Ichikawa et al. [120] paid attention to the oscillation mode with $(m_{\text{LB}}, N_{\text{LB}}) = (1, 1)$, and they proposed the following equation for the resonance frequency:

$$f_R = \frac{1}{2\pi H} \sqrt{\frac{2}{B}\left(\frac{8\sigma}{D\rho} - \frac{\nu^2}{H^2 B}\right)}, \tag{41}$$

where B is the ratio of the moving mass to the total mass of the liquid bridge and takes a value between 0.5 and 1. The dimensionless pulsation and the resonance frequency for the present liquid bridges are summarized in Table 9. In this table, the values related to Sanz and Diez [119] are read graphically from their ω_R^*-A_r curves. The values of f_R for $(m_{\text{LB}}, N_{\text{LB}}) = (1, 1)$ and $(0, 2)$ agree well with second- or third-peak frequencies in Figs. 60a-2 and 61a-2, i.e., $f = 6.73\,\text{Hz}$ for $A_r = 0.5$ and $f = 3.31\,\text{Hz}$ and $6.84\,\text{Hz}$ for $A_r = 1.00$. Additionally, for $A_r = 1.00$, there is

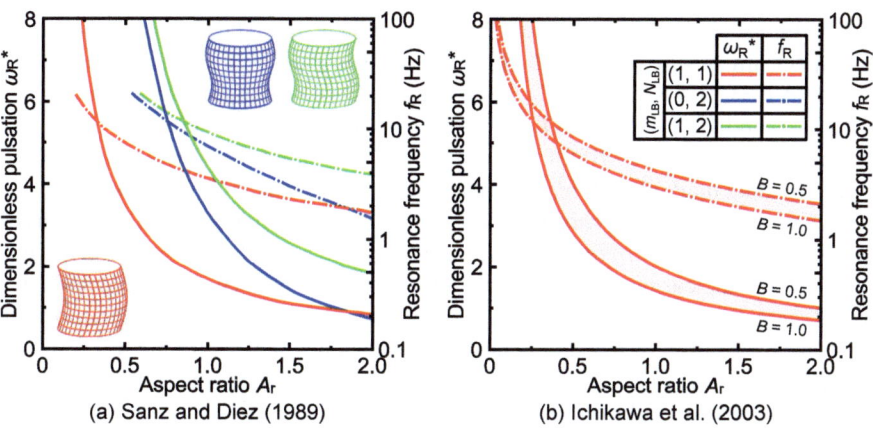

Fig. 63 Relationship between aspect ratio A_r and dimensionless pulsation ω_R^* (see *solid lines* and the left vertical axis) and resonance frequency f_R (see *dashed-dotted lines* and the left vertical axis) analyzed by **a** Sanz and Diez [119] and **b** Ichikawa et al. [120]. Results with *red*, *blue*, and *green* indicate oscillation mode with $(m_{LB}, N_{LB}) = (1, 1)$, $(0, 2)$, and $(1, 2)$, respectively. Illustrations of the liquid bridge in **a** represent schematics of oscillation modes

Table 9 Dimensionless pulsation ω_R^* and resonance frequency f_R for the liquid bridges of 5-cSt silicone oil with $D = 10$ mm, $V_r = 1$, and $A_r = 0.50$ and 1.00 evaluated with the models proposed by Sanz and Diez [119] and Ichikawa et al. [120]

	(m_{LB}, N_{LB})	$A_r = 0.50$		$A_r = 1.00$	
		ω_R^*	f_R (Hz)	ω_R^*	f_R (Hz)
Sanz and Diez [119]	(1, 1)	3.56	7.44	1.67	3.49
	(0, 2)	>10	>15	3.28	6.84
	(1, 2)	>10	>15	4.42	9.24
Ichikawa et al. [120]	(1, 1)	2.83–4.00	5.91–8.35	1.41–2.00	2.95–4.18

also a small peak of amplitude spectrum of δ_{DSD} at $f = 9.72$ Hz, which is close to the resonance frequency of oscillation mode with $(m_{LB}, N_{LB}) = (1, 2)$. These agreements between peak and resonance frequencies suggest that the high-frequency components of the DSD shown in Figs. 60a-2 and 61a-2 are due to g-jitter.

To return to the DSD due to thermocapillary convection, the change in the amplitude spectrum of δ_{DSD} with ΔT for $A_r = 1.00$ is shown in Fig. 64a: (1) steady state at $\Delta T = 11.4$ K ($\Delta T/\Delta T_c = 0.97$) and (2)–(4) oscillatory states at $\Delta T_c =$ 14.5, 15.7, and 16.8 K ($\Delta T/\Delta T_c = 1.24, 1.34$, and 1.44, respectively). The measurement position of the MIDM is approximately 0.5 mm below the hot disk. In this figure, the data marked with *black circles* are related to the internal convection, and those marked with *red* and *blue diamonds* are related to g-jitter. Firstly, let us focus on the results with a *black circle*. There is no clear peak at $f < 1$ Hz for $\Delta T < \Delta T_c$ (Fig. 64a-1), while a clear peak appears around $f = 0.2$ Hz in each amplitude spectrum for $\Delta T > \Delta T_c$ (Figs. 64a-2 to a-4), because of the onset of oscillatory thermocapillary convection. Figures 64b and c show the dependence of peak frequency f_{peak} and peak amplitude $A_{\text{DSD}}(f_{\text{peak}})$ (i.e., the magnitude of amplitude spectrum at f_{peak}) on ΔT (see the left vertical axis for the data with a *black circle*). As in previous studies [18, 39], the convective oscillation frequency increases linearly with ΔT. On the other hand, the amplitudes of DSD at each f_{peak} fit well with the power function of the form $A_{\text{DSD}} = 0.026\sqrt{\Delta T - 11.98}$. From this function, the value of ΔT at which A_{DSD} becomes zero can be found as 11.98 K, which is very close to the critical temperature difference $\Delta T_c = 11.7$ K measured from the flow and temperature observation. The DSD amplitude A_{DSD} increases linearly with $\sqrt{\Delta T}$ (or in other word, the squared amplitude A_{DSD}^2 increases linearly with ΔT), indicating that the flow becomes an oscillatory state via a supercritical Hopf bifurcation [39, 85, 121].

As shown in Fig. 64a, there are two other peaks of the amplitude spectrum due to g-jitter: one is at ~ 3.25 Hz and the other is at ~ 6.69 Hz. These frequencies agree well with the resonance frequencies of the liquid bridge with $(m_{\text{LB}}, N_{\text{LB}}) =$ (1, 1) and (0, 2), respectively. In Fig. 64, the data related to the former oscillation mode are indicated by *red diamonds*, and those to the later oscillation mode are indicated by *blue diamonds* (see the right vertical axis for them). In contrast to the DSD induced by the oscillatory thermocapillary convection, these peak frequencies are insensitive to ΔT (Fig. 64b), and corresponding DSD amplitudes and ΔT show no (or weak) correlation (Fig. 64c).

The axial distribution of $A_{\text{DSD}}(f)$ for (a) $f = 0.25$ Hz and (b) $f = 7.44$ Hz for $A_r = 0.50$ are shown in Fig. 65. The experimental conditions are the same as those in Fig. 60. In this result, attention is paid to the components having the same frequencies as (a) the oscillatory thermocapillary convection and (b) the resonance oscillation mode of $(m_{\text{LB}}, N_{\text{LB}}) = (1, 1)$, and the RMS values of $A_{\text{DSD}}(f)$ in the range of $\pm 10\%$ of these frequencies are given. Owing to the small field of view, the DSD along the entire free surface cannot be measured simultaneously; therefore, this result was obtained by traversing the MIDM. This is the reason for the discontinuities in the plots. The distribution of A_{DSD} for 0.25 Hz in Fig. 65a, which is induced by the internal oscillatory thermocapillary convection, shows two maxima: the first near the hot corner at $z \approx 4$ mm and the second near the cold corner at $z \approx 1$ mm. Such double-maxima pattern of DSD agrees qualitatively with the numerical result for $\text{Pr} = 4.38$ under zero-gravity condition reported by Kuhlmann and Nienhüser [109], who showed typical bulging and sinking near the hot and cold corners. Ferrera et al. [114] and Carrión et al. [112] also reported the distribution of DSD amplitude having

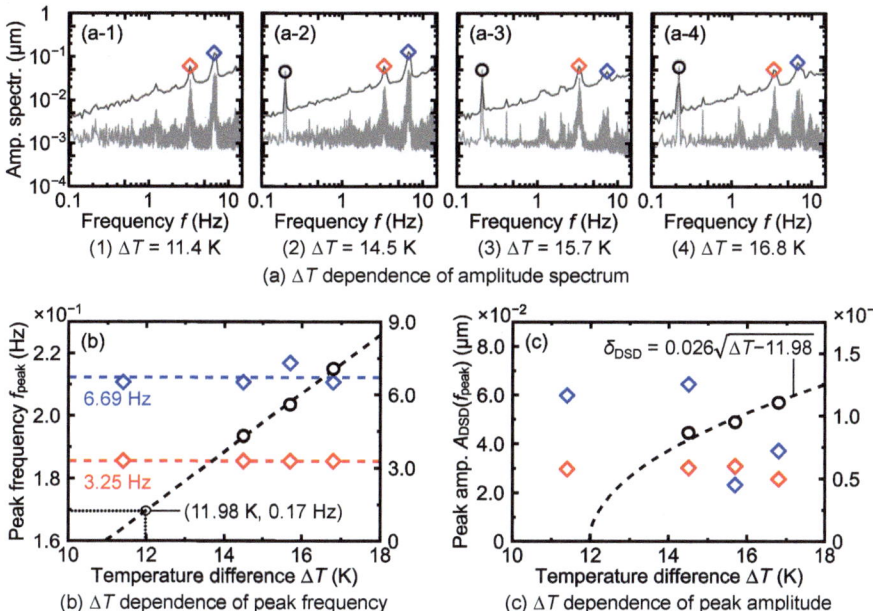

Fig. 64 a Amplitude spectrum of dynamic free-surface deformation (*light gray*) and 1/12 octave-band filtered amplitude spectrum (*dark gray*) for four different temperature conditions. **b, c** Dependence of peak frequency f_{peak} and peak amplitude $A_{\text{DSD}}(f_{\text{peak}})$ on ΔT. *Black circles* indicate the oscillation due to the thermocapillary-convection instability (see the left vertical axis), while *red* and *blue diamonds* indicate the g-jitter induced oscillations with $(m_{\text{LB}}, N_{\text{LB}}) = (1, 1)$ and $(0, 2)$, respectively (see the right vertical axis)

two maxima, although their magnitudes are not equivalent and the deformation near the hot corner is significant. We note that their aspect ratios are $A_r = 0.6$ and 0.615, respectively, and V_r is slightly lower than the unity. Additionally, their liquid bridges were deformed due to gravity. In contrast to the result for 0.25 Hz, the distribution of A_{DSD} for 7.44 Hz in Fig. 65b, which is induced by g-jitter, shows a single maximum at $z \approx 3.5$ mm, indicating the axial oscillation mode of $N_{\text{LB}} = 1$. Unfortunately, since the present MIDM was fixed in the azimuthal direction, the azimuthal oscillation mode m_{LB} cannot be determined from this result.

Figure 66 shows the distribution of $A_{\text{DSD}}(f)$ for $A_r = 1.00$: (a) the oscillatory thermocapillary convection-induced DSD and (b) the g-jitter-induced DSD, wherein the experimental conditions are the same as those in Fig. 61. The frequency considered in Fig. 66a is $f = 0.18$ Hz. The distribution of A_{DSD} for this frequency reaches its maximum at $z \approx 6.5$ mm and decreases monotonically towards the hot disk. This single-maximum pattern is similar to the numerical result reported by Kuhlmann and Nienhüser [109] for Pr = 4.38 under normal-gravity condition. Since the distribution of DSD reflects the oscillatory behavior of thermocapillary convection inside the liquid bridge, the difference in the characteristics of DSD shown in Figs. 65a and 66a suggests the difference in the axial flow structures between $A_r = 0.50$ and 1.00

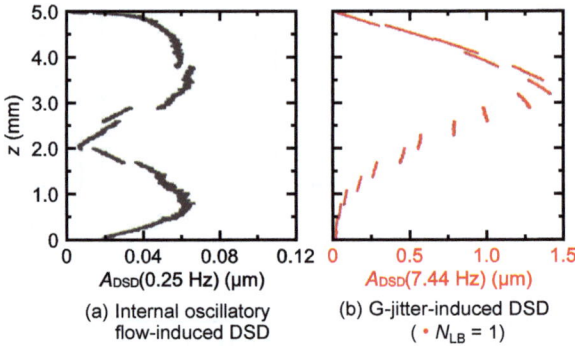

Fig. 65 Axial distributions of DSD amplitude induced by **a** oscillatory thermocapillary convection and **b** g-jitter for $A_r = 0.50$. (Reprinted by permission from Springer Nature: Springer Nature Yano et al. [16], Copyright Springer Science+Business Media B.V., part of Springer Nature 2018. Modifications have been made in this figure with the permission from the publisher. All rights reserved)

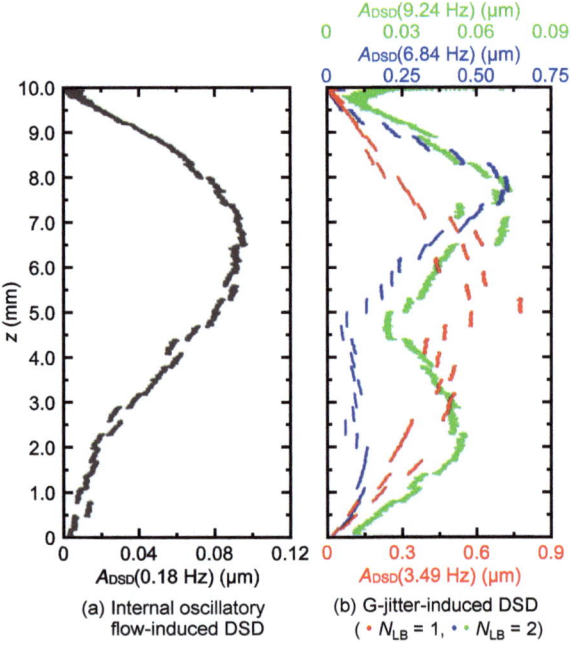

Fig. 66 Axial distributions of DSD amplitude induced by **a** oscillatory thermocapillary convection and **b** g-jitter for $A_r = 1.00$. Original data for $f = 0.18, 3.49$, and 9.24 Hz were reported by Yano et al. [16]

in spite of the same oscillation mode (i.e., RW type with $m = 1$). The amplitude of DSD due to thermocapillary-convection instability for $D = 10$ mm is on the order of $0.1\,\mu$m for both $A_r = 0.50$ and 1.00, which is close to or rather smaller than the results for $D = 5$–6 mm measured in the terrestrial experiments by Kanashima et al. [10], Ferrera et al. [114], and so on.

The frequencies considered in Fig. 66b are 3.49, 6.84, and 9.24 Hz, which correspond to the resonance frequencies of the oscillation modes of $(m_{LB}, N_{LB}) =$ (1, 1), (0, 2), and (1, 2), respectively (see the same color data and axes). As with the result in Fig. 65b, the distribution of A_{DSD} for 3.49 Hz indicates the oscillation with $N_{LB} = 1$. On the other hand, since there are two local maxima at $z \approx 2.5$ and 7.5 mm, the results for 6.84 Hz and 9.24 Hz indicates the oscillation with $N_{LB} = 2$. The distribution of A_{DSD} for 6.84 Hz is reminiscent of the results of Ferrera et al. [114] and Carrión et al. [112] that A_{DSD} has larger peak near the hot disk and smaller peak near the cold disk; however, the reason for such distribution remains unclear. Additionally, the oscillation at 9.24 Hz is insignificant because the amplitude of DSD for this frequency is an order of magnitude smaller than the others. It follows from the present DSD measurements that g-jitter strongly affects the behavior of the liquid bridge, and the liquid bridge formed in the ISS oscillates with various frequencies and resonance modes. Fortunately, the DSD caused by the thermocapillary-convection instability and that caused by g-jitter have obviously different frequency bands; therefore, they can be separated with appropriate filtering.

8 Transition to Chaos and Turbulence

Ichiro Ueno

This section introduces the thermal-flow fields in the half-zone liquid bridge under rather high thermocapillary effect. One realizes time-dependent 'oscillatory' convection beyond the threshold for the onset of primary instability. By further increasing the thermocapillary effect, it has been indicated by terrestrial and microgravity experiments that the thermal-flow field exhibits transition to chaotic and turbulent states. Evaluation of the chaotic states is conducted via various approaches of time-series chaos analysis.

8.1 Introduction

After the transition from the time-independent 'steady' state to the time-dependent 'oscillatory' one beyond the onset condition for the primary instability [18, 55, 58], the thermal-flow field in the half-zone (HZ) liquid bridges exhibits several regimes accompanying with different travelling-wave-type and standing-wave-type convections [17–19]. Under a fixed geometry of the liquid bridge, the temperature difference ΔT is a free parameter to determine the intensity of the thermocapillary effect in terms of $\mathrm{Ma} = |\gamma_T| \Delta T L / \rho \nu \kappa$, where γ_T is the temperature coefficient of the surface tension γ, L is the characteristic length of the liquid bridge, ρ, ν, and κ are the density, the kinematic viscosity, and the thermal diffusivity of the test liquid. The threshold for the primary instability is generally described by the critical Marangoni number $Ma_c^{(1)}$. The threshold in terms of the temperature difference to determine $Ma_c^{(1)}$ is called the critical temperature difference ΔT_c. A 'distance' in Ma from the threshold for the primary instability is generally measured by so-called over critical parameter $\epsilon = (Ma - Ma_c^{(1)})/Ma_c^{(1)}$ [121].

It must be emphasized that one has to face a severe difficulty in realizing high Ma especially in the terrestrial experiments. Due to the static pressure by the gravity, it is quite difficult to realize large-size liquid bridge without unavoidable deformation. This is a critical reason why liquid bridges of $O(10^{-3}\,\text{m})$ has been commonly employed in different research groups (e.g., Chun and Wuest [122]; Preisser et al. [17]; Cao et al. [123]; Petrov et al. [124]; Hirata et al. [125]; Kamotani et al. [97]; Ueno et al. [19]). Because there exists a severe limitation in increasing L, one has to increase ΔT to increase Ma for test liquid concerned, which results in inevitable evaporation [19, 27, 122]. In the following, knowledge accumulated by the previous research will be reviewed from a pioneering work conducted by Velten et al. [34] to a series of microgravity experiments.

8.2 Terrestrial Experiments

Pattern map of the thermal-flow field in the HZ liquid bridges was illustrated in wide ranges of Ma and Γ [2] with various kinds of test liquids by Velten et al. [34]. They categorized the induced thermal-flow fields into periodic and nonperiodic states. Frank and Schwabe [36] then investigated the route to the chaotic state by increasing Ma: By illustrating so-called return map from the time series of the temperature near the free surface, they showed the thermal-flow fields exhibit transitions from steady, time dependent, quasi-periodic, and to chaotic states. Shevtsova et al. [126] investigated the variation of the thermal-flow field via a series of numerical simulation, and illustrated the route to the chaotic state.

Ueno et al. [19] indicated several regimes of the thermal-flow fields by monitoring the behavior of the suspended particles. They found that there exist two different regimes for each traveling-wave-type (or rotating) and standing-wave-type (or pulsating) state (Fig. 67). The pseudo-phase space was reconstructed by considering the time-delayed D-dimensional coordinate system v_D defined as follows;

$$v_\xi = (T(t), T(t+\tau), T(t+2\tau), \cdots, T(t+(D-1)\tau)), \tag{42}$$

where τ is the delay time. They further measured the variations of the thermal-flow field quantitatively by evaluating the correlation dimension D_c and the maximum Lyapunov exponent λ_1 from the time-delayed coordinate system (Fig. 68). The correlation dimension shows a geometric nature of the trajectory in the phase space, and the Lyapunov exponents indicate a character of the time-dependent dynamics of the system. These evaluations clearly indicate the difference the thermal-flow field

---

[2] In their article the aspect ratio was defined as $A = H/2R$, thus $\Gamma = 2A$.

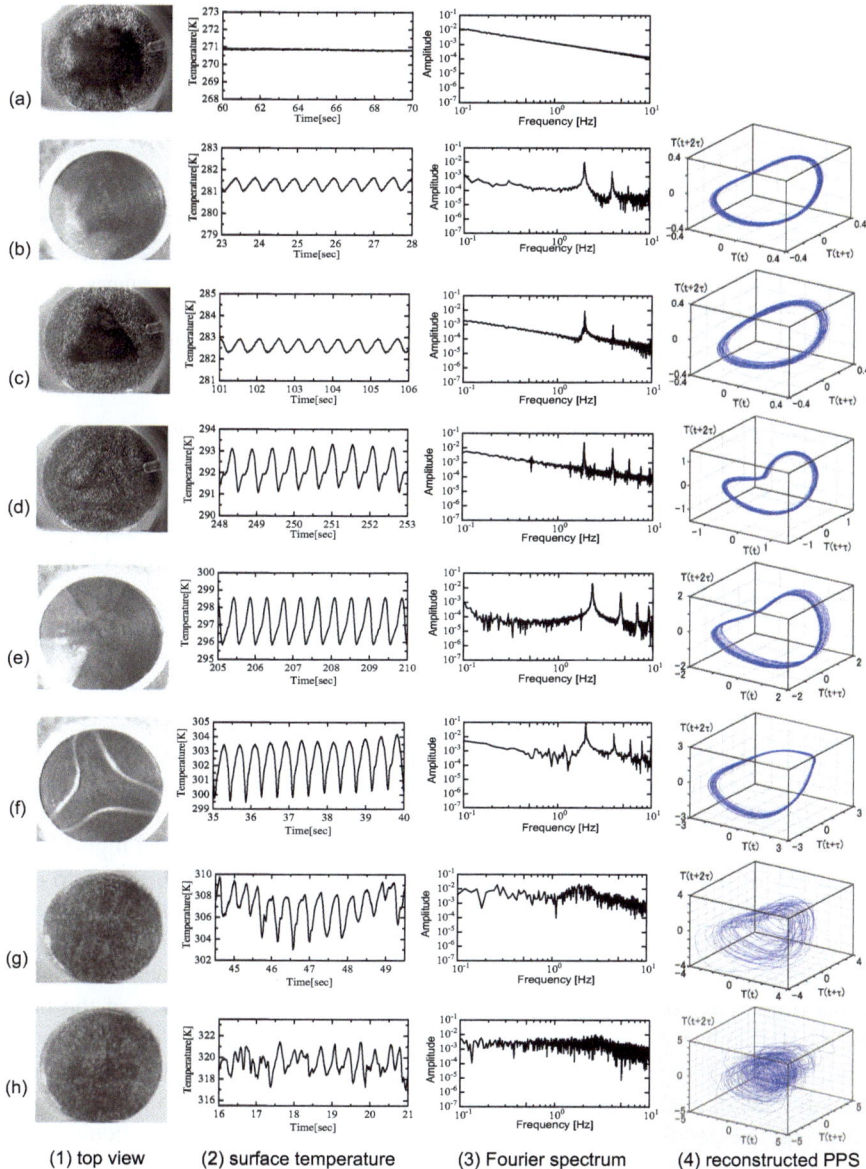

(1) top view   (2) surface temperature   (3) Fourier spectrum   (4) reconstructed PPS

Fig. 67 Flow regimes of **a** Rg1: steady flow, **b** Rg2: pulsating flow I, **c** Rg3: rotating flow I, **d** Rg4: transition, **e** Rg5: pulsating flow II, **f** Rg6: rotating flow II, **g** Rg7: chaotic flow I, and **h** Rg8: chaotic flow II (turbulence). Column (1) indicates top view of the flow field, (2) time series of the surface temperature variation, (3) its Fourier spectrum, and (4) reconstructed pseudophase space (PPS). Experimental conditions in Γ, ΔT and corresponding Ma for **a**–**h** are as follows: **a** (0.64, 28.5 K, 2.6×10^4), **b** (0.64, 34.2 K, 3.2×10^4), **c** (0.64, 36.6 K, 3.4×10^4), **d** (0.62, 69.7 K, 6.4×10^4), **e** (0.68, 67.5 K, 6.2×10^4), **f** (0.64, 83.7 K, 7.7×10^4), **g** (0.64, 90.7 K, 8.4×10^4), and **h** (0.64, 103.9 K, 9.6×10^4) in a cold environment or the ambient temperature $\Delta T_\infty = -20$ K. Reprinted from Ueno et al. [19], with the permission of AIP Publishing. All rights reserved

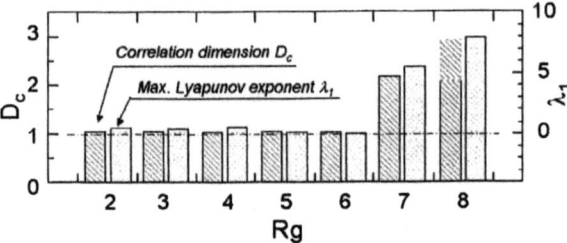

Fig. 68 Correlation dimension and maximum Lyapunov exponent λ_1 for flow regimes. The flow regimes 'Rg' and their experimental conditions are the same as given in the caption for Fig. 67. Reprinted from Ueno et al. [19], with the permission of AIP Publishing. All rights reserved

before and after the onset of the chaotic flow state. It is noted, as aforementioned, that the highly nonlinear thermal-flow field is hardly realized under normal gravity condition without encountering the undesignated problem of the test-liquid evaporation. This is due to the severe limitation in the characteristic length L of the liquid-bridge formation against the static pressure: One has to inevitably apply large ΔT in order to realize high Ma.

8.3 Microgravity Experiments

Long-duration microgravity environment enables us to perform a series of experiments by applying high Ma with lower ΔT if one employs large-scale liquid bridge [127–129]. Another significant advantage of such environment is that high-viscous liquid can be used as the test liquid while keeping ΔT relatively small. Sato et al. [127] and Matsugase et al. [128] conducted a series of microgravity experiments on the thermocapillary-induced convection under Ma far beyond $Ma_c^{(1)}$ or under large ϵ in the liquid bridges of high Pr (Pr > 100). Figure 69 illustrates typical examples of the time series of the temperature deviation over the free surface of the liquid bridge of $\Gamma = 2.5$ ($R = 25$ mm) and Pr = 206.8 for 25 °C. This experiment was conducted in the forth series of MEIS (MEIS-IV) in 2010. It is noted that the right and left ends of each frame correspond to the hot and cold disks to sustain the liquid bridge, respectively. Another note is that a partial surface of the liquid bridge is mimicked to visualize the surface temperature deviation. Column (a) shows the variation under $\epsilon = 1.2$, slightly beyond the threshold for the primary instability. Relatively cold region emerges near the hot disk, and propagates almost normal to the axial direction at a constant propagation speed. When one increases Ma to realize higher ϵ, relatively cold region also emerges near the hot disk (Column (b)). A sharp cold line travels from the region near the hot disk toward the cold end as well, but its propagation direction varies from that under lower ϵ. Such oblique propagation of the thermal wave over the free surface is commonly observed in the tall liquid bridges of high Pr [14, 93, 101]. Under higher ϵ, the thermal waves over the free

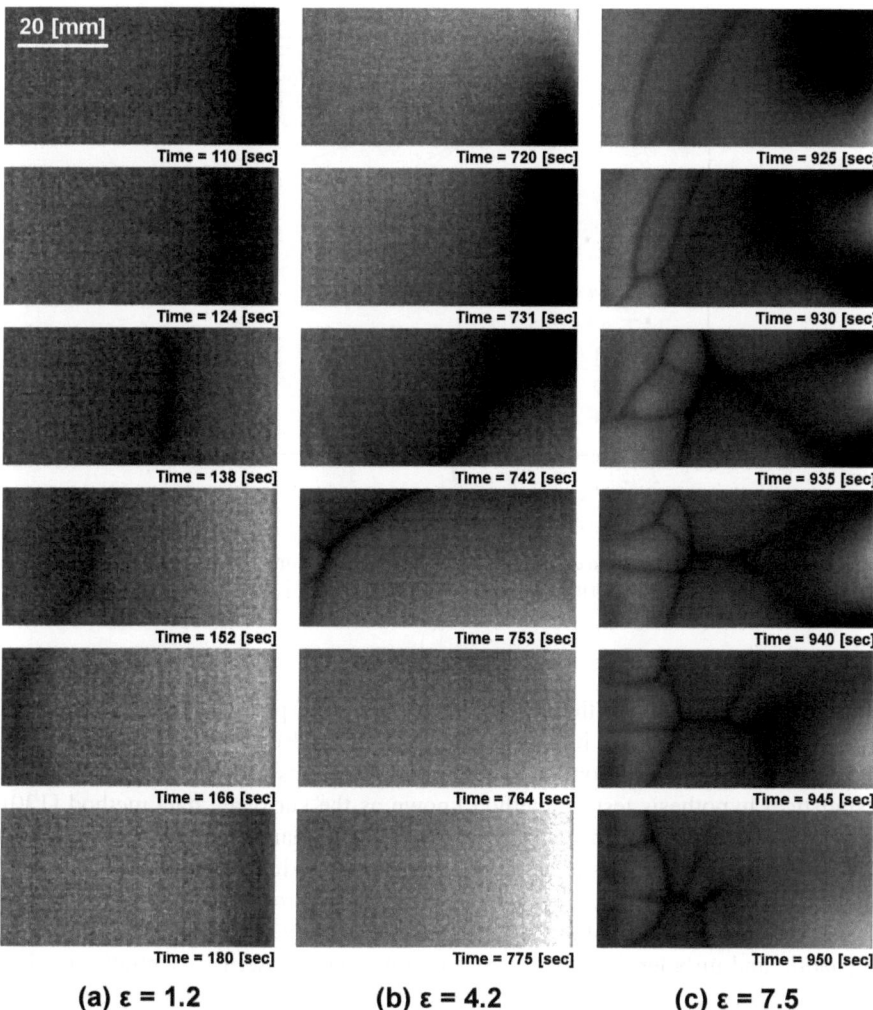

Fig. 69 Time series of surface-temperature deviation evaluated by IR images in liquid bridge of $(\Gamma, \text{Pr}, R) = (2.5, 206.8, 25\,\text{mm})$ under $(\text{Ma}, \epsilon) = $ **a** $(4.37 \times 10^4, 1.2)$, **b** $(10.5 \times 10^4, 4.2)$ and **c** $(17.2 \times 10^4, 7.5)$. Reprinted from Sato et al. [127], with the permission of Springer Nature. All rights reserved

surface become significantly complex (Column (c)). A cold line as seen in lower ϵ (Columns (a) and (b)) breaks apart to exhibit branch-like structure. One finds more finer structures especially near the cold-end region.

After a qualitative investigation on the non-linear thermocapillary-driven convection in the half-zone liquid bridges [127], a quantitative investigation was conducted. Matsugase et al. [128] obtained the time series of the surface temperature at the middle height of the liquid bridge to evaluate the transition process of the flow fields. They

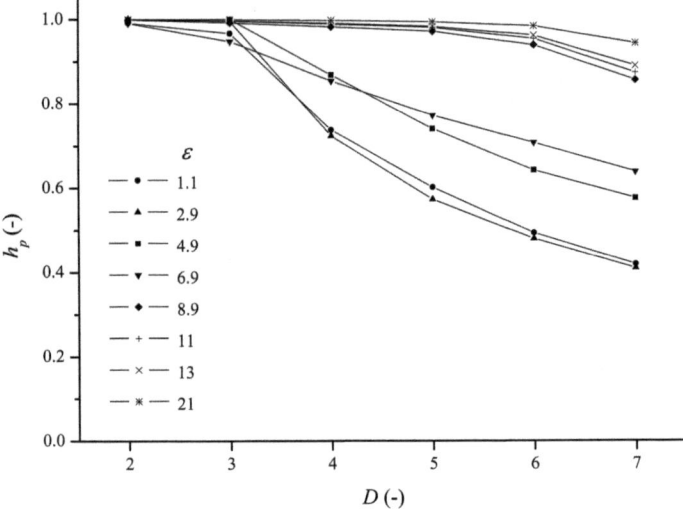

Fig. 70 Variation of permutation entropy h_p against embedding dimension D in liquid bridge of $(\Gamma, \text{Pr}, R) = (2.0, 112, 25\,\text{mm})$. Reprinted from Matsugase et al. [128], with permission from Elsevier. All rights reserved

indicated the transition of the thermal-flow field from periodic, to quasi-periodic, and finally to chaotic states by Fourier analysis. They also employed the time-series chaos analysis: Intrinsic nonlinearity of the time series was verified by applying a statistical hypothesis testing method known as the surrogate data method [130]. They successfully indicated a clear boundary of the dynamics underlying in the system between the deterministic and stochastic processes in ϵ. They further evaluated the permutation entropy h_p (Fig. 70), the translation error as well as the maximum Lyapunov exponent λ_1 to illustrate the transition process of the thermal-flow field to chaotic and turbulent states. A long-duration fine dataset is essentially required to conduct precise evaluations of those measures. Not only fine microgravity environment but also a long enough period of microgravity condition is indispensable for investigations on the thermal-flow transition processes. One finds that a Chinese group evaluated the correlation dimension and λ_1 in a more volatile liquid bridge in a microgravity experiment on the Tiangong 2 [129].

8.4 Other Topics in Thermal-Flow Transition

It has been known that the thermal-flow field in low-Pr liquid bridges exhibits primary and secondary instabilities by increasing Ma [18, 56, 132–134] before the transition to chaotic and turbulent states. That is, the thermal-flow field exhibits a transition from two-dimensional time-independent convection to two-dimensional time-dependent one via the primary instability, and then to three-dimensional time-dependent one via the secondary instability. Motegi et al. [135] investigated such twofold instabilities by Floquet analysis. In high-Pr liquid bridges, on the contrary, it had been discussed only primary instability in the transition process of the thermal-flow field [17–19], notwithstanding that a quasi-periodic state has been monitored [19, 36]. Ogasawara et al. [131] firstly indicated the secondary instability for high-Pr liquid bridge via numerical simulation. They focused on the thermal-flow field in a rather low liquid bridge, in which the azimuthal wave number after the onset of the primary instability is of $m = 3$. The thermal-flow field beyond the primary

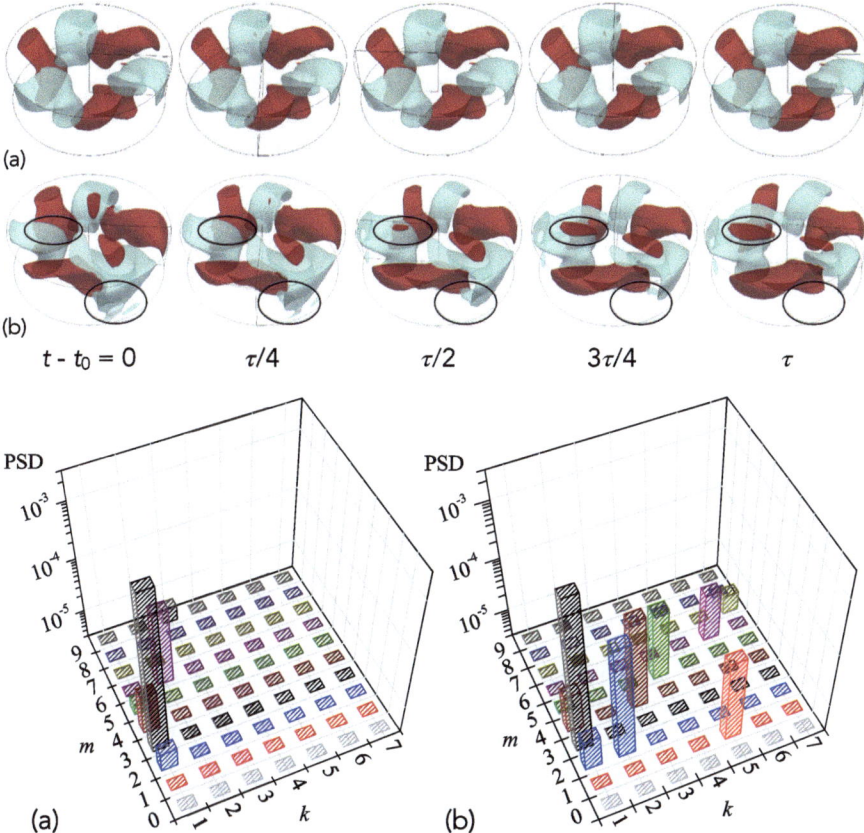

Fig. 71 Panel (1): Time series of temperature deviation $\hat{T}$ in liquid bridge of $(\Gamma, \text{Pr}) = (0.7, 4)$ in rotating frame of reference. Red and sky-blue surfaces correspond to the isosurfaces of $\hat{T} = \pm 0.06$, respectively. Panel (2): Power spectral density of azimuthal wave number $m(k)$ for kth component $\hat{T}(k)$ at $(r, z) = (1/\Gamma, 1/2)$. Common conditions for **a** and **b** are Re = 3000 < $\text{Re}_c^{(2)}$ and **b** 3500 > $\text{Re}_c^{(2)}$, respectively, where $\text{Re}_c^{(2)}$ represents the critical Reynolds number for secondary instability. Reprinted from Ogasawara et al. [131], with permission from the Japan Society for Mechanical Engineers. All rights reserved

instability but prior to the secondary instability is spatially and temporally periodic (Fig. 71(1-a)). The structure seems rigidly fixed in the rotating frame of reference [136–138]. After the onset of the secondary instability, their fundamental structure remains but perfect rigidity is broken: A rigid-body-like motion in the frame of reference is not formed anymore (see regions indicated by circles in Fig. 71(1-b)). They extract the variation of the thermal-flow structures before/after the secondary instability via the proper orthogonal decomposition (POD). In the oscillatory states before the secondary instability, the thermal-flow field consists of the primary component with the fundamental modal structure ($m = 3$ in this case) and those with higher harmonics of the primary components (Fig. 71(2-a)). After the onset of secondary instability, extra components emerge in the thermal-flow field (Fig. 71(2-b)); those components consist of the the fundamental azimuthal modal structures, which are different from higher harmonics of the primary flow field ($m = 2$ and 4 in this case). The onset condition for the secondary instability $Ma_c^{(2)}$ is determined by monitoring the development of the energy of the newly arisen components, which agrees well with the prediction through Floquet modes via linear stability analysis.

References

1. Wakatsuki T, Nishikawa W, Kobayashi R (2009) Preparation status of payload operations for the first experiment in JEM. Trans JSASS Space Tech Japan 7(ists26):Th_15–Th_20, https://doi.org/10.2322/tstj.7.Th_15
2. Onishi M, Sakai Y, Tomobe T, Nagashima T, Takada T (2010) Development of experiment equipment and racks of the JEM-PM (Japanese Experiment Module-Pressurized Module). IHI Eng Rev 43(1):6–15. https://www.ihi.co.jp/en/technology/review_library/2010/43_02.html, (Accessed 16 Feb 2025; 16:42:55)
3. JAXA (2014) Handbook for use of laboratory in Kibo. http://iss.jaxa.jp/kibo/library/fact/data/pmhandbook_en.pdf, (Accessed 16 Feb 2025; 16:42:55)
4. Yano T, Nishino K, Matsumoto S, Ueno I, Komiya A, Kamotani Y, Imaishi N (2018) Overview of "Dynamic Surf" project in Kibo—dynamic behavior of large-scale thermocapillary liquid ridges in microgravity. Int J Microgravity Sci Appl 35(1):350102. https://doi.org/10.15011/jasma.35.1.350102
5. Shin-Etsu Co, Ltd (2014) Silicone fluid KF-96 performance test results. https://www.shinetsusilicone-global.com/catalog/pdf/kf96_e.pdf, (Accessed 16 Feb 2025; 16:42:55)
6. Nishino K, Yano T, Kawamura H, Matsumoto S, Ueno I, Ermakov MK (2015) Instability of thermocapillary convection in long liquid bridges of high Prandtl number fluids in microgravity. J Cryst Growth 420:57–63. https://doi.org/10.1016/j.jcrysgro.2015.01.039
7. Yano T (2015) Study of instability and flow structure of marangoni convection in high prandtl number liquid bridges. dissertation. Ph.D. dissertation, Yokohama National University, http://hdl.handle.net/10131/9257, (in Japanese, Accessed 16 Feb 2025; 16:42:55)
8. Kawaji M, Ahmad W, DeJesus JM, Sutharshan B, Lorencez C, Ojha M (1993) Flow visualization of two-phase flows using photochromic dye activation method. Nucl Eng Des 420(1–2):343–355. https://doi.org/10.1016/0029-5493(93)90111-L
9. Kawaji M (1998) Two-phase flow measurement using a photochromic dye activation technique. Nucl Eng Des 184(2–3):379–392. https://doi.org/10.1016/S0029-5493(98)00210-6
10. Kanashima Y, Nishino K, Yoda S (2005) Effect of g-jitter on the thermocapillary convection experiment in ISS. Microgravity Sci Technol 16(1–4):285–289. https://doi.org/10.1007/BF02945992

11. Takeda Y (1991) Development of an ultrasound velocity profile monitor. Nucl Eng Des 126(2):277–284. https://doi.org/10.1016/0029-5493(91)90117-Z
12. Kawamura H, Ueno I (2006) Review on thermocapillary convection in a half-zone liquid bridge with high Pr fluid: Onset of oscillatory convection, transition of flow regimes, and particle accumulation structure. In: Savino R (ed) Surface tension-driven flows and applications. Research Signpost, Kerala, India, pp 1–24
13. Kawamura H, Nishino K, Matsumoto S, Ueno I (2010) Space experiment of Marangoni convection on international space station. In: International heat transfer conference, Washington, DC, USA, 8–13 Aug 2010. https://doi.org/10.1115/IHTC14-23346
14. Kawamura H, Nishino K, Matsumoto S, Ueno I (2012) Report on microgravity experiments of Marangoni convection aboard International Space Station. J Heat Transfer 134(3):031005. https://doi.org/10.1115/1.4005145
15. Yano T, Nishino K, Kawamura H, Ueno I, Matsumoto S, Ohnishi M, Sakurai M (2011) Space experiment on the instability of Marangoni convection in large liquid bridge–MEIS-4: effect of Prandtl number. J Phys: Conf Ser 327(1):012029. https://doi.org/10.1088/1742-6596/327/1/012029
16. Yano T, Nishino K, Matsumoto S, Ueno I, Komiya A, Kamotani Y, Imaishi N (2018) Report on microgravity experiments of dynamic surface deformation effects on Marangoni instability in high-Prandtl-number liquid bridges. Microgravity Sci Technol 30(5):599–610. https://doi.org/10.1007/s12217-018-9614-9
17. Preisser F, Schwabe D, Scharmann A (1983) Steady and oscillatory thermocapillary convection in liquid columns with free cylindrical surface. J Fluid Mech 126:545–567. https://doi.org/10.1017/S0022112083000324
18. Leypoldt J, Kuhlmann HC, Rath HJ (2000) Three-dimensional numerical simulation of thermocapillary flows in cylindrical liquid bridges. J Fluid Mech 414:285–314. https://doi.org/10.1017/S0022112000008570
19. Ueno I, Tanaka S, Kawamura H (2003) Oscillatory and chaotic thermocapillary convection in a half-zone liquid bridge. Phys Fluids 15(2):408–416. https://doi.org/10.1063/1.1531993
20. Kawamura H, Ono Y, Ueno I (2001) Transition and modal structure of oscillatory Marangoni convection in liquid bridge. Trans JSME, Ser B 67(658):1466–1473. https://doi.org/10.1299/kikaib.67.1466. (in Japanese)
21. Lappa M, Savino R, Monti R (2000) Influence of buoyancy forces on Marangoni flow instabilities in liquid bridges. Int J Numer Meth Heat Fluid Flow 10(7):721–749. https://doi.org/10.1108/09615530010350444
22. Fujimura K (2013) Linear and weakly nonlinear stability of Marangoni convection in a liquid bridge. J Phys Soc Jpn 82(7):074401. https://doi.org/10.7566/JPSJ.82.074401
23. Romanò F, Kuhlmann HC, Ishimura M, Ueno I (2017) Limit cycles for the motion of finite-size particles in axisymmetric thermocapillary flows in liquid bridges. Phys Fluid 29(9):093303. https://doi.org/10.1063/1.5002135
24. Schwabe D, Tanaka S, Mizev A, Kawamura H (2006) Particle accumulation structure in time-dependent thermocapillary flow in a liquid bridge under microgravity. Microgravity Sci Technol 18(3–4):117–127. https://doi.org/10.1007/BF02870393
25. Nishimura M, Ueno I, Nishino K, Kawamura H (2005) 3D PTV measurement of oscillatory thermocapillary convection in half-zone liquid bridge. Exp Fluids 38(3):285–290. https://doi.org/10.1007/s00348-004-0885-0
26. Wang J, Wu D, Duan L, Kang Q (2017) Ground experiment on the instability of buoyant-thermocapillary convection in large-scale liquid bridge with large Prandtl number. Int J Heat Mass Transf 108:2107–2119. https://doi.org/10.1016/j.ijheatmasstransfer.2016.12.095
27. Tanaka S, Kawamura H, Ueno I, Schwabe D (2006) Flow structure and dynamic particle accumulation in thermocapillary convection in a liquid bridge. Phys Fluids 18(6):067103. https://doi.org/10.1063/1.2208289
28. Schwabe D, Hintz P, Frank S (1996) New features of thermocapillary convection in floating zones revealed by tracer particle accumulation structures (PAS). Microgravity Sci Technol 9(3):163–168

29. Schwabe D, Frank S (1999) Particle accumulation structure (PAS) in the toroidal thermocapillary vortex of a floating zone-Model for a step in planet-formation? Adv Space Res 23(7):1191–1196. https://doi.org/10.1016/S0273-1177(99)00181-7
30. Matsumoto S, Nishino K, Ueno I, Yano T, Kawamura H (2014) Marangoni Experiment in Space. Int J Microgravity Sci Appl 31(Suppl2014):S51–S79. https://repository.exst.jaxa.jp/dspace/handle/a-is/16349, (in Japanese, Accessed 16 Feb 2025; 16:42:55)
31. Sakata T, Terasaki S, Saito H, Fujimoto S, Ueno I, Yano T, Nishino K, Kamotani Y, Matsumoto S (2022) Coherent structures of m = 1 by low-Stokes-number particles suspended in a half-zone liquid bridge of high aspect ratio: Microgravity and terrestrial experiments. Phys Rev Fluids 7(1):014005. https://doi.org/10.1103/PhysRevFluids.7.014005
32. Fukuda Y, Ogasawara T, Fujimoto S, Eguchi T, Motegi K, Ueno I (2021) Thermal-flow patterns of $m = 1$ in thermocapillary liquid bridges of high aspect ratio with free-surface heat transfer. Int J Heat Mass Transf 173:121196. https://doi.org/10.1016/j.ijheatmasstransfer.2021.121196
33. Schwabe D, Scharmann A (1979) Some evidence for the existence and magnitude of a critical Marangoni number for the onset of oscillatory flow in crystal growth melts. J Cryst Growth 46(1):125–131. https://doi.org/10.1016/0022-0248(79)90119-2
34. Velten R, Schwabe D, Scharmann A (1991) The periodic instability of thermocapillary convection in cylindrical liquid bridges. Phys Fluids A 3(2):267–279. https://doi.org/10.1063/1.858135
35. Kuhlmann HC (1999) Thermocapillary convection in models of crystal growth. Springer, Heidelberg. https://doi.org/10.1007/BFb0109562
36. Frank S, Schwabe D (1997) Temporal and spatial elements of thermocapillary convection in floating zones. Exp Fluids 23:234–251. https://doi.org/10.1007/s003480050107
37. Melnikov DE, Shevtsova VM, Legros JC (2004) Onset of temporal aperiodicity in high Prandtl number liquid bridge under terrestrial conditions. Phys Fluids 16(5):1746–1757. https://doi.org/10.1063/1.1699135
38. Yano T, Mabuchi Y, Yamaguchi M, Nishino K (2022) Internal flow structure and dynamic free-surface deformation of oscillatory thermocapillary convection in a high-Prandtl-number liquid bridge. Exp Fluids 63:95. https://doi.org/10.1007/s00348-022-03453-2
39. Schwabe D (2005) Hydrodynamic instabilities under microgravity in a differentially heated long liquid bridge with aspect ratio near the Rayleigh-limit: Experimental results. Adv Space Res 36(1):36–42. https://doi.org/10.1016/j.asr.2005.02.085
40. Schwabe D (2005) Hydrothermal waves in a liquid bridge with aspect ratio near the Rayleigh limit under microgravity. Phys Fluids 17(11):112104. https://doi.org/10.1063/1.2135805
41. Hu WR, Shu JZ, Zhou R, Tang ZM (1994) Influence of liquid bridge volume on the onset of oscillation in floating zone convection I experiments. J Cryst Growth 142(3–4):379–384. https://doi.org/10.1016/0022-0248(94)90349-2
42. Masud J, Kamotani Y, Ostrach S (1997) Oscillatory thermocapillary flow in cylindrical columns of high Prandtl number fluids. J Thermophys Heat Transfer 11(1):105–111. https://doi.org/10.2514/2.6207
43. Antar BN, Nuotio-Antar VS (1993) Fundamentals of low gravity fluid dynamics and heat transfer. CRC Press, Boca Raton, Florida. https://doi.org/10.1201/9781351072182
44. Slobozhanin LA, Perales JM (1993) Stability of liquid bridges between equal disks in an axial gravity field. Phys Fluids A: Fluid Dyn 5(6):1305–1314. https://doi.org/10.1063/1.858567
45. Ferziger JH, Peric M (2002) Computational methods for fluid dynamics, 3rd edn. Springer, Heidelberg. https://doi.org/10.1007/978-3-642-56026-2
46. Nishino K, Kato H, Torii K (2000) Stereo imaging for simultaneous measurement of size and velocity of particles in dispersed two-phase flow. Meas Sci Technol 11(6):633–645. https://doi.org/10.1088/0957-0233/11/6/306
47. Yano T, Nishino K, Kawamura H, Ueno I, Matsumoto S (2012) Effect of liquid-bridge shape on the instability of Marangoni convection in space experiment. In: Proceedings of the 15th international symposium on flow visualization, Minsk, Belarus, 25–28 June 2012, http://www.itmo.by/en/conferences/abstracts/isfv_15/, (Accessed 16 Feb 2025; 16:42:55)

48. Kuhlmann HC, Nienhüser C, Rath HJ, Yoda S (2002) Influence of the volume of liquid bridge on the onset of three-dimensional flow in thermocapillary liquid bridges. Adv Space Res 29(4):639–644. https://doi.org/10.1016/S0273-1177(01)00665-2
49. Shevtsova VM, Mojahed M, Melnikov DE, Legros JC (2003) The choice of the critical mode of hydrothermal instability in liquid bridge. In: Narayanan R, Schwabe D (eds) Interfacial fluid dynamics and transport process. Springer, Berlin Heidelberg, pp 241–262. https://doi.org/10.1007/978-3-540-45095-5_12
50. Shevtsova V, Mialdun A, Kawamura H, Ueno I, Nishino K, Lappa M (2011) Onset of hydrothermal instability in liquid bridge: experimental benchmark. Fluid Dyn Mater Process 7(1):1–27. https://doi.org/10.3970/fdmp.2011.007.001
51. Sakurai M, Ohishi N, Hirata A (2007) Oscillatory thermocapillary convection in a liquid bridge: part 1–1g experiments. J Cryst Growth 308(2):352–359. https://doi.org/10.1016/j.jcrysgro.2007.07.026
52. Kang Q, Wu D, Duan L, Hu L, Wang J, Zhang P, Hu W (2019) The effects of geometry and heating rate on thermocapillary convection in the liquid bridge. J Fluid Mech 881:951–982. https://doi.org/10.1017/jfm.2019.757
53. Kang Q, Wu D, Duan L, Zhang J, Zhou B, Wang J, Han Z, Hu L, Hu W (2020) Space experimental study on wave modes under instability of thermocapillary convection in liquid bridges on Tiangong-2. Phys Fuids 32(3):034107. https://doi.org/10.1063/1.5143219
54. Kuhlmann HC, Rath HJ (1993) Hydrodynamic instabilities in cylindrical thermocapillary liquid bridge. J Fluid Mech 247:247–274. https://doi.org/10.1017/S0022112093000461
55. Wanschura M, Shevtsova VM, Kuhlmann HC, Rath HJ (1995) Convective instability mechanisms in thermocapillary liquid bridges. Phys Fluids 7(5):912–925. https://doi.org/10.1063/1.868567
56. Rupp R, Müller G, Neumann G (1989) Three-dimensional time dipendent modelling of the marangoni convection in zone melting configurations for GaAs. J Cryst Growth 97(1):34–41. https://doi.org/10.1016/0022-0248(89)90244-3
57. Chen G, Lizée A, Roux B (1997) Bifurcation analysis of the thermocapillary convection in cylindrical liquid bridges. J Cryst Growth 180(3–4):638–647. https://doi.org/10.1016/S0022-0248(97)00259-5
58. Levenstam M, Amberg G, Winkler C (2001) Instabilities of thermocapillary convection in a half-zone at intermediate Prandtl numbers. Phys Fluids 13(4):807–816. https://doi.org/10.1063/1.1337063
59. Hibiya T, Nagafuchi K, Shiratori S, Yamane N, Ozawa S (2008) Attempt to study Marangoni flow of low-Pr-number fluids using a liquid bridge of silver. Adv Space Res 41(12):2107–2111. https://doi.org/10.1016/j.asr.2007.04.106
60. Jurisch M, Loser W (1990) Analysis of periodic non-rotational W striations in Mo single crystals due to nonsteady thermocapillary convection. J Cryst Growth 102(1–2):214–222. https://doi.org/10.1016/0022-0248(90)90904-Y
61. Han JH, Sun ZW, Dai LR, Xie JC, Hu WR (1996) Experiment on the thermocapillary convection of a mercury liquid bridge in a floating half zone. J Cryst Growth 169(1):129–135. https://doi.org/10.1016/0022-0248(96)00263-1
62. Cröll A, Kaiser Th, Schweizer M, Danilewsky AN, Lauer S, Tegetmeier A, Benz KW (1998) Floating-zone and floating-solution-zone growth of GaSb under microgravity. J Cryst Growth 191(3):365–376. https://doi.org/10.1016/S0022-0248(98)00215-2
63. Schwabe D, Velten R, Scharmann A (1990) The instability of surface tension driven flow in models for floating zones under normal and reduced gravity. J Cryst Growth 99(2):1258–1264. https://doi.org/10.1016/S0022-0248(08)80117-0
64. Simic-Stefani S, Kawaji M, Yoda S (2006) Onset of oscillatory thermocapillary convection in acetone liquid bridges: the effect of evaporation. Int J Heat Mass Transf 49(17–18):3167–3179. https://doi.org/10.1016/j.ijheatmasstransfer.2006.01.042
65. Watanabe T, Melnikov DE, Matsugase T, Shevtsova V, Ueno I (2014) The stability of a thermocapillary-buoyant flow in a liquid bridge with heat transfer through the interface. Microgravity Sci Technol 26(1):17–28. https://doi.org/10.1007/s12217-014-9367-z

66. Kamotani Y, Ostrach S, Vargas M (1984) Oscillatory thermocapillary convection in a simulated floating-zone configuration. J Cryst Growth 66(1):83–90. https://doi.org/10.1016/0022-0248(84)90079-4
67. Shevtsova VM, Mojahed M, Legros JC (1999) The loss of stability in ground based experiments in liquid bridges. Acta Astronaut 44(7–12):625–634. https://doi.org/10.1016/S0094-5765(99)00071-5
68. Hayashida H, Matsumoto S, Natsui H, Yoda S, Imaishi N (2004) Experimental study on transition to oscillatory thermocapillary flow in the half-zone liquid bridge of low Prandtl number fluid. J Jpn Soc Microgravity Appl 21(4):281–284. https://doi.org/10.15011/jasma.21.4.281, (in Japanese)
69. Melnikov DE, Shevtsova V, Yano T, Nishino K (2015) Modeling of the experiments on the Marangoni convection in liquid bridges in weightlessness for a wide range of aspect ratio. Int J Heat Mass Transf 87:119–127. https://doi.org/10.1016/j.ijheatmasstransfer.2015.03.016
70. Monti R, Fortezza R, Mannara G (1988) Results of the TEXUS 14-B flights experiment on a floating zone. First approach towards telescience in fluid science. Acta Astronaut 17(11–12):1221–1228. https://doi.org/10.1016/0094-5765(88)90011-2
71. Monti R, Fortezza R (1991) The scientific results of the experiment on oscillatory Marangoni flow performed in telescience on TEXUS 23. Microgravity Q 1(3):163–171
72. Hirata A, Nishizawa S, Imaishi N, Yasuhiro S, Yoda S, Kawasaki K (1993) Oscillatory Marangoni convection in a liquid bridge under microgravity by utilizing TR-IA sounding rocket. J Jpn Soc Microgravity Appl 10(4):241–250. https://doi.org/10.15011/jasma.10.4.241
73. Kawamura H, Saita K, Nishino K, Yamamoto M, Yoda S, Nakamura T, Morita TS, Kawasaki K, Tamaoki H (1997) Three-dimensional measurement of Marangoni convection in a liquid bridge under microgravity conditions in the TR-IA-4 sounding rocket. J Jpn Soc Microgravity Appl 14(1):34–41. https://doi.org/10.15011/jasma.14.1_34
74. Carotenuto L, Castagnolo D, Albanese C, Monti R (1998) Instability of thermocapillary convection in liquid bridges. Phys Fluids 10(3):555–565. https://doi.org/10.1063/1.869583
75. Savino R, Monti R (1996) Oscillatory Marangoni convection in cylindrical liquid bridges. Phys Fluids 8(11):2906–2922. https://doi.org/10.1063/1.869070
76. Altenbuchner L, Ettl J, Hörschgen M, Jung W, Kirchhartz R, Stamminger A, Turner P (2012) MORABA—overview on DLR's mobile rocket base and projects. In: Proceedings of the SpaceOps 2012 conference, Stockholm, Sweden, 11–15 June 2012. https://doi.org/10.2514/6.2012-1272497
77. Albanese C, Carotenuto L, Castagnolo D, Ceglia E, Monti R (1995) An investigation on the "Onset" of oscillatory marangoni flow. Adv Space Res 16(7):87–94. https://doi.org/10.1016/0273-1177(95)00140-A
78. Kamotani Y, Ostrach S (1998) Theoretical analysis of thermocapillary flow in cylindrical columns of high Prandtl number fluids. J Heat Transfer 120(3):758–764. https://doi.org/10.1115/1.2824346
79. Nishino K, Kasagi N, Hirata M (1989) Three-dimensional particle tracking velocimetry based on automated digital image processing. J Fluid Eng 111(4):384–391. https://doi.org/10.1115/1.3243657
80. Yano T, Nishino K, Kawamura H, Ueno I, Matsumoto S, Ohnishi M, Sakurai M (2012) 3-D PTV measurement of Marangoni convection in liquid bridge in space experiment. Exp Fluids 53(1):9–20. https://doi.org/10.1007/s00348-011-1136-9
81. Yano T, Nishino K (2019) Flow visualization of axisymmetric steady Marangoni convection in high-Prandtl-number liquid bridges in microgravity. Int J Microgravity Sci Appl 36(2):360202–1–8. https://doi.org/10.15011/jasma.36.2.360202
82. Marquardt DW (1963) An algorithm for least-squares estimation of nonlinear parameters. J Soc Ind Appl Math 11(2):431–441. https://doi.org/10.1137/0111030
83. Nishino K, Kawamura H, Emori T, Iijima Y, Kawasaki K, Makino K, Yoda S, Kawasaki H (1998) Simultaneous observation of three-dimensional fluid flow and surface temperature of unsteady Marangoni convection in a liquid bridge. J Jpn Soc Microgravity Appl 15(3):158–164. https://doi.org/10.15011/jasma.15.3.158, (in Japanese)

84. Villers D, Platten JK (1992) Coupled buoyancy and Marangoni convection in acetone: experiments and comparison with numerical simulations. J Fluid Mech 234:487–510. https://doi.org/10.1017/S0022112092000880
85. Chernatinsky VI, Birikh RV, Briskman VA, Schwabe D (2002) Thermocapillary flows in long liquid bridges under microgravity. Adv Space Res 29(4):619–624. https://doi.org/10.1016/S0273-1177(01)00652-4
86. Yano T, Nishino K, Kawamura H, Ueno I, Matsumoto S (2015) Instability and associated roll structure of Marangoni convection in high Prandtl number liquid bridge with large aspect ratio. Phys Fluids 27(2):024108. https://doi.org/10.1063/1.4908042
87. Popovich AT, Hummel RL (1967) A new method for non-disturbing turbulent flow measurements very close to a wall. Chem Eng Sci 22(1):21–25. https://doi.org/10.1016/0009-2509(67)80100-3
88. Nishino K (2003) A review of the Marangoni convection experiment in TR-IA #6. J Jpn Soc Microgravity Appl 20(3):232–236. https://doi.org/10.15011/jasma.20.3.232, (in Japanese)
89. Suzuki R (2010) Surface velocity measurement of oscillatory marangoni flow in a liquid bridge by photochromic method. Master's thesis, Yokohama National University (in Japanese)
90. Kamotani Y, Wang L, Hatta S, Wang A, Yoda S (2003) Free surface heat loss effect on oscillatory thermocapillary flow in liquid bridges of high Prandtl number fluids. Int J Heat Mass Transf 46(17):3211–3220. https://doi.org/10.1016/S0017-9310(03)00098-X
91. Yasnou V, Gaponenko Y, Mialdun A, Shevtsova V (2018) Influence of a coaxial gas flow on the evolution of oscillatory states in a liquid bridge. Int J Heat Mass Transf 123:747–759. https://doi.org/10.1016/j.ijheatmasstransfer.2018.03.016
92. Xun B, Li K, Hu WR, Imaishi N (2011) Effect of interfacial heat exchange on thermocapillary flow in a cylindrical liquid bridge in microgravity. Int J Heat Mass Transf 54(9–10):1698–1705. https://doi.org/10.1016/j.ijheatmasstransfer.2011.01.026
93. Fujimoto S, Ogasawara T, Ota A, Motegi K, Ueno I (2019) Effect of heat loss on hydrothermal wave instability in half-zone liquid bridges of high Prandtl number fluid. Int J Microgravity Appl 36(2):360204. https://doi.org/10.15011/jasma.36.2.360204
94. Melnikov DE, Shevtsova VM (2014) The effect of ambient temperature on the stability of thermocapillary flow in liquid column. Int J Heat Mass Transf 74:185–195. https://doi.org/10.1016/j.ijheatmasstransfer.2014.02.058
95. Romanò F, Kuhlmann HC (2019) Heat transfer across the free surface of a thermocapillary liquid bridge. Tech Mech 39(1):72–84. https://doi.org/10.24352/UB.OVGU-2019-008
96. Yano T, Nishino K, Ueno I, Matsumoto S, Kamotani Y (2017) Sensitivity of hydrothermal wave instability of Marangoni convection to the interfacial heat transfer in long liquid bridges of high Prandtl number fluids. Phys Fluids 29(4):044105. https://doi.org/10.1063/1.4979721
97. Kamotani Y, Wang L, Hatta S, Selver R, Yoda S (2001) Effect of free surface heat transfer on onset of oscillatory thermocapillary flow of high Prandtl number fluid. J Jpn Soc Microgravity Appl 18(4):283–288. https://doi.org/10.15011/jasma.18.4.283
98. Wang A, Kamotani Y, Yoda S (2007) Oscillatory thermocapillary flow in liquid bridge of high Prandtl number fluid with free surface heat gain. Int J Heat Mass Transf 50(21–22):4195–4205. https://doi.org/10.1016/j.ijheatmasstransfer.2007.02.035
99. Shitomi N, Yano T, Nishino K (2019) Effect of radiative heat transfer on the flow structure of thermocapillary convection in high-Prandtl-number liquid bridges in microgravity. Int J Heat Mass Transf 133:405–415. https://doi.org/10.1016/j.ijheatmasstransfer.2018.12.119
100. Yano T, Nishino K (2020) Numerical study on the effects of convective and radiative heat transfer on thermocapillary convection in a high-Prandtl-number liquid bridge in weightlessness. Adv Space Res 66:2047–2061. https://doi.org/10.1016/j.asr.2020.07.009
101. Ueno I, Kawasaki H, Watanabe T, Motegi K, Kaneko T (2014) Hydrothermal-wave instability and resultant flow patterns induced by thermocapillary effect in a half-zone liquid bridge of high aspect ratio. In: Proceedings of the international heat transfer conference 15, Kyoto, Japan, 10–15 Aug 2014. https://doi.org/10.1615/IHTC15.fcv.009489
102. Kawamura H, Tagaya E, Hoshino Y (2007) A consideration on the relation between the oscillatory thermocapillary flow in a liquid bridge and the hydrothermal wave in a thin liquid layer. Int

J Heat Mass Transf 50(7–8):1263–1268. https://doi.org/10.1016/j.ijheatmasstransfer.2006. 09.035
103. Xu JJ, Davis SH (1984) Convective thermocapillary instabilities in liquid bridges. Phys Fluids 27(5):1102–1107. https://doi.org/10.1063/1.864756
104. Ryzhkov II (2011) Thermocapillary instabilities in liquid bridges revisited. Phys Fluids 23(8):082103 1–6. https://doi.org/10.1063/1.3627150
105. Ermakov MK, Ermakova MS (2004) Linear-stability analysis of thermocapillary convection in liquid bridges with highly deformed free surface. J Cryst Growth 266(1–3):160–166. https://doi.org/10.1016/j.jcrysgro.2004.02.041
106. Ermakov MK (2015) Private communication
107. Landau LD, Lifshitz EM (1987) Fluid mechanics, 2nd edn. Course of theoretical physics, vol 6. Pergamon Press, Oxford. https://doi.org/10.1016/C2013-0-03799-1
108. Shevtsova V, Gaponenko YA, Nepomnyashchy A (2013) Thermocapillary flow regimes and instability caused by a gas stream along the interface. J Fluid Mech 714:644–670. https://doi.org/10.1017/jfm.2012.519
109. Kuhlmann HC, Nienhüser Ch (2002) Dynamic free-surface deformations in thermocapillary liquid bridges. Fluid Dyn Res 31(2):103–127. https://doi.org/10.1016/S0169-5983(02)00090-4
110. Montanero JM, Ferrera C, Shevtsova VM (2008) Experimental study of the free surface deformation due to thermal convection in liquid bridges. Exp Fluids 45(6):1087–1101. https://doi.org/10.1007/s00348-008-0529-x
111. Yang S, Liang R, He J (2016) Oscillating characteristic of free surface from stability to instability of thermocapillary convection for high Prandtl number fluids. Int J Heat Fluid Flow 61:298–308. https://doi.org/10.1016/j.ijheatfluidflow.2016.05.001
112. Carrión LM, Herrada MA, Montanero JM (2020) Influence of the dynamical free surface deformation on the stability of thermal convection in high-Prandtl-number liquid bridges. Int J Heat Mass Transf 146:118831. https://doi.org/10.1016/j.ijheatmasstransfer.2019.118831
113. Shu JZ, Yao YL, Zhou R, Hu WR (1994) Experimental study of free surface oscillations of a liquid bridge by optical diagnostics. Microgravity Sci Technol 7(2):83–89
114. Ferrera C, Montanero JM, Mialdun A, Shevtsova VM, Cabezas MG (2008) A new experimental technique for measuring the dynamical free surface deformation in liquid bridges due to thermal convection. Meas Sci Technol 19(1):015410. https://doi.org/10.1088/0957-0233/19/1/015410
115. ANSI (2009) ANSI/ASA S1.11-2004 (R2009) Octave-band and fractional-octave-band analog and digital filters. https://webstore.ansi.org/standards/asa/ansiasas1112004r2009 (Accessed 16 Feb 2025; 16:42:55)
116. Penley NJ, Schafer CP, Bartoe JDF (2002) The international space station as a microgravity research platform. Acta Astronaut 50(11):691–696. https://doi.org/10.1016/S0094-5765(02)00003-6
117. Barger LK, Flynn-Evans EE, Kubey A, Walsh L, Ronda JM, Wang W, Wright KP, Czeisler CA (2014) Prevalence of sleep deficiency and hypnotic use among astronauts before, during and after spaceflight: an observational study. Lancet Neurol 13(9):904–912. https://doi.org/10.1016/S1474-4422(14)70122-X
118. Goto M, Murakami K, Ookuma H (2011) Experiment environment and habitat vibration in "KIBO". J Jpn Soc Microgravity Appl 28(1):8–12. https://doi.org/10.15011/jasma.28.1.8 (in Japanese)
119. Sanz A, Diez JL (1989) Non-axisymmetric oscillations of liquid bridges. J Fluid Mech 205:503–521. https://doi.org/10.1017/S0022112089002120
120. Ichikawa N, Kawaji M, Misawa M, Psofogiannakis G (2003) Resonance behavior of a liquid bridge caused by horizontal vibrations. J Jpn Soc Microgravity Appl 20(4):292–300. https://doi.org/10.15011/jasma.20.4.292
121. Shiomi J, Kudo M, Ueno I, Kawamura H, Amberg G (2003) Feedback control of oscillatory thermocapillary convection in a half-zone liquid bridge. J Fluid Mech 496:193–211. https://doi.org/10.1017/S0022112003006323

122. Chun CH, Wuest W (1978) A micro-gravity simulation of the Marangoni convection. Acta Astronaut 5(9):681–686. https://doi.org/10.1016/0094-5765(78)90047-4
123. Cao ZH, Xie JC, Tang ZM, Hu WR (1991) The influence of buoyancy on the onset of oscillatory convection in a half floating zone. Adv Space Res 11(7):163–166. https://doi.org/10.1016/0273-1177(91)90276-P
124. Petrov V, Schatz M, Muehlner KA, VanHook SJ, McCormick WD, Swift JB, Swinney HL (1996) Nonlinear control of remote unstable states in a liquid bridge convection experiment. Phys Rev Lett 77(18):3779–3782. https://doi.org/10.1103/PhysRevLett.77.3779
125. Hirata A, Nishizawa S, Sakurai M (1997) Experimental results of oscillatory Marangoni convection in a liquid bridge under normal gravity. J Jpn Soc Microgravity Appl 14(2):122–129. https://doi.org/10.15011/jasma.14.2.122
126. Shevtsova VM, Melnikov DE, Legros JC (2003) Multistability of oscillatory thermocapillary convection in a liquid bridge. Phys Rev E 68(6):066311. https://doi.org/10.1103/PhysRevE.68.066311
127. Sato F, Ueno I, Kawamura H, Nishino K, Matsumoto S, Ohnishi M, Sakurai M (2013) Hydrothermal wave instability in a high-aspect-ratio liquid bridge of Pr > 200. Microgravity Sci Technol 25(1):43–58. https://doi.org/10.1007/s12217-012-9332-7
128. Matsugase T, Ueno I, Nishino K, Ohnishi M, Sakurai M, Matsumoto S, Kawamura H (2015) Transition to chaotic thermocapillary convection in a half zone liquid bridge. Int J Heat Mass Transf 89:903–912. https://doi.org/10.1016/j.ijheatmasstransfer.2015.05.041
129. Wang J, Wu D, Duan L, Kang Q (2020) Transition to chaos of buoyant-thermocapillary convection in large-scale liquid bridges. Microgravity Sci Technol 32(2):217–227. https://doi.org/10.1007/s12217-019-09770-2
130. Theiler J, Eubank S, Longtin A, Galdrikian B, Farmer JD (1992) Testing for nonlinearity in time series: the method of surrogate data. Physica D 58(1–4):77–94. https://doi.org/10.1016/0167-2789(92)90102-S
131. Ogasawara T, Motegi K, Hori T, Ueno I (2019) Secondary instability induced by thermocapillary effect in half-zone liquid bridge of high Prandtl number fluid. Mech Eng Lett 5(19):1–10. https://doi.org/10.1299/mel.19-00014
132. Levenstam M, Amberg G (1995) Hydrodynamical instabilities of thermocapillary flow in a half-zone. J Fluid Mech 297:357–372. https://doi.org/10.1017/S0022112095003132
133. Imaishi N, Yasuhiro S, Akiyama Y, Yoda S (2001) Numerical simulation of oscillatory Marangoni flow in half-zone liquid bridge of low Prandtl number fluid. J Cryst Growth 230(1–2):164–171. https://doi.org/10.1016/S0022-0248(01)01332-X
134. Takagi K, Otaka M, Natsui H, Arai T, Yoda S, Yuan Z, Mukai K, Yasuhiro S, Imaishi N (2001) Experimental study on transition to oscillatory thermocapillary flow in a low Prandtl number liquid bridge. J Cryst Growth 233(1–2):399–407. https://doi.org/10.1016/S0022-0248(01)01538-X
135. Motegi K, Fujimura K, Ueno I (2017) Floquet analysis of spatially periodic thermocapillary convection in a low-Prandtl-number liquid bridge. Phys Fluids 29(7):074104. https://doi.org/10.1063/1.4993466
136. Kuhlmann HC, Mukin RV, Sano T, Ueno I (2014) Structure and dynamics of particle-accumulation in thermocapillary liquid bridges. Fluid Dyn Res 46(4):041421. https://doi.org/10.1088/0169-5983/46/4/041421
137. Gotoda M, Toyama A, Ishimura M, Sano T, Suzuki M, Kaneko T, Ueno I (2019) Experimental study of coherent structures of finite-size particles in thermocapillary liquid bridges. Phys Rev Fluids 4:094301. https://doi.org/10.1103/PhysRevFluids.4.094301
138. Oba T, Toyama A, Hori T, Ueno I (2019) Experimental study on behaviors of low-Stokes number particles in weakly chaotic structures induced by thermocapillary effect within a closed system with a free surface. Phys Rev Fluids 4(10):104002. https://doi.org/10.1103/PhysRevFluids.4.104002

Open Access This chapter is licensed under the terms of the Creative Commons Attribution 4.0 International License (http://creativecommons.org/licenses/by/4.0/), which permits use, sharing, adaptation, distribution and reproduction in any medium or format, as long as you give appropriate credit to the original author(s) and the source, provide a link to the Creative Commons license and indicate if changes were made.

The images or other third party material in this chapter are included in the chapter's Creative Commons license, unless indicated otherwise in a credit line to the material. If material is not included in the chapter's Creative Commons license and your intended use is not permitted by statutory regulation or exceeds the permitted use, you will need to obtain permission directly from the copyright holder.

Surface-Tension Related Flows in Microgravity and Microscale: Liquid Bridges

Ichiro Ueno, Mitsuru Ohnishi, Satoshi Matsumoto, and Suguru Uemura

Abstract This chapter introduces examples of thermocapillary-driven flows and resultant phenomena in the liquid bridges under microgravity conditions and/or in microscale in spatial dimensions. In the first section, a unique phenomenon by tiny particles suspended in the half-zone liquid bridges, known as 'particle accumulation structures (PASs)' is introduced. The second section describes the effect of thermocapillary-flow on the pinching of the liquid bridges. In the third section, an application of the liquid-bridge pinching for the formation of nano-liter droplet is presented.

1 Experimental Study on Coherent Structures by Particles Suspended in Half-Zone Thermocapillary Liquid Bridges

Ichiro Ueno

1.1 Introduction

The accumulation of suspended particles in the thermocapillary-driven convection in the half-zone liquid bridges was firstly found by [65]. This unique phenomenon, known as particle accumulation structure (PAS) after [65], occurs in the traveling-

I. Ueno (✉)
Tokyo University of Science, Chiba, Japan
e-mail: ich@rs.tus.ac.jp

M. Ohnishi
Japan Aerospace Exploration Agency, Tokyo, Japan
e-mail: ohnishi.mitsuru@jaxa.jp

S. Matsumoto
Japan Aerospace Exploration Agency, Ibaraki, Japan
e-mail: matsumoto.satoshi@jaxa.jp

S. Uemura
Hokkaido University, Hokkaido, Japan
e-mail: uem@eng.hokudai.ac.jp

© The Author(s) 2025
H. Kawamura et al. (eds.), *Thermocapillary Convection in Microgravity*,
Fluid Mechanics and Its Applications 139,
https://doi.org/10.1007/978-981-96-2991-6_7

type oscillatory convection in the liquid bridges of high Prandtl-number fluids. Here the Prandtl number is described as $\mathrm{Pr} = \nu/\kappa$, where ν and κ are the kinematic viscosity and thermal diffusivity of the test liquid, respectively. High-Pr condition is essential for so-called hydrothermal wave (HTW) instability in geometries of thin liquid films [73] and liquid bridges [89]. Tanaka et al. [77] demonstrated that the particles accumulate to form a closed structure in the liquid bridge, which exhibits spatial structure with integer wave number in azimuthal direction [66, 77]. The PAS apparently rotates without varying its shape when observed from above (Fig. 1). That is, the PAS is a coherently formed structure, or the coherent structure [59], realized in the liquid bridge in the rotating frame of reference. The wave number m corresponds to that raised in the thermal-flow field induced in the high-Prandtl-number liquid bridges of finite height [25, 34, 47, 55, 82, 88]. This azimuthal wave number of the symmetric structure about the axis depends on the liquid-bridge shapes defined by the aspect ratio $\Gamma = H/R$ [55, 82, 88] and the volume ratio $V/V_0 = V/(\pi R^2 H)$ [94], where H and R are the distance between the disks and the disk radius, respectively, and V is the volume of the liquid bridge. Here V_0 indicates the volume of cylinder, of which the end disks correspond to the end surfaces. The PAS has the same azimuthal wave number as the oscillatory convection by the hydrothermal wave instability [15, 49, 66, 77, 80, 90]. The PAS of smaller m is formed in the liquid bridges of higher Γ [66, 77]. After finding of the PAS by Schwabe et al. [65], the PASs of moderate m ($2 \leq m \leq 5$) have been hitherto investigated actively by experiments [1, 13, 15, 38, 45, 49, 77, 80, 83] and by numerical simulations [6, 27, 32, 33, 36, 37, 40–42, 58, 59, 67, 76]. It is worth noting that the traveling-wave-type convection is one of the thermal-flow states realized in the three-dimensional time-dependent ('oscillatory') convection after the onset of the primary instability through an increase in the intensity of the thermocapillary effect [34, 55, 82, 88]. Two different types of oscillatory convection [34, 82] exist before the secondary instability [50]; the standing-wave-type convection and the traveling-wave-type one. In the former, a pair of thermal waves propagate over the free surface with the same amplitude and propagation speed but in the opposite azimuthal directions. In the latter, a single thermal wave propagates in an azimuthal direction. The intensity of the thermocapillary effect is generally described by the Reynolds number $\mathrm{Re} = |\gamma_T| \Delta T L/(\rho \nu^2)$, where γ_T is the temperature coefficient of surface tension γ, L is the characteristic length of the system, ρ is the density of the test liquid. In this geometry, the temperature difference between the end disks that sustain the liquid bridge, ΔT, is the governing factor for varying the thermocapillary effect. The intensity of the thermocapillary effect is described as the Marangoni number $\mathrm{Ma} = |\gamma_T| \Delta T H/(\rho \nu \kappa) = \mathrm{RePr}$.

Tanaka et al. [77] indicated two distinct types of PAS of the same m, which are named as spiral loops 1 (SL1) and 2 (SL2) by varying the intensity of the thermocapillary effect. Typical examples of these structures for $m = 3$ are depicted in Fig. 1. These are the snapshots observed through a transparent top rod; the white dots correspond to the particles suspended in the liquid bridge. The PAS consists of the major structure so-called 'blade' [77]: From the perspective of an observer who is viewing from above in the laboratory frame, the PAS seems rotating without changing its shape as the rigid structure as aforementioned. The azimuthal direction

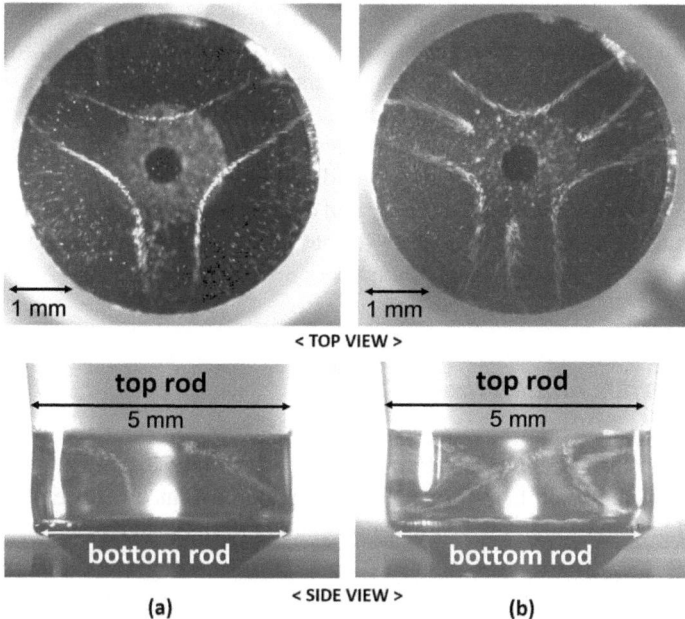

Fig. 1 Snapshots of particle accumulation as viewed from above (top row) and from side (bottom row) in liquid bridge of Pr = 28.6, $\Gamma = 0.68$ ($R = 2.5$ mm), and $V/V_0 = 1.0$: **a** SL1 PAS under Ma = 4.7×10^4 and **b** SL2 PAS under Ma = 5.8×10^4. The direction of the PAS rotation or the traveling-wave-type oscillatory convection by the hydrothermal wave instability is counterclockwise for both cases in the top row, and is from right to left in the bottom row. Reprinted from Toyama et al. [80] with the permission of Springer Nature. All rights reserved

of the rotation of the PAS is the same as that of the thermal wave over the free surface. The number of the blades matches to the azimuthal wave number m of the oscillatory convection after the onset of the primary instability. The tip of each blade corresponds to the trajectory of the particles traveling near the free surface toward the cold end from the hot end in the case of $\gamma_T < 0$. These particles then penetrate into the central region of the liquid bridge and rise toward the hot end by following the return flow. As the particles approach the hot end, they change their direction toward the free surface to form the tip of the adjacent blade. The net azimuthal direction of the particles forming the PAS is opposite to that of the PAS itself. The SL1-PAS represents the fundamental structure; if one follows the radial position of the blade (r) while varying the azimuthal position (θ), $\partial r/\partial \theta$ becomes greater than zero from the central region toward the tip, and becomes less than zero from the tip toward the central region. The SL2-PAS exhibits an additional loop structure near the tip. When the particles that form the PAS travel near the free surface, they exhibit a sharp additional rotation near the free surface before penetrating into the central region of the liquid bridge. Most existing research has focused on the SL1-PAS, because it emerges at lower Ma than the SL2 [66, 77, 80]. In other words, a larger ΔT is required to realize the SL2-PAS, which leads to a severe and inevitable issue of evaporation of the test liquid when performing the experiment under the normal gravity condition. To conduct the terrestrial experiments with the liquid bridges, one has to prepare the liquid bridge whose characteristic length L is small enough to prevent the deformation of the liquid bridge due to the static pressure [25], whereas a large ΔT must be imposed between the both ends of the liquid bridge to realize high Re with small L. This is one of the primary and critical reasons why the research on the SL2-PAS has been limited comparing to that on the SL1-PAS. One can find some research on the SL2-PAS through the numerical simulation [58].

1.2 Materials and Methods

The geometry of interests is quite simple as aforementioned; one prepares a set of cylindrical disks placed face-to-face at a designated distance between the both end surfaces. There exist a number of research dealing with the half-zone liquid bridges [4, 5, 8, 16, 17, 20, 21, 31, 53, 55, 64, 70, 71, 79, 82, 84–86]. The transparent rod, as introduced by Hirata et al. [16, 17], allows the observation of particle behaviors through it. This technique has been widely adopted by various research groups. An example of the experimental set up is illustrated in Fig. 2 after Oba et al. [49]. A liquid is sustained between the top rod, made of sapphire, and the bottom one, made of aluminum. The end surfaces of both rods are finished to have the same radius R. The top rod is heated by an electric heater wound around the rod. The temperature of the rod is measured by the thermocouple and controlled by the PID controller. The bottom rod is connected to the base block with drilled channels. The block is cooled by the coolant pumped from the constant temperature bath. To suppress undesignated natural convection around the liquid bridge, an external shield is often installed coaxially to the liquid bridge. Infrared (IR) camera is commonly used to measure the surface temperature. One might have to prepare a tiny window for the IR camera installed to the external shield. The particles behaviours are observed by camera(s). If one has two cameras, simultaneous observation is applicable through the top rod and the external shield. Oba et al. [49] applied three-dimensional particle tracking velocimetry (3-D PTV) with a conventional algorithm, so that a cubic beam splitter was installed above the top rod to realize simultaneous observation via multiple cameras with different incident angles to the central axis of the liquid bridge. It is noted that the characteristic time for the oscillatory convection becomes shorter as the characteristic length becomes smaller [61, 78], thus high-speed camera(s) might be necessary for precise observations.

Variety of test liquids have been employed as high Pr fluids. Silicone oil has been widely used after [8] because of its transparency for the visible light and well-known thermal properties including almost linear variation of the surface tension against the temperature. Because of the high temperature to realize thermocapillary-driven convection under high Ma, evaporation of the test liquid becomes a common

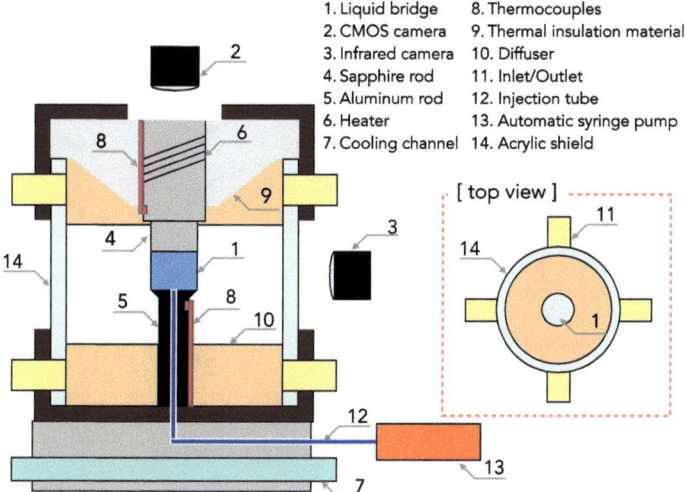

Fig. 2 Cross-sectional view of the experimental apparatus (not to scale). The liquid bridge is heated from above in this configuration. A narrow ZnSe window is placed at a portion of the external shield for the infrared camera to detect the surface temperature through the shield. The CMOS camera for the side view and the displacement sensor are omitted. Reprinted figure with permission from Gotoda et al. [15]. Copyright 2019 by the American Physical Society. All rights reserved

and severe problem to conduct the experiment as aforementioned. A liquid supply system through the rod has been installed to keep V/V_0 constant [14, 15, 37, 38, 49]. One finds the apparatus with similar mechanism for microgravity experiments [66, 95–97].

1.3 Thermocapillary-Driven Convection and Coherent Structures

1.3.1 General View

Physical mechanism for particle accumulation has been discussed over a decade; three major models have been proposed so far. At the early stages following the discovery of the PAS, Schwabe et al. [66] posited that the particles were amassed near the free surface due to the Marangoni effect. Through the fine experimental observation by using the test liquid of $Pr = 15$, they found that the tip of the PAS blade is located on the colder region of the thermal wave over the free surface. They considered that the thermocapillary-driven flow drives the particles approaching the free surface toward low-temperature band due to the negative temperature coefficient of the surface tension. Upon returning to the internal region of the liquid bridge, the particles follow a narrow band of the stream lines to form the PAS. It was indicated, however, that such spatial correlation between the PAS tip and the relatively cold

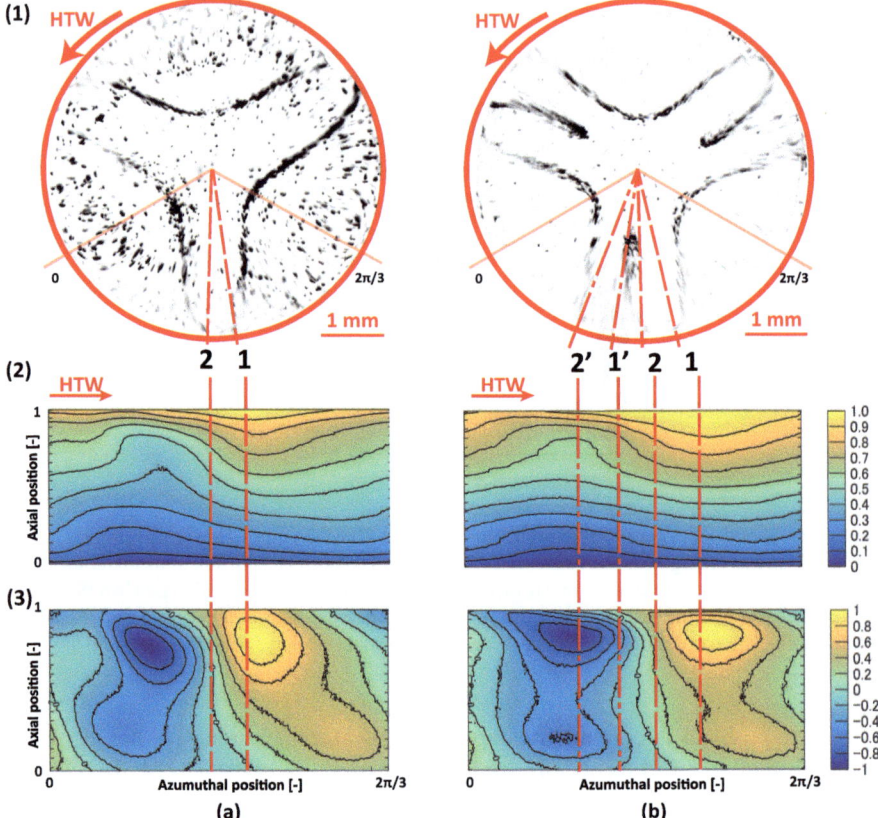

Fig. 3 Correlation between PAS and surface temperatures for **a** SL1 PAS under Ma = 4.7×10^4 and **b** SL2 PAS Ma = 5.8×10^4 in liquid bridge of Pr = 28.6, $\Gamma = 0.68$ ($R = 2.5$ mm), and $V/V_0 = 1.0$: The row (1) illustrates the top views of the PASs, and the rows (2) and (3) the absolute temperature and the temperature deviation over the free surface in a range of $0 \leq \theta \leq 2\pi/3$ (as defined in the row (1)), respectively. Reprinted from Toyama et al. [80] with the permission of Springer Nature. All rights reserved

band over the free surface is not always realized; Toyama et al. [80] illustrated the spatial correlations between the SL1- and SL2 PASs and the thermal wave over the free surface (see Fig. 3). They conducted a simultaneous observation of the surface temperature from the side and the particle motion inside the liquid bridge through the top rod, and reconstructed those distributions in the rotating frame of reference. It was indicated that the tip of the PAS blade locates between the cold and hot bands in the liquid bridge of higher Pr than that studied in [66]. This experimental work reveals that the particles are not collected by the thermocapillary effect over the free surface.

Two other models are so-called (i) 'phase locking model' [56] and (ii) 'particle–free-surface interaction model' [18, 26]. Pushkin et al. [56] proposed a model based on the 'phase locking' between the flow field and the particle motion; they suggested that a PAS would be formed by 'synchronization' between the particle turnover motion due to the basic flow in the liquid bridge and the azimuthal convective motion due to the hydrothermal wave instability. Hofmann and Kuhlmann [18] proposed a model by considering density-matched particles that PAS would be formed through the transfer of particles that 'collide' with the free surface to specific streamlines. They illustrated the presence of closed stream tubes in the flow in the rotating frame, and that particles accumulate on these stream tubes. It was considered that an effect of the finite particle size would force transfers of particle from one streamline to another. There has been active discussion in 2010s [27, 28, 37, 39, 42, 58, 60]. It is indispensable and of great importance to accumulate comprehensive knowledge with experimental and numerical approaches with finer spatial-temporal resolution.

When considering the process of the PAS formation, particularly when evaluating the formation time [66, 77], it is necessary to measure the accumulation; how much the accumulation is realized. Several measures have been introduced such as 'contrast' [14] and '$K(t)$ parameter' [40]. It was indicated that the formation time of SL1 PAS in the HTW of $m = 3$ is of the order of the thermal diffusion time by the terrestrial experiments [13]. It must be noted that, however, it is rather impossible in the experiments to disperse the particles without disturbing the thermal-flow field in the liquid bridge [1, 13]. We need supportive data via fine numerical simulation precisely solving the particles behaviour even close to the free surface and walls [60].

In the following, knowledges on the PAS in fully developed states will be introduced.

1.3.2 Particle Pathlines to Form Coherent Structures

Particle path lines

The PAS seems rotating azimuthally without changing its shape as a rigid structure at a constant angular velocity [77] as previously remarked. When viewed in the rotating frame of reference, the PAS exhibits a coherent structure within the liquid bridge [15, 30, 49]. Figure 4 provides typical examples of the path lines of the particles forming (a) a triangle particle depletion zone and (b) the SL1 PAS in (i) the rotating frame of reference and (ii) in the laboratory frame observed from above [15]. The images are obtained by accumulating the snapshots for 7 s or 10 periods. In the laboratory frame (column (ii)), the particles appear to scatter inside the liquid bridge except the particle depletion zone. At an appropriate ΔT, the particles gather along closed orbits to form coherent structures in the frame of reference rotating with the traveling-wave-type oscillatory convection due to the hydrothermal wave instability ((b) in column (i)). Mukin and Kuhlmann [39] indicated that the PAS emerges after the particle transfer from the chaotic streamlines to other streamlines located inside of Kolmogorov-Arnold-Moser (KAM) tori, and that such particle transfer is realized by the interaction with the free surface. These KAM tori are three-dimensional closed stream tubes in the liquid bridge, and the spatial structures of the particle

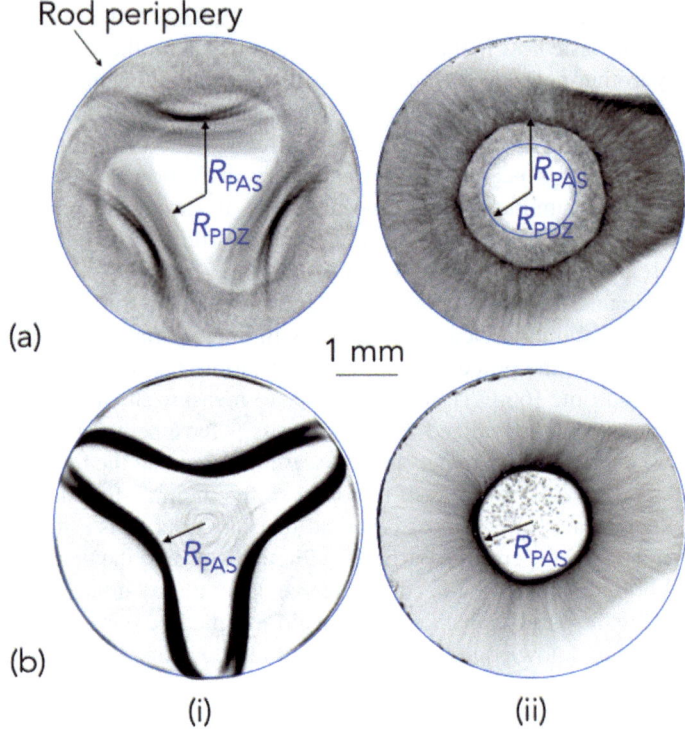

Fig. 4 Top views of pathlines of particles of 15 μm in diameter **a** before and **b** after PAS formation in (i) rotating frame of reference and in (ii) laboratory frame by averaging over 7 s (10 times fundamental periods of HTW) in liquid bridge of $Pr = 28.6$, $\Gamma = 0.64$ ($R = 2.5$ mm), and $V/V_0 = 1.0$: **a** $Ma = 3.13 \times 10^4$ and **b** $Ma = 4.30 \times 10^4$. Note that white regions at the top and bottom right in the images in the column (ii) are the regions where the light from the light source never irradiates in the liquid bridge: The light is supplied to the liquid bridge through the free surface from the left in these images and one cannot avoid the refraction of the light due to the curvature of the free surface. Those regions become vague and are not apparent by integrating images for a long period in the rotating frame of reference as shown in the column (i). Reprinted figure with permission from Gotoda et al. [15]. Copyright 2019 by the American Physical Society. All rights reserved

accumulations well match to those of the KAM tori such as T_3^3, T_3^9 and T_{core} [30].

Pattern map

The particle accumulation depends on the intensity of the thermocapillary effect; the occurring condition is described as a function of Ma [66]. Figure 5 illustrates a typical example of the variation of the particle distribution in the rotating frame of reference for the particles of different sizes [15]. The formation, development and decay of the coherent structures are observed with increasing Ma. Once these coherent structures form, the width of the PAS varies as a function of Ma, as seen in

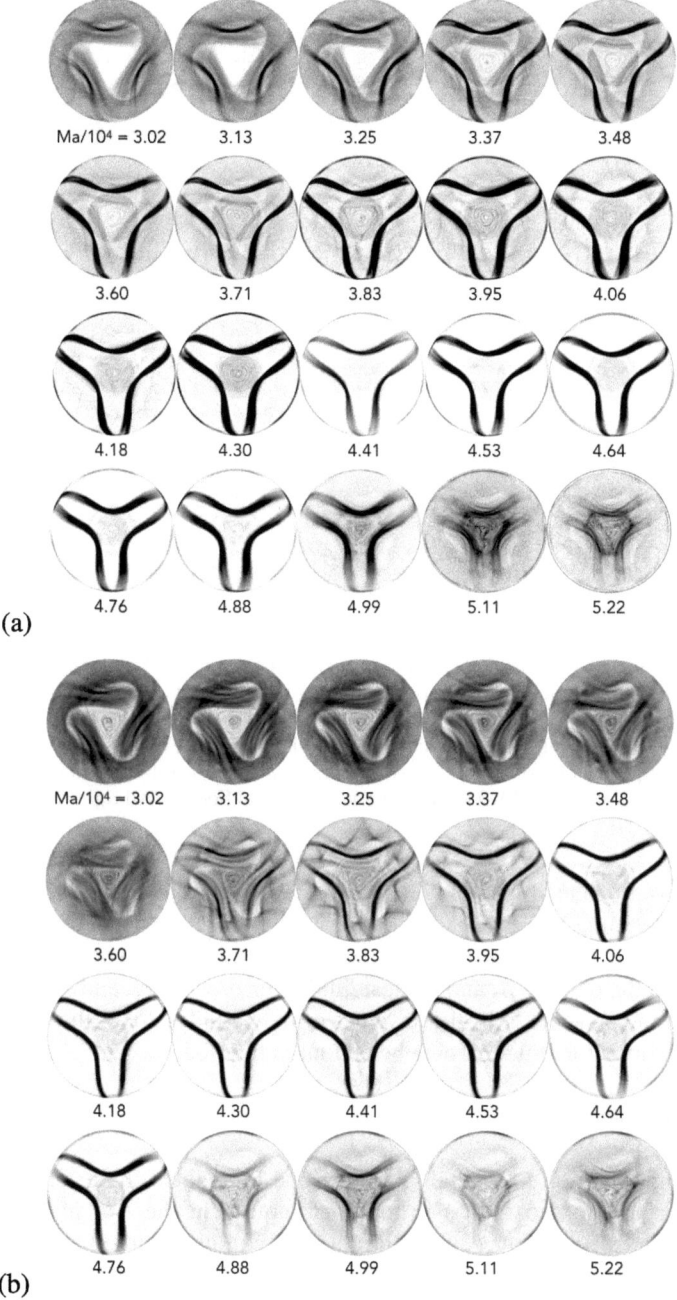

Fig. 5 Variations of coherent structures by the particles as a function of Ma for **a** $d_p = 15\,\mu\text{m}$ and **b** $d_p = 30\,\mu\text{m}$. Conditions except Ma are the same as indicated in Fig. 4. The azimuthal direction of the HTW is counterclockwise for all images. Reprinted figure with permission from Gotoda et al. [15]. Copyright 2019 by the American Physical Society. All rights reserved

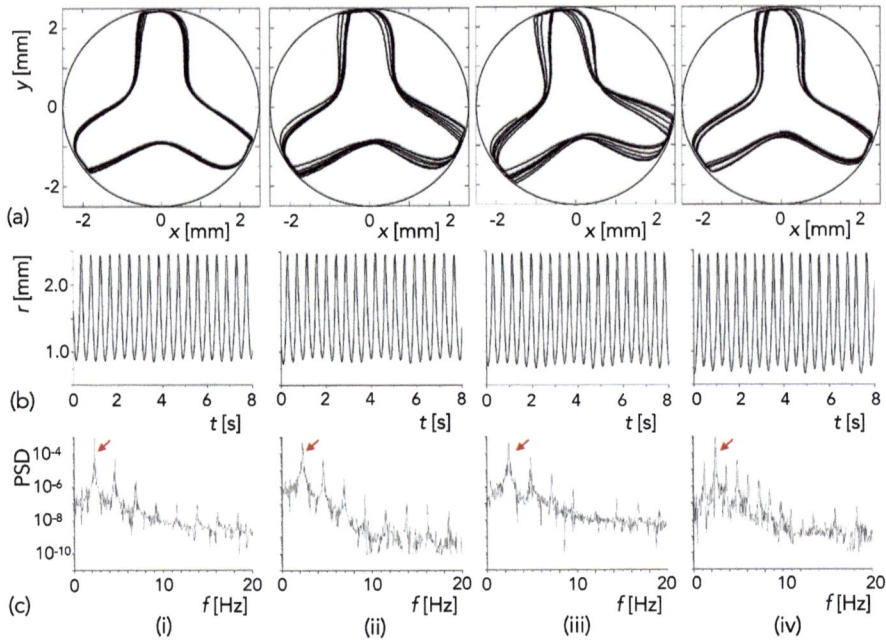

Fig. 6 **a** Reconstructed SL1 PAS configurations in rotating frame of reference by tracking a single particle forming PAS, **b** time series of particle radial positions, and **c** its power spectral density (PSD) for Ma values of (i)–(iv) 3.3×10^4, 3.7×10^4, 4.2×10^4, 4.6×10^4, respectively. Conditions are the same as indicated in Fig. 4. The particle diameter is of 15 μm. Arrows in row (**c**) indicate the fundamental frequency of the particle turnover motion f_0. Reprinted figure with permission from Gotoda et al. [15]. Copyright 2019 by the American Physical Society. All rights reserved

the KAM tori predicted by employing the modelled convection field [40]. Note that, even under the same Ma, the shape of the coherent structure depends on the particle size or corresponding Stokes number St $= \varrho d_p^2/(18H^2)$ [18], where ϱ is the density ratio between the particle ρ_p and the test fluid ρ_f or $\varrho = \rho_p/\rho_f$, and d_p is the particle diameter. In the present study, those values of the particles are of the order of 10^{-5}. In the following, individual particle behaviour is focused.

Particle behaviours

When paying attention to a single particle, one finds their characteristics to form the PAS. In addition to the basic turnover motion in the r–z plain due to the thermocapillary-driven convection, the particle exhibits azimuthal motion in the oscillatory convection, as described by Tanaka et al. [77]. The net azimuthal direction of the particle motion is opposite to that of the PAS itself. When the particle forms the PAS, the turnover motion itself varies as Ma [15]: The variation of the particle trajectory from (i) strictly periodic, (ii)–(iii) quasi-periodic and (iv) period doubled by varying Ma (Fig. 6a). Such variation is clearly demonstrated by monitoring the particle position in r (frame (b)) and its Fourier spectrum (frame (c)). When paying attention to the absolute minimum of the particle position in r (corresponding to R_{PAS} as indicated in Fig. 4), the variation of the minimum position of the particle in r depends on Ma. In the case of (i), the minimum position is almost constant. With increasing Ma for (ii) and (iii), it exhibits periodic modulations. With further increasing Ma to realize the period-doubled PAS [39], the minimum position of the traveling particle in r exhibits a large oscillation, having alternative values of 0.7 and 0.8 mm each time in this case, as the particles penetrate to the deepest position in the liquid bridge. A fundamental frequency $f_0 = 2.4$ Hz and its subharmonics $f_0/2 = 1.2$ Hz (the frame (c)–(iv)) are observed; this subharmonic frequency corresponds to the period-doubled trajectory of the PAS in the rotating frame of reference. It must be emphasized that the oscillatory convection itself is strictly periodic and not modulated under these conditions, as indicated by Toyama et al. [80] under Ma for the SL1 and SL2 PASs, and even under slightly higher Ma with no stable PASs. Through the tracking a single particle forming the PAS, the frequency ratio between the particle turnover motion f_p and the HTW f_{HTW} remains almost constant (about 1.6 in these series of the experiments) under the condition where the PAS is fully formed, and this ratio is independent of Ma.

To depict the particles' trajectory to form the PAS, one can monitor the Poincaré section. Figure 7 illustrates some example of the Poincaré section at $z = H/2$. When one tracks the particle forming the PAS, the particle passes several fixed areas in r–θ plain. In the case of the PAS in the HTW of $m = 3$, one finds three pairs of the area; a pair consists of the area locating near the free surface ('A' in the frame) and near the liquid-bridge centre ('B'). The area 'A' near the free surface corresponds to the trajectory near the tip of the PAS blade passing the plain downward, and the area 'B' near the centre corresponds to the trajectory where the particle follows the return flow inside the liquid bridge from the region near the cold disk toward the hot disk. One finds finite sizes of the areas. Note that these experimental results were obtained by employing rather big particles ($d_p = 30\,\mu$m). One would measure not only the size but also their shape in the Poincaré section, if one employs finer particles with finer resolution of the observation system, in order to make comparison with those of the KAM tori [40].

Through such observation, one can find some particle departing from a certain trajectory to another (see Fig. 8); Top frame illustrates an example of the Poincaré section detected at $z = 0.81 \pm 0.09$ mm under Ma $= 5.2 \times 10^4$ (same as row (c) in Fig. 7 but in the different experimental run). Frames (a)–(c) depict the sections obtained in the same experimental run to track a single particle under the same condition but in different successive periods. Frame (d) illustrates the whole and original Poincaré section obtained in a single experimental run for 73.2 s, that is, the sum of (a) to (c) corresponds to the original section. From these successive sections, it is evident that a single particle does not stay to settle at a particular structure, but rather switches among specific structures. In this specific case, the particle first migrates from (a) the core, then (b) stays for a while on a structure wrapping the core, and finally (c) settles on the PAS. Each structure bears a resemblance to the KAM torus realized in the thermocapillary-driven convection in the half-zone liquid bridge;

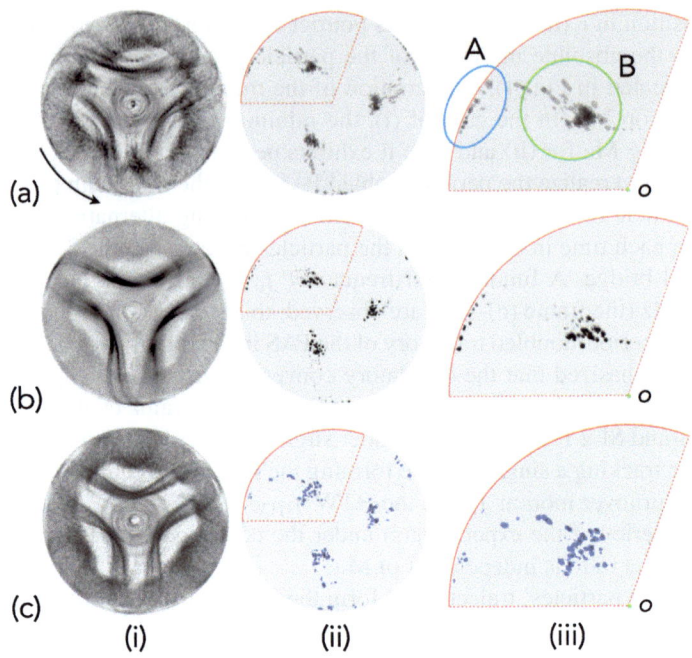

Fig. 7 PASs observed through top rod in rotating frame of reference under Ma = **a** 4.1×10^4, **b** 4.7×10^4, and **c** 5.2×10^4 in liquid bridge of Pr = 28.5, $\Gamma = 0.68$ ($R = 2.5$ mm), and $V/V_0 = 1.0$ with suspended particles of 30 μm in diameter: Columns (i) and (ii) indicate the projected images, and Poincaré sections at midheight of liquid bridge, respectively. Images for column (i) are obtained by by integrating 500 frames (for about 8.3 s). Images for column (ii) **a** and **b** are obtained by integrating 3900 frames (for almost 65 s), and **c** 2340 frames (for almost 39 s). Column (iii) shows a zoomed view of $1/3 = 1/m$ region of image in (ii). The fundamental frequencies f_0 are of **a** 1.41 Hz, **b** 1.42 Hz, and **c** 1.44 Hz. The rotating direction of the HTW in this figure is counter-clockwise in the laboratory frame. Reprinted from Yamaguchi et al. [90] with permission of The Japan Society of Microgravity Application. All rights reserved

the structure known as the toroidal core resembles one of the tori T_{core}, whereas the structure wrapping the core resembles T_3^9 [30]. Additionally, the PAS in the HTW of $m = 3$ also resembles T_3^3. The middle and bottom frames of the figure illustrate the corresponding temporal variations of r- and θ-positions of a particle on Poincaré section. In the variation of r-position, the corresponding KAM torus is indicated. The section (a) illustrates the paths of the particle on the core, whereby the particle is attracted by T_{core} first. Subsequently, the particles switches to the attractor T_3^9 in the section (b). This torus, however, does not keep attracting the particle; the particle migrates between the attractors T_3^9 and T_{core}. Finally, in this case, the particle is attracted to T_3^3 to keep forming the PAS (the section (c)). Scenario of the particle migration is not thus far predictable; each torus becomes unstable to keep attracting particle under high Ma. It was indicated that the particle would not be able to stay in a long period especially at the core, which might reflect that the torus T_{core} in the high-Pr liquid bridge becomes weakly chaotic [58]. Further research is needed

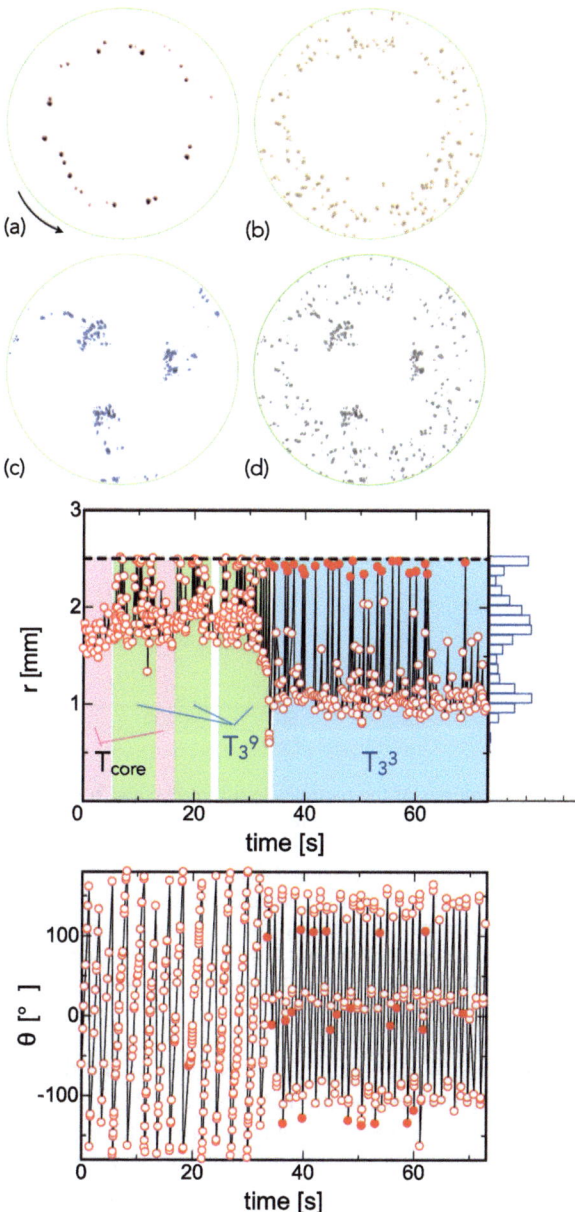

Fig. 8 (Top) Poincaré section at midheight of liquid bridge obtained by integrating **a** $0\,\text{s} \leq t \leq 6.1\,\text{s}$, **b** $6.1\,\text{s} \leq t \leq 33.4\,\text{s}$, **c** $33.4\,\text{s} \leq t \leq 73.2\,\text{s}$, and **d** total images of the section for 73.2 s detected at $z = 0.81 \pm 0.09$ mm under $\text{Ma} = 5.2 \times 10^4$ (same as row (**c**) in Fig. 7 but in the different experimental run). (middle) and (bottom) Corresponding temporal variations of r- and θ-positions of a particle on Poincaré section. Histogram on the right of top graph illustrates the particle number density measured with a constant interval $\Delta r = 0.1$ mm in the observation period. The direction of HTW is counter-clockwise in the laboratory frame. The conditions are the same as those in Fig. 7. Reprinted from Yamaguchi et al. [90] with permission of The Japan Society of Microgravity Application. All rights reserved

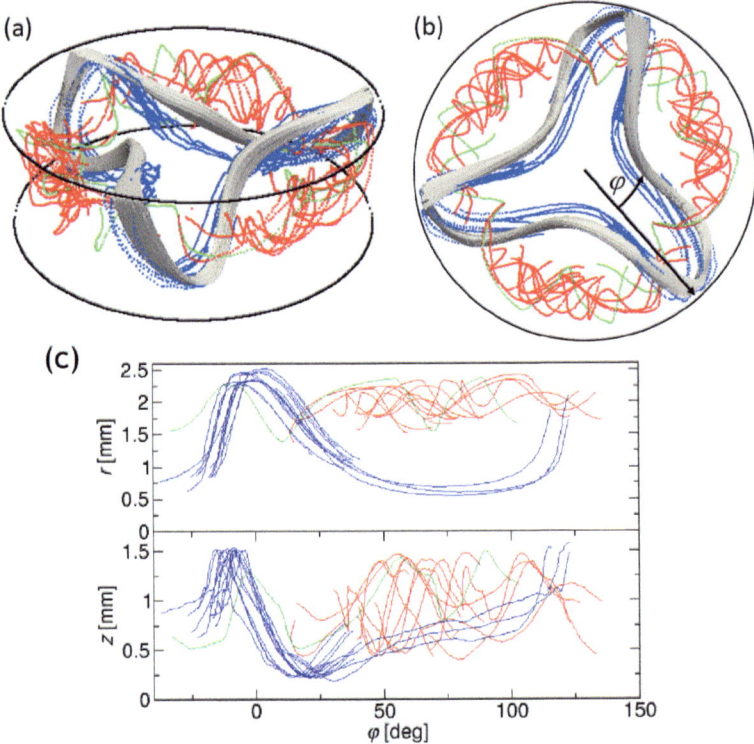

Fig. 9 Reconstructed trajectories of the particles in **a** bird-eye and **b** top views, respectively, in liquid bridge of Pr = 28.6, $\Gamma = 0.68$ ($R = 2.5$ mm), and $V/V_0 = 1.0$ with suspended particles of 30 μm in diameter: The figures consist of the trajectories of the particles on PAS (14 trajectories in blue), on the core (16 in red) and on the structure wrapping the core (3 in green). The direction of HTW is clockwise, and the net direction of the particles on PAS is counter-clockwise (in positive φ direction) in the laboratory frame. Bundles of pathlines to form PAS obtained by the numerical simulation [57] are also drawn in grey. Frame (c): distributions of all trajectories in $r - \varphi$ (top) and $z - \varphi$ (bottom) plane. Reprinted figure with permission from Oba et al. [49]. Copyright 2019 by the American Physical Society. All rights reserved

to describe the attractivity of each structure of KAM tori and the mechanism on the interaction between the particles and the KAM tori.

It is of utmost importance and indispensable to discern the three-dimensional behaviour of the particles to comprehend the mechanism of the PAS formation and the correlation with the KAM tori. It is, however, quite difficult especially under the normal gravity condition because of the small size of the liquid bridge with deformed free surface. The time-dependent deformation of the free surface in the oscillatory convection prevent the precise measurement of the particles with the visible light through the free surface. In such a sense, microgravity conditions provide some ideal environment since one can employ larger liquid bridges as noted in various studies [22, 23, 25, 35, 47, 63, 87, 91–93, 95–97]. Under the normal gravity, some efforts have been conducted to realize three-dimensional measurements of

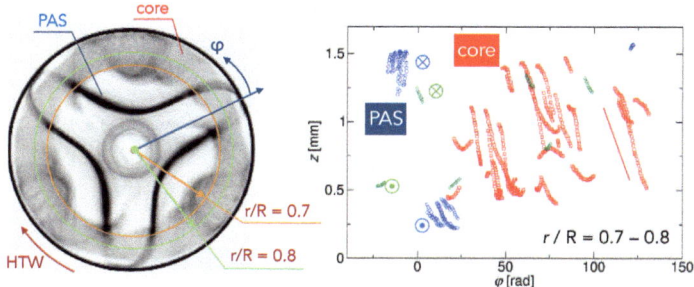

Fig. 10 (Left) Coherent structure under the same condition as shown in Fig. 9, and (right) spatial distributions of paths in z-φ space in $0.7 \leq r/R < 0.8$. In the right frame, the particles attracted to T_3^3, T_{core} and T_3^9 are drawn in blue, red and green, respectively. Arrow indicates the direction of motions of the particles concerned. The direction in r of the particle motions is also indicated; ⊙ indicates the motion toward the free surface (or in positive r direction), and ⊗ that toward the center axis of the liquid bridge (or in negative r direction)

the particles inside the liquid bridge of $O(10^{-3}\,\text{m})$ through the transparent top rod [1, 45, 46]. Oba et al. [49] accomplished the three-dimensional reconstruction by implementing the three-dimensional particle velocimetry (3-D PTV) with a classical algorithm (see Fig. 9): Frames (a) and (b) illustrate the reconstructed trajectories of the particles in the bird-eye and top views, respectively, and frame (c) shows the distributions of all trajectories in r-φ (top) and z-φ (bottom) planes. The figures in Frames (a) and (b) consist of the trajectories of the particles on the PAS (14 trajectories in blue), on the core (16 in red) and on the structure wrapping the core (3 in green). Note that the trajectories were obtained in $1/3$ region of the liquid bridge in azimuthal direction by the experiments, and the same results are plotted repeatedly with a phase difference of $\pm 2\pi/3$ in the rest of the region. This process was based on the three-fold rotational symmetry structure of the flow field. Because of the high frequency of the PAS in the small-size liquid bridge, Oba et al. [49] employed synchronized high-speed cameras to track the motion of the particles forming the various coherent structures as a function of time.

Obtaining the three-dimensional behaviour of the particles allows for monitoring their motions within designated areas (see Fig. 10). The particles form various structures that resemble various KAM tori, such as T_3^3, T_3^9 and T_{core}, as indicated by the model flow [40]. As previously described, finer tracking with smaller particles is indispensable to conduct a comprehensive comparison between the KAM tori and the coherent structures more in detail.

Coherent structures in HTW of $m = 1$

The azimuthal wave number of the oscillatory convection is contingent upon the geometric characteristics of the liquid bridge, especially on the aspect ratio Γ. Because of the gravity effect as well as the Rayleigh limit, the primary focus of the investigation has been on the oscillatory convection with $2 \leq m \leq 5$. Note that the experiments with shorter liquid bridge for larger m ($m \geq 5$) involve critical problem of the evapo-

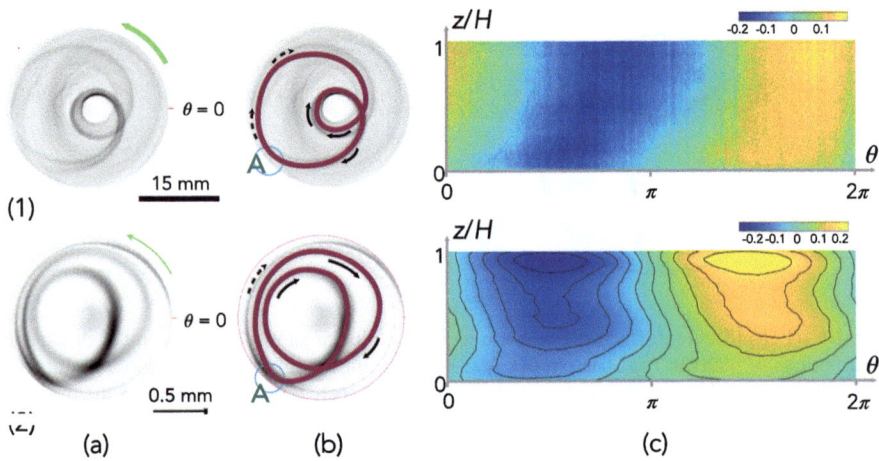

Fig. 11 **a** Coherent structure in rotating frame of reference observed through hot disk, **b** its schematics, and **c** temperature-deviation distribution over free surface for (1) microgravity and (2) terrestrial experiments. The coherent-structure images are obtained by integrating the particle images for (1) four fundamental periods or 131.4 s and for (2) ten fundamental periods or 3.8 s. Arrows in the row **a** indicate the direction of the traveling wave. Solid and dashed arrows in the row **b** indicate the upward ($\partial z/\partial t \geq 0$) and downward ($\partial z/\partial t \leq 0$) motions of the particles forming the PAS, respectively. Circle in the row **b** illustrates the position 'A' where the coherent structure seems locating closest to the free surface. Reprinted figure with permission from Sakata et al. [61]. Copyright 2022 by the American Physical Society. All rights reserved

ration; one has to increase ΔT to realize higher Ma despite of smaller characteristic length L. Therefore, it is rather difficult to find any knowledge accumulated by the experimental approaches on the PAS in the HTW of $m = 1$ and $m \geq 5$.

As for the PAS in the HTW of $m = 1$ as well as $m \geq 5$, only a limited number of research has been conducted through the terrestrial experiments: Yazawa and Kawamura [98] prepared two sets of the end disks whose diameters were of 10 mm and 20 mm, and realized the SL1-PAS of various m up to 20 in the liquid bridges whose Pr is of 28 and 68.

Sasaki et al. [62] and Schwabe et al. [66] examined the PAS in the HTW of $m = 1$ in the liquid bridge of moderate Pr. Schwabe et al. [66] indicated that the existing range of the PAS in the HTW of $m = 1$ against Γ and Ma is much narrower comparing to those of the PAS in the HTW of $2 \leq m \leq 4$. It was found that two types of PAS in the HTW of $m = 1$ exists by varying Ma [62]. It must be noted that the PAS in the HTW of $m \geq 2$ also exhibits two types with increasing Ma (SL1- and SL2 PASs as introduced), but the variation of PAS in the HTW of $m = 1$ is different from those. When one observes from above, the particles form the closed path with a spiral structure inside the liquid bridge for the SL PAS under lower Ma. Such a spiral structure corresponds to the helical motion following the return flow in the central region. As for the SL PAS under higher Ma, on the other hand, the particles form the closed path without such a spiral structure. That is, this type of PAS realized at higher Ma exhibits a simpler structure. It seems opposite to the correlation for the

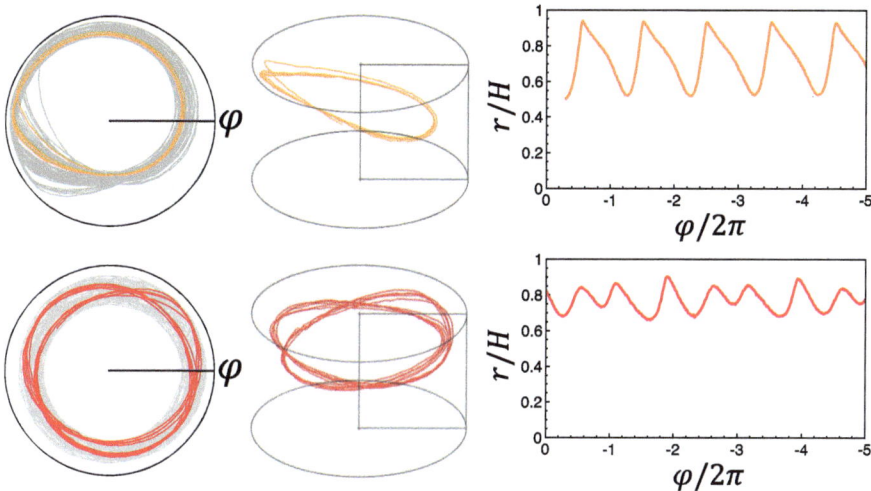

Fig. 12 Coherent particle trajectories emerged simultaneously under Ma $= 3.3 \times 10^4$ in the liquid bridge of $\Gamma = 1.6$ and $V/V_0 \sim 1$. Top and bottom: trajectories that close after a single revolution and two-fold revolutions, respectively, about the z axis. Reprinted figure with permission from Noguchi and Ueno [48]. Copyright 2023 by the American Physical Society. All rights reserved

SL1- and SL2 PASs against Ma. Schwabe et al. [66] also realized the SL PAS with a spiral structure in the central region of the liquid bridge. The author's group at Tokyo University of Science recently demonstrated the SL PAS with a spiral structure in the liquid bridge of Pr $= 28.6$, and succeeded in the reconstruction in the rotating frame of reference (see Fig. 11) [61]. This PAS in the HTW of $m = 1$ is realized under Ma $= 1.86 \times 10^4$ or $\Delta T = 20$ K in the liquid bridge of $\Gamma = 2.0$ ($R = 0.75$ mm) and $V/V_0 = 0.8$. The induced oscillatory convection seems corresponding to the type of HTW(b) indicated by the linear stability analysis [12]. The SL PAS in the HTW of $m = 1$ with a spiral structure was finely reproduced in the liquid bridge of Pr $= 8$ by Capobianchi and Lappa [7] by the numerical simulation. They indicated the correlation between the coherent structure and the KAM torus. It must be noticed that they reproduced the coherent structure only by the particle lighter than the test liquid, or, $\varrho = \rho_p/\rho_l < 1$. The terrestrial experiments [61, 62, 66, 78] have revealed that the particles of $\varrho > 1$ realize the coherent structure in the HTW of $m = 1$ as those of $m \geq 2$ [1, 13, 15, 45, 49, 66, 77]. As for the SL PAS in the HTW of $m = 1$ without a spiral structure, it was numerically predicted by Barmak et al. [2] and subsequently demonstrated experimentally by Noguchi and Ueno [48].

Knowledge on the PAS in the HTW of $m = 1$ has been gradually accumulated (see Fig. 12) [48]. Through the preceding works [15, 48, 49, 66, 77, 80], it has been indicated that the azimuthal wave number of coherent structure, m_{PAS}, matches the azimuthal wave number of HTW, m_{HTW}, for $m_{\text{HTW}} \geq 2$. The 'PAS in the HTW of $m = 1$ with a spiral structure' [61, 78], however, does not follow the correlation $m_{\text{PAS}} = m_{\text{HTW}}$ that has been hitherto recognized. Sensui et al. [68] proposed a rational wave number for the coherent structures, $m_{\text{PAS}} = q/p$, by considering the coherent structure closing in p-time azimuthal revolutions after q-time turnover by the particles, where p and q are the mutually prime positive integers. Based on this definition, the PAS with a spiral in the HTW of $m_{\text{HTW}} = 2$, as demonstrated by experiments [61, 78] and predicted by numerical simulation [7], is successfully categorized as $m_{\text{PAS}} = 1/2$. Note that this definition also works perfectly for the cases of PASs of $m_{\text{PAS}} \geq 1$. Further, Sensui et al. [68] predicted various types of coherent structures of $m_{\text{PAS}} = q/p$ in imaginary traveling-wave flow, some of which were experimentally demonstrated [24, 48]. Since a high correlation in spatial structure between the coherent structures and the KAM tori has been indicated [2, 30, 48, 49], exploring of coherent structures would lead to unveiling the thermal flow fields in tall liquid bridges.

1.4 Concluding Remarks

Coherent structures formed by particles suspended in half-zone thermocapillary liquid bridges via experimental approaches were introduced in this article. Since Schwabe et al. [65] found this unique phenomenon so-called particle accumulation structure (PAS), general knowledge on its shape, occurring conditions, and formation processes has been accumulated. By focusing on the spatial-temporal behaviours of particles forming the PAS of the azimuthal wave number of $m = 3$, the correlation between the particle behaviour and the ordered flow structures, known as the Kolmogorov-Arnold-Moser tori, was illustrated as function of the intensity of the thermocapillary effect and the particle size. Recent works on the PAS in the HTW of $m = 1$, through the experimental and numerical approaches, were briefly introduced. A rational wave number for the coherent structures, $m_{\text{PAS}} = q/p$, was proposed by Sensui et al. [68] by considering the coherent structure closing in p-time azimuthal revolutions after q-time turnover by the particles, where p and q are the mutually prime positive integers. This definition works perfectly not only for the coherent structures in complex shape in the HTW of $m_{\text{HTW}} = 1$ [48, 61, 78] but also those in the HTW of $m_{\text{PAS}} = m_{\text{HTW}} \geq 1$. Knowledge accumulated via ground-based research on this unique phenomenon would lead to future research including microgravity experiments on the International Space Station (ISS) by the project 'Japanese European Research Experiments on Marangoni Instability (JEREMI)' [29, 69].

2 Dynamics of Liquid Bridge

Mitsuru Ohnishi and Satoshi Matsumoto

2.1 Introduction

Liquid bridges, anchored at both ends by solid rods, exhibit characteristic limitations in their axial length, known as the Plateau-Rayleigh limit [54, 75]. Under quasi-static conditions, free from gravitational effects and devoid of liquid motion, a liquid bridge with an initial radius R_0 becomes unstable at a critical length $L = 2\pi R_0$, where the capillary pressure advantage undergoes a transition. This critical point marks the intersection of the two principal curvatures of the free surface. The shape of the static interface is determined by the volume-constrained Young-Laplace equation.

A pinned liquid bridge exhibits countless infinite shape instabilities as its length increases. The shapes correspond to eigenmodes and occur at eigenvalues in the bifurcation analysis. The instability at the Plateau-Rayleigh limit is often observed the first such eigenmode.

In microgravity environments, large liquid bridges can form making the phenomenon easier to observe. Increasing the inter-disk distance leads to the development of instability, resulting in the rupture of the liquid bridge. The behavior of the breakup process is expected to vary when temperature differences exist at the free surface of the liquid bridges. To simulate the breakup (pinch-off) process of these liquid bridges, numerical analysis is performed. This intriguing phenomenon can be observed during microgravity experiments conducted on the International Space Station (ISS).

2.2 Numerical Method

The numerical analysis dealt with a two-dimensional incompressible fluid with an interface corresponding to a liquid bridge. The improved SOLA-VOF (Solution Algorithm-Volume of Fluid) method [44] is employed to capture the interface shape that evolves moment by moment.

The governing equations of continuity, momentum, and energy are expressed as follows [51];

$$\frac{\partial u}{\partial x} + \frac{\partial v}{\partial y} + \frac{\xi \cdot u}{x} = 0 \tag{1}$$

$$\frac{\partial u}{\partial t} + u\frac{\partial u}{\partial x} + v\frac{\partial u}{\partial y} = -\frac{1}{\rho}\frac{\partial P}{\partial x} + v\left[\frac{\partial^2 u}{\partial x^2} + \frac{\partial^2 u}{\partial y^2} + \frac{\xi}{x}\left(\frac{\partial u}{\partial x} - \frac{u}{x}\right)\right] \tag{2}$$

$$\frac{\partial v}{\partial t} + u\frac{\partial v}{\partial x} + v\frac{\partial v}{\partial y} = -\frac{1}{\rho}\frac{\partial P}{\partial y} + v\left(\frac{\partial^2 v}{\partial x^2} + \frac{\partial^2 v}{\partial y^2} + \frac{\xi}{x}\frac{\partial v}{\partial x}\right) \tag{3}$$

$$\frac{\partial T}{\partial t} + u\frac{\partial T}{\partial x} + v\frac{\partial T}{\partial y} = \kappa\left(\frac{\partial^2 T}{\partial x^2} + \frac{\partial^2 T}{\partial y^2} + \frac{\xi}{x}\frac{\partial T}{\partial x}\right) \tag{4}$$

where, u and v represent the velocity in the x and y directions, respectively, P is the pressure, T is the temperature, v is the kinematic viscosity, and κ is the thermal diffusivity. The coordinate system is determined by the value of ξ, where $\xi = 0$ corresponds to Cartesian geometry, and $\xi = 1$ corresponds to cylindrical geometry.

Table 1 Thermophysical properties of fluid

Properties	Unit	Value
Density, ρ	kg/m^3	950
Surface tension, σ	mN/m	20.6
Temperature coefficient of surface tension, σ_T	mN/(m K)	-6.2×10^2
Kinematic viscosity, ν	m^2/s	2.0×10^5
Thermal diffusivity, κ	m^2/s	9.7×10^8

The VOF (Volume of Fluid) fraction equation is to track the volume fractions of different fluid phases within a computational domain. It is an essential equation for representing and tracking the interfaces between liquid and gas in a numerical simulation. The VOF fraction equation is used to determine how much of each fluid occupies a given computational cell.

The VOF fraction equation can be expressed as:

$$\frac{\partial F}{\partial t} + u \frac{\partial F}{\partial x} + v \frac{\partial F}{\partial y} = 0 \tag{5}$$

where F is the fluid volume, and this equation is used to calculate the shape of the interface.

As the boundary conditions on the fluid surface, following equations are used;

$$P = \mu \frac{\partial u_n}{\partial n} + \sigma \left(\frac{1}{R} + \frac{\xi}{R_c} \right) \tag{6}$$

$$\mu \left(\frac{\partial u_n}{\partial m} + \frac{\partial u_m}{\partial n} \right) = \frac{\partial \sigma}{\partial m} \tag{7}$$

$$\sigma = \sigma_0 - \sigma_T (T - T_0) \tag{8}$$

where u_n is the normal velocity, u_m is the tangential velocity, μ is the viscosity, σ is the surface tension, R is the curvature of the surface, R_c is the curvature of the surface in the case of cylindrical coordinate, σ_T is the temperature coefficient of surface tension, and indices 'n' and 'm' represent the normal direction and the tangential direction to the fluid surface, respectively. These boundary equations show the stress balances in both directions on the fluid surface.

The Incomplete LU decomposition Conjugate Gradient (ILUCG) method, which is fast and numerically stable, is used to solve the pressure Poisson equation. The second-order Euler-leap flog method is used to time derivative term.

Thermophysical properties of silicone oil fluid is listed in Table 1.

2.3 Pinch-Off Experiment and Comparison with Simulation

Chapter "Microgravity Experiments in Kibo Onboard the International Space Station" of this book provides a detailed account of microgravity experiments. Here, the pinch-off experiment is focusing on the pinch-off experiment conducted aboard the International Space Station (ISS). The experiment involved the formation of a liquid bridge with a 30 mm radius and a length of 63.5 mm. The volume ratio, representing the ratio of liquid volume to the gap volume between the supporting disks, was set at 0.71. To induce a temperature difference, the experiment involved imposing variations at both ends of the liquid bridge. The progressively narrowing liquid bridge is achieved by extracting liquid from the center of the cold-side rod. The time series illustrating the evolution of the liquid bridge shape is presented in Figs. 13 and 14. Importantly, a neck in the liquid bridge emerges, biased toward the heating disk due to the thermocapillary force. Ultimately, the liquid bridge undergoes pinch-off, resulting in the formation of a satellite droplet. In Figs. 13 and 14, the upper photos depict experimentally visualized snapshots captured by the CCD camera, while the lower figures display the numerical results. The numerical analysis results accurately reproduce the pinch-off phenomenon observed experimentally.

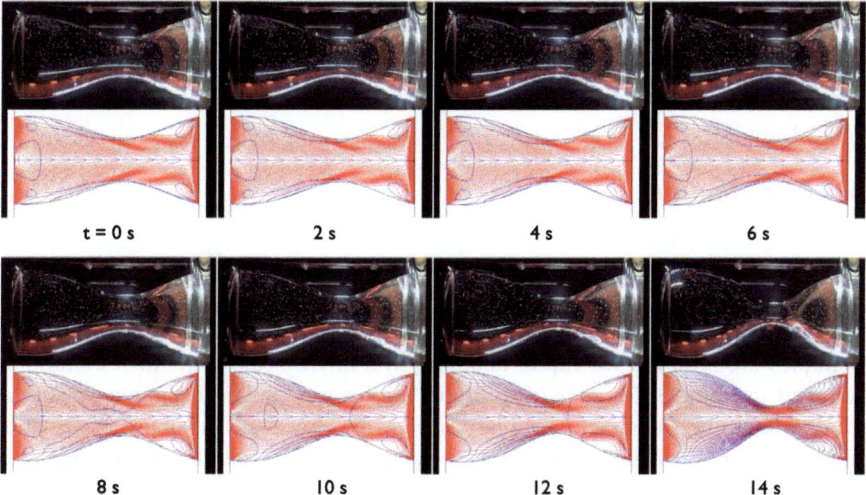

Fig. 13 Time series of liquid bridge shape compared with microgravity experiment and numerical simulation (early stage before pinch off). Above: experiment (Courtesy by JAXA. All rights reserved), Bottom: numerical simulation

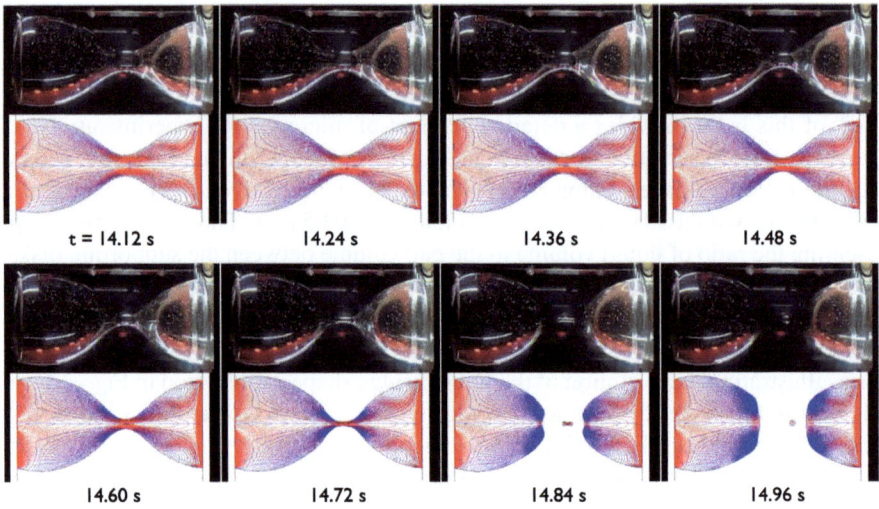

Fig. 14 Time series of liquid bridge shape compared with microgravity experiment and numerical simulation (later stage before/after pinch off). Above: experiment (Courtesy by JAXA. All rights reserved), Bottom: numerical simulation

3 Nano Liter Droplet Formation and Its Volume Control by Thermocapillary Effect

Suguru Uemura

3.1 Introduction

Manipulation of micro droplets has attracted a great deal of attention in the field of chemistry, bio and life sciences [9, 72, 74]. Sample size reduction takes advantage of high reaction rate, high throughput and space saving. For these reasons, new techniques are required for liquid handling in a micro scale. The most commonly employed techniques for preparing liquid microspots on solid planner substrates include inkjet, microcontact, and pin printing [3]. However, its robustness is often compromised by delicate sample surface interactions, gas bubbles and particulates. For these reasons, we have applied a simple technique using a small pin for nano liter droplet formation. When the pin is dipped into the liquid and pulled up, the formed meniscus collapses (snap-off) and a droplet remains on the pin end face. A number of studies which focused on the meniscus shape and droplet formation process have been performed experimentally and numerically [11, 19, 52], however, only few studies focused on the resulted droplet volume [81].

In the present study, the droplet volume on the pin end face and its reproducibility were measured. Furthermore, in order to control the droplet volume, the thermocap-

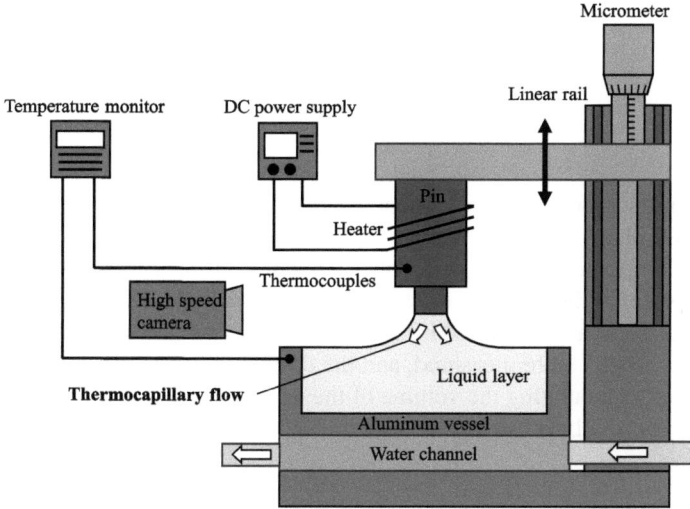

Fig. 15 Experimental apparatus

illary flow was induced by imposing a temperature difference on the free surface of the meniscus.

3.2 Experiments

The experimental apparatus is shown in Fig. 15. The pin was made of sapphire glass (Al_2O_3 single crystal) whose end face was mirror finished and its edge was kept sharp. Heater was wrapped around the pin and gave a temperature difference on the free surface. The pin was moved up and down by attaching on a liner stage. The linear stage was pulled up by rotating a micrometer. The pulling speed was kept less than 0.01 mm/s to avoid effect of the pull-up speed for the droplet formation phenomena. Test fluids were pure water or silicone oil (2, 5, 10, 100, 1000 cSt, KF-96, Shin-Etsu Co., Ltd.). The test fluid was kept in the aluminum vessel. The rim of the vessel was chemically coated to prevent overflow of the test fluid. When the thermocapillary flow was induced, a temperature gradient was imposed on the free surface by heating the rod and cooling the vessel.

The initial surface shape of liquid layer was set approximately flat, that is, the mean curvature was close to zero. The snap-off instant was observed with a high-speed camera (FASTCAM-APX RS, Photron Ltd.), which was placed at side of the test section.

The droplet volumes were characterized by the Bond number Bo and the capillary number Ca, defined as follows.

$$\text{Bo} = \frac{\Delta \rho g R^2}{\sigma_0} \quad (9)$$

$$\text{Ca} = \frac{\mu U}{\sigma_0} = \frac{\mu}{\sigma_0} \frac{|\sigma_T| \Delta T}{\mu} = \frac{|\sigma_T| \Delta T}{\sigma_0} \quad (10)$$

Considering the thermocapillary effect, Marangoni velocity ($U = |\sigma_T| \Delta T / \mu$) was employed for characteristic velocity.

All droplet formations were performed in 5 replicates in order to obtain the droplet volume reproducibility at each experimental condition. The droplet volumes were calculated directly from captured images. The diameter d and height h of the bottom circle of the droplet were measured, and the droplet volume V was calculated from the formula for calculating the volume of the partial sphere. In addition, the droplet volume was normalized by the pin radius.

$$\beta = \frac{V}{R^3} \quad (11)$$

3.3 Results and Discussion

Droplet formation process without thermocapillary flow The first experiment of droplet formation has been performed without thermocapillary flow. The snap-off of the meniscus and droplet formation processes were observed through a high-speed camera (shown in Fig. 16). As pulling up the pin, the meniscus deformed slowly until $t = -500$ ms. The interface highly deforms at a certain height and progresses to the snap-off of the meniscus (around $t = -50$ ms to $t = 0$ ms). Neck diameter of the meniscus become increasingly smaller and eventually snap-off occurs. After the snap-off, a droplet is formed at the pin end face with damping the shape oscillation. The final droplet volume at $t = 40$ ms was 640 nL.

A satellite droplet, of which volume was approximately 2.5 nL, emerged just after the snap-off. The satellite droplet either bounced on the droplet of the end face or merged without bouncing [10, 43]. Further research on the criteria of the uncertain satellite droplet behavior is needed, however, influences on the main droplet volume and its reproducibility are very small because the satellite droplet is significantly smaller than the main droplet.

At $t = 0$ ms, a cone shape was formed near the pin end face as shown in dashed line [11]. Measured cone volume was about the same as the final droplet volume. Therefore it has been found that the resulted droplet volume is dominated by the cone volume, i.e., the interface shape at the instant of snap-off.

Fig. 16 Highlight of the snap-off without thermocapillary flow. Test fluid was 2-cSt silicone oil. Pin radius $R = 1$ mm. Bo = 0.47. Frame rate was 20000 fps. The t is time, with snap-off occurrence set as zero

3.3.1 Droplet Volume

The droplet volume has been measured with varying the test fluid and pin diameter. Figure 17 shows nondimensional droplet volume β versus Bo. Average values are plotted on the graph. The droplet volume is reproducible with an uncertainty of 1–2%. For Bo < 0.01, the nondimensional droplet volume keeps constant value $\beta = 1.1$. It means that the droplet takes similar figure because the β indicates shape coefficient of the droplet. For Bo > 0.01, on the other hand, the β decreases with increasing the Bond number. Figure 18 shows that the droplet shape on Bo = 1.11 becomes flat as compared to Bo = 4.0×10^{-4}.

Considering the cone shape at the instant of snap-off ($t = 0$ ms in Fig. 16), the cone volume is approximated as follows.

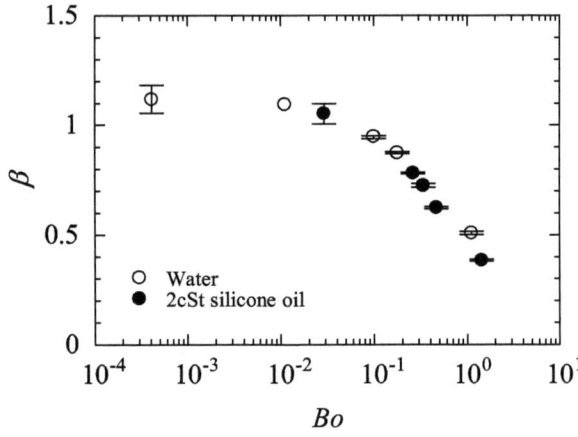

Fig. 17 Nondimensional droplet volume versus Bond number. The error bars represent ±1 standard deviation

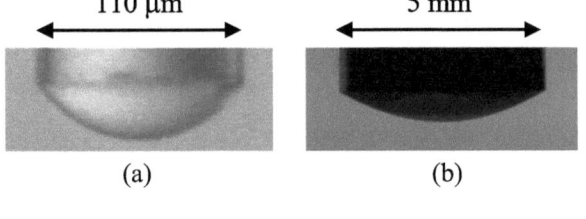

Fig. 18 Comparison of the water droplet shape on **a** Bo = 4.0×10^4 and **b** Bo = 1.11

$$V_{\text{cone}} = \frac{1}{3}\pi R^2 H_{\text{cone}} \qquad (12)$$

Equation (12) can be substituted for Eq. (11) because the cone volume is almost the same volume as the resulted droplet volume. Equation (13) indicates that the nondimensional droplet volume β clearly describes not only droplet shape coefficient but also nondimensional cone height.

$$\beta = \frac{1}{R^3} V_{\text{cone}} = \frac{1}{3}\pi \frac{H_{\text{cone}}}{R} \qquad (13)$$

Consequently, Fig. 17 also indicates that the nondimensional cone height at the instant of snap-off decreases with increasing the Bond number.

3.3.2 Volume Control with Using the Thermocapillary Effect

To control the droplet volume, the thermocapillary flow was induced on the meniscus interface. Pin radius $R = 1.0$ mm and silicone oil were employed. Figure 19 shows droplet formation process which was observed through a high-speed camera. The snap-off process, satellite droplet emergence and final droplet formation were almost as same as the experimental result without thermocapillary flow from a macroscopic

Fig. 19 Highlight of the snap-off with thermocapillary flow. Test fluid was 2-cSt silicone oil. Pin radius $R = 1$ mm. Bo = 0.67, Ca = 0.38. Frame rate was 20000 fps. The t is time, with snap-off occurrence set as zero. White and gray arrows indicate thermocapillary flow (white arrow) and return flow (gray arrow) directions, respectively

viewpoint. Significant difference was occurrence of the dewetting on the pin end face as shown in Fig. 19. The contact line slipped on the pin end face from $t = -10$ ms to $t = 0$ ms. The dewetting was a temporary phenomenon since the contact line rewetted the entire pin end face after the snap-off.

We consider that the dewetting was caused by the thermocapillary flow. The dewetting was observed not only the side view but also top view with employing a transparent pin and seeding tracer particles. Before the snap-off, the induced thermocapillary flow involved return flow in the central part of the meniscus in order to satisfy the continuity. The return flow was blocked by the meniscus constriction at the instant of snap-off, and on the other hand, the thermocapillary effect still induced the flow on the free surface. As the result, the thermocapillary flow drags the contact line on the pin end face.

Fig. 20 Schematics of profiles at instant of snap-off without (left, Ca = 0.01, Bo = 0.47) and with thermocapillary effect (right, Ca = 0.38, Bo = 0.67). Hatching represents conical shape approximation region

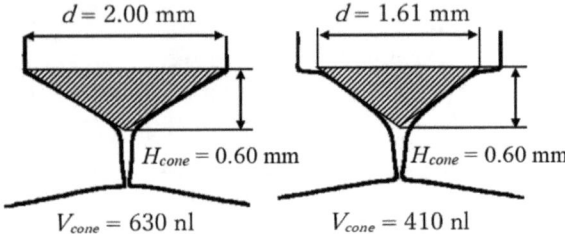

Interface shape at the instant of snap-off without and with thermocapillary flow is illustrated in Fig. 20. The base radius with thermocapillary flow became smaller compared to the cone shape without thermocapillary flow. On the other hand, the cone heights were the same value between the two experiments. That is, the cone shape was sharpened by the dewetting. The final droplet volume was 410 nL on Ca = 0.38 (ΔT = 82.5 K), that is approximately 35% volume reduction compared to the droplet volume without thermocapillary flow.

3.3.3 Volume Reduction by the Thermocapillary Flow

The temperature difference ΔT has been varied to investigate the droplet volume dependence on the capillary number (see Fig. 21). Average values are plotted as a function of Ca for various kinematic viscosities. The droplet volume decreased almost linearly with increasing Ca, and the gradient becomes steeper as the viscosity increases. The droplet volume was reproducible with an uncertainty of 1–2%. Reproducibility of the droplet volume was independent of Ca. These results suggest that the droplet volume can be controlled with using the thermocapillary effect with a high accuracy.

Fig. 21 Non-dimensional droplet volume versus capillary number

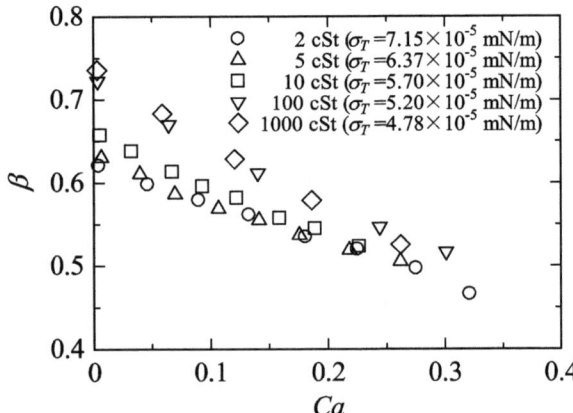

3.4 Conclusion

Small pin dispensing method has been applied for nano liter droplet formation. Snap-off process was observed through a high-speed camera. The droplet volume which was measured by captured images was characterized by the Bond number Bo. The droplet volume was reproducible with an uncertainty of 1–2% under the present experimental condition. At the instant of the snap-off, cone shape emerged on the pin end face and its shape governed the resulted droplet volume. It is found that the nondimensional droplet shape strongly depends on Bo since nondimensional droplet height determined by Bo.

The thermocapillary flow has been induced in order to control the droplet volume. The droplet volume was decreased linearly with increasing the capillary number Ca (i.e., the intensity of the thermocapillary flow). The volume reproducibility and decreasing tendency (linearity) did not depend on the viscosity under the present experimental condition. The droplet volume reduction is caused by the sharpening of the cone shape. The thermocapillary flow causes dewetting of the pin end faces, as a result, only the cone radius is reduced at the instant of snap-off, while the height is not affected by the thermocapillary flow.

Acknowledgements The experimental research on the PAS in the HTW of $m = 1$ through the terrestrial experiments [61] as indicated in Sect. 1 were supported by the Japan Society for the Promotion of Science (JSPS) through Challenging Research (Exploratory) (project number: 20K20977). Special thanks are owed to Prof. Johan Roeraade of KTH for his valuable comments and discussion on the experiment as indicated in Sect. 3. The contributor SU has been supported by Japan Society for the Promotion Science (JSPS) Fellowship.

References

1. Abe Y, Ueno I, Kawamura H (2009) Dynamic particle accumulation structure due to thermocapillary effect in noncylindrical half-zone liquid bridge. Ann NY Acad Sci 1161(1):240–245. https://doi.org/10.1111/j.1749-6632.2008.04073.x
2. Barmak I, Romanò F, Kuhlmann HC (2021) Finite-size coherent particle structures in high-Prandtl-number liquid bridges. Phys Rev Fluids 6:084301
3. Basaran OA (2002) Small-scale free surface flows with breakup: drop formation and emerging applications. Am Instit Chem Eng AIChE J 48:1842–1848
4. Cao ZH, Xie JC, Tang ZM, Hu WR (1991a) The influence of buoyancy on the onset of oscillatory convection in a half floating zone. Adv Space Res 11(7):163–166. https://doi.org/10.1016/0273-1177(91)90276-P
5. Cao ZH, You XT, Tang ZM, Hu WR (1991b) Experimental investigation of thermocapillary convection in half floating zone. Adv Space Res 11(7):229–232. https://doi.org/10.1016/0273-1177(91)90287-T
6. Capobianchi P, Lappa M (2020) Particle accumulation structures in noncylindrical liquid bridges under microgravity conditions. Phys Rev Fluids 5:084304
7. Capobianchi P, Lappa M (2021) On the influence of gravity on particle accumulation structures in high aspect-ratio liquid bridges. J Fluid Mech 908:A29

8. Chun CH, Wuest W (1978) A micro-gravity simulation of the Marangoni convection. Acta Astronaut 5:681–686
9. Darhuber AA, Troian SM (2005) Principles of microfluidic actuation by modulation of surface stresses. Annu Rev Fluid Mech 37:425–455
10. Dell'Aversana P, Neitzel GP (2004) Behavior of noncoalescing and nonwetting drops in stable and marginally stable states. Exp Fluids 36:299–308
11. Eggers J (1997) Nonlinear dynamics and breakup of free-surface flows. Rev Mod Phys 69(3):865–930
12. Fujimoto S, Ogasawara T, Ota A, Motegi K, Ueno I (2019) Effect of heat loss on hydrothermal wave instability in half-zone liquid bridges of high Prandtl number fluid. Int J Microgravity Sci Appl 36(2):2019p360204. https://doi.org/10.15011/jasma.36.2.360204
13. Gotoda M, Sano T, Kaneko T, Ueno I (2015) Evaluation of existence region and formation time of particle accumulation structure (PAS) in half-zone liquid bridge. Eur Phys J Spec Top 224:299–307. https://doi.org/10.1140/epjst/e2015-02361-7
14. Gotoda M, Melnikov DE, Ueno I, Shevtsova V (2016) Experimental study on dynamics of coherent structures formed by inertial solid particles in three-dimensional periodic flows. Chaos 26:073106
15. Gotoda M, Toyama A, Ishimura M, Sano T, Suzuki M, Kaneko T, Ueno I (2019) Experimental study of coherent structures of finite-size particles in thermocapillary liquid bridges. Phys Rev Fluids 4:094301. https://doi.org/10.1103/PhysRevFluids.4.094301
16. Hirata A, Nishizawa S, Sakurai M (1997a) Experimental results of oscillatory marangoni convection in a liquid bridge under normal gravity. https://doi.org/10.15011/jasma.14.2.122
17. Hirata A, Sakurai M, Ohishi N (1997b) Effect of gravity on marangoni convection in a liquid bridge. https://doi.org/10.15011/jasma.14.2.130
18. Hofmann E, Kuhlmann HC (2011) Particle accumulation on periodic orbits by repeated free surface collisions. Phys Fluids 23(7):072106
19. Huh C, Scriven L (1969) Shapes of axisymmetric fluid interfaces of unbounded extent. J Colloid Interface Sci 30(3):323–337
20. Irikura M, Arakawa Y, Ueno I, Kawamura H (2005) Effect of ambient fluid flow upon onset of oscillatory thermocapillary convection in half-zone liquid bridge. Microgravity Sci Technol 16:176–180
21. Kamotani Y, Wang L, Hatta S, Wang A, Yoda S (2003) Free surface heat loss effect on oscillatory thermocapillary flow in liquid bridges of high prandtl number fluids. Int J Heat Mass Transf 46(17):3211–3220
22. Kang Q, Wu D, Duan L, Hu L, Wang J, Zhang P, Hu W (2019) The effects of geometry and heating rate on thermocapillary convection in the liquid bridge. J Fluid Mech 881:951–982
23. Kang Q, Wu D, Duan L, Zhang J, Zhou B, Wang J, Han Z, Hu L, Hu W (2020) Space experimental study on wave modes under instability of thermocapillary convection in liquid bridges on Tiangong-2. Phys Fluids 32:034107. https://doi.org/10.1063/1.5143219
24. Kato K, Sensui S, Noguchi S, Kurose K, Ueno I (2024) Experimental study on coherent structures by small particles suspended in high aspect-ratio ($\Gamma = 2.5$) thermocapillary liquid bridges of high Prandtl number. Eur Phys J Spec Top
25. Kawamura H, Nishino K, Matsumoto S, Ueno I (2012) Report on microgravity experiments of Marangoni convection aboard international space station. J Heat Transfer 134:031005
26. Kuhlmann HC, Hofmann E (2011) The mechanics of particle accumulation structures in thermocapillary flows. Eur Phys J Spec Top 192:3–12
27. Kuhlmann HC, Muldoon FH (2012) Particle-accumulation structures in periodic free-surface flows: inertia versus surface collisions. Phys Rev E 85:046310
28. Kuhlmann HC, Muldoon FH (2013) On the different manifestations of particle accumulation structures (PAS) in thermocapillary flows. Eur Phys J Spec Top 219:59–69
29. Kuhlmann HC, Lappa M, Melnikov DE, Mukin RV, Muldoon FH, Pushkin DO, Shevtsova VM, Ueno I (2014a) The jeremi-project on thermocapillary convection in liquid bridge part a: particle accumulation structures. Fluid Dyn Mater Process 10(1):1–10

30. Kuhlmann HC, Mukin RV, Sano T, Ueno I (2014) Structure and dynamics of particle-accumulation in thermocapillary liquid bridges. Fluid Dyn Res 46:041421
31. Lan CW, Kim YJ, Kou S (1990) A half-zone study of marangoni convection in floating-zone crystal growth under microgravity. J Cryst Growth 104:801–808
32. Lappa M (2013a) Assessment of the role of axial vorticity in the formation of particle accumulation structures in supercritical Marangoni and hybrid thermocapillary-rotation-driven flows. Phys Fluids 25:012101. https://doi.org/10.1063/1.4769754
33. Lappa M (2013) On the variety of particle accumulation structures under the effect of g-jitters. J Fluid Mech 726:160–195
34. Leypoldt J, Kuhlmann HC, Rath HJ (2000) Three-dimensional numerical simulation of thermocapillary flows in cylindrical liquid bridges. J Fluid Mech 414:285–314
35. Matsugase T, Ueno I, Nishino K, Ohnishi M, Sakurai M, Matsumoto S, Kawamura H (2015) Transition to chaotic thermocapillary convection in a half zone liquid bridge. Int J Heat Mass Transfer 89:903–912. https://doi.org/10.1016/j.ijheatmasstransfer.2015.05.041
36. Melnikov DE, Pushkin DO, Shevtsova VM (2011) Accumulation of particles in time-dependent thermocapillary flow in a liquid bridge: modeling and experiments. Eur Phys J Spec Top 192:29–32
37. Melnikov DE, Pushkin DO, Shevtsova VM (2013) Synchronization of finite-size particles by a traveling wave in a cylindrical flow. Phys Fluids 25:092108. https://doi.org/10.1063/1.4821291
38. Melnikov DE, Watanabe T, Matsugase T, Ueno I, Shevtsova V (2014) Experimental study on formation of particle accumulation structures by a thermocapillary flow in a deformable liquid column. Microgravity Sci Technol 26:365–374
39. Mukin RV, Kuhlmann HC (2013) Topology of hydrothermal waves in liquid bridges and dissipative structures of transported particles. Phys Rev E 88:053016
40. Muldoon FH, Kuhlmann HC (2013) Coherent particulate structures by boundary interaction of small particles in confined periodic flows. Phys D 253(15):40–65
41. Muldoon FH, Kuhlmann HC (2014) Different particle-accumulation structures arising from particle-boundary interactions in a liquid bridge. Int J Multiph Flow 59:145–159
42. Muldoon FH, Kuhlmann HC (2016) Origin of particle accumulation structures in liquid bridges: particle-boundary-interactions versus inertia. Phys Fluids 28:073305
43. Neitzel GP, Dell'Aversana P (2002) Noncoalescence and nonwetting behavior of liquids. Annu Rev Fluid Mech 34(1):267–289
44. Nichols BD, Hirt CW, Hotchkiss RS (1980) SOLA-VOF: a solution algorithm for transient fluid flow with multiple free boundaries. Technical report, Los Alamos National Lab. (LANL), Los Alamos, NM, United States, https://doi.org/10.2172/5122053
45. Niigaki Y, Ueno I (2012) Formation of particle accumulation structure (PAS) in half-zone liquid bridge under an effect of thermo-fluid flow of ambient gas. https://doi.org/10.2322/tastj.10.Ph_33
46. Nishimura M, Ueno I, Nishino K, Kawamura H (2005) 3D PTV measurement of oscillatory thermocapillary convection in half-zone liquid bridge. Exp Fluids 38:285–290
47. Nishino K, Yano T, Kawamura H, Matsumoto S, Ueno I, Ermakov MK (2015) Instability of thermocapillary convection in long liquid bridges of high Prandtl number fluids in microgravity. J Crystal Growth 420:57–63. https://doi.org/10.1016/j.jcrysgro.2015.01.039
48. Noguchi S, Ueno I (2023) Spatial-temporal behaviors of low-stokes-number particles forming coherent structures in high-aspect-ratio liquid bridges by thermocapillary effect. Phys Rev Fluids 8:114002. https://doi.org/10.1103/PhysRevFluids.8.114002
49. Oba T, Toyama A, Hori T, Ueno I (2019) Experimental study on behaviors of low-stokes number particles in weakly chaotic structures induced by thermocapillary effect within a closed system with a free surface. Phys Rev Fluids 4:104002. https://doi.org/10.1103/PhysRevFluids.4.104002
50. Ogasawara T, Motegi K, Hori T, Ueno I (2019) Secondary instability induced by thermocapillary effect in half-zone liquid bridge of high Prandtl number fluid. Mech Eng Lett 5(19):00014. https://doi.org/10.1299/mel.19-00014

51. Ohnishi M, Yoshihara S, Azuma H (1992) Computer simulation of thermocapillary motion with surface deformation. Microgravity Q 2:17–28
52. Padday JF, Pétré G, Rusu CG, Gamero J, Wozniak G (1997) The shape, stability and breakage of pendant liquid bridges. J Fluid Mech 352:177–204
53. Petrov V, Schatz MF, Muehlner KA, VanHook SJ, McCormick WD, Swift JB, Swinney HL (1996) Nonlinear control of remote unstable states in a liquid bridge convection experiment. Phys Rev Lett 77:3779–3782
54. Plateau JAF (1873) Statique Expérimentale et Théorique des Liquides Soumis aux Seules Forces Molécu-laires. Gauthier-Villars, Paris
55. Preisser F, Schwabe D, Scharmann A (1983) Steady and oscillatory thermocapillary convection in liquid columns with free cylindrical surface. J Fluid Mech 126:545–567
56. Pushkin DO, Melnikov DE, Shevtsova VM (2011) Ordering of small particles in one-dimensional coherent structures by time-periodic flows. Phys Rev Lett 106:234501
57. Romanò F, Kuhlmann HC. Private communications
58. Romanò F, Kuhlmann HC (2018) Finite-size lagrangian coherent structures in thermocapillary liquid bridges. Phys Rev Fluids 3:094302
59. Romanò F, Kuhlmann HC (2019) Finite-size coherent structures in thermocapillary liquid bridges. Int J Microgravity Sci Appl 36:2019p360201
60. Romanò F, Kuhlmann HC, Ishimura M, Ueno I (2017) Limit cycles for the motion of finite-size particles in axisymmetric thermocapillary flows in liquid bridges. Phys Fluids 29:093303. https://doi.org/10.1063/1.5002135
61. Sakata T, Terasaki S, Saito H, Fujimoto S, Ueno I, Yano T, Nishino K, Kamotani Y, Matsumoto S (2022) Coherent structures of m = 1 by low-Stokes-number particles suspended in a half-zone liquid bridge of high aspect ratio: Microgravity and terrestrial experiments. Phys Rev Fluids 7:014005. https://doi.org/10.1103/PhysRevFluids.7.014005
62. Sasaki Y, Tanaka S, Kawamura H (2005) Particle accumulation structure in thermocapillary convection of small liquid bridge. In: Proceedings of 6th Japan/China workshop on microgravity sciences
63. Sato F, Ueno I, Kawamura H, Nishino K, Matsumoto S, Ohnishi M, Sakurai M (2013) Hydrothermal wave instability in a high-aspect-ratio liquid bridge of Pr >200. Microgravity Sci Technol 25(1):43–58
64. Schwabe D, Mizev AI (2011) Particles of different density in thermocapillary liquid bridges under the action of travelling and standing hydrothermal waves. Eur Phys J Spec Top 192:13–27
65. Schwabe D, Hintz P, Frank S (1996) New features of thermocapillary convection in floating zones revealed by tracer particle accumulation structure (PAS). Microgravity Sci Technol 9:163–168
66. Schwabe D, Mizev AI, Udhayasankar M, Tanaka S (2007) Formation of dynamic particle accumulation structures in oscillatory thermocapillary flow in liquid bridges. Phys Fluids 19:072102
67. Seki T, Tanaka S, Kawamura H (2005) Numerical simulation of particle accumulation structure in oscillatory thermocapillary convection of a liquid bridge. In: Proceedings of thermal engineering conference (in Japanese). Japan Society for Mechanical Engineers, pp 169–170. https://doi.org/10.1299/jsmeted.2005.169
68. Sensui S, Noguchi S, Kato K, Ueno I (2024) Coherent structures formed by small particles in traveling-wave flows. Phys Rev E 110:015101. https://doi.org/10.1103/PhysRevE.110.015101
69. Shevtsova V, Gaponenko Y, Kuhlmann HC, Lappa M, Lukasser M, Matsumoto S, Mialdun A, Montanero JM, Nishino K, Ueno I (2014) The jeremi-project on thermocapillary convection in liquid bridges part b: overview on impact of co-axial gas flow. Fluid Dyn Mater Process 10(2):197–240
70. Shevtsova VM, Mojahed M, Legros JC (1999) The loss of stability in ground based experiments in liquid bridges. Acta Astronaut 44(7–12):625–634
71. Shiomi J, Kudo M, Ueno I, Kawamura H, Amberg G (2003) Feedback control of oscillatory thermocapillary convection in a half-zone liquid bridge. J Fluid Mech 496:193–211
72. Sjödahl J, Kempka M, Hermansson K, Thorsén A, Roeraade J (2005) Chip with twin anchors for reduced ion suppression and improved mass accuracy in MALDI-TOF mass spectrometry. Anal Chem 77(3):827–832

73. Smith MK, Davis SH (1983) Instabilities of dynamic thermocapillary liquid layers part 1 convective instabilities. J Fluid Mech 132:119–144
74. Stone HA, Stroock AD, Ajdari A (2004) Microfluidics toward a lab-on-a-chip. Ann Rev Fluid Mech 36:381–411. https://doi.org/10.1146/annurev.fluid.36.050802.122124
75. Strutt JW VI (1879) On the capillary phenomena of jets. Proc R Soc London 29(196–199):71–97. https://doi.org/10.1098/rspl.1879.0015
76. Takatsuka M, Tanaka S, Ueno I, Kawamura H (2002) Dynamic particle accumulation structure of Marangoni convection in liquid bridge—2. numerical simulation. In: Proceedings of thermal engineering conference (in Japanese), Japan Society for Mechanical Engineers, pp 307–308. https://doi.org/10.1299/jsmeptec.2002.0_307
77. Tanaka S, Kawamura H, Ueno I, Schwabe D (2006) Flow structure and dynamic particle accumulation in thermocapillary convection in a liquid bridge. Phys Fluids 18:067103. https://doi.org/10.1063/1.2208289
78. Terasaki S, Sensui S, Ueno I (2023) Thermocapillary-driven coherent structures by low-stokes-number particles and their morphology in high-aspect-ratio liquid bridges. Int J Heat Mass Transfer 203:123772. https://doi.org/10.1016/j.ijheatmasstransfer.2022.123772
79. Tiwari S, Nishino K (2010) Effect of confined and heated ambient air on onset of instability in liquid bridges of high pr fluids. Fluid Dyn Mater Process (FDMP) 6:109–136
80. Toyama A, Gotoda M, Kaneko T, Ueno I (2017) Existence conditions and formation process of second type of spiral loop particle accumulation structure (SL-2 PAS) in half-zone liquid bridge. Microgravity Sci Technol 29(4):263–274. https://doi.org/10.1007/s12217-017-9544-y
81. Uemura S, Stjernstrom M, Sjodahl J, Roeraade J (2006) Picoliter droplet formation on thin optical fiber tips. Langmuir 22:10272–10276
82. Ueno I, Tanaka S, Kawamura H (2003) Oscillatory and chaotic thermocapillary convection in a half-zone liquid bridge. Phys Fluids 15(2):408–416. https://doi.org/10.1063/1.1531993
83. Ueno I, Abe Y, Noguchi K, Kawamura H (2008) Dynamic particle accumulation structure (PAS) in half-zone liquid bridge—reconstruction of particle motion by 3-D PTV -. Adv Space Res 41:2145–2149
84. Ueno I, Kawazoe A, Enomoto H (2010) Effect of ambient-gas forced flow on oscillatory thermocapillary convection of half-zone liquid bridge. Fluid Dyn Mater Process (FDMP) 6(1):99–108. https://doi.org/10.3970/fdmp.2010.006.099
85. Velten R, Schwabe D, Scharmann A (1991) The periodic instability of thermocapillary convection in cylindrical liquid bridges. Phys Fluids A 3(2):267–279
86. Wang A, Kamotani Y, Yoda S (2007) Oscillatory thermocapillary flow in liquid bridges of high prandtl number fluid with free surface heat gain. Int J Heat Mass Transf 50:4195–4205
87. Wang J, Wu D, Duan L, Kang Q (2020) Transition to chaos of buoyant-thermocapillary convection in large-scale liquid bridges. Microgravity Sci Technol 32:217–227
88. Wanschura M, Shevtsova VM, Kuhlmann HC, Rath HJ (1995) Convective instability mechanisms in thermocapillary liquid bridges. Phys Fluids 7(5):912–925
89. Xu JJ, Davis SH (1984) Convective thermocapillary instabilities in liquid bridges. Phys Fluids 27:1102–1107. https://doi.org/10.1063/1.864756
90. Yamaguchi K, Hori T, Ueno I (2019) Long-term behaviors of a single particle forming a coherent structure in thermocapillary-driven convection in half-zone liquid bridge of high prandtl-number fluid. https://doi.org/10.15011/jasma.36.2.360203
91. Yano T, Nishino K, Kawamura H, Ueno I, Matsumoto S, Ohnishi M, Sakurai M (2011) Space experiment on the instability of Marangoni convection in large liquid bridge—MEIS-4: effect of Prandtl number. J Phys Conf Ser 327:012029
92. Yano T, Nishino K, Kawamura H, Ueno I, Matsumoto S, Ohnishi M, Sakurai M (2012) 3-D PTV measurement of Marangoni convection in liquid bridge in space experiment. Exp Fluids 53:9–20
93. Yano T, Nishino K, Kawamura H, Ueno I, Matsumoto S (2015) Instability and associated roll structure of Marangoni convection in high Prandtl number liquid bridge with large aspect ratio. Phys Fluids 27:024108. https://doi.org/10.1063/1.4908042

94. Yano T, Maruyama K, Matsunaga T, Nishino K (2016) Effect of ambient gas flow on the instability of Marangoni convection in liquid bridges of various volume ratios. Int J Heat Mass Transfer 99:182–191. https://doi.org/10.1016/j.ijheatmasstransfer.2016.03.085
95. Yano T, Nishino K, Ueno I, Matsumoto S, Kamotani Y (2017) Sensitivity of hydrothermal wave instability of Marangoni convection to the interfacial heat transfer in long liquid bridges of high Prandtl number fluids. Phys Fluids 29:044105
96. Yano T, Nishino K, Matsumoto S, Ueno I, Komiya A, Kamotani Y, Imaishi N (2018a) Overview of "Dynamic Surf" project in Kibo-dynamic behavior of large-scale thermocapillary liquid bridges in microgravity. Int J Microgravity Sci Appl 35:350102. https://doi.org/10.15011/jasma.35.1.350102
97. Yano T, Nishino K, Matsumoto S, Ueno I, Komiya A, Kamotani Y, Imaishi N (2018) Report on microgravity experiments of dynamic surface deformation effects on Marangoni instability in high-Prandtl-number liquid bridges. Microgravity Sci Technol 30:599–610
98. Yazawa S, Kawamura H (2010) Experiment of Marangoni convection in a liquid bridge with a low to medium aspect ratio. In: Proceedings of national heat transfer symposium of Japan (in Japanese), p SP409

Open Access This chapter is licensed under the terms of the Creative Commons Attribution 4.0 International License (http://creativecommons.org/licenses/by/4.0/), which permits use, sharing, adaptation, distribution and reproduction in any medium or format, as long as you give appropriate credit to the original author(s) and the source, provide a link to the Creative Commons license and indicate if changes were made.

The images or other third party material in this chapter are included in the chapter's Creative Commons license, unless indicated otherwise in a credit line to the material. If material is not included in the chapter's Creative Commons license and your intended use is not permitted by statutory regulation or exceeds the permitted use, you will need to obtain permission directly from the copyright holder.

Surface-Tension Related Flows in Microgravity and Microscale: Hanging Droplet, Thin Films, and Positive Surface Tension Temperature Coefficient (Self-rewetting Fluids)

Ichiro Ueno, Suguru Shiratori, Anselmo Cecere, Raffaele Savino, and Yoshiyuki Abe

Abstract Thermocapillary-driven convection in geometries different from the half-zone liquid bridges is introduced. The first section illustrates an example in the case of the hanging droplet, a droplet placed on a heated disk whereas a cold disk is installed near the droplet head. The second section introduces another example in the case of the thin film, especially focusing on the solutal Marangoni effect in coating processes. The third section illustrates an example of the application. Effects of the temperature dependence of the surface tension on the thermal devices are introduced, with focusing on the liquids with positive temperature coefficient of surface tension, or "self-rewetting fluids."

I. Ueno (✉)
Tokyo University of Science, Chiba, Japan
e-mail: ich@rs.tus.ac.jp

S. Shiratori
Tokyo City University, Tokyo, Japan
e-mail: sshrator@tcu.ac.jp

A. Cecere · R. Savino
Università degli studi di Napoli, Naples, Italy
e-mail: anselmo.cecere@unina.it

R. Savino
e-mail: rasavino@unina.it

Y. Abe
Prometek, Habère-Lullin, France

1 Thermocapillary-Driven Convection in Hanging Droplet

Ichiro Ueno

1.1 Introduction

A number of experiments on Marangoni convection, or thermocapillary-driven convection, have been conducted under the microgravity as well as normal gravity conditions in order to understand its influences on crystal growth in the floating-zone (FZ) method [19–22, 29]. The half-zone (HZ) geometry has been employed for these fundamental researches; coaxial cylindrical rods with a designated temperature difference are prepared, and a liquid is bridged between the rods by the surface tension and the wettability. A 'half' portion of the full-zone liquid bridge is mimicked by this geometry.

Napolitano et al. [58] found unexpectedly a unique phenomenon during their on-orbit experiment on thermocapillary-driven convection in the HZ liquid bridge in the Space Shuttle. After an accidental breakage of the liquid bridge to form semi-spherical droplets sitting on the rods with a certain temperature difference, they tried to re-bridge the liquid by moving one rod to make a contact between the droplets, but in vain. Some experimental and numerical researches were carried out to indicate the mechanism of such a non-coalescing phenomenon between the face-to-face droplets [25, 55, 101] following that report. This unique phenomenon is caused by the thermocapillary-driven convection on the surface of the droplets; the thermocapillary effect on the heated droplet drags the ambient gas into the region between the droplets' heads, and that effect on the cooled droplet drags out the gas from that region. As a result, the entrained gas forms lubricating gas film between the heads of the droplets to prevent from their coalescing.

Ueno et al. [101] were interested in the flow structure inside the non-coalescence droplets, and indicated the suspended particles in the heated droplet accumulate on a single line like a whip. That is, the particles flow down along the free surface of the droplet, and rise up in the central region toward the heated rod. Each particle radially spreads with a constant phase difference of particle motion toward the free surface in each azimuthal plane. Then the structure by the particles does not change its global shape and 'seems' to rotate like a rigid structure under a constant azimuthal velocity (see Fig. 1 (I)). This unique behavior of the suspended particles is similar to a phenomenon known as the particle accumulation structure (PAS) in HZ liquid bridge of high Prandtl number ($\mathcal{P}$) fluid after Schwabe et al. [85]. It has been known that the PAS emerges in a traveling-wave-type time-dependent 'oscillatory' flow [34, 86, 96, 99] induced by the hydrothermal wave instability in high-$\mathcal{P}$ liquid bridge [30, 69, 106, 113] as indicated in the previous chapter. Such unique particle behaviors nor the induced flow patterns due to the thermocapillary effect, however, had not been reported in the geometry of the hanging droplet before Takakusagi and Ueno [95] to the best of our knowledge. To elucidate particles behaviors inside a droplet brings important knowledge on not only controlling the dispersion of the particles but also processes of crystal growths. Such geometry has been employed especially in the crystal growth processes of proteins and gels known as a hanging drop method [24, 44, 48, 54, 66]. Spontaneous formations of unique structures by particles accumulation have potential applications not only to the high-quality crystal growth, but also to techniques of mixing and stirring particles without using external forces in a closed system of a hanging droplet.

In this section, we introduce experimental results on the flow patterns induced by thermocapillary effect and resultant particles behaviors with a geometry of a single hanging droplet of a high-$\mathcal{P}$ fluid under normal- and microgravity conditions. A hanging droplet formed on a heated rod facing downward is employed as the target geometry instead of face-to-face geometry; a bottom rod, cooled to impose a designated temperature difference between the rods, is placed just beneath the droplet tip (Fig. 1 (II)). The flow fields in the hanging droplet and the behaviors of suspended particles are discussed as functions of the volume and shape of the droplet, and of the temperature difference between the both ends of the droplet. To detect spatio-temporal correlation between the unique particle behaviors and the induced flow fields, three-dimensional particle tracking is conducted to reconstruct time series of particle positions inside the droplet as well as the surface temperature observation with an infrared (IR) camera.

Results under microgravity conditions [107] were accumulated in the series of on-orbit experiments, called Marangoni Experiment in Space (MEIS), which were conducted as the first scientific experiments on the Japanese Experiment Module 'Kibo' aboard the International Space Station (ISS) since 2008. In the MEIS project, five series of experiments, MEIS-1, MEIS-2, MEIS-3, MEIS-4 and MEIS-5 were carried out in the Japan fiscal year of 2008, 2009, 2011, 2010 and 2012, respectively. During the experiments in the MEIS-1, a hanging droplet on a heated rod was realized and the suspended particles in the heated droplet accumulate on a single line like a whip as observed in a small-scale droplet [95]. Another series of experiments on hanging droplet were carried out in MEIS-3: Similar flow regimes in the droplet to those under normal gravity conditions were observed.

1.2 Methods

Apparatus for conducting the series of experiments on the hanging droplet is basically the same as for the half-zone liquid bridges as illustrated in the Sect. 1 in Chapter "Surface-Tension Related Flows in Microgravity and Microscale: Liquid Bridges". Coaxial cylindrical rods of R in radius are placed as each end surface is face-to-face located. A droplet of a designated volume is formed on the surface of the top rod facing downward. The top rod is heated to maintain its temperature at T_h. The coaxial bottom rod is cooled to keep the temperature at T_c. The temperature at the tip of the droplet must be different from the cooled rod temperature. The tip temperature, however, is hardly detected in this system. Thus, the temperature difference is defined as this formula as the matter of practical convenience. The distance between the droplet tip

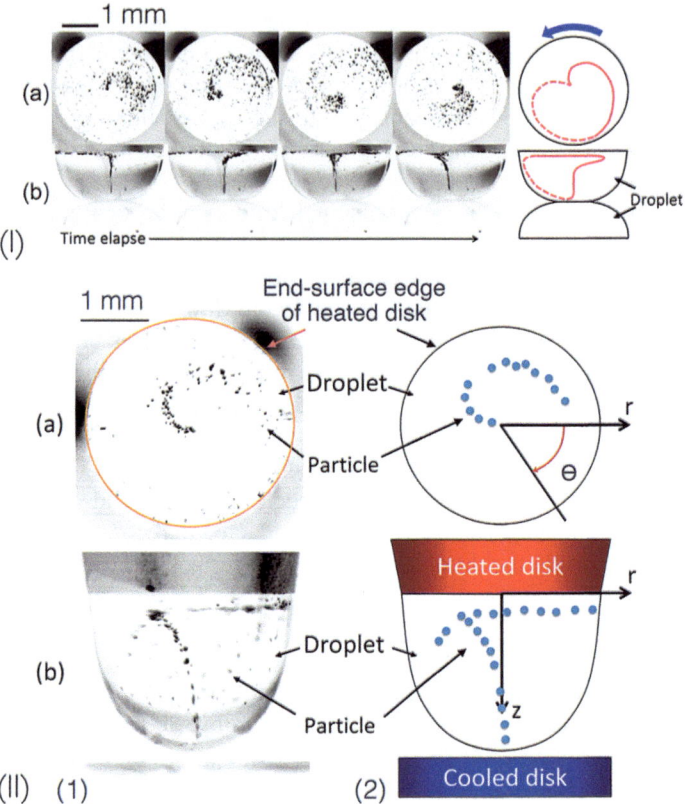

Fig. 1 (I) Time series of snapshots of PAS in face-to-face droplets (first four frames from the left) and its schematic view (right) corresponding to the second frame of the time series; **a** top view and **b** side view. (II) Target geometry of a hanging droplet shown as (1) snapshots and (2) their schematic views with the $r - \theta - z$ coordinate system; **a** top view and **b** side view. The PAS propagates azimuthally in counter-clockwise for both cases. Reprinted with permission from Takakusagi and Ueno [95]. Copyright 2017 American Chemical Society. All rights reserved

and the end surface of the bottom rod is able to be varied by traversing the bottom rod vertically. The distance is kept as small as possible by checking from the side for any droplets of different volumes. The height of the droplet H is also evaluated from the side view.

The intensity of the induced flow is described via the Marangoni number $\mathcal{M}$:

$$\mathcal{M} = \frac{|\gamma_T|\Delta T L}{\rho \nu \kappa} = \mathcal{R}\mathcal{P}, \tag{1}$$

where ρ, ν and κ are the density, kinematic viscosity and thermal diffusivity of the test liquid, respectively, $\gamma_T = \partial \gamma / \partial T$ is a temperature coefficient of surface tension, ΔT is the temperature difference to which the droplet is exposed ($\Delta T = T_h - T_c$), L is the characteristic length of the target geometry, $\mathcal{R}$ is the Reynolds number defined as $(|\gamma_T|\Delta T L)/(\rho \nu^2)$ and $\mathcal{P}$ is the Prandtl number defined as ν/κ.

Two different characteristic lengths are introduced to define $\mathcal{M}$ for the droplet of V in volume: The one is $L = V/R^2$ introduced by Takakusagi and Ueno [95], and the other is $L' = VH/R^3 = L\Gamma$. The latter is defined with considering the effect of droplet shape under the normal gravity condition. Corresponding $\mathcal{M}$s are defined as $\mathcal{M} = \mathcal{M}(L)$, and $\mathcal{M}' = \mathcal{M}(L')$, respectively. The hanging-droplet volume V is evaluated from the sideview images [100, 102]: The positions of the free surface (or perimeter) are detected at each z position, then the volume of the droplet is evaluated by accumulating thin cylindrical 'disks' of one pixel in height under an assumption that the shape of the droplet is axisymmetric.

In the series of the on-ground experiments, the end-surface radius of the top rod, R, is of 1.5 mm. The temperatures of both rods, T_h and T_c, are measured by thermocouples. Through the series of experiments, T_c is kept at 20.0(5) °C. The temperature difference to which the droplet is exposed is defined as $\Delta T = T_h - T_c$. The volume of the droplet V is varied from 5.0 μL to 15.0 μL. The droplet height H and droplet volume V were correlated due to the normal gravity: The droplet height increases almost linearly as a function of the droplet volume. The shape of the droplet is described by aspect ratio Γ defined as H/R. Test fluids are 2- and 5-cSt silicone oils (KF96L-2cs and KF96L-5cs, Shin-Etsu Chemical Co., Ltd., Japan), whose Prandtl number $\mathcal{P}$ at 25 °C is of 28.1 and 68.4, respectively. Gold-coated cross-linking acrylic particles are examined as the test particles. Two different sizes are utilized, whose d_p and the density ρ_p of the particles are $(d_p, \rho_p) = (15\,\mu\text{m}, 1.80 \times 10^3\,\text{kg/m}^3)$ and $(30\,\mu\text{m}, 1.49 \times 10^3\,\text{kg/m}^3)$. The particle motions are monitored by two CMOS cameras with 512 pixel × 512 pixel at a frame rate of 60 frame per second (fps) (FASTCAM- 512PCI, Photron, Inc., Japan); one from the top through the top rod and the other from the side. The surface temperature is simultaneously measured with an infrared (IR) camera with 320 pixel × 240 pixel at a frame rate of 60 fps with a temperature resolution of 0.05 K at 30 °C (Thermography R300, NEC Avio Infrared Technologies Co., Ltd., Japan) with a closeup lens (TVC-2100UB, NEC Avio Infrared Technologies Co., Ltd., Japan).

For the on-orbit experiments, a hanging droplet of the silicone oil is formed between two coaxial disks of 15.0 mm in radius, and its ambient is filled with Argon gas. The heated disk is made from transparent sapphire to facilitate observation of the internal flow fields, and the heating is done by a transparent indium tin oxide (ITO) film coating on its surface and in contact with the liquid. The cooled disk is made from aluminum and its temperature is controlled by a Peltier device. The temperature of the cooled disk is maintained at 20 °C. Three finger-sized black/white CCD cameras are mounted near the heated disk, and another CCD camera is placed for side view. The cameras have image sensors of 768 pixel × 494 pixel, and a frame rate at 30 fps. An IR camera having a wavelength sensitivity of 8 μm to 14 μm is used to visualize surface temperature from position shifted by $\pi/2$ from the side-view camera in azimuthal direction. Test fluids for the on-orbit experiments are 5- and 20-cSt silicone oils of 67 and 207 in $\mathcal{P}$ at 25 °C, respectively. Details of experimental apparatus are described in references [46, 47, 114–116]. Test particles are the same kind but in different size as the on-ground experiments, whose diameter and density are of $(180\,\mu\text{m}, 1.30 \times 10^3\,\text{kg/m}^3)$.

1.3 Flow Patterns

Figure 2 illustrates typical example of time series of snapshots of particle motions and surface temperature deviation from the averaged temperature in time during the fundamental period of the hydrothermal wave obtained in the on-ground experiments. The top views and side views are shown in (a) and (b), respectively. The distributions of the surface temperature and its deviation from the averaged field are illustrated in (c) and (d), respectively. It is noted that the side views are detected at the azimuthal position $\theta = \pi/2$, and the surface temperature at $\theta = 0$. When ΔT or V is relatively low, that is, $\mathcal{M}$ is lower than the threshold, an axisymmetric time-independent 'steady' flow emerges. Note that the flow never exhibits the transition to any 'oscillatory' flows in any droplets of $\Gamma \lesssim 1.0$ by increasing ΔT even up to 60 K in the present system [95]. To realize oscillatory flows in small-Γ droplet, a large temperature difference must be added between the both ends of the droplet. It must be noted that the evaporation of the test fluid becomes inevitable in the case of $\Delta T > 60$ K. The particle motions in the hanging droplet are categorized based on the thermal-flow fields of (1) axisymmetric 'Steady' flow state and (2)–(4) three-dimensional 'oscillatory' flow states. The latter state can be further separated into (2) 'Standing wave' and (3)–(4) 'Traveling wave,' judging from behavior of particles in the droplet as well as the temperature variation over the droplet surface. It is found that two kinds of the particle accumulation structures are realized in the Traveling wave; they are named as (3) 'PAS1' for the simpler structure and (4) 'PAS2' for more complicated structure. The PAS in the hanging droplet is always observed whenever the traveling hydrothermal wave occurs. This is the uniquely different point comparing to the case of the half-zone liquid bridge. In the half-zone liquid bridge, the PAS emerges under certain limited conditions in the Marangoni number as well as the aspect ratio [34, 36, 62, 96, 99].

In the Steady flow state (row (1) in Fig. 2), any particles never exhibit azimuthal motion during one turn-over motion in certain $r - z$ plane. The temperature deviation over the free surface exhibits quite tiny temporal variations; the maximum variation of $|\hat{T}|$ of $\lesssim 0.03$ K, smaller than the resolution of the IR camera of ± 0.05 K. The temperature deviation shows negligible changes against time. It must be noted again that the particles settle to the bottom in the droplet under the condition $\Delta T = 0$ because of the normal gravity. Once ΔT is increased, the particles follow the thermocapillary-driven flow. The particles slide down near the free surface from the region around the edge of top rod to the tip of the droplet. Then they rise up from the bottom toward the top rod through the central region of the droplet, and they spread radially from the central region of the droplet toward the free surface in certain $r - \theta$ plane. This is the basic patterns of the particles due to the thermocapillary-driven basic flow inside the droplet. The streamlines near the free surface become denser all along to the tip of the droplet, where is the coldest part over the free surface.

As $\mathcal{M}$ is increased, three-dimensional oscillatory 'Standing wave' emerges beyond a threshold ΔT_c or $\mathcal{M}_c$. In the oscillatory flow regimes, there exist time-dependent thermal-flow components due to the hydrothermal wave instability in addition to the basic flow. In the 'Standing wave' state (row (2) in Fig. 2), the particles start swinging alternatively from side to side. In this state, the surface temperature deviation exhibits an alternate behavior: The colder spot and the hotter one appear alternately. There is a pair of a node and an antinode in the hydrothermal wave realized in the droplet (that is, the azimuthal mode number $m = 1$ as seen in the high-aspect-ratio liquid bridges [31, 72, 97]). Dashed line in the schematic image of the top view in Fig. 2 (2) indicates the plane of the node. As the particles travel in the droplet, they tend to gather in the middle region of the antinode over the free surface. This region corresponds to the coldest spot over the free surface. Once the particles are gathered to flow in the coldest spot, they do not move azimuthally over the free surface in the fully developed state. The particles fall down to the tip of the droplet, and rise up toward the hot disk almost along the center axis of the droplet following the flow of the relatively colder fluid. This returning flow makes the opposite side of the droplet cooled to the prior coldest spot, and brings the particles to the opposite side of the droplet. Thus, thermal wave behaves as the 'Standing wave,' and they do not move in azimuthal direction. Through a series of experiments, the azimuthal position of node plane appears randomly when ΔT exceeds the threshold, and it never changes its position while the 'Standing wave' emerges in the droplet. This is a reflection of a uniformity of the geometrical and thermal-fluid boundary conditions of the apparatus.

As $\mathcal{M}$ is further increased, the flow field exhibits a transition to the 'Traveling wave' state (row (3) in Fig. 2). The rotating behavior of the particle accumulation structure (PAS) like a rigid structure without any dynamic deformation is quite similar to that in the liquid-bridge geometry [36, 50, 61, 62, 86, 96, 97, 99]. The surface temperature deviation also seems to rotate as a rotating wave with a propagation angle, that is, the azimuthal phase of the hydrothermal wave depends on the vertical coordinate and that the dependence of the phase on z is approximately linear. In Fig. 2 (3), the curved particle line (PAS) and the thermal wave over the free surface travel in the counter-clockwise direction when one observes from above the droplet. Such 'oscillating' flow patterns beyond the threshold $\mathcal{M}_c$ accompanying with thermal wave over the free surface correspond to the hydrothermal wave predicted by Smith and Davis [91] in the liquid films and by others [30, 69, 106, 113] in liquid bridges as aforementioned. It is noted that the azimuthal direction of the traveling behavior is not predictable through the conduction of the experiments. As $\mathcal{M}$ is increased more, the particle accumulation structure exhibits a transition from 'PAS1' to 'PAS2' in the regime of the Traveling wave. In the PAS2 state, although the flow field still remains periodic and the thermal wave over the free surface travels in azimuthal direction without any significant difference comparing to PAS1 case, the particles disperse, and there seem to exist multiple pathlines inside the droplet. It is noted that, as discussed latter, this PAS2 has a single closed structure and never deforms its shape in rotation; that is, the PAS2 seems to rotate as a rigid body as well as the PAS1. As imposing much larger temperature difference, the flow field becomes a chaotic flow.

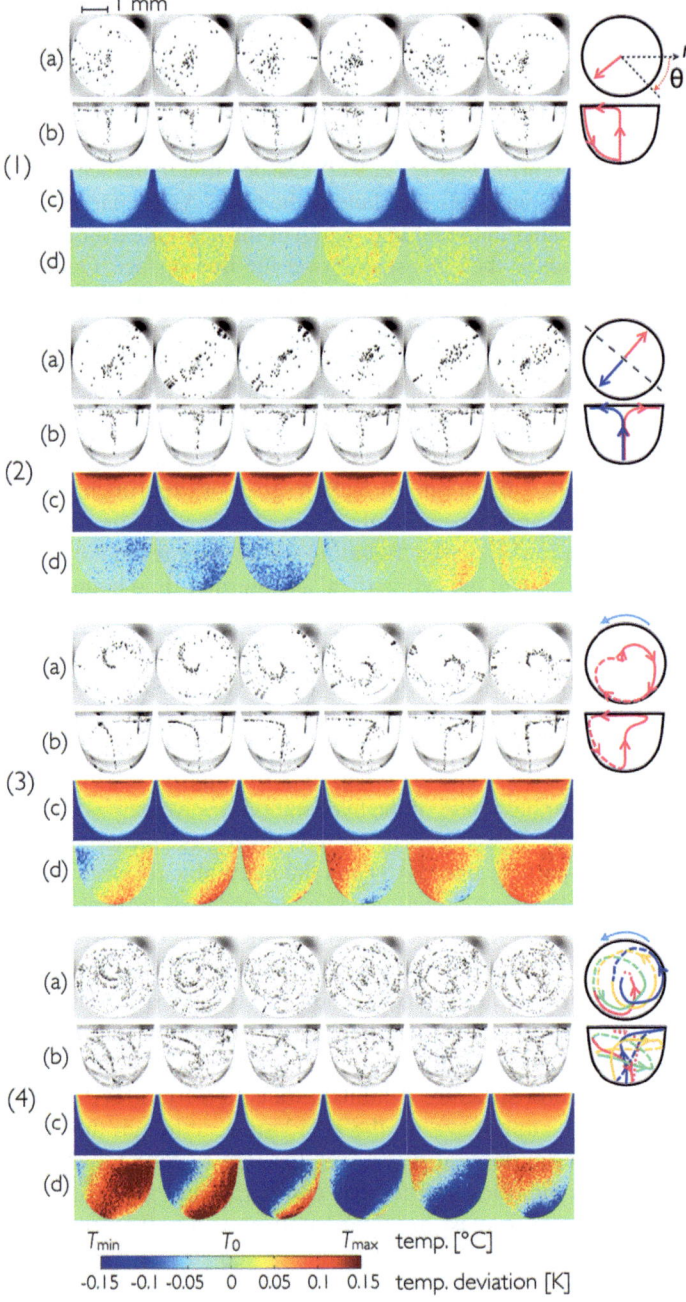

◀**Fig. 2** Time series of snapshots of each flow regimes in the droplet of 5 cSt silicone oil (whose volume and aspect ratio are of 11 μL and of 1.5, respectively) in all flow regimes; (1) Steady flow, (2) Standing wave, (3) PAS1 and (4) PAS2 in Traveling wave; **a** top view, **b** side view, **c** and **d** distributions of surface temperature and its deviation from averaged field, respectively. Note that the camera for side view is located at the azimuthal position $\theta = \pi/2$ and the IR camera at $\theta = 0$. Experimental conditions of $(\Delta T, \mathcal{M})$ in each flow regime are (1) (10.2 K, 0.99×10^4), (2) (20.3 K, 2.1×10^4), (3) (21.7 K, 2.3×10^4), and (4) (31.7 K, 3.6×10^4). Time intervals Δt are of (1) 0.80 s, (2) 0.178 s, (3) 0.172 s, and (4) 0.126 s, respectively. The solid line shown in schematic of (3) is a part of the PAS line inside the droplet, and the dashed line is that along the droplet surface. The solid lines shown in schematic of (4) are a part of the PAS line located in front half of the droplet, and the dashed lines are those in rear half of the droplet. Combinations of (T_{min}, T_0, T_{max}) in °C of the color bar for rows **c** are (1) (24.0, 25.5 and 27.0), (2) (29.5, 32.5 and 35.5), (3) (30.2, 33.7 and 37.2), and (4) (37.0, 42.0 and 47.0), respectively. Reprinted with permission from Takakusagi and Ueno [95]. Copyright 2017 American Chemical Society. All rights reserved

1.4 Particle Accumulation

In the PAS1 state, as described above, the particles disperse again in the droplet, but form a whip-like structure that 'seems' to rotate. The trajectory of an individual particle in the hydrothermal wave state accompanying PAS1 lies approximately in a plane of $\theta = $ const. in a single turnover period of the particle motion. Such behavior is similar to that in the case of the Steady flow. However, upon rising from the cold bottom spot to the hot wall near the z axis the particle can move to another plane at another azimuth. This behavior is most likely due to the symmetry breaking of the hydrothermal wave with azimuthal wave number $m = 1$. The essence of rotating whip-like structure is that each particle steadily travels in each $r - z$ plane but with a constant phase difference from neighboring particles as shown in Fig. 3a. The square marks indicates the positions of different particles at the same instance t [s] $= t_0$, and the circles at $t = t_0 + \Delta t$ ($\Delta t = 0.33$ s). The solid red and blue lines indicate the apparent structures with the particles at $t = t_0$ and $t = t_0 + \Delta t$, respectively. At a certain instant, particles exist on a curved structure (also see the square marks and the solid line in the schematics of the top view illustrated in Fig. 3b). After a time elapsed, the particles travel almost straight outward to keep the shape of the apparent structure (the circle marks and the solid line in the same frame). That is, a particle on a $r - z$ plane at $\theta_1 + \Delta\theta$ in azimuthal direction exhibits the same motion in the $r - z$ plane as the other particle at θ_1 with a time difference of $\Delta\theta/(2\pi f_0)$, where f_0 is the fundamental frequency of the traveling hydrothermal wave. Thus the PAS seems to rotate even though each particle travels in each $r - z$ plane.

In the PAS2 state, the particle behaviors become much more complicated: The particles exhibit azimuthal motion in a region of $r > 0$ in addition to the basic circular movement in a $r - z$ plane. The apparent structure of the PAS2 consists of a single closed line as that in the PAS1. This particle on the PAS2 travels in inner area of the droplet between the surface and the center line in addition to the basic trajectory of the PAS1. One can see a folded structure in which the particle come close to the free surface several times at different θ before rising up near the center line of the

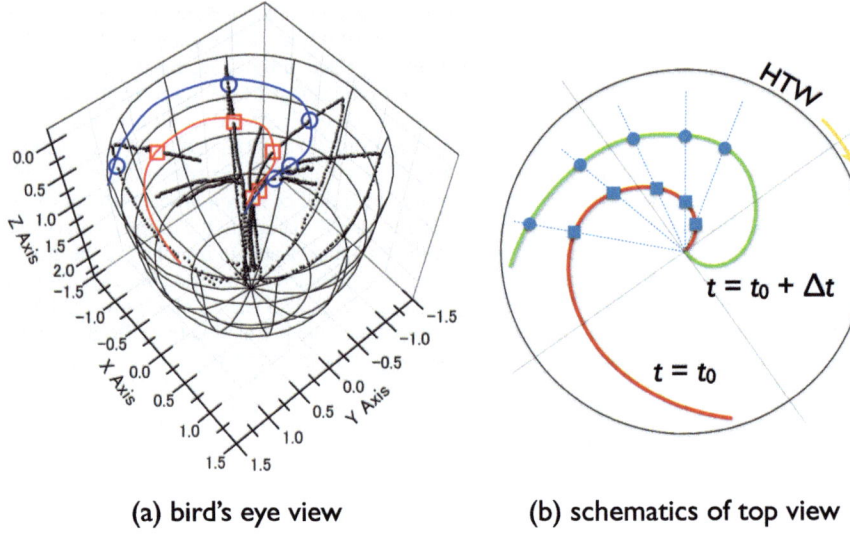

(a) bird's eye view   (b) schematics of top view

Fig. 3 Essence of rotating whip-like structure: **a** bird's eye view of the particle path lines and corresponding PAS1 at different instants (squares at $t = t_0$ and circles at $t = t_0 + \Delta t$ ($\Delta t = 0.33$ s). Each particle steadily travels in each $r - z$ plane but with a constant phase difference from neighboring particles. **b** Schematic top view of typical particles traveling near end-surface (at $z \sim 0$) on the PAS1 at two different instants. Reprinted with permission from Takakusagi and Ueno [95]. Copyright 2017 American Chemical Society. All rights reserved

droplet: The PAS2 has a structure elongated and deformed from the PAS1 under higher thermocapillary effect. So that the particle comes closer to the free surface three times in a single cycle, whereas only once in the case of the PAS1. The behavior of the particle on the PAS2 is similar in quality to that of the particle on the 'SL-2 PAS' [96, 99] as reconstructed by Niigaki and Ueno [59]. Knowledge on the flow structure and accompanying particle behaviors driven by the higher thermocapillary effect, however, has not been intensively accumulated.

Axisymmetric 'Steady flow' state and PAS1 in the 'Traveling wave' state are also realized in the on-orbit experiments in MEIS-3 as shown in Fig. 4 (1) and (2), respectively. In PAS1, the particles scatter azimuthally after the transition, and travel in different $r - z$ plane with a constant phase difference with respect to the fundamental frequency of traveling HTW, f_0. That is, considering two particles travelling in $\theta = \theta_1$ and $\theta = \theta_1 + \Delta\theta$, each particle exhibits the same basic turnover motion as seen in Steady flow, but with a phase difference of $\Delta\theta/(2\pi f_0)$ as observed in the elongated droplets ($\Gamma > 1$) on-ground experiments (see Fig. 3).

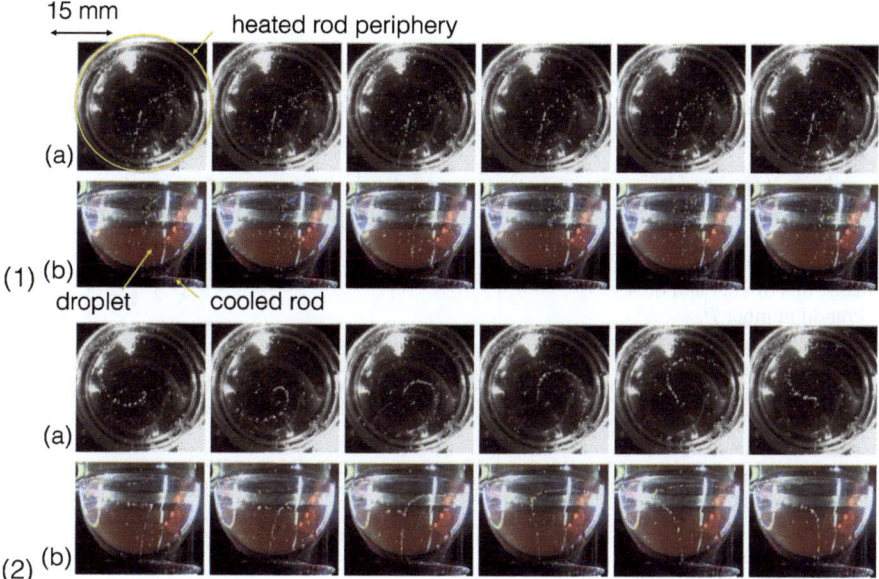

Fig. 4 Time series of snapshots in droplet of 20 cSt silicone oil (whose volume and aspect ratio are of 8.4 cc and of 1.1, respectively) in (1) Steady flow and in (2) Traveling wave realized in the 'Kibo' aboard the International Space Station; **a** top view and **b** side view. Experimental condition of ΔT in each regime is (1) 23.7 K and (2) 39.4 K, and time intervals Δt is 5.5 s and 2.2 s, respectively. The radius of the disk is of 15 mm. Reprinted from Watanabe et al. [107] with permission from Elsevier. All rights reserved

1.5 Onset of Oscillatory Convection

Flow field exhibits a transition from 2-D steady flow to 3-D time-dependent one, as described in this section, when ΔT exceeds the threshold ΔT_c. One needs less temperature difference to realize the oscillatory flow as increasing Γ. And, one needs less temperature difference under the constant Γ to realize the flow transition for larger droplet. Figure 5a indicates the transition condition in term of the critical Marangoni number $\mathcal{M}_c$. The critical Marangoni number is evaluated by putting the critical temperature difference ΔT_c and the characteristic length $L = L_1 = V/R^2$ into the Eq. (1) [95]. The transition conditions in $\mathcal{M}_c$ exhibit almost uniform behavior despite of the difference of the droplet size and $\mathcal{P}$ of the test fluid. The threshold $\mathcal{M}_c$ decreases monotonically as Γ. This indicates that the flow becomes unstable as the Γ increases; the longer the distance in the direction normal to the temperature difference becomes, the more unstable the induced flow becomes. One would notice that there exist slight dispersions in the $\mathcal{M}_c$ in varying R for the results in the terrestrial experiments. This can be explained by the variation of the droplet shape; the droplet shape does not have a similarity in varying R under the normal gravity condition while the droplet volume are varied in proportion to R^3 [95]. The hanging droplet

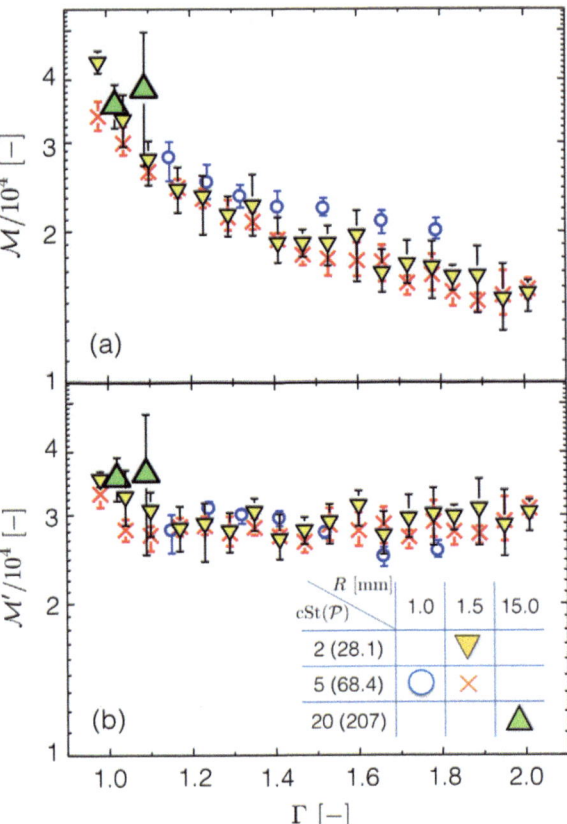

Fig. 5 Critical Marangoni number $\mathcal{M}_c$ as a function of aspect ratio Γ; Marangoni number defined as using different characteristic length **a** $\mathcal{M}_c = \mathcal{M}_c$ ($L = L_1 = V/R^2$) and **b** $\mathcal{M}'_c = \mathcal{M}_c$ ($L = L' = L_1\Gamma$). Test fluids for on-ground experiments are 2- and 5-cSt silicone oils of 28.1 and 68.4 in Prandtl number $\mathcal{P}$, respectively, and those for the on-orbit experiments are 5- and 20-cSt silicone oils of 67 and 207 in $\mathcal{P}$, respectively. The values of the Prandtl number are measured at 25 °C. Reprinted from Watanabe et al. [107] with permission from Elsevier. All rights reserved

distorts its shape of larger Γ in case of larger V and/or larger R; the droplet becomes deformed to be in catenary shape due to its own weight under the normal gravity conditions.

Another characteristic length $L = L' = VH/R^3 = L_1\Gamma$ is thus introduced to take account of the shape effect on the transition condition by $\mathcal{M}'_c = \mathcal{M}_c(L')$ (Fig. 5b). Results obtained in the MEIS-3 are also plotted in Fig. 5a, b. It is noted again that the on-orbit experiments on the hanging droplet enable us to avoid any static deformation due to the gravity (see Fig. 4). Thus the aspect ratio of the droplet in the on-orbit experiment would be different from that under the normal gravity. Nevertheless, the critical value of the flow transition becomes almost constant against the aspect ratio $\Gamma = H/R$ in spite of the differences of R and/or $\mathcal{P}$. From the above, the $\mathcal{M}_c$ is found to be significantly affected by the shape of the droplets, which is influenced by the gravity, and not sensitive to the droplet size. It must be noted that the data at $\Gamma = 1.1$ by the on-orbit experiment has relatively larger range of uncertainty. This is due to an unavoidable larger step of the temperature-difference variation to explore the transition point [107].

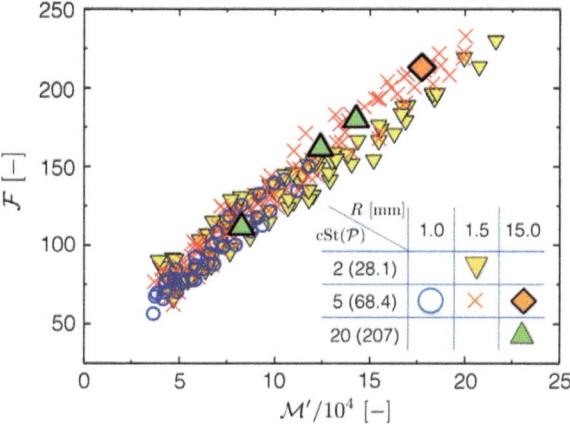

Fig. 6 Non-dimensional frequency $\mathcal{F}$ in the oscillatory flow as a function of Marangoni number $\mathcal{M}'$. Test fluids for on-ground experiments are 2- and 5-cSt silicone oils of 28.1 and 68.4 in Prandtl number $\mathcal{P}$, respectively, and those for the on-orbit experiments are 5- and 20-cSt silicone oils of 67 and 207 in $\mathcal{P}$, respectively. The values of the Prandtl number are measured at 25 °C. Reprinted from Watanabe et al. [107] with permission from Elsevier. All rights reserved

In the oscillatory flow regimes, the flow becomes time-dependent with a fundamental frequency f_0. The fundamental frequency is linearly increased as ΔT at each Γ, and its gradient against $\mathcal{M}'$ becomes larger as Γ. To illustrate comprehensive characterization of the oscillatory flows in ranges of droplet size and $\mathcal{P}$ of the test fluids, a non-dimensional frequency $\mathcal{F}$ is introduced defined as follows, and plot against $\mathcal{M}'$ in Fig. 6.

$$\mathcal{F} = \frac{f_0 H^2}{\kappa}. \tag{2}$$

In the figure, the results obtained in the on-orbit experiments, MEIS-1 and MEIS-3, are also plotted as the diamond and triangle, respectively, in addition to the on-ground experiments. The correlation between $\mathcal{F}$ and $\mathcal{M}'$ is approximately describe as $\mathcal{F} \sim 9.43 \times 10^{-4} \mathcal{M}' + 34.4 \; \{\mathcal{M}' \mid \mathcal{M}' \geq \mathcal{M}'_c\}$, where $\mathcal{M}'_c$ is the critical Marangoni number as shown in Fig. 5. These data are in good agreement with the terrestrial experiments in spite of rather large differences of R and $\mathcal{P}$. The fundamental frequency f_0 of the oscillatory flows in the hanging droplet in a wide range of $\mathcal{M}'$ is well described via a non-dimensional formula independently of the droplet size and $\mathcal{P}$ of the fluid.

1.6 Concluding Remarks

Thermocapillary-driven convection and resultant particle behaviors in the hanging droplets are introduced with varying the aspect ratio and the size of the droplet, and the Prandtl number $\mathcal{P}$ of the test fluid through the on-ground and on-orbit experiments. A droplet is hung on a heated cylindrical rod facing downward, and another rod cooled is placed just beneath the droplet, then we expose the droplet to a designated temperature difference between the rods in order to induce thermocapillary-driven convection inside the droplet due to the non-uniform temperature distribution over the free surface. The regimes of convection fields are categorized from the steady flow to the oscillatory flows through the observation of the suspended particle motion and the surface temperature variation with varying the temperature difference ΔT. With increasing ΔT, the flow exhibits a transition from 2-D steady to 3-D time-dependent 'oscillatory' flows, as seen in the half-zone liquid bridges [65, 106] and thin liquid films [67, 91]. In the range of the present conditions of the experiments, the flow always exhibits the oscillatory flows of distinct azimuthal modal structures with $m = 1$ in azimuthal wave number in spite of the variation of the droplet shape. This is a unique aspect of this geometry comparing to the flow fields in the different geometries such as liquid bridge [45, 61, 72, 88, 97] and evaporating droplet.

The suspended particles gather to form a closed structure inside the droplet in the Steady flow state and the Traveling wave flow in the oscillatory state. In the Steady flow state, the particles form a closed ring at a certain azimuthal position, that is, in a $r - z$ plane, and the ring-like structure never changes their azimuthal position as far as the flow is in the two-dimensional steady regime. In the Traveling wave flow states, the ring-like structure in the Steady flow state stretches to deform in the azimuthal direction but keeping the closed structure. Two kinds of the particle accumulation structures are realized: 'PAS1' under lower $\mathcal{M}$ with the simpler structure, and 'PAS2' under higher $\mathcal{M}$ with complicated structure. The PASs in the hanging droplet are always observed whenever the traveling hydrothermal wave occurs. This is the uniquely different point comparing to the case of the half-zone liquid bridges [7, 31, 35, 36, 61, 62, 72, 86, 96, 97, 99]. Onset conditions of the oscillatory flows in the hanging droplets are also introduced. Transition point is well described by Marangoni number $\mathcal{M}' = \mathcal{M}\Gamma$, which measures the effect of the droplet aspect ratio $\Gamma = H/R$. The fundamental frequency of the oscillatory flow is linearly increased as the Marangoni number $\mathcal{M}'$, which is well described via a non-dimensional formula independent of the droplet size and the Prandtl number $\mathcal{P}$ of the test fluid.

2 Thin Liquid Film

Suguru Shiratori

2.1 Overview

The Marangoni effect also plays a significant role in thin liquid film flows, to which we will review in this section. Coating process of thin liquid films has enormous practical applications in the micro-fabrication techniques for a wide range of industrial manufacturing, for instance, semiconductors, micro-electro mechanical systems (MEMS), color filters of image sensors and displays, and anti-reflection films for optical lenses. For semiconductor devices, coated photoresist films are removed in the middle of the fabrication process, whereas for MEMS devices, anti-reflection films for optical lenses, or color filters of image sensors, some parts of photoresist films are used as final structures of the devices. Therefore, the design of devices determines the film thickness, which often exceeds the order of $10\,\mu$m, and its uniformity is desired at a high level. However, liquid film suffers from various thickness undulations due to physical phenomena during the coating process. The better-known thickness undulations are "striation" and "edge-bead", which are shown in Fig. 7.

Striation is typical thickness undulation, which appears during the spin-coating as radial spoke-like patterns all over the film. Figure 7a shows a thickness distribution measured after the spin coating of the mixure of the epoxy resin and xylene solvent. The mechanism of striation has been investigated and, Marangoni-Bénard instability is considered to play a fundamental role [8–11, 23, 37–39, 49, 90]. Edge-bead is another typical thickness undulation, which appears as a thick ridge along the substrate periphery (Fig. 7b). The substrate region eroded by edge beads is not usable, thus they result in the loss of product yield. Shiratori and Kubokawa [89] investigated edge-bead generation for the case where the bead had a double-peaked shape in the direction away from the substrate periphery, and proposed a simple explanation for the mechanism of the double peak.

Such thickness variations must be avoided or suppressed in industry. To this end, numerical simulations are often used to find the optimal coating conditions, which allows thickness undulations to be minimized. In previous research on thin liquid films, many numerical methods have been developed [27, 33, 68, 87, 94, 108, 117].

In the following subsections, the generation mechanism for these thickness undulation will be explained after the mathematical formulation of the models.

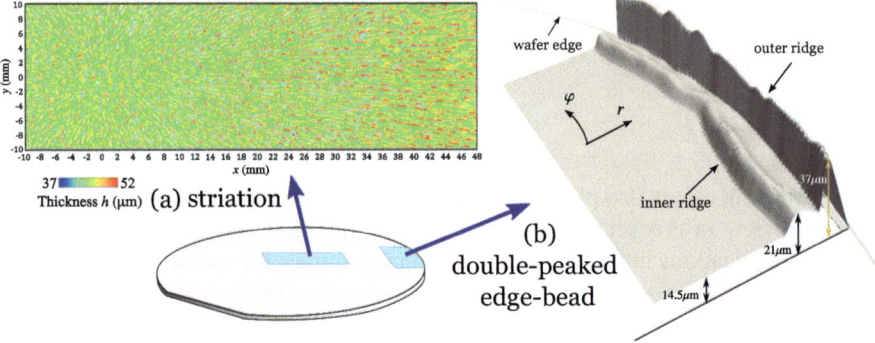

Fig. 7 Thickness distributions which indicate the striation (**a**) and the double-peaked edge-bead (**b**) in the films generated by the spin-coating of a mixture solution which consists of epoxy resin and xylene solvent. Panel **a**: Reprinted from Shiratori et al. [90] with permission from Elsevier. All rights reserved. Panel **b**: Reprinted from Shiratori and Kubokawa [89] with the permission of AIP Publishing. All rights reserved

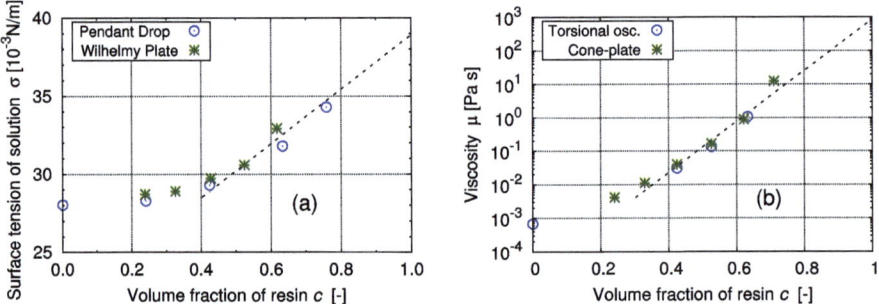

Fig. 8 Physical properties of the mixture of epoxy resin (EPHE-3150) and xylene solvent as a function of the resin fraction c. **a** Surface tension measured by a pendant drop method (circles) and by the method of Wilhelmy plate (stars). **b** Viscosity measured by the vibrational viscometer (circles) and by the rotational viscometer (stars). Reprinted from Shiratori and Kubokawa [89] with the permission of AIP Publishing. All rights reserved

2.2 Mathematical Formulation of the Liquid Film Flows

Physical properties and evaporation rate

In the industrial coating process, the liquid is consist of the resin and volatile solvent. Due to the evapolation of the solvent, the concentration of the resin in the solution has spatial variation, which leads the surface tension variation along the free surface. For many combinations of solvent-resin mixtures, the surface tension becomes larger for the larger concentration of the resin, as shown in Fig. 8a. Therefore, the solutal

Marangoni effect (solutocapillary effect) often be dominant rather than the thermal Marangoni (thermocapillary effect). Besides the surface tension, the viscosity, diffusivity, evaporation rate are also dependent on the concentration. In the rest of this subsection, the concentration is described as the volume fraction of resin c. The surface tension $\sigma(c)$, viscosity $\mu(c)$, diffusivity $D(c)$, and evaporation rate $E(c, x)$ are modeled as

$$\sigma(c) = \sigma_0 + \sigma_c (c - c_b), \tag{3}$$
$$\mu(c) = \mu_0 \exp\left[K_\mu (c - c_b)\right], \tag{4}$$
$$D(c) = D_0 \exp\left[-K_D (c - c_b)\right], \tag{5}$$
$$E(c, x) = E_0 (1 - \chi(c)) \mathcal{F}(x) \mathcal{G}(\Omega), \tag{6}$$

where c_b is the reference concentration, σ_0, μ_0, D_0 are the reference values for each properties at $c = c_b$, and $\sigma_c = \partial \sigma / \partial c$ is a concentration coefficient of the surface tension. The exponents K_μ and K_D are empirical constants which are determined from the measurements. The evaporation of the solvent is considered to be dependent on not only the concentration but also the spatial location x and the rotation rate Ω. E_0 is the constant and has unit of velocity. $\chi(c)$ stands for the concentration dependence, where the Raoult's law can be assumed as

$$\chi(c) = \frac{(1 - c)(\rho_s/M_s)}{c(\rho_r/M_r) + (1 - c)(\rho_s/M_s)}, \tag{7}$$

where M_s and M_r are the molar mass of the solvent and resin, respectively. ρ_s and ρ_r are The function $\mathcal{F}(x)$ denotes the spatial dependence, whereas $\mathcal{G}(\Omega)$ the dependence on the rotation rate. These functions are often selected as

$$\mathcal{F}(x) = \left[1 - \frac{r^2}{R}\right]^{\frac{1}{2}}, \tag{8}$$
$$\mathcal{G}(\Omega) = \left(1 + \frac{\Omega}{[\text{rad/s}]}\right)^{\frac{1}{2}}, \tag{9}$$

where R and r are the radius of the substrate and radial length from the substrate center, respectively.

Longwave approximation with Marangoni effect

Generally, the liquid film flows have large difference of length scales in between horizontal and vertical directions. By applying some approximations, the governing equations for the liquid film flows can be much simplified. In the following subsections, such approximations are briefly explained. The detailed formumlation can be found e.g. in Oron et al. [63] and Craster and Matar [18].

Let us consider to nondimensionalize the Navier-Stokes equation using separate length scales for the horizontal (L) and vertical (h_0) directions. Because the length ratio $\varepsilon = h_0/L$ is sufficiently small ($\varepsilon \ll 1$), the terms of 2nd or higher order in ε can be neglected. After applying the lubrication approximation, which is also called longwave approximation, the velocity fields in the liquid film can be written as

$$\frac{\partial p}{\partial x} + \frac{\partial \phi}{\partial x} = \mu \frac{\partial^2 u}{\partial z^2}, \tag{10}$$

$$\frac{\partial p}{\partial z} + \frac{\partial \phi}{\partial z} = 0, \tag{11}$$

where u is the horizontal velocity, p is the pressure, and ϕ is the centrifugal potential

$$\phi = -\frac{1}{2}\rho r^2 \Omega^2. \tag{12}$$

For the boundary conditions, the following no-slip and the Marangoni effect are considered:

$$u = 0 \quad \text{at} \quad z = 0, \tag{13}$$

$$\mu \frac{\partial u}{\partial z} = \sigma_c \frac{\partial c}{\partial x} \quad \text{at} \quad z = h. \tag{14}$$

Because Eq. (11) means the pressure is uniform along the layer-wise direction, p is determined from the pressure jump across the free surface (Laplace pressure), which is given by $\Delta p = -\sigma \kappa$ from the curvature of the surface κ.

$$\mu \frac{\partial u}{\partial z} = \int \phi_x dz = \phi_x z + c_1, \tag{15}$$

$$\mu u = \int (\phi_x z + c_1) \, dz = \frac{1}{2}\phi_x z^2 + c_1 z + c_2, \tag{16}$$

where ϕ_x is the gradient of the pressure and centrifugal potential $\phi_x = \partial p/\partial x + \partial \phi/\partial x$. The constants c_1 and c_2 can be determined from the boundary conditions Eqs. (13) and (14) as $c_2 = 0$ and $c_1 = \partial \sigma/\partial x - \phi_x h$. From Eq. (11), the pressure is uniform along the thickness direction, and it is determined from the surface curvature as

$$\Delta p = -\sigma \kappa = -\sigma \frac{\partial^2 h}{\partial x^2}. \tag{17}$$

Integrating Eq. (10), subject to boundary conditions Eqs. (13) and (14), leads to the horizontal velocity

$$u = \frac{1}{\mu}\frac{\partial (p+\phi)}{\partial x}\left(\frac{1}{2}z^2 - hz\right) + \frac{\sigma_c}{\mu}\frac{\partial c}{\partial x}, \tag{18}$$

where the pressure gradient $\partial p/\partial x$ can be obtained by differentiating Eq. (17). Integrating Eq. (18) gives the flux Q as

$$Q = \int_0^h u\,dz = \frac{\sigma h^3}{3\mu}\frac{\partial^3 h}{\partial x^3} + \frac{\sigma_c h^2}{2\mu}\frac{\partial c}{\partial x} + \frac{\rho r \Omega^2 h^3}{3\mu}. \tag{19}$$

Their first term on the right hand side of Eq. (19) stands for the flux driven by the Laplace pressure due to surface curvature, the second term by the solutocapillary effect, and the last term by the centrifugal force. From the mass conservation and the kinematic condition for the interface, the time evolution of the height function h can be written as

$$\frac{\partial h}{\partial t} = -\frac{\partial Q}{\partial x} - E, \tag{20}$$

where E is the evaporation rate. If the spatial thickness variation is sufficiently small and we are only interested in horizontally averaged field, Eq. (20) can be reduced as

$$\frac{\partial h}{\partial t} = -\frac{2\rho\Omega^2 h^3}{3\mu} - E. \tag{21}$$

Concentration field

General mass transfer equation can be written as

$$\frac{\partial c}{\partial t} + (\boldsymbol{u}\cdot\nabla)c = \nabla\cdot(D(c)\nabla c), \tag{22}$$

where $D(c)$ is the concentration-dependent diffusivity.

When the Sherwood number $\mathrm{Sh} = E_0 h_0/D_0$ is sufficiently small, the concentration is assumed as uniform along the depth direction (so-called "well-mixed" assumption [109]). Hence, only the depth-averaged concentration $c = c(r, t)$ is considered. Here, the concentration c is defined as the volume fraction of resin, which can be interpreted as a ratio of an imaginary thickness of the resin over the total hight of the film [28]. Such a concentration field satisfies the mass conservation

$$\frac{\partial(ch)}{\partial t} = -\frac{1}{r}\frac{\partial}{\partial r}(rQ_{\text{res}}), \tag{23}$$

where Q_{res} denotes the convective and diffusive flux of resin given by

$$Q_{\text{res}} = -D(c)h\frac{\partial c}{\partial r} + Qc. \tag{24}$$

Using the identity $\partial_t(ch) = c\partial_t h + h\partial_t cm$, the evolution equation for the concentration field can be obtained as

$$\frac{\partial c}{\partial t} = -\frac{Q}{r}\frac{\partial c}{\partial r} + \frac{1}{rh}\frac{\partial}{\partial r}\left(D(c)rh\frac{\partial c}{\partial r}\right) + \frac{Ec}{h}. \tag{25}$$

The last term acts to increase the resin fraction at faster rate in thinner h region.

For larger Sherwood number cases, the above-mentioned well-mixed assumption is no longer valid and the concentration variation in thickness direction must be considered. If the spatial thickness variation is sufficiently small and we are only interested in horizontally averaged field, the evolution equation for the concentration field can be reduced as one dimensional form:

$$\frac{\partial c}{\partial t} = \frac{\partial}{\partial z}\left(D(c)\frac{\partial c}{\partial z}\right). \tag{26}$$

Although this equation is rather simple, the spatial domain may varies in time, for instance in the spin-coat problem. To simply apply the boundary conditions for the time-varying spatial domain, it is better to transform the governing equations to shrinking coordinate. The detailed treatment of the transformed is found in Shiratori et al. [90].

2.3 Double-Peaked Edge-Bead

Edge-bead is a thick ridge that appears along the substrate periphery. The substrate region eroded by edge beads is not usable, thus they result in the loss of product yield. Shiratori and Kubokawa [89] investigated edge-bead generation for the case where the bead had a double-peaked shape in the direction away from the substrate periphery, and proposed a simple explanation for the mechanism of the double peak (Fig. 9).

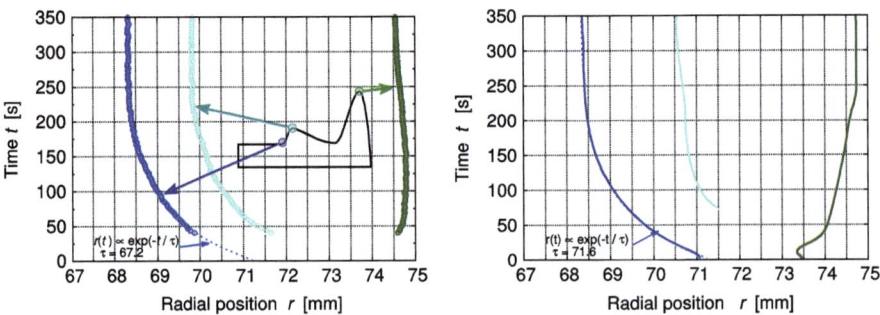

Fig. 9 Time evolution of selected thickness extrema tracked from the spatio-temporal thickness variation. Dashed line is an exponential decay function fitted to the radial movement of the inner bead. Reprinted from Shiratori and Kubokawa [89] with the permission of AIP Publishing. All rights reserved

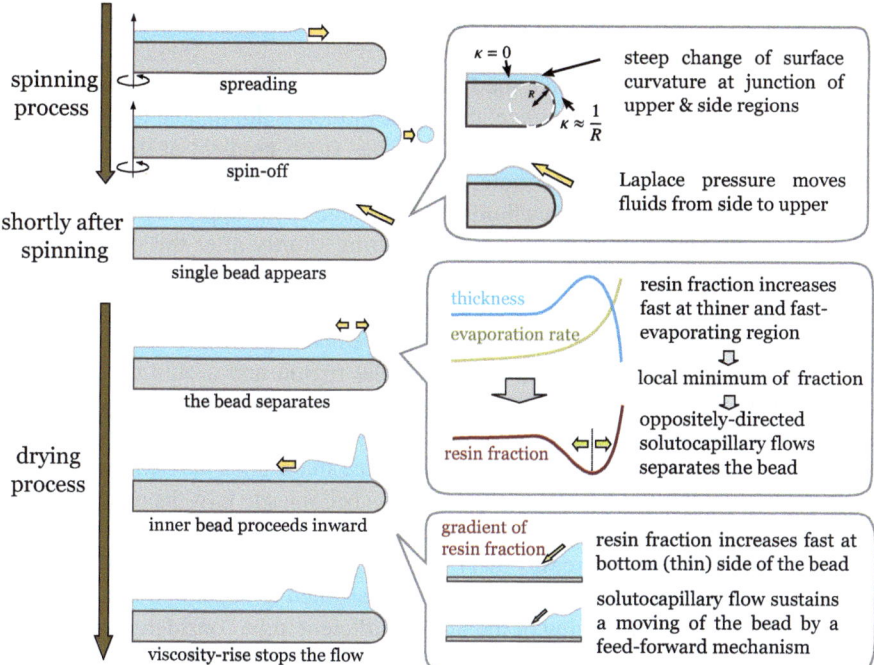

Fig. 10 Mechanism how the double-peaked edge-bead is formed. Reprinted from Shiratori and Kubokawa [89] with the permission of AIP Publishing. All rights reserved

Figure 10 shows a picture of the mechanism how the double-peaked edge-bead forms. In the followings, the detail is explained in the order of event. At the beginning of the spin-coat process, a specified volume of the liquid is dispensed on the center of the substrate. Then it is spread by the centrifugal force, and the liquid flows radially outward across the substrate (Fig. 10a). After the wavefront of the liquid film reached the outer edge of the substrate, the liquid proceeds to the side region of the substrate periphery. In the experimental observation, the liquid region was reached on the backside of the substrate at shortly after the spinning process. When the liquid volume supplied in the side region becomes a certain amount, the surface tension cannot sustain the liquid, thus the part of the liquid breaks up and flies off to the outside of the substrate (Fig. 10b). Depending on the flow rate and the rotation speed, several patterns of breakup are observed; e.g. the spoke-like continuous jets or a number of droplets. Lefebvre [52] has observed and summarised this breakup phenomena (see Fig. 34 in Chapter "Thermocapillary Convection in Liquid Bridges of Finite Length" his book). At low flow rates, the liquid film on the periphery breaks up to droplets, whereas at higher flow rates, filaments are formed along the periphery, which subsequently breakup to droplets due to the Rayleigh instability. The length scale of the discharged droplet can roughly be estimated as order of the substrate thickness d. After each breakup, the part of the liquid is expected to remain on the side region of the substrate.

When the spinning is stopped, the force balance suddenly changes due to a loss of the centrifugal force. At this stage, the liquid motion is dominated by the Laplace pressure. Since the curvature of the substrate has steep gradient at the junction of upper and side regions of the substrate, Laplace pressure acts to reduce the curvature distribution, and it quickly moves fluids from the side to the upper region (Fig. 10c). In this way, the single edge-bead is formed during shortly after the spinning-stop. In the order to determine the radial extent of the bead, Shiratori and Kubokawa [89] investigated the fluid motion on the curved substrate based on the formulation of Weidner et al. [108] and Roy et al. [68] who have investigated the similar problem. The substrate is considered to have a upper flat region and a side region curved with a constant curvature. At the moment of the spin-stopping, the liquid film is assumed uniform along the flat region of the substrate, whereas relatively thick film is assumed on the side region. Such a initial surface shape deforms rapidly due to the Laplace pressure. In the short time, the thickness becomes to have its minimum near the junction of the upper and the side region. For the typical case of $h_0 = 50\,\mu\text{m}$, the minimum thickness decreases to 1/5 of its initial value during the time of 1 s. Therefore, at this point the fluid region can be assumed as separated into the upper and the side regions. The drying of the film is sufficient to be considered only for the upper region on the flat substrate.

Hereafter, let us consider the effect of drying and the corresponding time scale. In terms of evaporation, two different effects on a concentration field should be noticed; one is a spatial distribution of evaporation rate $\mathcal{F}(x)$. According to the spatial dependence of the evaporation rate Eq. (8), the solvent evaporates faster in the outer region of the substrate. This leads to a gradient of resin fraction which increases outward. Another effect is caused by a thickness distribution; the increase in resin fraction due to solvent evaporation is not uniform, even if the evaporation is uniform. According to the last term (Ec/h) in Eq. (25), the change rate of c is inversely proportional to the thickness h [28]. Hence, when the film thickness has a bead-like profile, the resin fraction increases at a faster rate in the thin region relative to the thick region. For the single bead, which is formed due to the Laplace pressure, both the above two effects come into play in the concentration distribution. A resulting concentration field becomes to have a minimum extrema, which causes two oppositely-directed solutocapillary flow (Fig. 10d). This is the reason why the bead separates.

Once the bead is separated, the outer bead moves outward rapidly and almost stays at the outermost position. In contrast, the inner bead moves rather slow. The movement of the inner bead can be explained in terms of the above-described thickness effect on the change rate of concentration; the resin fraction increases at a faster rate in the thin region relative to the thick region. Because there exists a relatively thin region at the further inside of the inner bead, a gradient of the concentration is formed so that the resin fraction becomes higher along inward direction. This concentration gradient leads the solutocapillary flow which moves the inner bead inward. After the bead moved slightly, there can be found the similar structure at the moved position.

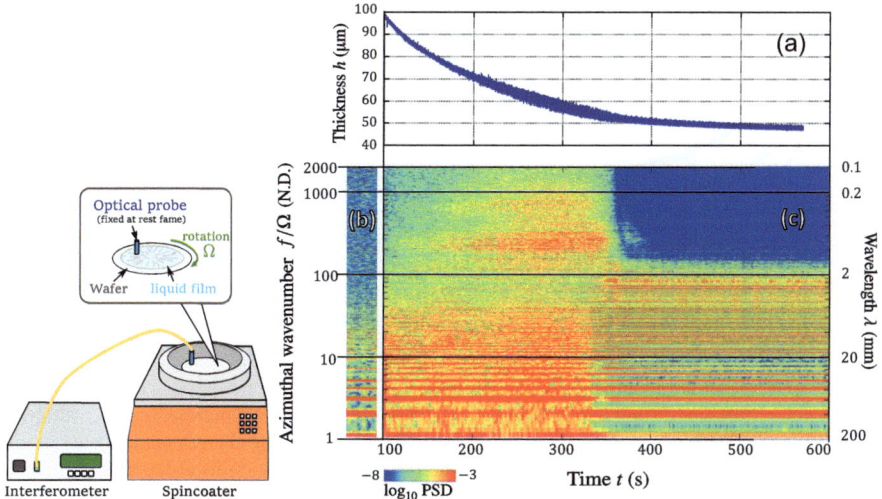

Fig. 11 Spatio-temporal thickness variation measured during spin coating with the rotation rate $\Omega = 50$ rpm. **a** Time evolution of thickness. **b** Spectrogram of intensity for the case of bare substrate. **c** Spectrogram of thickness variation of (**a**). In the spectrograms, the horizontal axes stand for time t. The left vertical axes are the frequency normalized by the rotation rate, which means the wavenumber in the unit rotation. The right vertical axes are the wavelength, which is converted from the frequency. The color contours indicate a logarighmic intensity of the signals for the corresponding time and frequency. Reprinted from Shiratori et al. [90] with permission from Elsevier. All rights reserved

Therefore the movement is sustained due to a kind of feed-forward mechanism, in which the bead carries a source of driving by itself (Fig. 10e). As time advances, the flow is gradually suppressed with the increase of viscosity, and the movement of the bead becomes vanishingly slow after some time (Fig. 10f). The time constant for this decay is expected to be scaled using a exponent of viscosity K_μ in Eq. (4).

2.4 Striations in Spincoating

Striation is another typical thickness undulation, which appears during spin-coating as shown in Fig. 7a. As mentioned above, the generation mechanism of striation has been explained by the Marangoni-Bénard instability under the rotation field. Shiratori et al. [90] investigated the formation process of striations by measuring spatio-temporal thickness variations during a spin coating for mixtures of epoxy resin and xylene solvent. The thickness was measured by an interferometer, whose optical probe was fixed in the rest frame, whereas the substrate and the film were rotated. In this way, the optical probe drew a trajectory of a circle in the rotating frame. Therefore, the recorded time-series data of the thickness contains information on the azimuthal distribution and time variation. Figure 11a shows the time evolution

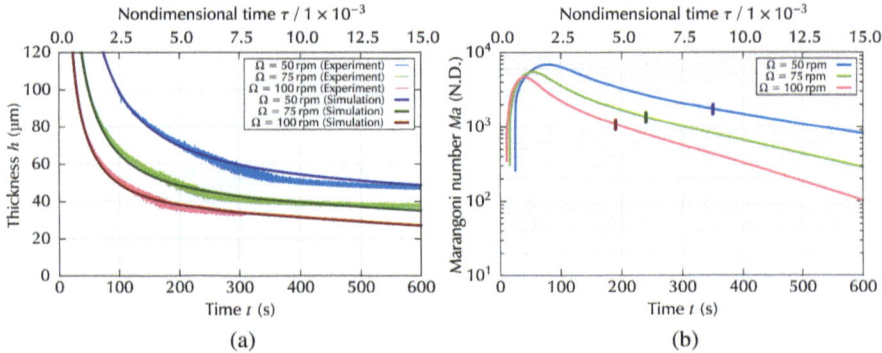

Fig. 12 a Time evolution of the thickness obtained by the numerical simulations (thick lines) and the optical measurements (thin lines). b Time evolution of the Marangoni number for three different rotation rates Ω. The vertical lines are time instants when the short-wavelength components vanish in the experiments. Reprinted from Shiratori et al. [90] with permission from Elsevier. All rights reserved

of thickness. Time $t = 0$ was determined as when the rotation started. At the time when the thickness is able to be detected ($t \approx 100$ s), time variation of the thickness does not show monotonic decrease but oscillating behaviour. By applying a time-frequency analysis of a short-time Fourier transform (STFT) to the recorded thickness data, the azimuthal wavelength and its time variation can be extracted. In STFT, the total time signal was divided into shorter segments of equal lengths. Then the Fourier transform was computed separately on each shorter segment. Figure 11b, c are the time-frequency spectrograms for the intensity of reflected light in the bare-substrate rotation (b) and for thickness variation (c). From Fig. 11a it can be noticed that the amplitude of the thickness variation increased during the time range of $t = 200$ s ~ 300 s. During the same time range in Fig. 11c, the components of shorter wavelength $\lambda < 2$ mm became drastically strong. It can be suggested that these time-varying components were corresponding to the striation, and its generation seems to start in the early phase of spin coating. It should be noted that the intensity of shorter wavelength components suddenly becomes negligibly small at $t \approx 350$ s. In Fig. 11a, the amplitude of the thickness variation also suddenly becomes small.

To explain the reason for this sudden vanishment of shorter wavelength components, Shiratori et al. [90] also performed a numerical simulation based on the physical model described in Sect. 2.2. Figure 12a shows time evolution of thickness obtained by their numerical simulations for cases of three different rotation rate. In the same figure, the experimental results are plotted together for comparison. For all cases the behaviour of thickness decrease predicted by the numerical simulation showed good agreement with the mean thickness of the experimental observations. This agreement means that physical model and the parameters applied have some degree of validity, and the discussions based on the numerical simulations using this model are deemed significant.

Using this numerical results, Shiratori et al. [90] predict the transient Marangoni number during spin-coating as shown in Fig. 12b. They calculated the solutal Marangoni number which is defined as

$$\text{Ma}^S = \frac{\partial \sigma}{\partial c} \frac{\Delta c h}{\mu D}. \qquad (27)$$

For all cases of the rotation rate, the Marangoni number rapidly increases at an early stage. This is because the concentration difference Δc is quite small at the early stage. In their simulation the initial concentration is assumed to be uniform; thus, it requires a certain amount of time for the concentration field to develop. The Marangoni number starts to decrease taking its maximum at the time $t = 50$–80 s. This decrease in Ma is mainly caused by a decrease in thickness h.

In the experiment, the shorter wavelength components suddenly vanished in the middle of the process. This sudden spectrum change can be explained by the disappearance of the disturbance corresponding to one of the bifurcations. By comparison between this sudden wavelength change and numerical results, the Marangoni number corresponding to the possible bifurcation can be predicted in the range of $1000 < \text{Ma} < 1800$.

These results suggest that we can suppress the striations by controlling the coating conditions and/or by appropriate selecting the solvent so that the Marangoni number become lower than the above value range.

3 Positive Surface Tension Temperature Coefficient (Self-rewetting Fluids)

Anselmo Cecere, Raffaele Savino, and Yoshiyuki Abe

3.1 Self-rewetting Fluids

Binary mixtures of water and long-chain alcohols like butanol, pentanol, hexanol and heptanol show an anomalous nonlinear behavior of the surface tension with temperature; as it is depicted in Fig. 13, in some conditions and contrary to ordinary liquids, the interfacial force increases with increasing temperature resulting in a thermocapillary flows at liquid/vapor interfaces that moves liquid towards hotter regions. Furthermore, a differential evaporation along the interface gives an additional surface tension gradient that increases the reverse Marangoni flow. For these reasons, such mixtures are usually referred in literature as Self-ReWetting Fluids (SRWFs).[1]

---

[1] Terminology "self-rewetting fluids" was named by Prof. Kawamura (Tokyo University of Science, at that time) after the oral presentation of "Microgravity experiments on phase change of self-wetting fluids" delivered at the session chaired by Prof. Kawamura during an ECI Conference (Davos, Switzerland, Sep. 14–19, 2003).

SRWFs were proposed as advanced working fluids for heat transfer applications for the first time from Abe et al. [4] and Savino and Monti [74], and have been investigated both numerically and experimentally as working fluids for several heat transfer devices. Interferometric systems and flow visualization methods have been extensively used to investigate the fundamental physical mechanisms involved in the heat and mass transfer process of SRWF mixtures. Experiments have been carried out also in short term weightlessness of microgravity platforms, while experiments in space environment are foreseen to be performed on the International Space Station (ISS) where the Marangoni effect can be observed in relatively large systems. In this chapter main results obtained in the last ten years will be revised. The Sect. 3.2 summarized experimental data on surface tension carried out by several research groups. In the Sect. 3.3 fundamental study on SRWF are presented and finally in Sect. 3.4 through Sect. 3.6 boiling characteristics, main applications on heat pipe and future space experiments planned to be performed in space are described.

3.2 Surface Tension

The surface tension of alcohol aqueous solutions was first investigated theoretically [104, 105] and experimentally [64]. In case of alcohols with high number of carbon atoms (≥ 4), they observed an unexpected presence of a minimum in the surface tension-temperature curve. After that, measurements have been carried out by several research groups with the aid of pendant drop technique [41, 80], maximum bubble pressure method [17, 93] and Wilhelmy plate tensiometer [112]. Figure 14 shows experimental results obtained for water/butanol (5 and 6 wt%), water/pentanol (1.5 and 2 wt%) water/hexanol (0.6 wt%) and water/heptanol (0.05 and 0.1 wt%) (values are normalized with respect to the values of the surface tension σ_{ref} at room temperature (30 °C)).

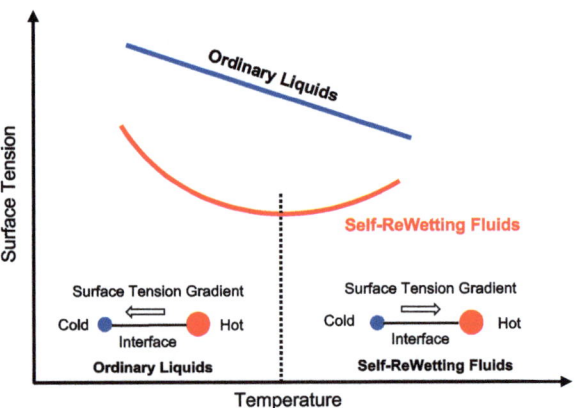

Fig. 13 Surface tension behaviour of ordinary liquids and self-rewetting fluids

For all cases investigated when the temperature increases above a certain point, the surface tension of the aqueous solution will increase with the temperature. Above this point, which depends on concentration and alcohol type, and in particular, on the number of carbon atoms as theoretically explained in the pioneering works [104, 105], the interfacial force shows a reverse gradient of surface tension that is reversed by the temperature rise. Even if all the measurements carried out agree on this behavior, some discrepancy shall be noted. For some mixtures, the minimum position in the surface tension curve and the intensity of the surface tension derivative are not yet fully understood. It is worth to mention that, as shown in Fig. 15, many additional efforts have been carried out to improve the self-rewetting effects.

Savino et al. [80] proposed the so-called Self-ReWetting brines based on potassium hydroxide and acetic acid (FD-40) and potassium formate (FP-40) with the aim to extend the operative temperature range of such mixture and overcome the relatively high freezing point of the water. They observed the same surface tension anomaly also in case of ternary mixtures. These authors also employed silica, gold, and single-wall nano-horns indicate as Self-ReWetting nanofluids in Fig. 15 on the

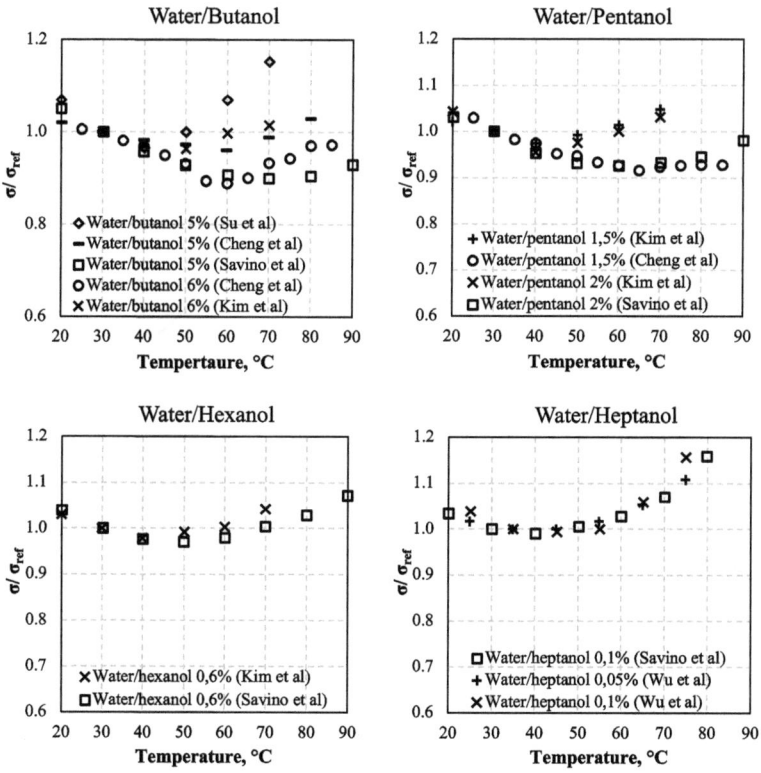

Fig. 14 Surface tension measurements for water alcohol mixtures obtained with the aid of different optical techniques

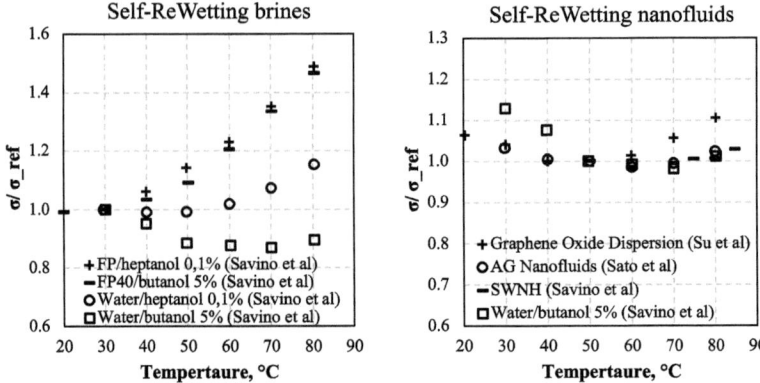

Fig. 15 Self rewetting brine and nanofluids

right (in this case data are normalized with respect to the values σ_{ref} at temperature of 50 °C) [82] and they found not only an anomaly in the surface tension (SWNH) but also in freezing point depression, enhanced thermal conductivity, and wetting improvement. Sato et al. [73] employed silver nanoparticles (AG Nanofluids) of different aspect ratio showing the self-rewetting behaviour with an enhanced thermal conductivity and viscosity. Su et al. [93] proposed a new self-rewetting nanofluid by mixing n-butanol aqueous solution with graphene oxide dispersion solution. They show the positive temperature dependence of the surface tension, was rather enhanced by the addition of graphene.

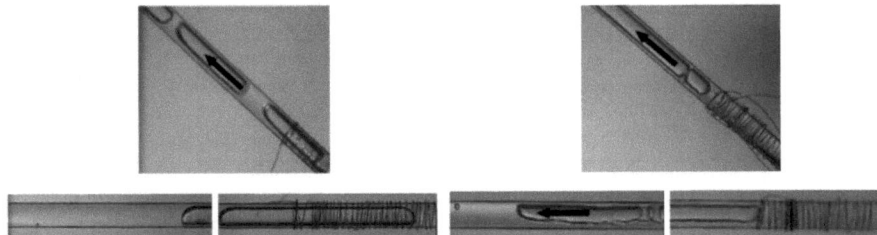

Fig. 16 Slug boiling behaviour in water (left) and SRWF (right) under different inclination with respect to the gravity vector. Reprinted from Savino et al. [78] with permission from Elsevier. All rights reserved

3.3 Interface Behavior with a Positive Temperature Coefficient

Savino and Paterna [75] carried out theoretical developments and numerical simulations on effects of Marangoni flow on heat pipe performances. Notably, their work demonstrated that using binary mixtures with an inverse Marangoni effect, the maximum heat transfer rate corresponding to the dryout limit in grooved heat pipes can be strongly improved in comparison to common ordinary liquids. In parallel, a series of flow visualization experiments [78] have been performed with glass capillary tubes with a heating section at one end and a cooling section at another end and partially filled with both ordinary liquids and SRWFs, see Fig. 16. Experimental results show that when capillaries are inclined of 45° with respect to the gravity vector, in case of the capillary filled with SRWF (on the right in Fig. 16) the size of vapour slugs generated at the heating region is much smaller and a very wavy interface of the vapour slugs can be observed in the condensation region, which implies a very strong liquid downward flow. For the capillary filled with pure water (on the left in Fig. 16) no significant differences are evident in the merging behaviour of vapour slugs but more liquid in the condensation region and less liquid in the evaporation region with respect to the heat pipe filled with SRWF are present. Similar results are described also in gravity-assisted configuration (thermosiphon).

When the wickless heat pipes are in the horizontal configuration dryout conditions are established in the evaporator of the capillary filled with pure water. On the contrary, the heat pipe filled with SRWF continues to operate, slug vapour bubbles are continuously generated and move from right to left toward the condenser, where the condensed liquid is spontaneously supplied at the evaporator.

Figure 17 shows the Marangoni flows in a thin liquid layer film in presence of a horizontal thermal gradient observed by means of interferometric system by Cecere et al. [12] with the aim to investigate fluid behavior, temperature, and concentration distributions along the liquid vapour interface.

Figure 17a shows the streaklines in case of an ordinary liquid (ethanol). The phase map obtained through interferometric analysis in Fig. 17a shows a strong capillary flow at liquid vapour interface that moves the liquid from the hot region (right) to the cold region (left). Similar experiments carried out with a SRWF (water/butanol 5 wt%) but with two different wavelengths (blue and red in Fig. 17b shows a contrary behaviour. In this case, the distortion of the isotherms, as well as of the isoconcentration lines towards the hot side, show the existence of a reverse Marangoni flow along liquid-vapour interface. Due to evaporation-condensation processes the alcohol concentration decreases from the cold to the hot side and both thermocapillary and solutal effects are established along the liquid vapour interface pushing liquid from cold to the hotter region of the experimental cell. The experiment proved that in case of SRWF, both thermocapillary and solutocapillary forces acted in the same direction along a liquid-vapor interface.

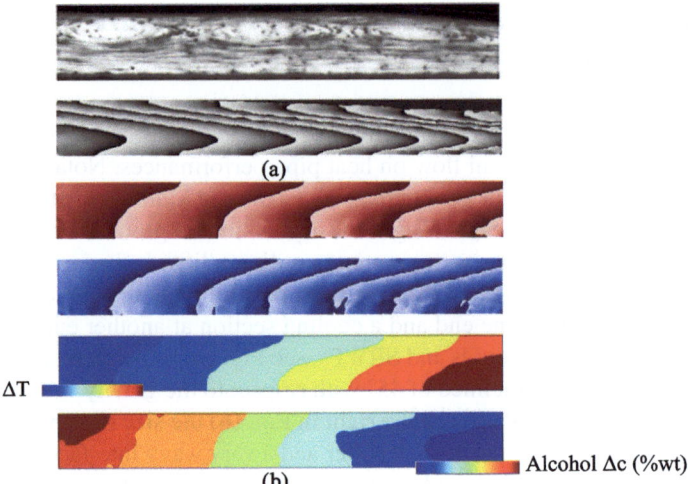

Fig. 17 Marangoni flows in a thin liquid layer film in presence of a horizontal thermal gradient (left: cold side, right: hot side) observed with the aid of a two-wavelength Mach-Zehnder interferometer in case of **a** ordinary liquid and **b** a SRWF. $\Delta T = 10$ K, alcohol $\Delta c = 1.2\%$. Reprinted from Cecere et al. [12] with the permission of Springer Nature. All rights reserved

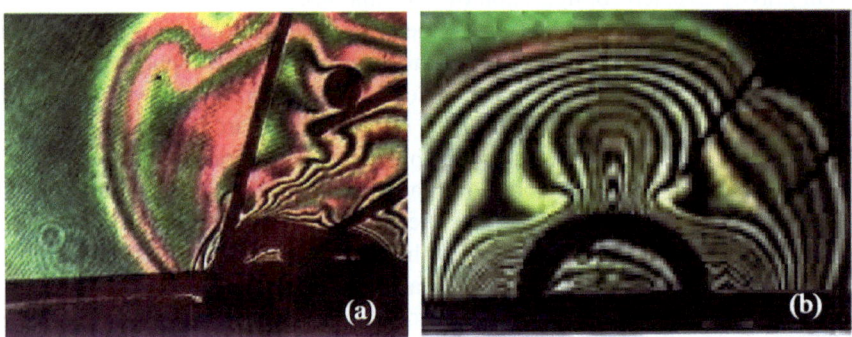

Fig. 18 Interferograms of two-wavelength Mach-Zehnder interferometer of single bubble in microgravity. **a** CFC-113, **b** CFC-112/CFC-12

3.4 Boiling Characteristics of Self-rewetting Fluids

Fundamental difference in the boiling characteristics between pure liquids and non-azeotropic binary liquids was studied in microgravity available in drop shaft [56] with the aid of a two-wavelength interferometer by Abe and Iwasaki [1]. Figure 18 compares the interferograms of single bubbles for subcooled CFC-113 and CFC-12/CFC-112 (CFC12: 6.2 wt%) observed in microgravity. As depicted, a strong thermocapillary flow was developed around the bubble in CFC-113. In contrast to this, the layers of thermal boundary and concentration boundary were developed around the bubble in CFC-12/CFC-112 due to the preferential evaporation of CFC-12. The experimental results were afterward qualitatively verified by a numerical simulation for the influence of non-condensable gas in boiling conducted by Wu and Dhir [110].

In the case of SRWFs, both concentration and temperature profiles developed along bubbles induce an intense Marangoni flow around the bubbles, which are considered to enhance the boiling heat transfer.

Pool boiling heat transfer of binary mixtures in microgravity was first studied by Abe et al. [2] with a plate heater, in which two non-azeotropic binary compositions for aqueous solutions of ethanol (11.3, 27.3 wt%) were employed. The boiling heat transfer coefficients for both compositions on 50 mm × 50 mm glass plate showed 20–60% enhancement in microgravity compared with those in the normal gravity. From the observation from the rear side of a transparent heater, they attributed the boiling enhancement to a vigorous Marangoni flow at the bubble/heater interface induced by the preferential evaporation of ethanol-rich component. For SRWFs, pool boiling in microgravity was studied by Abe et al. [3], but the experiments were focused on the bubble and the flow behavior at a spot heater on a glass plate, and no heat transfer data were acquired.

A number of experimental studies with a wire heat source have been conducted for SRWFs and other alcoholic aqueous solutions. van Wijk et al. [103] reported 2.5 times higher CHF (Critical Heat Flux) for 1-butanol aqueous solutions and 3 times higher CHF of 1-pentanol aqueous solutions comparing with that for pure water. Shoji and his coworker confirmed nearly three times higher CHF for 1-butanol aqueous solution comparing with that for water [60]. In high pressure conditions, Kawaji and his coworkers noticed an appreciable increase in the CHFs for SRWFs comparing with water [57]. Hu and coworkers have been conducting extensive heat transfer studies on SRWFs including pool boiling with a wire heat source. For 5 wt% butanol aqueous solution, they reported 1.91 times higher CHF for that of water [42], and they also studied for 0.1 wt% heptanol aqueous solution and reported 2.45 times higher CHF [43]. In both solutions, considerably higher heat transfer coefficients than water were observed in the entire boiling region.

Although higher CHFs and higher boiling heat transfer coefficients than those for water have been confirmed for pool boiling with wire heat source, studies on pool boiling with using plate heaters are rather limited and the results were somewhat inconsistent. With using copper block heaters, Ohta and his coworkers [70] and Abe et al. [4] observed nearly the same CHFs and slightly higher nucleate boiling heat transfer coefficients for high-carbon alcoholic aqueous solutions in comparison with water. On the other hand, McGillis and Carey [53] and Sakashita et al. [71] reported 2.1 timers higher and 1.7 times higher CHFs for 2-propanol aqueous solutions, respectively. 2-propanol aqueous solutions were also studied in Sakai et al. [70] and Abe et al. [4], but they did not observe higher CHFs than water. Note that the experiments by McGillis and Carey [53] were conducted in a reduced pressure, 1 kPa, and the heater sizes in McGillis and Carey [53] and Sakashita et al. [71] were smaller than Sakai et al. [70] and Abe et al. [4], which may imply the influence of heater size in the CHFs.

Figure 19 compares the nucleate boiling behavior of water and 1-butanol aqueous solution at relatively low heat fluxes on a copper block heater [4]. As shown clearly, the bubble sizes and nucleation site densities are considerably different. Figure 20 summarizes the results reported in Abe et al. [4] with the correlation by Stephan and Abdelsalam [92] for water.

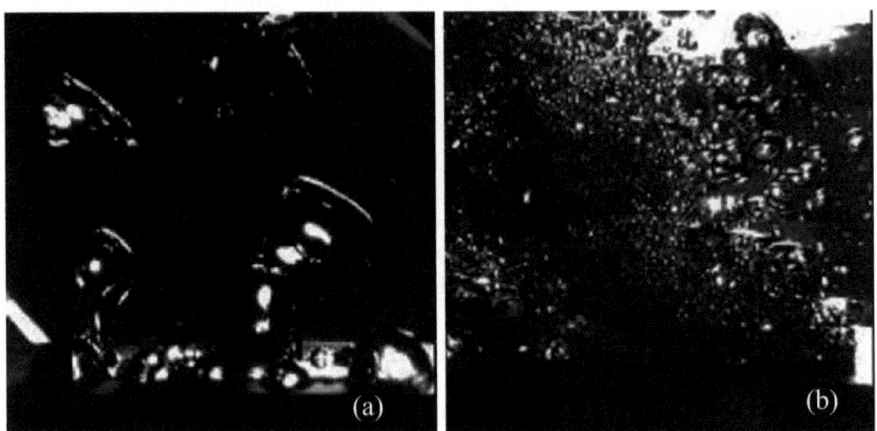

Fig. 19 Nucleate boiling behavior on copper block heater of 20 mm in diameter. **a** Water ($q = 82 \, \text{kW/m}^2$), **b** 1-Butanol aq. sol. 6.0 wt% ($q = 66 \, \text{kW/m}^2$)

Table 1 Experiments on heat pipe filled with SRWFs

References	HP technology	SRWF	Main results
Abe et al. [3, 4, 6], Savino et al. [76, 77]	Wickless heat pipe	1-butanol (1.5)	Faster liquid return towards the evaporator in μg conditions
Abe et al. [5], di Francescantonio et al. [26], Savino et al. [77, 79]	Wicked heat pipe	Binary (water and alcohols) ternary mixtures	Enhanced dryout limit
Savino et al. [40], Fumoto et al. [32]	Oscillating heat pipe	Binary (water and alcohols) SRWF nanofluids	Enhanced dryout limit, lower thermal resistance
Wu [111]	Loop heat pipe	Binary (water and alcohols)	Higher critical heat load and lowest thermal resistance

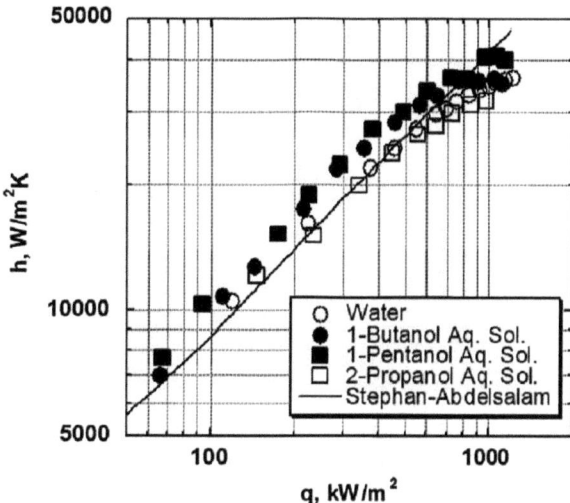

Fig. 20 Boiling curves for various fluids (1-butanol 6.0 wt%, 1-pentanol 2.0 wt%, 2-propanol 4.8 wt%)

3.5 Research and Application in Heat Pipe Technology

Although the idea of heat pipes with alcoholic aqueous solutions as a working fluid was already proposed and studied [98], the Marangoni effect in the solutions was out of the scope of these studies. Kuramae and Suzuki first paid their attention to the Marangoni effect of dilute ethanol aqueous solution in wickless heat pipe in microgravity [51]. For these devices, the inverse Marangoni effect, caused by the increase of surface tension with temperature, provides an additional mechanism that forces liquids to return from the condenser to the evaporator, other than forces like capillary and gravitational. Several experiments have been carried out considering wickless, wicked, loop and oscillating heat pipe, see Table 1.

Surface tension as well as thermocapillary effects play a more important role in heat and mass transfer processes when the Bond number (Bo) becomes smaller, i.e., with the reduction in scale or gravity. A comparative study on heat transfer performances of wickless heat pipes in low gravity conditions was carried out also by Savino et al. [76, 77] and Abe et al. [3, 6]. Infrared experimental results show that in case of heat pipe filled with binary mixture the thermal performances were almost the same in normal and low gravity conditions, whereas the heat pipe filled with ordinary liquids became less efficient in reduced gravity. Based on this result, the interesting concept of a wickless, flexible, inflatable and deployable ultralight-weight radiator was also demonstrated. Figure 21b shows the polyimide single heat pipe panel; when the fluid is in the channel, because of the inner pressure, the panel is spontaneously inflated, deployed, and finally rolled out. Inflation and deployment test of polyimide panels was carried out in a vacuum chamber on board a parabolic flight. Figure 21 compares the IR images taken in normal gravity (Fig. 21a) and in low gravity (Fig. 21c) for two wickless single heat pipe panels filled with water and

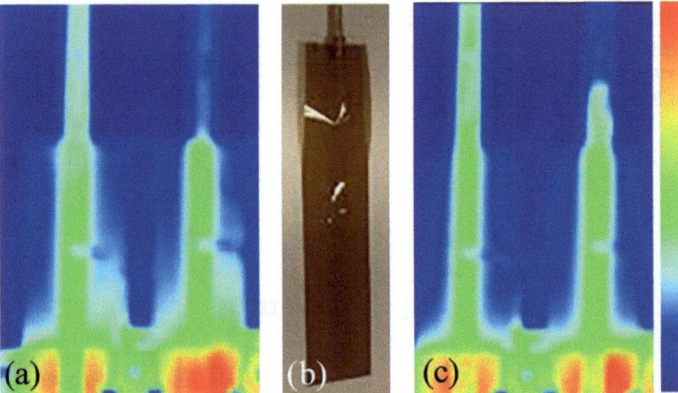

Fig. 21 Thermographic images of polyimide single heat pipe panel (**b**); **a** normal gravity (left: water, right: 1-butanol 1.5 wt% aqueous solution); **c** low gravity (left: water, right: 1-butanol 1.5 wt% aqueous solution)

SRWF. In the case of the panel filled with SRWF (right on the infrared image (a) and (c)) the temperature at the evaporation region decreased and the temperature distribution along the channel became more uniform in low gravity. In contrast, the maximum temperature of the water heat pipe (left on the infrared image (a) and (c)) increased during the microgravity phase.

Thorough and extensive study wicked heat pipes with SRWFs were initiated and conducted by Abe et al. [5], di Francescantonio et al. [26], and Savino et al. [77, 79]. Figure 22 compares the thermal performance of various working fluids in 8 mm-diameter heat pipes and 4 mm-diameter heat pipes of 250 mm in length with a composite wick (shallow grove + wire mesh) in the horizontal orientation. As shown in the comparisons, the dryout limits of heat pipes were improved significantly with using SRWFs. On the other hand, the thermal resistance is nearly the same or even higher than water heat pipes due to the temperature difference between the bubble point and dew point in non-azeotropic compositions.

The behavior of the thermal resistance is opposite to the results of oscillating heat pipes studied by Hu et al. [40], they confirmed a significant improvement in thermal resistance comparing with water oscillating heat pipes. Fumoto et al. [32] investigated the thermal performances of water and 1-butanol in a micro pulsating heat pipe composed of 20 parallel channels made of a copper capillary tube with an internal diameter of 0.8 mm. Results show that 1-butanol aqueous solution had higher effective thermal conductivity in all investigated thermal loading regimes, because of the stable generated oscillations of the working fluid. Wu [111] proposed to overcome the hydrophobic property of PTFE wick structure used in loop heat pipe with SRWF, getting better thermal performances with 6 wt% butanol aqueous solution.

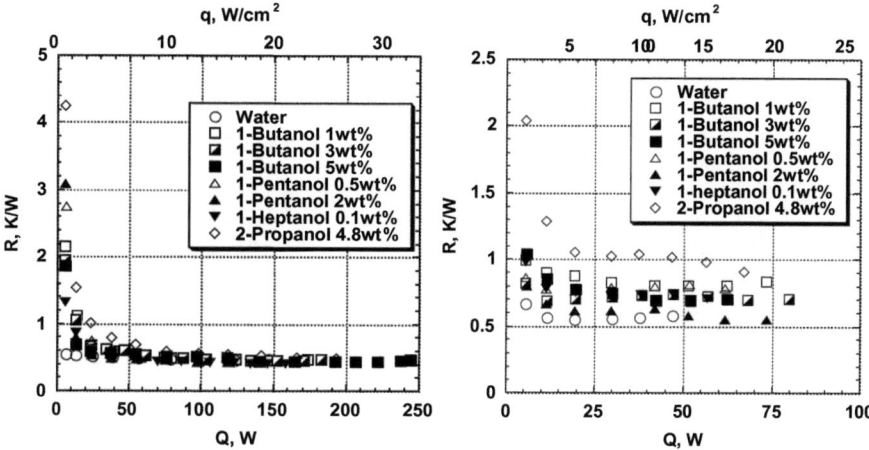

Fig. 22 Thermal performance of wicked heat pipes with different working fluids; **a** 8 mm in diameter, **b** 4 mm in diameter. Reprinted from Savino et al. [79] with permission from Elsevier. All rights reserved

3.6 Low Gravity and Space Experiments

Experiments on SRWFs are foreseen to be performed in space. A facility named Heat Transfer Host 1 will be developed by the European Space Agency to carry out experiments on several heat transfer phenomena. Figure 23 shows the preliminary breadboard developed by Redwire Space nv Belgium. Experiments will be observed from a top window by a common infrared and optical diagnostic setup, Fig. 23a, c.

SRWF experiment intends to enable contributions to the understanding of the basic fluid dynamic and physical chemical mechanisms in multi-component two-phase systems and particularly of the interplay between phase change, heat and mass transfer, surface properties in evaporation-based heat transfer devices like heat pipes [81, 83]. Figure 23d shows the experimental concept of the SRWF experimental test container while Fig. 23b shows the preliminary breadboard developed during the Phase B. A metallic groove with a V-shaped cross section is filled with a sample liquid and heated and cooled at the two opposite sides by heaters and Peltier elements mounted below the evaporator and condenser sections, respectively. Based on the surface properties of the liquid a meniscus is formed inside the groove. Due to evaporation and condensation processes, the liquid meniscus radius changes along the channel. The pressure difference due to the different meniscus curvature between evaporator and condenser generates the capillary pumping for the liquid flow directed from the condenser to the evaporator section.

Preliminary experiments have been carried out also in short term weightlessness of a parabolic flight campaigns with the aim to understand its fundamental physics. Cecere et al. [15, 16] investigated the capillary-driven two-phase flow in a water/butanol solution under normal and reduced gravity conditions on board the

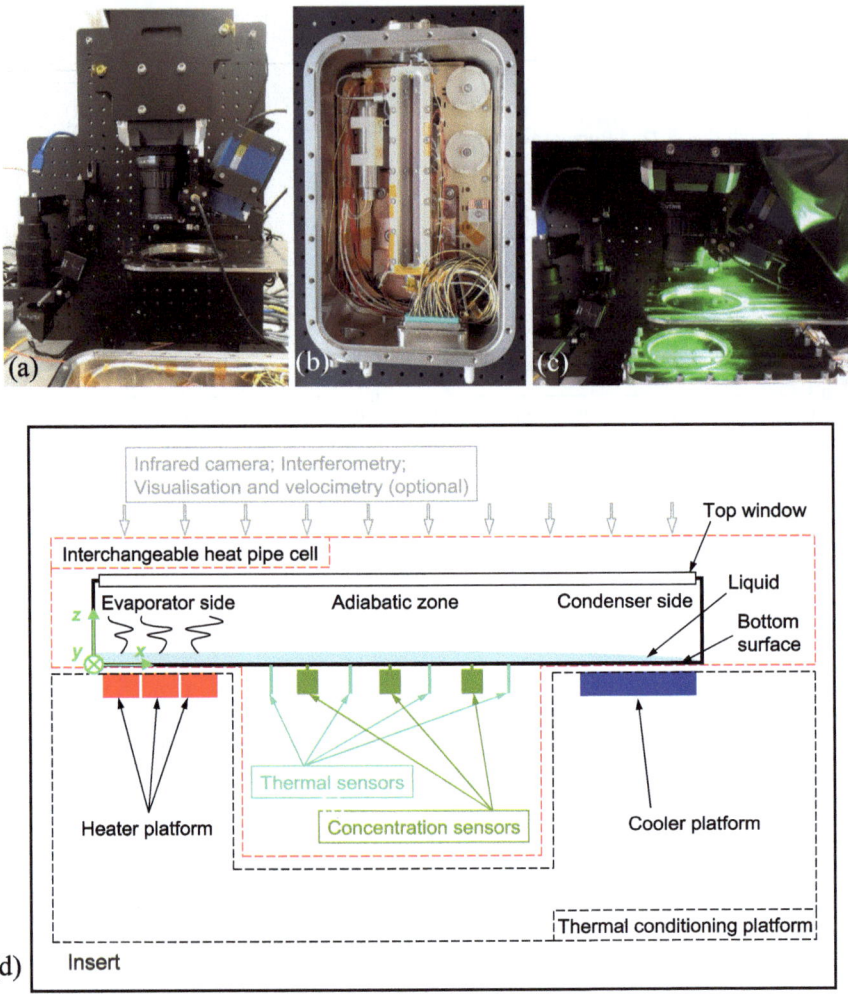

Fig. 23 SRWF experiment onboard the International Space Station (photos **a**, **b**, **c** courtesy of ESA and Redwire Space nv Belgium. All rights reserved). Panel **d**: Reprinted from Cecere et al. [16] with the permission of Springer Nature. All rights reserved

'zero-g' plane of the European Space Agency. The experimental setup, shown in Fig. 24a) is based on a cell with a transparent top window enabling the visualization of the liquid distribution during the parabolic manoeuvres. The evaporation rate is regulated changing the power input in a range between 0 and 30 W. Optical diagnostic devices is based on a LED illumination device proposed by Savino et al. [84] and Cecere et al. [14] where the liquid in the groove is visualized thanks to the presence of LED light reflections along the gas/liquid interface.

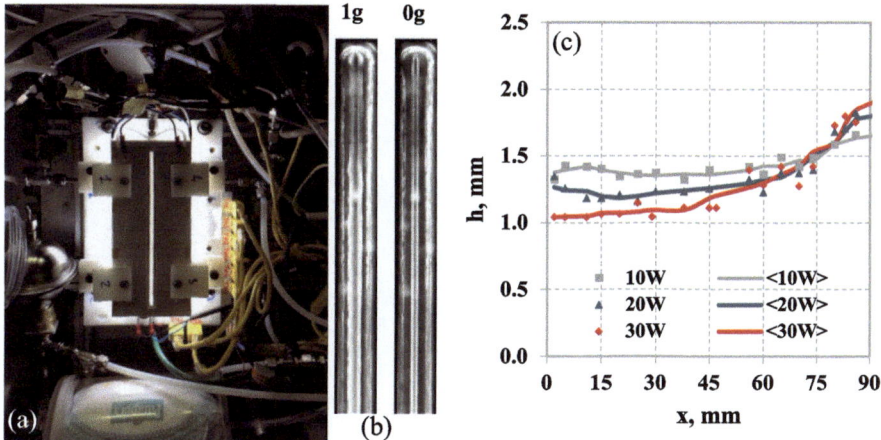

Fig. 24 SRWF experiment onboard the ESA Zero-g plane. **a** experimental setup, **b** images of the liquid film under normal and low gravity conditions, and **c** liquid film thickness under lower gravity conditions for different power inputs. Panels **b** and **c**: Reprinted from Cecere et al. [16] with the permission of Springer Nature. All rights reserved

Figure 24b shows images acquired during normal (1g) and low gravity (0g) conditions. During the microgravity phase because of the absence of the hydrostatic pressure, the liquid thickness in the groove and the curvature radius of the meniscus are reduced and the white bands reflected from the meniscus become thinner and thinner. Experimental results show that due to the capillary force, during the microgravity phase, the liquid film remains confined in the groove and as shown in Fig. 24c, increasing the power level, the liquid height at the evaporator decreases. The experiment was simulated by means of CFD analysis modelled considering the meniscus shape detected with the CCD camera. A good agreement is found between numerical and experimental results. The computed viscous pressure loss allows to isolate the effect of gravity on the capillary flow of water/butanol system and to analyse the reverse Marangoni effect which contrary to the ordinary liquid is directed from cold to the hotter region aiding the basic capillary flow and pumping the liquid to the evaporator.

Additionally, SRWF mixtures will be tested also in a flat Pulsating Heat Pipe (PHP) in a second insert. Figure 25 shows the oscillating flow patterns of a flat plate pulsating heat pipe filled with an ordinary liquid (water), and a SRWF observed by Cecere et al. [13] under variable gravity conditions on board a 'Zero-g' plane. In absence of gravity the average temperature at the evaporator section increases and a dryout of the evaporator is observed. For both fluids, the only way to decrease the temperature of the evaporator zone under microgravity conditions is the oscillation of liquid plugs between the evaporator and the condenser sections.

Figure 25a shows a view of the PHP (left) and the associated temperatures (right) during the aircraft manoeuvre for water and a power input of 200 W. Under micro-

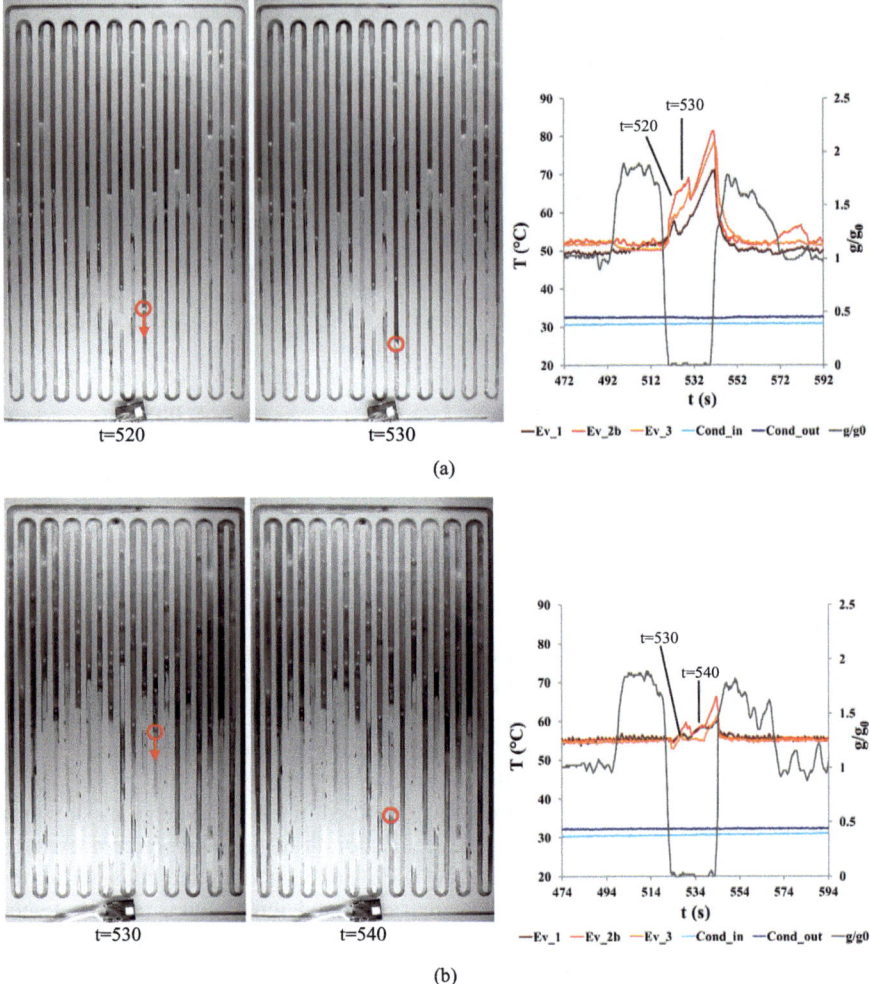

Fig. 25 Flow pattern and associated temperature (right) under microgravity condition for water (**a**) and water-butanol mixture (**b**). Power input: 200 W. Reprinted from Cecere et al. [13] with permission from Elsevier. All rights reserved

gravity conditions, at $t = 520$ s a liquid plug moves progressively towards the evaporator zone. The plug reaches a superheated evaporator zone ($t = 530$ s) triggering a mass transfer mechanism in the whole PHP due to the vapor pressure increase during its rapid evaporation. The associate heat transfer from slug/plug motion is responsible of the temperature oscillation shown in Fig. 25 (on the right). When the liquid plug oscillations stop, it prevents the energy transfer between the hot and the cold side and results in increasing temperatures. In case of water, during the 20 s of microgravity conditions the fluctuations are stopped, and the liquid accumulates at the condenser section. Instead, in case of the SRWF, the liquid re-wet several times the evaporator zone triggering the oscillating regime (Fig. 25b). Liquid plugs in different channels move towards the hot side (lower part in Fig. 25), where they evaporate triggering the oscillatory cycle. This vaporization leads to a new liquid vapor distribution of the fluid inside the PHP, and to decrease the temperature. In the short time of the microgravity period, the evaporator never becomes completely dry as it appears for pure water, and this is a very promising results in view of the space experiment.

3.7 Concluding Remark

The addition of high-carbon alcohols in water drastically changes the surface tension behavior inducing a rise of the surface tension with temperature. Such anomalous behaviour is observed not only for water-based two-component solutions, i.e., SRWFs, but also for complicated aqueous solutions such as brines (self-rewetting brines) and mixture with suspended nanoparticle (nano self-rewetting fluids). Pool and flow boiling as well as vaporization from a thin liquid layer film experiments have been reviewed. Experiments proved that both thermocapillary and solutocapillary forces acted in the same direction along the liquid-vapor interface of SRWFs. For pool boiling heat transfer with wire heat sources, the CHF is appreciably increased in SRWFs when compared with water. In the case of plate heaters, however, experimental results by different authors are rather contradictory. Such mixtures have been extensively used as working fluids in several heat transfer devices. In either pulsating or ordinary heat pipes, the employment of SRWFs results in an extended dryout limit and pulsating heat pipes exhibit even an improved thermal resistance, in comparison to those containing water as a working fluid.

Acknowledgements The study introduced in Sect. 1 was partially financially supported by Grant-in-Aid for Scientific Research (B) (21360101 and 24360078) and by a Grant-in-Aid for Challenging Exploratory Research (16K14176) from the Japan Society for the Promotion of Science (JSPS). The work in Sect. 3 was in part financially supported by the European Space Agency through the Microgravity Application Program (contract numbers: 4000115115/15/NL/PG, 4000128640).

References

1. Abe Y, Iwasaki A (1999) Observation of vapor bubble of non-azeotropic binary mixture in microgravity with a two-wavelength interferometer. In: Proceedings of the 5th ASME/JSME joint thermal engineering conference, AJTM 99–6418
2. Abe Y, Oka T, Mori YH, Nagashima A (1994) Pool boiling of a non-azeotropic binary mixture under microgravity. Int J Heat Mass Transf 37(16):2405–2413
3. Abe Y, Iwasaki A, Tanaka K (2004) Microgravity experiments on phase change of self-rewetting fluids. Ann N Y Acad Sci 1027(1):269–285
4. Abe Y, Iwasaki A, Tanaka K (2005) Thermal management with self-rewetting fluids. Microgravity Sci Technol 16:148–152

5. Abe Y, Tanaka K, Mochizuki M, Sato M, di Francescantonio B, Savino R (2006a) Heat pipes with self-rewetting fluids. In: Proceedings of the 8th international heat pipe symposium, pp 76–81
6. Abe Y, Tanaka K, Nakagawa M, di Francescantonio N, Savino R, Iwasaki A (2006b) Flexible wickless heat pipes radiator with self-rewetting fluids. In: 9th AIAA/ASME Joint thermophysics and heat transfer conference, No AIAA 2006–3105
7. Barmak I, Romanò F, Kuhlmann HC (2021) Finite-size coherent particle structures in high-Prandtl-number liquid bridges. Phys Rev Fluids 6:084301
8. Birnie DP III (1997) Combined flow and evaporation during spin coating of complex solutions. J Non-Crystaline Solids 218:174–178. https://doi.org/10.1016/S0022-3093(97)00141-5
9. Birnie DP III (2001) Rational solvent selection strategies to combat striation formation during spin coating of thin films. J Mater Res 16(4):1145–1154. https://doi.org/10.1557/JMR.2001.0158
10. Birnie DP III (2013) A model for drying control cosolvent selection for spin-coating uniformity: the thin film limit. Langmuir 29:9072–9078. https://doi.org/10.1021/la401106z
11. Birnie DP III, Hau SK, Kamber DS, Kaz DM (2005) Effect of ramping-up rate on film thickness for spin-on processing. J Mater Sci 16:715–720. https://doi.org/10.1007/s10854-005-4973-6
12. Cecere A, Paola R, Savino R, Abe Y, Carotenuto L, Vaerenbergh S (2011) Observation of Marangoni flow in ordinary and self-rewetting fluids using optical diagnostic systems. Eur Phys J Spec Top 192:109–120
13. Cecere A, De Cristofaro D, Savino R, Ayel V, Sole-Agostinelli T, Marengo M, Romestant C, Bertin Y (2018a) Experimental analysis of a flat plate pulsating heat pipe with self-rewetting fluids during a parabolic flight campaign. Acta Astronaut 147:454–461
14. Cecere A, De Cristofaro D, Savino R, Boveri G, Raimondo M, Veronesi F, Oukara F, Rioboo R (2018b) Visualization of liquid distribution and dry-out in a single-channel heat pipes with different wettability. Exp Thermal Fluid Sci 96:234–242
15. Cecere A, Mungiguerra S, Di Martino G, Savino R (2018c) Self-rewetting capillary flow under evaporation and condensation processes in parabolic flight conditions. In: Proceedings of the international astronautical congress (Bremen, Oct 2018), p Code 147415
16. Cecere A, Di Martino GD, Mungiguerra S (2019) Experimental investigation of capillary-driven two-phase flow in water/butanol under reduced gravity conditions. Microgravity Sci Technol 31(4):425–434
17. Cheng KK, Park C (2017) Surface tension of dilute alcohol-aqueous binary fluids: n-Butanol/water, n-Pentanol/water, and n-Hexanol/water solutions. Heat Mass Transf 53(7):2255–2263
18. Craster RV, Matar OK (2009) Dynamics and stability of thin liquid films. Rev Mod Phys 81:1131–1198. https://doi.org/10.1103/RevModPhys.81.1131
19. Cröll A, Müller W, Nitsche R (1986) Floating-zone growth of surface-coated silicon under microgravity. J Cryst Growth 79:65–70
20. Cröll A, Müller-Sebert W, Benz KW, Nitsche R (1991) Natural and thermocapillary convection in partially confined silicon melt zones. Microgravity Sci Technol 3:204–215
21. Cröll A, Dold P, Benz KW (1994) Segregation in Si floating-zone crystals grown under microgravity and in a magnetic field. J Cryst Growth 137:95–101
22. Cröll A, Kaiser T, Schweizer M, Danilewsky AN, Lauer S, Tegetmeier A, Benz KW (1998) Floating-zone and floating-solution-zone growth of GaSb under microgravity. J Cryst Growth 191:365–376
23. Daniels BK, Szmanda CR, Templeton MK, Trefonas P III (1986) Surface tension effects in microlithography—striations. In: Advances in resist technology and processing III, SPIE, vol 631. https://doi.org/10.1117/12.963641
24. Day J, McPherson A (1992) Macromolecular crystal growth experiments on international microgravity laboratory-1. Protein Sci 1:1254–1268
25. Dell'Aversana P, Tontodonato V, Carotenuto L (1997) Suppression of coalescence and of wetting: the shape of the interstitial film. Phys Fluids 9(9):2475–2485

26. di Francescantonio N, Savino R, Abe Y (2008) New alcohol solutions for heat pipes: Marangoni effect and heat transfer enhancement. Int J Heat Mass Transf 51(25):6199–6207
27. Diez JA, Kondic L (2002) Computing three-dimensional thin film flows including contact lines. J Comp Phys 183:274–306. https://doi.org/10.1006/jcph.2002.7197
28. Eres MH, Weidner DE, Schwartz LW (1999) Three-dimensional direct numerical simulation of surface-tension-gradient effects on the leveling of an evaporating multicomponent fluid. Langmuir 15:1859–1871. https://doi.org/10.1021/la980414u
29. Eyer A, Leiste H, Nitsche R (1985) Floating zone growth of silicon under microgravity in a sounding rocket. J Cryst Growth 71:173–182
30. Fujimura K (2013) Linear and weakly nonlinear stability of Marangoni convection in a liquid bridge. J Phys Soc Jpn 82:074401
31. Fukuda Y, Ogasawara T, Fujimoto S, Eguchi T, Motegi K, Ueno I (2021) Thermal-flow patterns of $m = 1$ in thermocapillary liquid bridges of high aspect ratio with free-surface heat transfer. Int J Heat Mass Transfer 173:121196. https://doi.org/10.1016/j.ijheatmasstransfer.2021.121196
32. Fumoto K, Sasa M, Okabe T, Savino R, Inamura T, Shirota M (2019) Research on heat transfer performance of the open-loop micro pulsating heat pipe with self-rewetting fluids. Microgravity Sci Technol 31(3):261–268
33. Gaskell PH, Jimack PK, Sellier M, Thompson HM (2006) Flow of evaporating, gravity-driven thin liquid films over topography. Phys Fluids 18:013601. https://doi.org/10.1063/1.2148993
34. Gotoda M, Sano T, Kaneko T, Ueno I (2015) Evaluation of existence region and formation time of particle accumulation structure (PAS) in half-zone liquid bridge. Eur Phys J Spec Top 224:299–307. https://doi.org/10.1140/epjst/e2015-02361-7
35. Gotoda M, Melnikov DE, Ueno I, Shevtsova V (2016) Experimental study on dynamics of coherent structures formed by inertial solid particles in three-dimensional periodic flows. Chaos 26:073106
36. Gotoda M, Toyama A, Ishimura M, Sano T, Suzuki M, Kaneko T, Ueno I (2019) Experimental study of coherent structures of finite-size particles in thermocapillary liquid bridges. Phys Rev Fluids 4:094301. https://doi.org/10.1103/PhysRevFluids.4.094301
37. Haas DE (2006) Predicting the uniformity of two-component, spin-deposited films. PhD thesis, University of Arizona. https://repository.arizona.edu/handle/10150/195952
38. Haas DE, Birnie DP III (2000) Real-time monitoring of striation development during spin-on-glass deposition. In: Proceedings of the American ceramic social symposium on sol-gel commercialization and application, vol 123, pp 133–138. https://doi.org/10.1051/jp3:1993253
39. Haas DE, Birnie DP III (2002) Evaluation of thermocapillary driving forces in the development of striations during the spin coating process. J Mater Sci 37:2109–2116. https://doi.org/10.1023/A:1015250120963
40. Hu Y, Zhang S, Li X, Wang S (2014) Heat transfer enhancement mechanism of pool boiling with self-rewetting fluid. Int J Heat Mass Transf 79:309–313
41. Hu Y, Zhang S, Li X, Wang S (2015) Heat transfer enhancement of subcooled pool boiling with self-rewetting fluid. Int J Heat Mass Transf 83:64–68
42. Hu Y, Chen S, Huang J, Song M (2018) Marangoni effect on pool boiling heat transfer enhancement of self-rewetting fluid. Int J Heat Mass Transf 127:1263–1270
43. Hu Y, Wang H, Song M, Huang J (2019) Marangoni effect on microbubble emission boiling generation during pool boiling of self-rewetting fluid. Int J Heat Mass Transf 134:10–16
44. Jancarik J, Kim SH (1991) Sparse matrix sampling: a screening method for crystallization of proteins. J Appl Crystallogr 24:409–411
45. Kato K, Sensui S, Noguchi S, Kurose K, Ueno I (2024) Experimental study on coherent structures by small particles suspended in high aspect-ratio ($\Gamma = 2.5$) thermocapillary liquid bridges of high Prandtl number. Eur Phys J Spec Top
46. Kawamura H, Ueno I, Ishikawa T (2002) Study of thermocapillary flow in a liquid bridge towards an on-orbit experiment aboard the ISS. Adv Space Res 29:611–618

47. Kawamura H, Nishino K, Matsumoto S, Ueno I (2012) Report on microgravity experiments of Marangoni convection aboard international space station. J Heat Transfer 134:031005
48. Koszelak S, Day J, Leja C, Cudney R, McPherson A (1995) Protein and virus crystal growth on international microgravity laboratory-2. Biophys J 69:13–19
49. Kozuka H, Ishikawa Y, Ashibe N (2004) Radiative striations of spin-coating films: surface roughness measurement and in-situ observation. J Sol-Gel Sci Tech 31:245–248. https://doi.org/10.1023/B:JSST.0000047996.53649.3d
50. Kuhlmann HC, Mukin RV, Sano T, Ueno I (2014) Structure and dynamics of particle-accumulation in thermocapillary liquid bridges. Fluid Dyn Res 46:041421
51. Kuramae M, Suzuki M (1993) Two-component heat pipes utilizing the Marangoni effect. J Chem Eng Jpn 26(2):230–231
52. Lefebvre AH (1988) Atomization and sprays. Taylor & Francis (chap 4), pp 127–136
53. McGillis WR, Carey VP (1996) On the role of Marangoni effects on the critical heat flux for pool boiling of binary mixtures. J Heat Transfer 118(2):103–109
54. Mikol V, Rodeau JL, Giegé R (1990) Experimental determination of water equilibration rates in the hanging drop method of protein crystallization. Anal Biochem 186:332–339
55. Monti R, Savino R, Lappa M, Tempesta S (1998) Behavior of drops in contact with pool surfaces of different liquids. Phys Fluids 10(11):2786–2796
56. Mori T, Goto K, Ohashi R, Sawaoka A (1993) Capabilities and recent activities of Japan Microgravity Center (JAMIC). Microgravity Sci Technol 5(4):238–42
57. Morovati M, Bindra H, Esaki S, Kawaji M (2011) Enhancement of pool boiling and critical heat flux in self-rewetting fluids at above atmospheric pressures. In: ASME/JSME 2011 8th thermal engineering joint conference, p T10246
58. Napolitano LG, Monti R, Rosso G (1986) Marangoni convection in one- and two-liquids floating zones. Naturwissenschaften 73:352–355
59. Niigaki Y, Ueno I (2012) Formation of particle accumulation structure (PAS) in half-zone liquid bridge under an effect of thermo-fluid flow of ambient gas. https://doi.org/10.2322/tastj.10.Ph_33
60. Nishiguchi S, Shoji M (2011) Boiling heat transfer of butanol aqueous solution-augmentation of critical heat flux. ASTM Int 8(6):JAL13452
61. Noguchi S, Ueno I (2023) Spatial-temporal behaviors of low-stokes-number particles forming coherent structures in high-aspect-ratio liquid bridges by thermocapillary effect. Phys Rev Fluids 8:114002. https://doi.org/10.1103/PhysRevFluids.8.114002
62. Oba T, Toyama A, Hori T, Ueno I (2019) Experimental study on behaviors of low-stokes number particles in weakly chaotic structures induced by thermocapillary effect within a closed system with a free surface. Phys Rev Fluids 4:104002. https://doi.org/10.1103/PhysRevFluids.4.104002
63. Oron A, Davis SH, Bankoff SG (1997) Long-scale evolution of thin liquid films. Rev Mod Phys 69(3):931–980. https://doi.org/10.1103/RevModPhys.69.931
64. Petre G, Azouni MA (1984) Experimental evidence for the minimum of surface tension with temperature at aqueous alcohol solution/air interfaces. J Colloid Interface Sci 98(1):261–263
65. Preisser F, Schwabe D, Scharmann A (1983) Steady and oscillatory thermocapillary convection in liquid columns with free cylindrical surface. J Fluid Mech 126:545–567
66. Provost K, Robert MC (1991) Application of gel growth to hanging drop technique. J Cryst Growth 110(1–2):258–264
67. Riley RJ, Neitzel GP (1998) Instability of thermocapillary-buoyancy convection in shallow layers part 1 characterization of steady and oscillatory instabilities. J Fluid Mech 359:143–164
68. Roy RV, Roberts AJ, Simpson ME (2002) A lubrication model of coating flows over a curved substrate in space. J Fluid Mech 454:235–261. https://doi.org/10.1017/S0022112001007133
69. Ryzhkov II (2011) Thermocapillary instabilities in liquid bridges revisited. Phys Fluids 23:082103
70. Sakai T, Yoshii S, Kajimoto K, Kobayashi H, Shinmoto Y, Ohta H (2011) Heat transfer enhancement observed in nucleate boiling of alcohol aqueous solutions at very low concentration. In: 14th international heat transfer conference, vol 1, pp 471–478

71. Sakashita H, Ono A, Nakabayashi Y (2010) Measurements of critical heat flux and liquid-vapor structure near the heating surface in pool boiling of 2-propanol/water mixtures. Int J Heat Mass Transf 53(7):1554–1562
72. Sakata T, Terasaki S, Saito H, Fujimoto S, Ueno I, Yano T, Nishino K, Kamotani Y, Matsumoto S (2022) Coherent structures of m = 1 by low-Stokes-number particles suspended in a half-zone liquid bridge of high aspect ratio: microgravity and terrestrial experiments. Phys Rev Fluids 7:014005. https://doi.org/10.1103/PhysRevFluids.7.014005
73. Sato M, Abe Y, Urita Y, Di Paola R, Cecere A, Savino R (2011) Thermal performance of self-rewetting fluid heat pipes containing dilute solutions of polymer-capped silver nanoparticles synthesized by microwave-polyol process. Int J Transp Phenom 12:339–345
74. Savino R, Monti R (2005) Heat pipes for space applications. Space Technol 25(1):59–61
75. Savino R, Paterna D (2006) Marangoni effect and heat pipe dry-out. Phys Fluids 18:118103
76. Savino R, Abe Y, Fortezza R (2008a) Comparative study of heat pipes with different working fluids under normal gravity and microgravity conditions. Acta Astronaut 63:24–34
77. Savino R, Di Paola R, Cecere A, Abe Y, Tanaka K, Nakagawa M, Saito M (2008b) Heat pipes with self-rewetting fluids for space applications. In: International conference on environmental systems, SAE international, No. 2008–01–1954
78. Savino R, Cecere A, Di Paola R (2009) Surface tension-driven flow in wickless heat pipes with self-rewetting fluids. Int J Heat Fluid Flow 30(2):380–388
79. Savino R, Cecere A, Di Paola R, Abe Y, Castagnolo D, Fortezza R (2009) Marangoni heat pipe: An experiment on board MIOsat Italian microsatellite. Acta Astronaut 65(11–12):1582–1592
80. Savino R, Di Paola R, Cecere A, Fortezza R (2010) Self-rewetting heat transfer fluids and nanobrines for space heat pipes. Acta Astronaut 67(9):1030–1037
81. Savino R, Abe Y, Castagnolo D, Celata GP, Kabov O, Kawaji M, Sato M, Tanaka K, Thome JR, Vaerenbergh SV (2011) Ground-based activities in preparation of SELENE ISS experiment on self-rewetting fluids. J Phys Conf Ser 327:012032
82. Savino R, Di Paola R, Gattia DM, Marazzi R, Antisari MV (2011) Self-rewetting fluids with suspended carbon nanostructures. J Nanosci Nanotechnol 11:8953–8958
83. Savino R, Cecere A, Vaerenbergh S, Abe Y, Pizzirusso G, Tzevelecos W, Mojahed M, Galand Q (2013) Some experimental progresses in the study of self-rewetting fluids for the SELENE experiment to be carried in the thermal platform 1 hardware. Acta Astronaut 89:179–188
84. Savino R, De Cristofaro D, Cecere A (2017) Flow visualization and analysis of self-rewetting fluids in a model heat pipe. Int J Heat Mass Transf 115:581–591
85. Schwabe D, Hintz P, Frank S (1996) New features of thermocapillary convection in floating zones revealed by tracer particle accumulation structure (PAS). Microgravity Sci Technol 9:163–168
86. Schwabe D, Mizev AI, Udhayasankar M, Tanaka S (2007) Formation of dynamic particle accumulation structures in oscillatory thermocapillary flow in liquid bridges. Phys Fluids 19:072102
87. Schwartz LW, Roy RV (2004) Theoretical and numerical results for spin coating of viscous liquids. Phys Fluids 16(3):569–584. https://doi.org/10.1063/1.1637353
88. Sensui S, Noguchi S, Kato K, Ueno I (2024) Coherent structures formed by small particles in traveling-wave flows. Phys Rev E 110:015101. https://doi.org/10.1103/PhysRevE.110.015101
89. Shiratori S, Kubokawa T (2015) Double-peaked edge-bead in drying film of solvent-resin mixtures. Phys Fluids 27(10):102105. https://doi.org/10.1063/1.4934670
90. Shiratori S, Kato D, Sugasawa K, Nagano H, Shimano K (2020) Spatio-temporal thickness variation and transient Marangoni number in striations during spin coating. Int J Heat Mass Transfer 154:119678. https://doi.org/10.1016/j.ijheatmasstransfer.2020.119678
91. Smith MK, Davis SH (1983) Instabilities of dynamic thermocapillary liquid layers part 1 convective instabilities. J Fluid Mech 132:119–144
92. Stephan K, Abdelsalam M (1980) Heat-transfer correlations for natural convection boiling. Int J Heat Mass Transf 23:73–87

93. Su X, Zhang M, Han W, Guo X (2015) Enhancement of heat transport in oscillating heat pipe with ternary fluid. Int J Heat Mass Transf 87:258–264
94. Sultan E, Boudaoud A, Amar MB (2005) Evaporation of a thin film: diffusion of the vapor and Marangoni instabilities. J Fluid Mech 543:183–202. https://doi.org/10.1017/S0022112005006348
95. Takakusagi T, Ueno I (2017) Flow patterns induced by thermocapillary effect and resultant structures of suspended particles in a hanging droplet. Langmuir 33(46):13197–13206
96. Tanaka S, Kawamura H, Ueno I, Schwabe D (2006) Flow structure and dynamic particle accumulation in thermocapillary convection in a liquid bridge. Phys Fluids 18:067103. https://doi.org/10.1063/1.2208289
97. Terasaki S, Sensui S, Ueno I (2023) Thermocapillary-driven coherent structures by low-stokes-number particles and their morphology in high-aspect-ratio liquid bridges. Int J Heat Mass Transfer 203:123772. https://doi.org/10.1016/j.ijheatmasstransfer.2022.123772
98. Tien CL, Rohani AR (1972) Theory of two-component heat pipes. J Heat Transfer 94(4):479–484
99. Toyama A, Gotoda M, Kaneko T, Ueno I (2017) Existence conditions and formation process of second type of spiral loop particle accumulation structure (SL-2 PAS) in half-zone liquid bridge. Microgravity Sci Technol 29(4):263–274. https://doi.org/10.1007/s12217-017-9544-y
100. Ueno I, Arima M (2007) Behavior of vapor bubble in subcooled pool. Microgravity Sci Technol 19:128–129. https://doi.org/10.1007/BF02915773
101. Ueno I, Miyauchi A, Kawamoto A (2006) What's happening in head-to-head droplets? behavior of particle suspended in non-coalescence droplets formed between coaxial rods with temperature difference. In: Proceedings of the 12th international symposium flow visualization cd-rom:paper211
102. Ueno I, Ando J, Koiwa Y, Saiki T, Kaneko T (2015) Interfacial instability of a condensing vapor bubble in a subcooled liquid. Eur Phys J Spec Top 224:415–424. https://doi.org/10.1140/epjst/e2015-02370-6
103. van Wijk W, Vos A, van Stralen S (1956) Heat transfer to boiling binary liquid mixtures. Chem Eng Sci 5(2):68–80
104. Vochten R, Petre G (1973) Study of the heat of reversible adsorption at the air-solution interface ii experimental determination of the heat of reversible adsorption of some alcohols. J Colloid Interface Sci 42(2):320–327
105. Vochten R, Petre G, Defay R (1973) Study of the heat of reversible adsorption at the air-solution interface I thermodynamical calculation of the heat of reversible adsorption of nonionic surfactants. J Colloid Interface Sci 42(2):310–319
106. Wanschura M, Shevtsova VM, Kuhlmann HC, Rath HJ (1995) Convective instability mechanisms in thermocapillary liquid bridges. Phys Fluids 7(5):912–925
107. Watanabe T, Takakusagi T, Ueno I, Kawamura H, Nishino K, Ohnishi M, Sakurai M, Matsumoto S (2018) Terrestrial and microgravity experiments on onset of oscillatory thermocapillary-driven convection in hanging droplets. Int J Heat Mass Transf 123:945–956
108. Weidner DE, Schwartz LW, Eley RR (1996) Role of surface tension gradients in correcting coating defects in corners. J Colloid Interface Sci 179:66–75. https://doi.org/10.1006/jcis.1996.0189
109. Weidner DE, Schwartz LW, Eres MH (1997) Simulation of coating layer evolution and drop formation on horizontal cylinders. J Colloid Interface Sci 187:243–258. https://doi.org/10.1006/jcis.1996.4711
110. Wu J, Dhir VK (2011) Numerical simulation of dynamics and heat transfer associated with a single bubble in subcooled boiling and in the presence of noncondensables. J Heat Transfer 133(4):041502
111. Wu SC (2015) Study of self-rewetting fluid applied to loop heat pipe. Int J Therm Sci 98:374–380
112. Wu SC, Lee TJ, Lin WJ, Chen YM (2017) Study of self-rewetting fluid applied to loop heat pipe with PTFE wick. Appl Therm Eng 119:622–628

113. Xu JJ, Davis SH (1984) Convective thermocapillary instabilities in liquid bridges. Phys Fluids 27:1102–1107. https://doi.org/10.1063/1.864756
114. Yano T, Nishino K, Kawamura H, Ueno I, Matsumoto S, Ohnishi M, Sakurai M (2011) Space experiment on the instability of Marangoni convection in large liquid bridge–MEIS-4: effect of Prandtl number. J Phys Conf Ser 327:012029
115. Yano T, Nishino K, Kawamura H, Ueno I, Matsumoto S, Ohnishi M, Sakurai M (2012) 3-D PTV measurement of Marangoni convection in liquid bridge in space experiment. Exp Fluids 53:9–20
116. Yano T, Nishino K, Kawamura H, Ueno I, Matsumoto S (2015) Instability and associated roll structure of Marangoni convection in high Prandtl number liquid bridge with large aspect ratio. Phys Fluids 27:024108. https://doi.org/10.1063/1.4908042
117. Yiantsios SG, Higgins BG (2006) Marangoni flows during driving of colloidal films. Phys Fluids 18:082103. https://doi.org/10.1063/1.2336262

Open Access This chapter is licensed under the terms of the Creative Commons Attribution 4.0 International License (http://creativecommons.org/licenses/by/4.0/), which permits use, sharing, adaptation, distribution and reproduction in any medium or format, as long as you give appropriate credit to the original author(s) and the source, provide a link to the Creative Commons license and indicate if changes were made.

The images or other third party material in this chapter are included in the chapter's Creative Commons license, unless indicated otherwise in a credit line to the material. If material is not included in the chapter's Creative Commons license and your intended use is not permitted by statutory regulation or exceeds the permitted use, you will need to obtain permission directly from the copyright holder.

Development of Fluid Dynamics Experiments in Kibo Aboard International Space Station and Beyond

Satoshi Matsumoto

Abstract The concluding chapter of this book introduces the International Space Station (ISS) and the Japanese Experiment Module "Kibo." It elaborates on the facilities and equipment utilized for conducting thermal capillary convection experiments on the International Space Station. The evolution of "Kibo" and its future prospects, particularly its role in technological advancements for future space exploration, are outlined. Finally, the application of surface tension-driven flows in upcoming human activities in space is explained.

1 International Space Station and Kibo Module

The space station is a habitable modular satellite (Fig. 1). The International Space Station program is a spectacular international cooperation project involving 15 countries around the world, including the United States, Russia, Europe, Canada, and Japan. It was proposed by the United States in 1984 and started with a call to other countries. This program aims to contribute to the development of science and technology by constructing a permanent crewed facility where astronauts will stay and conduct various space experiments and activities.

The International Space Station (ISS) is a platform that orbits in low Earth orbit approximately 400 km above the ground. The ISS is made up of many elements shown in Fig. 2. One of the pressurized modules to conduct scientific research and technology development is the Japanese Experiment Module (JEM) known as "Kibo" developed by Japan (Fig. 3). The largest element Kibo of the ISS consists of four major components: two main experiment facilities, which include pressurized modules and exposed facilities, a logistics module for storage, and a robotic arm. The Kibo docked with the ISS in 2008, and Japan began its activities in orbit as an international partner of the ISS along with the United States, Europe, and Russia.

S. Matsumoto (✉)
Japan Aerospace Exploration Agency, Tsukuba, Japan
e-mail: matsumoto.satoshi@jaxa.jp

Fig. 1 International space station (Courtesy of NASA)

The pressurized module is cylindrical in shape with an outer diameter of approximately 4.45 m, an inner diameter of approximately 4.22 m, and a total length of approximately 11.68 m. Inside the pressurized module, four "racks," which are structures for mounting equipment, are mounted on the same circumference. The racks are arranged in six rows in the axial direction of the pressurized section (five rows on the ceiling side), and a total of 23 racks can be mounted. Of these, 12 racks are allocated for experimental equipment (including two racks for experimental storage volume). The remaining 11 racks are for JEM system equipment. The crew's normal workspace (cabin), which is enclosed by racks and "closeout panels" (wall panels), is mostly a nearly square cross section of about 2.20 m on each side.

The thermocapillary convection experiments were performed in the fluid physics experiment rack called "Ryutai rack" located at JPM1A3 position in the Kibo pressurized module (Fig. 4).

2 Thermocapillary Experiment in Microgravity

Thermocapillary convection is a flow driven by a surface tension gradient caused by a temperature difference at the gas–liquid interface. The surface tension is generally higher on the side with lower temperature, and the fluid on the surface is pulled from the side with higher temperature to the side with lower temperature, creating a surface flow and resulting in internal flow. This phenomenon has been known for a

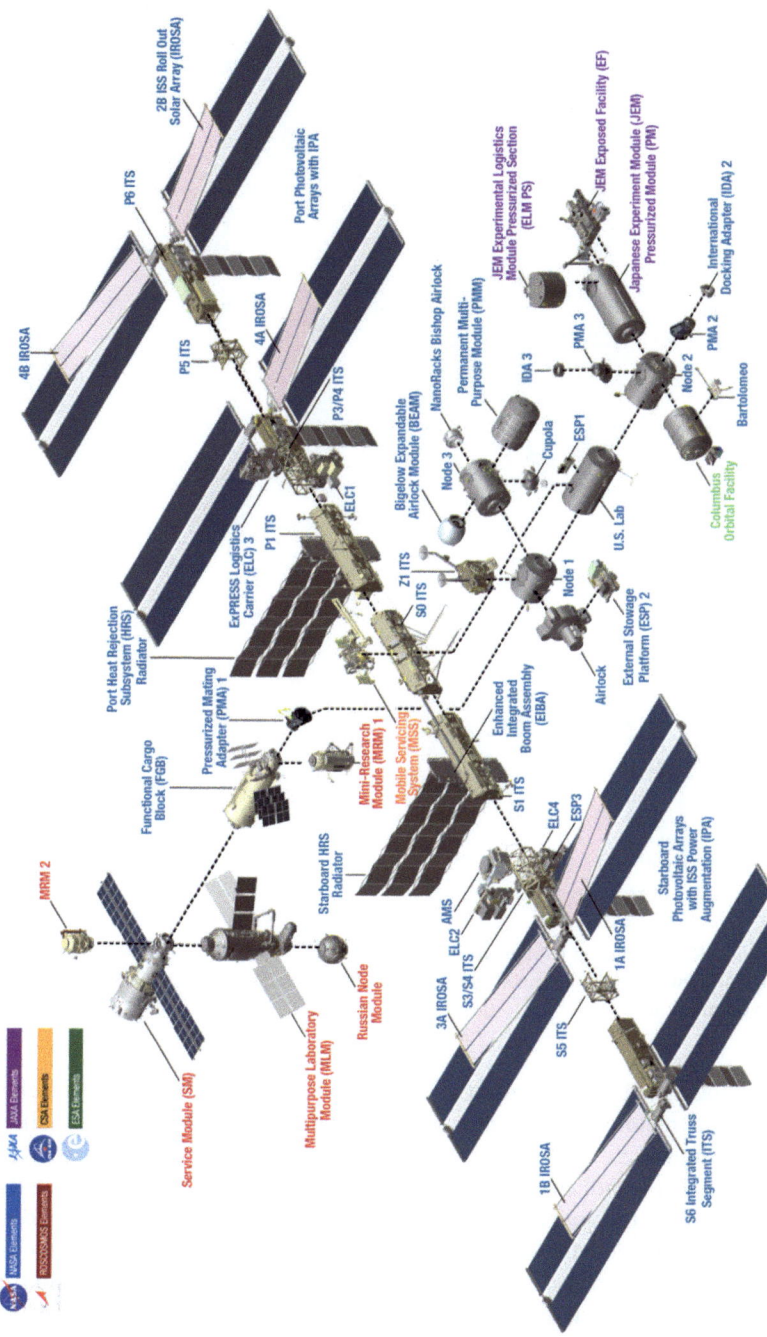

Fig. 2 Drawing of the international space station with all of the elements (Courtesy of NASA)

Fig. 3 Japanese experiment module "Kibo" (Courtesy of NASA)

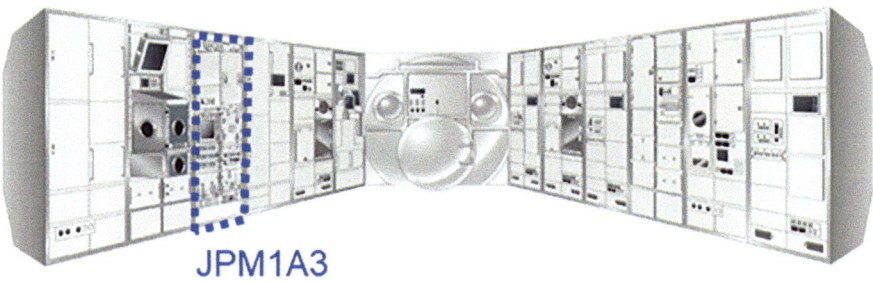

Fig. 4 Location of Ryutai Rack in JEM pressurized module (JPM1), Aft3 position (A3). This figure shows the interior view from the entrance side of JEM. The deck and ceiling side are not shown (©JAXA with permission)

long time. Although active observation of the phenomenon began in the seventeenth century, scientific treatment of the phenomenon did not begin until the latter half of the nineteenth century.

However, it wasn't until the mid-twentieth century that the complexity of the phenomenon was fully recognized theoretically. Experimental and theoretical studies were subsequently conducted from the 1950s. In the 1970s, experiments under microgravity environment led to a new focus on the phenomenon as a phenomenon in microgravity [5]. Surface-driven-flow is known to significantly affect the quality of crystals grown, which is a manufacturing method for semiconductor, alloy, and oxide crystals [4]. The microgravity experiment, Surface Tension Driven Convection Experiment (STDCE), conducted during the United States Microgravity Laboratory-I (USML-1) mission, utilized an open circular container configuration with a heated/cooled center part of the surface. This setup appears to simulate the Czochralski method used in crystal growth [2]. In the space experiment onboard the Kibo, we used a liquid bridge shape that simulated the floating zone method for crystal growth, and a simpler half-zone configuration to make the phenomenon easier to understand.

Three microgravity experiments on thermocapillary convection in a liquid bridge configuration were carried out on the Kibo, as listed below.

(1) "Chaos, Turbulence and its Transition Process in Marangoni Convection (MEIS)" Principal Investigators: Hiroshi Kawamura (Tokyo University of Science)/Koichi Nishino (Yokohama National University)
(2) "Spatio-temporal Flow Structure in Marangoni Convection," Principal Investigator: Shinichi Yoda (Japan Aerospace Exploration Agency)
(3) "Experimental Assessment of Dynamic Surface Deformation Effects in Transition to Oscillatory Thermocapillary Flow in Liquid Bridge of High Prandtl Number Fluid," Principal Investigators: Satoshi Matsumoto (Japan Aerospace Exploration Agency)/Yasuhiro Kamotani (Case Western Reserve University)

In August 2008, the memorable first experiment in the Kibo was "Chaotic and turbulent flows and their transition processes in Marangoni convection". The fact that this experiment was the first scientific experiment played a major role in the rapid progress of Japan's human space program.

Before the experiment, astronauts spent approximately 10 h over three days preparing for the experiment. During the experiment, scientists and ground operators monitored video images of flow and thermos-viewer and sensor values in real time, making temperature adjustments and fine-tuning settings as necessary. It involved a considerable amount of hectic work. By issuing commands and collecting data from the ground, we could adjust experimental conditions as required and proceed with the experiment.

In order to make the most of the limited experimental opportunities and to obtain results, we had positioned the above three experiments as a series of tasks for thermocapillary convection research in Japan, and were proceeding with the systematic acquisition of data. In thermocapillary convection, it is known that the flow transitions from steady flow to oscillatory flow, chaotic flow, and turbulent flow as in other convection phenomena by increasing the control parameter (temperature difference). However, experimental data on the conditions under which flow transitions occur are insufficient, and extensive accumulation of data is needed. In particular, liquid bridge with a diameter of over 5 mm are difficult to form on the ground because they are subject to large deformation and surface tension cannot support their own weight (Fig. 5). On the other hand, microgravity in space enables the formation of large liquid bridge with ideal shape, and space experiments provide a wide range of systematic data.

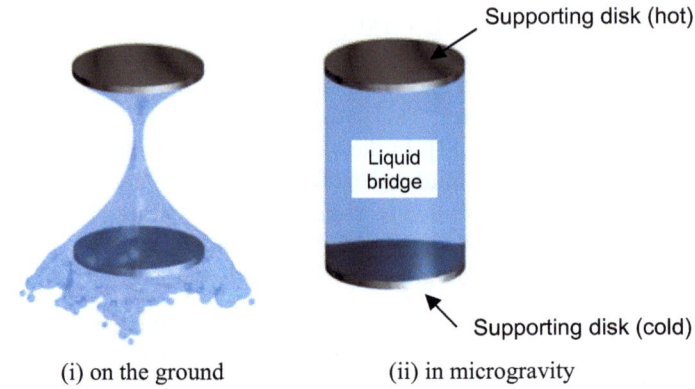

Fig. 5 Large liquid bridge formation in microgravity. (©JAXA with permission)

The driving force behind thermocapillary convection is the surface tension gradient, which is fundamentally distinct from that of density-driven convection, resulting from volumetric forces within a liquid. In a terrestrial gravity field, isolating thermocapillary convection for studying its dynamic behavior is challenging due to buoyant convection caused by thermal density differences. Therefore, a microgravity environment is necessary to maintain liquid bridge shape and to eliminate natural convection, allowing for the observation of pure thermocapillary convection. Furthermore, to accurately capture the transition process of thermocapillary convection, experiments spanning several hours are required. Thus, conducting experiments using Kibo, which offers a high-quality microgravity environment for extended durations, holds great significance.

Key features of microgravity experiments include: (1) the capability to form large liquid bridges, (2) an expanded parameter range, (3) absence of density-driven convection, and (4) prevention of liquid bridge deformation due to gravity. The prolonged microgravity environment facilitated by Kibo is conducive to systematically acquiring highly precise data.

A liquid bridge with diameters of 10, 30, or 50 mm is formed between heating and cooling disks, and thermocapillary convection is induced by applying a temperature difference across the bridge (Fig. 6). Convection patterns are observed using images, while temperatures at various locations are measured using thermocouples and platinum temperature sensors to collect data. Silicone oil with a kinematic viscosity ranging from 5 to 20 cSt (centistokes) was utilized as the working fluid. Silicone fluid is often used due to its chemical stability, non-toxic nature (essential for confined human spaces such as a space station), and transparency, which enables visualization of internal flows. Microparticles, sized between 30 and 200 μm in diameter, are added to the silicone oil to aid in visualizing internal flow dynamics.

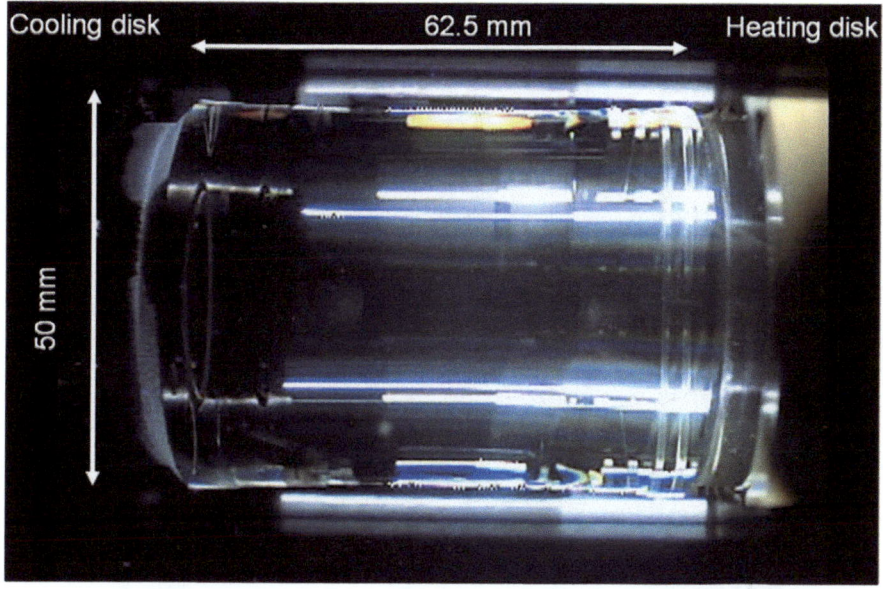

Fig. 6 Liquid bridge formed in space (diameter: 50 mm, height: 62.5 mm) (©JAXA with permission)

On the ISS, vibrations are generated by rotating objects (such as fans, hard disks, etc.) mounted on instruments, by the movement of structures like antennas and solar panels, and by the activities of astronauts. These disturbances, known as g-jitters, typically range from 10^{-6} to 10^{-4} g in magnitude. A liquid is held between supporting disks by its own surface tension, with its sides forming free surfaces. Consequently, it is highly sensitive to vibrations. Depending on the diameter and length of the liquid bridge, a liquid bridge measuring several tens of millimeters typically exhibits a natural frequency ranging from 0.1 to several Hz. This frequency range overlaps with the vibration frequencies induced by astronauts' activities. During astronaut activity periods, significant vibrations in large liquid bridges can occur, potentially leading to the collapse of the liquid bridge under worst-case scenarios. To mitigate this risk, liquid bridges are typically formed during astronauts' nighttime hours (usually from 21:30 to 6:00 GMT) when the vibration environment is relatively stable, enabling experiments to be conducted in an environment with minimal liquid bridge vibration.

3 Experimental Facility

Experiments are conducted using the Fluid Physics Experiment Facility (FPEF) and the Image Processing Unit (IPU) mounted on the Ryutai rack in Kibo (Fig. 7).

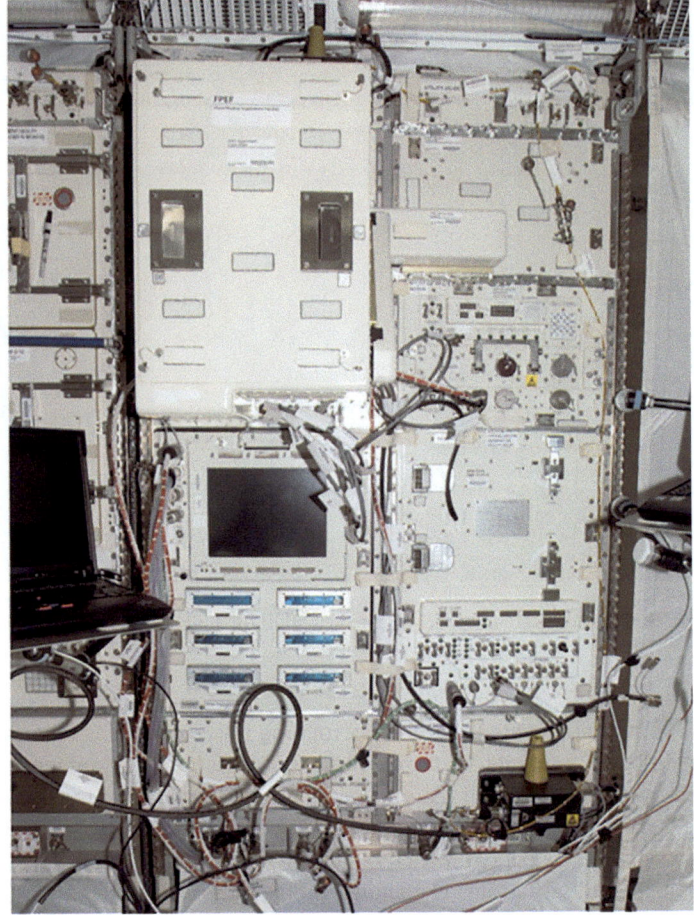

Fig. 7 Fluid rack installed in Kibo. Top left: FPEF, Bottom left: IPU (Courtesy of NASA)

The FPEF is designed to accommodate experimental inserts (Fig. 8) with functionalities tailored to the experiment's objectives. It can accommodate multiple research experiments by fabricating test specimens that align with the research.

(1) Fluid Physics Experiment Facility (FPEF)

The FPEF occupies the upper left quarter of the Ryutai rack and primarily provides the following functions to the experimental inserts:

- Main power supply
- Control of power supply to heaters, Peltier elements, etc.
- Input of sensor signals such as thermocouples and platinum resistance elements
- Analog and digital I/O
- Motor drivers

Fig. 8 Experimental insert (©JAXA with permission)

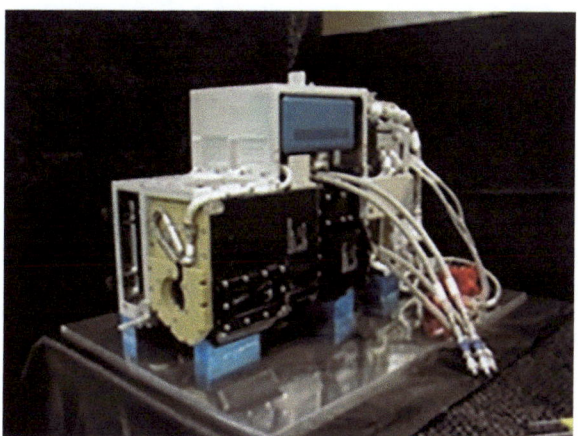

- Cooling water circulation
- Gas exhaust
- Argon gas supply
- CCD camera signal input.

The fluid physics experimental apparatus itself possesses functions for observing fluid behavior, including 3D-velocimetry from end of liquid bridge, observation of the liquid bridge side, and surface temperature distribution measurement.

A schematic diagram of the observation system is depicted in Fig. 9. The flow is observed from the top of the liquid bridge through a transparent heated disk, and images are captured by three CCD cameras positioned at varying angles and locations. These images undergo analysis using a particle tracking method to construct a three-dimensional flow field, providing insights into the flow structure within the liquid bridge and its temporal changes. Observations from the side of the liquid bridge facilitate the capture of the bridge shape and the flow appearance from a lateral perspective. Considering that thermocapillary convection exerts a significant driving force at the free surface, where velocity changes are notable, surface flow velocity can be measured using photochromic methods. Additionally, an infrared camera is employed to observe the surface temperature distribution.

(2) Experimental insert

The experimental insert measures 560 mm in length, 250 mm in width, and 360 mm in height, with a mass of approximately 36 kg. It comprises three primary components: (1) a frame structure, (2) a liquid-bridge forming section equipped with an internal mechanism for freely creating a liquid bridge, and (3) a sample cassette containing the experimental sample. These components are launched separately into orbit and assembled by astronauts once in space, integrating them into the FPEF (Fig. 10).

The liquid bridge formation section includes a heating disk and a cooling disk, with the length of the liquid bridge adjustable by moving the cooling disk. The heating disk, made of transparent sapphire, is essential for observation by a CCD

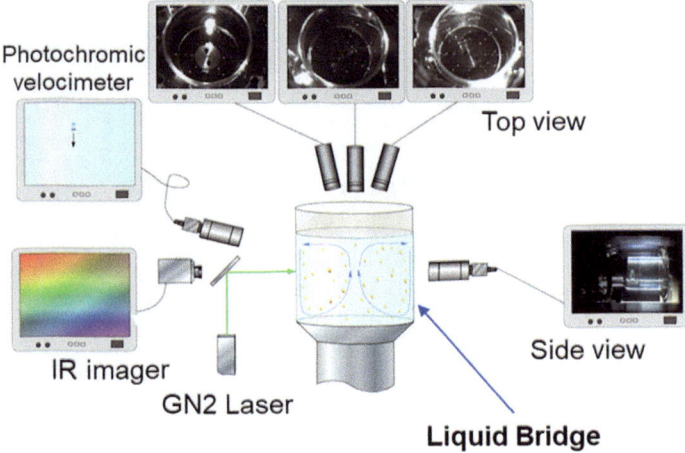

Fig. 9 Schematic diagram of the observation system

Fig. 10 Experimental specimen being set up by astronauts (©JAXA/NASA with permission)

camera from the top surface and boasts high thermal conductivity for temperature uniformity. It features patterned ITO (Indium-Tin-Oxide) film for resistance heating, with additional ITO film on the opposite surface for temperature measurement via its resistance temperature dependence. The disk's sharp 45° edge securely holds the liquid bridge. Technological advancements in uniformly patterning ITO film without blunting this edge encountered several technical challenges. However, overcoming these obstacles has led to the disk's excellent performance in space experiments. The cooling disk, made of aluminum, is cooled by a Peltier element.

The experimental sample is housed in a sample cassette (approximately 150 cc for a 30 mm diameter liquid bridge). As the cassette is assembled into the liquid bridge, a discharge port on the cooling disk's surface opens, supplying oil to both the heating and cooling disk surfaces. Corresponding to the cooling disk's movement, the bellows, with an effective diameter matching that of the liquid bridge in the cassette, contracts to supply the appropriate oil volume. An auxiliary pump facilitates fine-tuning the liquid bridge's shape at the 0.01 cc level, adjusting its slimming or fattening, thereby providing experimental data on the liquid bridge's shape dependence.

4 Experimental Operation

After the setup of the experimental hardware, it was remotely operated by commands from ground operators, and the functionality of the device had been verified. On the first day of the experiment, a 9.6 mm-long liquid bridge was formed, and oscillatory flow transitions were successfully observed by applying a temperature difference. The liquid bridge was formed very well, and thermocapillary convection was successfully observed by applying a temperature difference. Although bubbles were introduced into the liquid bridge, we succeeded in finding a method to remove the bubbles successfully during the next several experiments.

During experimental operation, the UI (user integration) position compiles the requests of researchers and progresses the experiment while conveying the requests to the experimental operation control staff. The liquid column Marangoni experiment is probably one of the most difficult experiments using the ISS. Because the cylindrical liquid is supported and held only by the disks at the top and bottom ends, it is necessary to make subtle adjustments when forming and storing the liquid column. Additionally, in order to obtain good data, it is essential to frequently send commands to control the device, such as adjusting the shape of the liquid column or changing the temperature profile in accordance with the phenomena observed in real time. Therefore, we sent an average of 150 commands in one day of the experiment. FPEF alone has 118 commands, and inhibit conditions are imposed on each command. The operation of checking each of them one by one, determining the instructions from the UI, and sending commands in rapid succession required a very sophisticated operation, but the fluid rack operation control personnel were unable to manage the heavy-duty operation. was carried out appropriately.

5 Perspectives of Kibo Utilization

The utilization of Kibo, which began in 2008, has involved numerous scientific experiments, educational missions, and humanities trials, and its status has changed over time in the following three phases. The first phase is the exploration of potential utilization, the second phase is the demonstration of the social and practical value of the utilization, and the third phase is the development into a socially established sustainable utilization.

(1) first phase

The initial stage of Kibo utilization was as a laboratory where Japan could proactively plan and implement the long-term use of the microgravity environment, which Japan had acquired for the first time. Based on the utilization of ISS equipment, experiments selected for the initial use of Kibo were implemented.

Furthermore, new application experiments are selected in a wide range of fields based on proposals based on the free ideas of researchers, advance understanding and elucidation of phenomena in the space environment and develop promising fields and areas for microgravity. The range of uses for scientific research has been expanded. In addition, to support these diverse uses, JAXA have developed new multipurpose experiment racks and other common infrastructure experiment equipment and have accumulated space experiment technology and know-how.

(2) second phase

The nation has provided opportunities for space experiments as one of the research tools for cutting-edge research that is being strategically promoted by the government. Advanced space experiment technologies have supported cutting-edge research utilization from a technological aspect.

As a private sector-led research and development utilization, the JAXA has been reviewing and enhancing its utilization menu, including the establishment of regular services, and has been contributing to private sector research and development activities by providing regular services that meet the needs of private companies.

(3) third phase

Ten years after its completion, the Japanese Experiment Module "Kibo" has provided opportunities for scientific research, the production of high-quality protein crystals, the release of nano-satellites, and other experiments. By providing space continuously and stably over a long period of time, the JEM is being established as a space "platform" for space environment utilization and technology demonstration, as well as for familiar space utilization. In addition, the space is being shifted from government-supported utilization to private enterprises. Self-invested utilization and independent utilization by academia are expanding.

Under these circumstances, private sector activities in Low Earth Orbit (LEO) (including ISS) are becoming more active both in Japan and abroad (especially in the U.S.). In order for Kibo to become a part of the social infrastructure and for LEO to continue to develop as a place for human economic activities after the ISS retirement, system and technology development will be promoted in cooperation with the private sector.

6 Use of Surface-Tension-Driven Flows in Future Human Activities in Space

Space activities are expanding into deep space, with missions targeting the Moon and Mars. With the stable utilization of space in low Earth orbit, space exploration has emerged as the next frontier. The United States is advancing the Artemis program, collaborating with international partners to establish a sustained presence on the Moon. This effort lays the groundwork for private companies to develop a lunar economy and eventually send humans to Mars. Our long-term goal is exploration. The Gateway program, a space station orbiting the Moon, is underway, and preparations are being made for lunar surface activities. The lunar orbiting space station (Gateway) provides a microgravity environment similar to the International Space Station. Conversely, the Moon and Mars surfaces exhibit gravitational acceleration of $1/6g$ and $1/3g$, respectively, presenting unique challenges. Liquid handling is crucial for human space activities.

The behavior of liquids in a normal gravity environment on Earth is naturally different in microgravity or low gravity, and gravity can be used as a driving force from the perspective of liquid transport, so it may be actively utilized. On the other hand, in micro-gravity and low-gravity environments where driving forces are reduced, fluid handling is required by other means.

Microfluidic processes on Earth, such as cooling water circulation in capillary-driven heat pipes, can maintain passive control if the system dimensions are small; in microgravity environments within spacecraft, similar control can be achieved due to reduced body forces. Although possible, it is used in systems with large characteristic dimensions such as capillary length. For example, storing and managing liquid fuel or cryogenic materials in units of 1 ton for space flight, or circulating large amounts of water in spacecraft life support systems, separation, etc. [1, 7].

One of the important geometrical structures in capillary fluid systems is edges (or internal angles, wedges, grooves, etc.). If the angle of the edge and the contact angle between the liquid and the wall are sufficiently small, the edge will be naturally filled with liquid. This edge wetting condition is an effective means for passive fluid transport and control. On the macroscale, it can be used to expel large amounts of fluid from one region of a vessel to another, or to passively transport immiscible fluids. It can be used for various purposes such as separating [6]. For example, to enhance fluid motion passively, the utilization of corner flow shows promise as a method. Additionally, thermo-capillarity can accelerate the movement of liquid within cavity pipes, as depicted in Fig. 11.

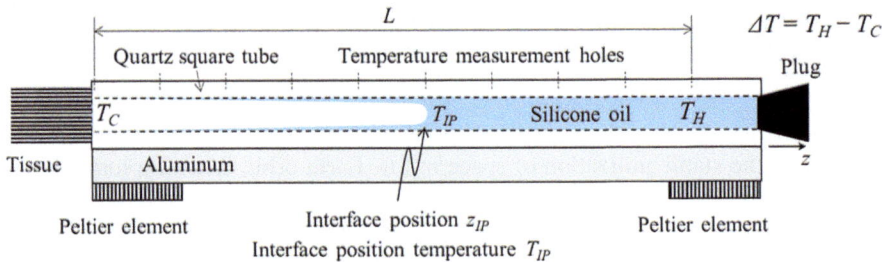

Fig. 11 Schematic of corner flow with thermo-capillarity in square pipe, own work

References

1. Chato DJ (2008) Cryogenic fluid transfer for exploration. Cryogenics 48:206–209. https://doi.org/10.1016/j.cryogenics.2008.03.013
2. Kamotani Y, Ostrach S, Pline A (1995) A thermocapillary convection experiment in microgravity. J Heat Transfer 117(3):611–618. https://doi.org/10.1115/1.2822621
3. Nixon D (2016) International space station–architecture beyond earth. Circa Press. ISBN 978-0-9930721-3-0
4. Schwabe D (1988) Surface-tension-driven flow in crystal growth melts. In: Superhard materials, convection, and optical devices. crystals, vol 11. Springer, Berlin, Heidelberg. https://doi.org/10.1007/978-3-642-73205-8_2
5. Walter HU (1984) Scientific results and accomplishments of the TEXUS programme, vol 11, pp 659–671. https://doi.org/10.1016/0094-5765(84)90050-X
6. Weislogel MM, Collicott SH (2004) Capillary rewetting of vaned containers: spacecraft tank rewetting following thrust resettling. AIAA J 42:2551–2561
7. Weislogel MM, Thomas EA, Graf JC (2009) A novel device addressing design challenges for passive fluid phase separations aboard spacecraft. Microgravity Sci Technol 21:257–268. https://doi.org/10.1007/s12217-008-9091-7

Open Access This chapter is licensed under the terms of the Creative Commons Attribution 4.0 International License (http://creativecommons.org/licenses/by/4.0/), which permits use, sharing, adaptation, distribution and reproduction in any medium or format, as long as you give appropriate credit to the original author(s) and the source, provide a link to the Creative Commons license and indicate if changes were made.

The images or other third party material in this chapter are included in the chapter's Creative Commons license, unless indicated otherwise in a credit line to the material. If material is not included in the chapter's Creative Commons license and your intended use is not permitted by statutory regulation or exceeds the permitted use, you will need to obtain permission directly from the copyright holder.

Index

A
Acquisition of Signal (AOS), 205, 206
Ambient airflow, 158
Amplitude equation, 99
Amplitude spectrum, 273–278
Anisotropic local stress, 44
Antisymmetric mode, 179
Aspect ratio, 84, 115, 180, 298, 335, 336, 339, 341, 342, 346
—A_r, 196, 210–220, 236, 261, 266, 268
Aspect ratio (A), 159
Aspect ratio (Γ), 166, 282
Aspect ratio (Γ_H), 180
Astronaut Doi, 12
Astronaut Hoshide, 12
Astronaut Wakata, 12
Azimuthal modal number (m), 182
Azimuthal mode number (m), 142, 209, 211, 213, 216, 222, 259
Azimuthal wavelength, 137
Azimuthal wave number, 82, 298, 299, 311, 314, 339, 344

B
Bakker's equation, 35, 38, 43
Basic flow, 178
Bénard, 55
Bénard cell, 122
Bénard-Pearson convection, 68
Biot number (Bi), 52, 60, 157, 261, 268
Boltzmann constant, 28
Bond number (Bo), 53, 221, 319, 321, 322, 325
Boundary conditions related to the thermocapillarity, 50

Boussinesq approximation, 57
Bubble, 207, 208, 211
—removal, 207, 208
Buoyancy, 55, 178, 212, 262

C
Camera
—calibration, 234
—parameter, 234
Capillarity, 5
Capillary
—number, 53, 55, 319, 324, 325
—pressure, 315
Capillary length, 54
Capillary number (Ca), 54, 272
Capillary rise, 5
Centrifugal force, 351
Chaos, 281–288
Chaotic flow, 116
Characteristic length, 298, 300, 312
Characteristic thermocapillary velocity, 231, 245
Coating process, 345
Co-flow, 78, 82, 92, 93
Co-flow direction, 241, 242, 254, 255, 257–260
Coherent particulate structure, 147
Coherent structure, 298
Cold disk, 196, 200, 205, 207, 248–253
Computational Fluid Dynamics (CFD), 251, 262–270
Concentration field, 349
Condensation, 359, 365
Conducting boundary, 59
Conductive heat transfer, 248, 261, 265

Constant temperature (isothermal) boundary, 59
Contact angle, 3, 19
Continuity equation, 262, 270
Control, 165
Control and suppression, 156
Control gain, 166
Convective heat transfer, 248, 261, 262, 265
Corner flow, 389
Correlation dimension (D_c), 282, 286
Counter-flow, 78, 82, 92, 93
Counter-flow direction, 241, 242, 255, 257–260
Critical
 —dimensionless oscillation frequency (F_c), 217, 218, 222, 223, 249, 251, 261
 —oscillation frequency (f_c), 210, 252
 —temperature difference (ΔT_c), 209, 210, 252, 282
Critical condition, 63
Critical Marangoni number (Ma_c), 114, 130, 157, 217, 218, 222, 223, 226–228, 249, 251, 261, 271, 282
Critical Rayleigh number (Re_c), 61, 62, 226
Critical wavelength, 62
Critical wavenumber, 62
Crystal growth, 7, 104
Czochralski method, 105

D
Damping performance (γ), 172
Deformation, 300, 310
Deterministic Lateral Displacement (DLD), 149
Diameter ratio (D_r), 221
Dimensionless
 —resonance angular oscillation frequency (ω_R^*), 276
Dimensionless angular oscillation frequency (ω^*), 258, 260
Dimensionless oscillation frequency (F), 217
Disconnected two neutral curves, 92, 96
Drifting Bénard-Marangoni cell, 240, 260
Dynamic free-Surface Deformation (DSD), 270–281

E
Edge-bead, 345, 346, 350–352
Effective characteristic length, 132
Emissivity (ε), 264, 265

Energy equation, 230, 262, 270
Entropy, 33
Equilibrium, 31
Equilibrium distance, 27, 29
Equilibrium MD, 38
Equilibrium state, 27, 30, 31
Evaporation, 300, 311, 336, 346, 347, 352, 355, 359, 361, 364, 365, 368
 —rate, 347, 349, 352, 366

F
Feedback control, 165
Finite volume method, 265
First law of thermodynamics, 18
Flat liquid-vapor interface, 36
Floating Zone (FZ) method, 105, 381
Floquet
 —analysis, 287
 —mode, 288
Fluid Physics Experiment Facility (FPEF), 194, 196, 205, 383
Fourier analysis, 274, 286
Fourier's law, 200
Free energy, 33
Frequency skip, 218, 223, 225, 236, 242, 256, 258, 261
Full-Zone (FZ), 177

G
Gateway program, 389
Ghost particle, 235
Gibbs free energy, 21
G-jitter, 13, 205, 275, 276, 278, 279, 281, 383
Greenwich Mean Time (GMT), 205
Growth rate, 61

H
Half zone liquid bridge, 105
Heat exchange, 156
Heat flux
 critical— (CHF), 361
Heat gain, 161
Heating output, 166
Heat loss, 157
Heat pipe, 6, 356, 359, 363–365
 grooved—, 359
 loop—, 362–364
 ordinary—, 369
 oscillating—, 362–364
 pulsating— (PHP), 364, 367, 369

Index

wicked—, 362–365
wickless—, 359, 362, 363
Heat transfer, 156
   —coefficient (h), 261, 268
   —ratio, 266, 268
Heat-transfer boundary, 59
Helmholtz free energy, 33, 35
Hopf bifurcation, 278
Horizontal Bridgeman method, 105
Hot
   —corner, 250, 271, 278
   —disk, 196, 200, 202
Hydrophobic, 4
Hydrothermal Wave (HTW), 77, 79, 83, 89, 90, 93, 166, 210, 215, 217, 242, 254–261
   —instability, 226, 242, 254, 268
Hydrothermal wave instability, 166

I

Image Processing Unit (IPU), 194, 196, 203, 205, 383
Indium-Tin-Oxide (ITO), 386
   —(film) heater, 201, 202
   —(film) sensor, 200, 202, 209, 210
Infinite liquid layer, 55
Infinitely long liquid column, 52
Infrared (IR)
   —camera, 202, 209, 254, 268, 270
   —image (imaging), 215, 216
Instability, 156, 315
   hydrothermal wave—, 298, 299, 303, 332, 336
   Marangoni-Bénard—, 345
   primary—, 298, 299
   Rayleigh—, 351
   secondary—, 298
Interaction stress, 41
Interface, 26
Interfacial heat transfer, 156, 241, 248–261
Interfacial tension, 24
Interferometer, 360
   Mach-Zehnder—, 360
Intermolecular interaction term, 40
International Space Station (ISS), 10, 11, 194, 199–201, 205, 275, 377
Interpretation of the surface tension, 30
Inverted scheme, 164, 166

J

Japanese Experiment Module (JEM), 377
Japan Standard Time (JST), 205

K

KAM theorem, 148
KAM torus, 148
Kibo, 104
Kinetic stress, 41

L

Laplace pressure, 348
Lennard-Jones potential, 30
Levenberg-Marquardt method, 234
Life science experiment, 11
Light-Emitting Diode (LED), 271
Linear dispersion relation, 88, 96
Linear Stability Analysis (LSA), 56, 58, 88, 158, 179, 226, 258, 259, 261, 262, 288
Liquid bridge, 52, 156, 196, 207
Liquid column, 52
Liquid-Vapor (LV) interface, 38, 42
Liquid-vapor interfacial, 44
Liquid-Vapor (LV) or Liquid-Gas (LG) surface tension, 22
Longitudinal roll, 91, 92
Longitudinal wave, 79
Longwave approximation, 348
Loss of Signal (LOS), 205, 206
Lubrication approximation, 348
Lyapunov exponent (λ_1), 282, 286

M

Magnitude ratio M_R, 184
Marangoni
   —Bénard instability, 345
   —effect, 345, 347, 348, 356, 363
   —flow, 359, 360
   —number, 51, 54, 55, 129, 130, 132, 298, 334, 336, 342–344, 354, 355
   —velocity, 51, 320
   critical—number, 129, 130, 217, 218, 341–343
   inverse—effect, 359, 363
   reverse—effect, 367
   reverse—flow, 355, 359
   solutal—effect, 346
   solutal—number, 355
Marangoni-Bénard convection, 68
Marangoni-Bénard instability, 353
Marangoni convection, 107
Marangoni flow, 8
Marangoni number (Ma), 51, 54, 66, 110, 157, 217, 228–230, 258, 270, 281

Marangoni Reynolds number, 51, 54
Marangoni velocity, 51
Marginal Marangoni number, 67, 77
Marginal (neutral) Rayleigh number, 62
Marginal Rayleigh number, 62, 63
Mars, 389
MAXUS 4, 120
Method of plane, 40
Microfluidic, 389
Microgravity, 13, 112, 114, 120
Microgravity environment, 10
Microgravity Measurement Apparatus (MMA), 13, 196, 273, 275
Micro-Imaging Displacement Meter (MIDM), 203, 271, 273, 276, 278
Mixed-mode PAS, 137
Modal analysis, 185
Mode number, 82, 141
Molecular Dynamics (MD), 23, 26
Molecular dynamics simulations, 38
Momentum equation, 262, 270
Moon, 389
Multimodal structure, 182
Multimode control, 170

N
Navier-Stokes equation, 348
Neutral stability curve, 217, 218, 227, 250, 256, 258–261
Non-inverted scheme, 166
Normal mode, 88
Normal stress, 42
Numerical analysis, 144
Numerical simulation, 145, 212, 226, 245, 256
Nusselt number (Nu), 55, 268

O
Octave-band, 274, 275
Over-critical parameters (ε), 170, 282

P
Particle Accumulation Structure (PAS), 116, 124, 135, 214, 297, 332, 336, 337, 344
Particle deficient zone, 145
Particle free zone, 172, 183
Partition disk, 158
Pearson, 64
Pearson-Bénard convection, 69
Peltier element, 200

Permutation entropy (h_p), 286
Phase inclination θ(m), 184
Phase separation, 26
Photochromic dye, 197, 243
—activation (PDA) technique, 202, 209, 243–248
—trace, 203, 210, 244
Physical experiment, 11
Pinch-off, 315
Plateau-Rayleigh instability, 52, 80
Plateau-Rayleigh limit, 314, 315
Platinum resistance thermometer (Pt) sensors, 200, 202
Power Spectral Density (PSD), 274
Prandtl number (Pr), 55, 127, 157, 197, 198, 211, 212, 225–228, 236, 249, 259, 270, 298, 317, 332, 334, 335, 342–344
PreDeterminant Circumferential Heating (PDCH), 165, 188
Preferred mode number, 187
Propagation angle, 79, 82
Propagation direction, 91, 93
Proper Orthogonal Decomposition (POD), 288
Pseudo-phase space, 282

R
Rack, 12
Radiative heat transfer, 262–270
Radiosity, 264
Raleigh-limit, 120
Rayleigh-Bénard convection, 55
Rayleigh number (Ra), 55, 178
Ray tracing method, 265
R-B convection, 55
Regularization, 87
Resonance
—frequency (f_R), 276, 281
Return flow, 71, 74, 79
Reynolds number (Re), 55, 179, 270, 298
Ridge, 345, 350
Ring heater, 177
Rocket experiments, 139
Roll structure, 236, 240–242, 250, 257, 260
Rotating Wave (RW), 97, 99, 209, 213, 222, 224, 255
RYUTAI rack, 13, 194, 196, 378

S
Satellite droplet, 317, 320, 322
Second law of thermodynamics, 18

Self-Rewetting Fluids (SRWFs), 355
Shape effect, 161
Sherwood number, 349
Silicone oil, 127, 197, 205, 243, 382
Single mode control, 165
Snap-off, 318–321
Solid-liquid and/or solid-gas interface, 22
Solutocapillary
—effect, 347, 349, 359
—flow, 352
—force, 359
Solutocapillary flow, 8
Space Acceleration Measurement System (SAMS), 13
Spin-coat, 351
Standing thermocapillary wave, 119
Standing Wave (SW), 97, 99, 133, 167, 209, 213, 222, 224, 255
Stefan-Boltzmann
—constant (σ_{SB}), 264
—law, 262
Stereoscopic vision, 233, 235
Stress anisotropy, 42
Striation, 345, 346, 353–355
STS-123, 12
STS-124, 12
STS-127, 12
Stuart-Landau equation, 98
Subcritical bifurcation, 99
Supercritical bifurcation, 99
Suppression, 165
Surface tension, 2, 3, 6, 18, 19, 38, 300, 316, 332, 346, 347, 351
    concentration coefficient of—, 347
    temperature coefficient of—, 298, 301, 316, 334
Surface-tension-driven flow, 389
Surface tension in terms of free energy, 19
Surface tension in thermodynamic equilibrium, 25
Surface-to-surface radiation, 253, 264
Surrogate data method, 286
Symmetric mode, 179

T
Tears of wine, 9
Telemetry, 205–207
Temperature coefficient of the surface tension, 21, 22, 50
Temperature dependence, 6
Temperature oscillation, 116
TEXUS, 114

Thermal
—conductivity (λ), 197, 200, 203, 248, 251
—diffusion, 208, 225
—diffusivity (κ), 197, 228
Thermal boundary condition, 53
Thermocapillarity, 48
Thermocapillarity condition, 8, 48, 81
Thermocapillary
—driven convection, 297, 300, 306, 307, 332, 344
—driven flow, 301, 336
—effect, 44, 298, 302, 304, 314, 320, 323, 324, 332, 333, 340, 347, 359
—flow, 8, 54, 69, 318–325
—force, 317, 359
Thermocapillary boundary condition, 53
Thermocapillary Convection, 44, 156, 107, 108
Thermocouple, 200, 203, 207, 209, 210, 229, 273
Thermodynamics, 25
Three Dimensional Particle Tracking Velocimetry (3-D PTV), 232–242, 256, 266
Time-delayed coordinate system, 282
Time-dependent thermocapillary flow, 114
TNSB, 197, 243
Toroidal core, 135
Tracer particle, 198, 209, 210, 213, 232, 235, 243
Transducer, 203, 207
Translation error, 286
Transverse wave, 79, 90
Travelling wave, 119, 133, 167
Turbulence, 135, 281–288

U
Ultrasonic (or Ultrasound) Velocity Profiler (UVP), 203

V
View factor (F_{ij}), 264, 266
Volume of Fluid (VOF)
—fraction equation, 316
solution algorithm (SOLA)- —method, 315
Volume ratio (V_R), 159, 210, 220–225, 236, 251, 298

W
Water droplets, 2
Wavenumber, 61
Weakly nonlinear, 97, 98, 100
Weightlessness, 11
Well-mixed assumption, 349

Y
Young-Laplace equation, 221, 222, 315
Young's equation, 4, 20, 21, 24

MIX
Papier aus verantwortungsvollen Quellen
Paper from responsible sources
FSC® C105338

If you have any concerns about our products,
you can contact us on
ProductSafety@springernature.com

In case Publisher is established outside the EU,
the EU authorized representative is:
**Springer Nature Customer Service Center GmbH
Europaplatz 3, 69115 Heidelberg, Germany**

Printed by Libri Plureos GmbH
in Hamburg, Germany